Representations of Algebras

CONTEMPORARY MATHEMATICS

705

Representations of Algebras

17th Workshop and International Conference on
Representation of Algebras
August 10–19, 2016
Syracuse University, Syracuse, New York

Graham J. Leuschke
Frauke Bleher
Ralf Schiffler
Dan Zacharia
Editors

AMS AMERICAN MATHEMATICAL SOCIETY

2010 *Mathematics Subject Classification*. Primary 13D09, 13F60, 14K02, 14E16, 16E10, 16G10, 16T30, 18E35, 18F30, 20G07, 20G10, 30F10.

Library of Congress Cataloging-in-Publication Data

Names: International Workshop and Conference on Representations of Algebras (17th : 2016 : Syracuse, N.Y.) | Leuschke, Graham J., 1973- editor. | Bleher, Frauke, 1968- editor. | Schiffler, Ralf, 1968- editor. | Zacharia, Dan (Dan I.), editor.

Title: Representations of algebras : 17th International Workshop and Conference on Representation of Algebras, August 10-19, 2016, Syracuse University, Syracuse, New York / Graham J. Leuschke, Frauke Bleher, Ralf Schiffler, Dan Zacharia, editors.

Description: Providence, Rhode Island : American Mathematical Society, [2018] — Series: Contemporary mathematics ; volume 705 | Includes bibliographical references.

Identifiers: LCCN 2017045514 | ISBN 9781470435769 (alk. paper)

Subjects: LCSH: Representations of algebras–Congresses. | Representations of groups–Congresses. | Algebra–Congresses. | AMS: Commutative algebra – Arithmetic rings and other special rings – Cluster algebras. msc | Algebraic geometry – Abelian varieties and schemes – Isogeny. msc | Algebraic geometry – Birational geometry – McKay correspondence. msc | Associative rings and algebras – Homological methods – Homological dimension. msc | Associative rings and algebras – Representation theory of rings and algebras – Representations of Artinian rings. msc | Associative rings and algebras – Hopf algebras, quantum groups and related topics – Connections with combinatorics. msc | Category theory; homological algebra – Abelian categories – Localization of categories. msc | Category theory; homological algebra – Categories and geometry – Grothendieck groups. msc | Group theory and generalizations – Linear algebraic groups and related topics – Structure theory. msc | Group theory and generalizations – Linear algebraic groups and related topics – Cohomology theory. msc | Functions of a complex variable – Riemann surfaces – Compact Riemann surfaces and uniformization. msc

Classification: LCC QA176 .I587 2016 | DDC 512/.2–dc23

LC record available at https://lccn.loc.gov/2017045514

Contemporary Mathematics ISSN: 0271-4132 (print); ISSN: 1098-3627 (online)

DOI: http://dx.doi.org/10.1090/conm/705

---

10 9 8 7 6 5 4 3 2 1 23 22 21 20 19 18

Contents

Preface

The seventeenth Workshop and International Conference on Representations of Algebras (ICRA 2016) was held August 10–19, 2016, at Syracuse University in Syracuse, NY, USA. The ICRA is a major event in the field with a long tradition, initiated at Carleton University in 1974. The ICRA is organized every other year at varying locations throughout the world; the previous iteration, ICRA 2014, was held in Sanya, China, and the next is scheduled to be held in Prague, Czech Republic, in August 2018. This was the first ICRA held in the United States.

The Scientific Advisory Committee for ICRA 2016 consisted of Luchezar Avramov (Nebraska, USA), Karin Erdmann (Oxford, UK), Birge Huisgen-Zimmermann (Santa Barbara, USA), Henning Krause (Bielefeld, Germany), Idun Reiten (NTNU, Norway), and Jerzy Weyman (Connecticut, USA). The Organizing Committee consisted of Frauke Bleher (Iowa, USA), Calin Chindris (Missouri, USA), Harm Derksen (Michigan, USA), Mark Kleiner (Syracuse, USA), Graham Leuschke (Syracuse, USA), Ralf Schiffler (Connecticut, USA), and Dan Zacharia (Syracuse, USA) (Chair).

Following the tradition of the conference series, ICRA 2016 was divided into two parts. The Workshop consisted of six short courses of three hours each, given by Claire Amiot (Grenoble), Michel Brion (Institut Fourier), Markus Linckelmann (City University London), Robert Marsh (Leeds), Julia Pevtsova (Washington), and Ryo Takahashi (Nagoya). This was followed by a five-day Conference with approximately 150 research presentations, 20 of which were plenary talks. ICRA 2016 had about 250 participants from 30 countries all over the world.

These Proceedings contain the notes of the workshop courses by Brion, Linckelmann and Takahashi as well as selected plenary conference talks and two additional invited contributions. The research areas of these notes range over associative rings and algebras, commutative algebra, algebraic geometry, category theory, group theory and combinatorics. We would like to express our appreciation to the authors who submitted contributions, as well as to the referees for their assistance.

The ICRA Prize, for outstanding work by a young mathematician working in the area of representations of algebras, was awarded to Greg Stevenson (Bielefeld), for his introduction of fundamental new concepts based on geometric ideas using the framework of triangulated categories, and to Qiu Yu (Chinese University of Hong Kong) for important contributions on the topological structure of Bridgeland stability conditions and a sequence of studies about Calabi-Yau and cluster categories. The Selection Committee consisted of Osamu Iyama, Idun Reiten, Jan Schröer, Jie Xiao, and Dan Zacharia (Chair).

We are happy to thank the National Science Foundation, the National Security Agency, the Syracuse University College of Arts & Sciences, and the Office of the

Vice President for Research for their generous financial support. We are also grateful to Syracuse University graduate students Rachel Gettinger, Casey Necheles, Eric Ottman, Robert Roy, and Joshua Stangle, as well as to Mathematics Department Administrative Assistant Kimberly Canino, for their invaluable assistance with the logistics of ICRA 2016.

Sadly, Ragnar-Olaf Buchweitz passed away before the publication of this volume, which contains one of his last papers. We dedicate this volume to his memory.

Contemporary Mathematics
Volume **705**, 2018
http://dx.doi.org/10.1090/conm/705/14204

Commutative algebraic groups up to isogeny. II

Michel Brion

ABSTRACT. This paper develops a representation-theoretic approach to the isogeny category $\underline{\mathcal{C}}$ of commutative group schemes of finite type over a field k, studied in our earlier work. We construct a ring R such that $\underline{\mathcal{C}}$ is equivalent to the category R-mod of all left R-modules of finite length. We also construct an abelian category of R-modules, R-$\widetilde{\text{mod}}$, which is hereditary, has enough projectives, and contains R-mod as a Serre subcategory; this yields a more conceptual proof of the main result of our earlier work, asserting that $\underline{\mathcal{C}}$ is hereditary. We show that R-$\widetilde{\text{mod}}$ is equivalent to the isogeny category of commutative quasi-compact k-group schemes.

CONTENTS

1. Introduction

In this paper, we develop a representation-theoretic approach to the isogeny category of commutative algebraic groups over a field k, studied in [**Br16**]. This abelian category, that we denote by $\underline{\mathcal{C}}$, is equivalent to the quotient of the abelian

2010 *Mathematics Subject Classification.* Primary 14L15, 16G10; Secondary 14K02, 16E10, 18E35, 20G07.

Key words and phrases. commutative algebraic group, isogeny, torsion pair, homological dimension.

category $\mathcal{C}$ of group schemes of finite type over k by the Serre subcategory $\mathcal{F}$ of finite k-group schemes. The main result of [**Br16**] asserts that $\underline{\mathcal{C}}$ is *hereditary*, i.e., $\mathrm{Ext}^i_{\underline{\mathcal{C}}}(G, H) = 0$ for all $i \geq 2$ and all $G, H \in \underline{\mathcal{C}}$. By a theorem of Serre (see [**Se60**, 10.1]) which was the starting point of our work, this also holds for the original category $\mathcal{C}$ when k is algebraically closed of characteristic 0. But for certain fields k, the extension groups in $\mathcal{C}$ can be non-zero in arbitrarily large degrees, as a consequence of [**Mi70**, Thm. 1].

To prove that $\underline{\mathcal{C}}$ is hereditary, the approach of [**Br16**] is similar to that of Serre in [**Se60**], generalized by Oort in [**Oo66**] to determine the extension groups in $\mathcal{C}$ when k is algebraically closed of positive characteristic. As $\underline{\mathcal{C}}$ is easily seen to be a finite length category, it suffices to check the vanishing of higher extension groups for all simple objects $G, H \in \underline{\mathcal{C}}$. These are the additive group, the simple tori and the simple abelian varieties, and one may then adapt the case-by-case analysis of [**Se60, Oo66**] to the easier setting of the isogeny category.

In this paper, we obtain a more conceptual proof, by constructing a ring R such that $\underline{\mathcal{C}}$ is equivalent to the category R-mod of all left R-modules of finite length; moreover, R-mod is a Serre category of an abelian category R-$\widetilde{\mathrm{mod}}$ of left R-modules, which is hereditary and has enough projectives. For a more precise statement, we refer to Theorem 3.5 in §3.2.5, which can be read independently of the rest of the paper. Our result generalizes, and builds on, the equivalence of the isogeny category of unipotent groups over a perfect field of positive characteristic with the category of modules of finite length over a localization of the Dieudonné ring (see [**DG70**, V.3.6.7]).

More specifically, the abelian category $\mathcal{C}$ has very few projectives: the unipotent groups in characteristic 0, and the trivial group otherwise (see [**Br16**, Thm. 2.9, Cor. 5.15])). This drawback was remedied in [**Se60**] by considering the abelian category $\widehat{\mathcal{C}}$ of *pro-algebraic groups*. If k is algebraically closed of characteristic 0, then $\widehat{\mathcal{C}}$ is hereditary, has enough projectives, and contains $\mathcal{C}$ as a Serre subcategory. We obtain a similar result for the isogeny category $\underline{\mathcal{C}}$ over an arbitrary field; this category has more projectives than $\mathcal{C}$ (e.g., the tori), but still not enough of them. We show that $\underline{\mathcal{C}}$ is a Serre subcategory of the isogeny category $\widetilde{\underline{\mathcal{C}}}$ of *quasi-compact group schemes*, which is a hereditary abelian category having enough projectives. In addition, $\widetilde{\underline{\mathcal{C}}}$ is equivalent to the category R-$\widetilde{\mathrm{mod}}$ mentioned above (see again §3.2.5 for a more precise statement). The quasi-compact group schemes, studied by Perrin in [**Pe75, Pe76**], form a restricted class of pro-algebraic groups, discussed in more details in §3.2.2.

Since $\underline{\mathcal{C}}$ is a length category, it is equivalent to the category of all left modules of finite length over a basic pseudo-compact ring A, which is then uniquely determined (this result is due to Gabriel, see [**Ga62**, IV.4] and [**Ga71**, 7.2]). The ring R that we construct is also basic, but not pseudo-compact; it may be viewed as a dense subring of A. Its main advantage for our purposes is that the above category R-$\widetilde{\mathrm{mod}}$ consists of R-modules but not of A-modules.

This paper is organized as follows. In Section 2, we study homological properties of abelian categories equipped with a torsion pair. This setting turns out to be very useful when dealing with algebraic groups, since these are obtained as extensions of groups of special types: for example, every connected algebraic group is an extension of an abelian variety by a affine algebraic group, and these are unique up to isogeny. The main result of Section 2 is Theorem 2.13, which explicitly describes certain abelian categories equipped with a torsion pair, in terms of modules over triangular matrix rings.

Section 3 begins with a brief survey of the structure theory for commutative algebraic groups, with emphasis on categorical aspects. We also treat in parallel the affine group schemes (which form the pro-completion of the abelian category of affine algebraic groups) and the quasi-compact group schemes. We then obtain our main Theorem 3.5 by combining all these structure results with Theorem 2.13. Next, after some auxiliary developments in Subsection 3.3, we study the finiteness properties of the spaces of morphisms and extensions in the isogeny category $\underline{\mathcal{C}}$. In particular, we show that $\underline{\mathcal{C}}$ is $\mathbb{Q}$-linear, Hom- and Ext-finite if and only if k is a number field (Proposition 3.15). The final Subsection 3.5 initiates the study of the indecomposable objects of $\underline{\mathcal{C}}$, by considering a very special situation: extensions of abelian varieties with prescribed simple factors by unipotent groups, over the field of rational numbers. Using a classical result of Dlab and Ringel on representations of species (see [**DR76**]), we obtain a characterization of finite representation type in that setting (Proposition 3.16).

Notation and conventions. All considered categories are assumed to be small. We denote categories by calligraphic letters, e.g., $\mathcal{X}, \mathcal{Y}$, and functors by boldface letters, e.g., $\mathbf{L}, \mathbf{R}$. By abuse of notation, we write $X \in \mathcal{X}$ if X is an object of $\mathcal{X}$. Also, we say that $\mathcal{X}$ contains $\mathcal{Y}$ if $\mathcal{Y}$ is a full subcategory of $\mathcal{X}$.

For any ring R, we denote by R-Mod the category of left R-modules, and by $R\text{-Mod}^{\mathrm{fg}}$ (resp. $R\text{-Mod}^{\mathrm{ss}}$, R-mod) the full subcategory of finitely generated modules (resp. of semi-simple modules, of modules of finite length).

2. A construction of hereditary categories

2.1. Two preliminary results. Let $\mathcal{C}$ be an abelian category. Recall that the *homological dimension* of $\mathcal{C}$ is the smallest non-negative integer $n =: \mathrm{hd}(\mathcal{C})$ such that $\mathrm{Ext}^{n+1}_{\mathcal{C}}(X, Y) = 0$ for all $X, Y \in \mathcal{C}$; equivalently, $\mathrm{Ext}^{m}_{\mathcal{C}}(X, Y) = 0$ for all $X, Y \in \mathcal{C}$ and all $m > n$. If there is no such integer, then $\mathrm{hd}(\mathcal{C})$ is understood to be infinite.

Also, recall that $\mathcal{C}$ is said to be *semi-simple* if $\mathrm{hd}(\mathcal{C}) = 0$; equivalently, every short exact sequence in $\mathcal{C}$ splits. If $\mathrm{hd}(\mathcal{C}) \leq 1$, then $\mathcal{C}$ is said to be *hereditary*.

We now record two easy lemmas, for which we could not locate appropriate references.

LEMMA 2.1. *The following conditions are equivalent for an abelian category $\mathcal{C}$ and a non-negative integer n:*

(i) $\mathrm{hd}(\mathcal{C}) \leq n$.
(ii) *The functor* $\mathrm{Ext}^{n}_{\mathcal{C}}(X, ?)$ *is right exact for any* $X \in \mathcal{C}$.
(iii) *The functor* $\mathrm{Ext}^{n}_{\mathcal{C}}(?, Y)$ *is right exact for any* $Y \in \mathcal{C}$.

PROOF. (i) $\Rightarrow$ (ii), (i) $\Rightarrow$ (iii) Both assertions follow from the vanishing of $\mathrm{Ext}^{n+1}_{\mathcal{C}}(X, ?)$ in view of the long exact sequence of Ext groups.

(ii) $\Rightarrow$ (i) Let $\xi \in \mathrm{Ext}^{n+1}_{\mathcal{C}}(X, Y)$ be the class of an exact sequence

$$0 \longrightarrow Y \longrightarrow X_{n+1} \longrightarrow \cdots \longrightarrow X_1 \longrightarrow X \longrightarrow 0$$

in $\mathcal{C}$. We cut this sequence in two short exact sequences

$$0 \longrightarrow Y \longrightarrow X_{n+1} \longrightarrow Z \longrightarrow 0, \quad 0 \longrightarrow Z \longrightarrow X_n \longrightarrow \cdots \longrightarrow X_1 \longrightarrow X \longrightarrow 0$$

with classes $\xi_1 \in \mathrm{Ext}^1_{\mathcal{C}}(Z,Y)$, $\xi_2 \in \mathrm{Ext}^n_{\mathcal{C}}(X,Z)$ respectively. Then ξ is the Yoneda product $\xi_1 \cdot \xi_2$. Since the natural map $\mathrm{Ext}^n_{\mathcal{C}}(X, X_{n+1}) \to \mathrm{Ext}^n_{\mathcal{C}}(X,Z)$ is surjective, there exists a commutative diagram with exact rows

$$\begin{array}{ccccccccccccc}
0 & \longrightarrow & X_{n+1} & \longrightarrow & X'_n & \longrightarrow & \cdots & \longrightarrow & X'_1 & \longrightarrow & X & \longrightarrow & 0 \\
 & & \downarrow & & \downarrow & & & & \downarrow & & \downarrow{\scriptstyle \mathrm{id}} & & \\
0 & \longrightarrow & Z & \longrightarrow & X_n & \longrightarrow & \cdots & \longrightarrow & X_1 & \longrightarrow & X & \longrightarrow & 0.
\end{array}$$

Also, we have a commutative diagram with exact rows

$$\begin{array}{ccccccccc}
0 & \longrightarrow & Y & \longrightarrow & X'_{n+1} & \xrightarrow{f'} & X_{n+1} & \longrightarrow & 0 \\
 & & \downarrow{\scriptstyle \mathrm{id}} & & \downarrow & & \downarrow{\scriptstyle f} & & \\
0 & \longrightarrow & Y & \longrightarrow & X_{n+1} & \xrightarrow{f} & Z & \longrightarrow & 0,
\end{array}$$

where the top exact sequence is split (as $\mathrm{id}_{X_{n+1}}$ yields a section of f'). Thus, $f^*(\xi_1) = 0$ in $\mathrm{Ext}^1_{\mathcal{C}}(X_{n+1}, Y)$. By concatenating both diagrams, we obtain a morphism of extensions

$$\begin{array}{ccccccccccccc}
0 & \longrightarrow & Y & \longrightarrow & X'_{n+1} & \longrightarrow & \cdots & \longrightarrow & X'_1 & \longrightarrow & X & \longrightarrow & 0 \\
 & & \downarrow{\scriptstyle \mathrm{id}} & & \downarrow & & & & \downarrow & & \downarrow{\scriptstyle \mathrm{id}} & & \\
0 & \longrightarrow & Y & \longrightarrow & X_{n+1} & \longrightarrow & \cdots & \longrightarrow & X_1 & \longrightarrow & X & \longrightarrow & 0.
\end{array}$$

Thus, ξ is also represented by the top exact sequence, and hence $\xi = f^*(\xi_1) \cdot \xi_2 = 0$. This completes the proof of (ii) $\Rightarrow$ (i).

A dual argument shows that (iii) $\Rightarrow$ (i). □

Next, recall that a subcategory $\mathcal{D}$ of an abelian category $\mathcal{C}$ is said to be a *Serre subcategory* if $\mathcal{D}$ is non-empty, strictly full in $\mathcal{C}$, and stable under taking subobjects, quotients and extensions.

LEMMA 2.2. *Let $\mathcal{C}$ be a hereditary abelian category, and $\mathcal{D}$ a Serre subcategory. Then $\mathcal{D}$ is hereditary.*

PROOF. In view of the assumption on $\mathcal{D}$, the natural map $\mathrm{Ext}^1_{\mathcal{D}}(X,Y) \to \mathrm{Ext}^1_{\mathcal{C}}(X,Y)$ is an isomorphism for any $X, Y \in \mathcal{D}$. So the assertion follows from Lemma 2.1. □

2.2. Torsion pairs. Throughout this subsection, we consider an abelian category $\mathcal{C}$ equipped with a *torsion pair*, that is, a pair of strictly full subcategories $\mathcal{X}$, $\mathcal{Y}$ satisfying the following conditions:

(i) $\mathrm{Hom}_{\mathcal{C}}(X,Y) = 0$ for all $X \in \mathcal{X}$, $Y \in \mathcal{Y}$.
(ii) For any $C \in \mathcal{C}$, there exists an exact sequence in $\mathcal{C}$

$$0 \longrightarrow X_C \xrightarrow{f_C} C \xrightarrow{g_C} Y_C \longrightarrow 0, \tag{2.1}$$

where $X_C \in \mathcal{X}$ and $Y_C \in \mathcal{Y}$.

Then $\mathcal{X}$ is stable under quotients, extensions and coproducts, and $\mathcal{Y}$ is stable under subobjects, extensions and products. Moreover, the assignment $C \mapsto X_C$ extends to an additive functor $\mathbf{R} : \mathcal{C} \to \mathcal{X}$, right adjoint to the inclusion. Dually, the assignment $C \mapsto Y_C$ extends to an additive functor $\mathbf{L} : \mathcal{C} \to \mathcal{Y}$, left adjoint to the inclusion (see e.g. [**BR07**, Sec. 1.1] for these results).

LEMMA 2.3. *Assume that $\mathcal{X}$, $\mathcal{Y}$ are Serre subcategories of $\mathcal{C}$. Then:*

(i) $\mathbf{R}$, $\mathbf{L}$ *are exact.*
(ii) $\mathrm{Ext}^n_{\mathcal{C}}(X,Y) = 0$ *for all* $X \in \mathcal{X}$, $Y \in \mathcal{Y}$ *and* $n \geq 1$.
(iii) $\mathrm{hd}(\mathcal{C}) \leq \mathrm{hd}(\mathcal{X}) + \mathrm{hd}(\mathcal{Y}) + 1$.

PROOF. (i) Let $C \in \mathcal{C}$ and consider a subobject $i : C_1 \hookrightarrow C$. Denote by $q : C \to C_2 := C/C_1$ the quotient map, and by $C_1 \cap X_C$ the kernel of the map $(q, g_C) : C \to C_2 \times Y_C$. Then $C_1 \cap X_C \hookrightarrow X_C$, and hence $C_1 \cap X_C \in \mathcal{C}$. Moreover, $C_1/C_1 \cap X_C \hookrightarrow C/X_C \cong Y_C$, and hence $C_1/C_1 \cap X_C \in \mathcal{Y}$. Thus, $C_1 \cap X_C = \mathbf{R}(C_1)$ and $C_1/C_1 \cap X_C = \mathbf{L}(C_1)$. So we obtain a commutative diagram of exact sequences

$$\begin{array}{ccccccccc} 0 & \longrightarrow & \mathbf{R}(C_1) & \xrightarrow{\mathbf{R}(i)} & \mathbf{R}(C) & \longrightarrow & \mathbf{R}(C)/\mathbf{R}(C_1) & \longrightarrow & 0 \\ & & \downarrow{\scriptstyle f_{C_1}} & & \downarrow{\scriptstyle f_C} & & \downarrow{\scriptstyle g} & & \\ 0 & \longrightarrow & C_1 & \xrightarrow{i} & C & \xrightarrow{q} & C_2 & \longrightarrow & 0. \end{array}$$

As we just showed, the left square is cartesian; it follows that g is a monomorphism. This yields a commutative diagram with exact rows and columns

$$\begin{array}{ccccccccc} & & 0 & & 0 & & 0 & & \\ & & \downarrow & & \downarrow & & \downarrow & & \\ 0 & \longrightarrow & \mathbf{R}(C_1) & \xrightarrow{\mathbf{R}(i)} & \mathbf{R}(C) & \longrightarrow & \mathbf{R}(C)/\mathbf{R}(C_1) & \longrightarrow & 0 \\ & & \downarrow{\scriptstyle f_{C_1}} & & \downarrow{\scriptstyle f_C} & & \downarrow{\scriptstyle g} & & \\ 0 & \longrightarrow & C_1 & \xrightarrow{i} & C & \xrightarrow{q} & C_2 & \longrightarrow & 0 \\ & & \downarrow & & \downarrow & & \downarrow & & \\ 0 & \longrightarrow & \mathbf{L}(C_1) & \longrightarrow & \mathbf{L}(C) & \longrightarrow & C_2/\mathrm{Im}(g) & \longrightarrow & 0 \\ & & \downarrow & & \downarrow & & \downarrow & & \\ & & 0 & & 0 & & 0 & & \end{array}$$

Moreover, $\mathbf{R}(C)/\mathbf{R}(C_1) \in \mathcal{X}$ and $C_2/\mathrm{Im}(g) \in \mathcal{Y}$ by our assumption on $\mathcal{X}$, $\mathcal{Y}$. It follows that $\mathbf{R}(C)/\mathbf{R}(C_1) = \mathbf{R}(C_2)$ and $C_2/\mathrm{Im}(g) = \mathbf{L}(C_2)$. Thus, $\mathbf{R}, \mathbf{L}$ are exact.

(ii) We first show that $\mathrm{Ext}^1_{\mathcal{C}}(X,Y) = 0$. Consider an exact sequence

$$0 \longrightarrow Y \longrightarrow C \longrightarrow X \longrightarrow 0$$

in $\mathcal{C}$. Then the induced map $\mathbf{R}(C) \to \mathbf{R}(X) = X$ is an isomorphism, since $\mathbf{R}$ is exact and $\mathbf{R}(Y) = 0$. Thus, the above exact sequence splits; this yields the assertion.

Next, we show the vanishing of any $\xi \in \mathrm{Ext}^n_{\mathcal{C}}(X,Y)$. For this, we adapt the argument of Lemma 2.1. Choose a representative of ξ by an exact sequence

$$0 \longrightarrow Y \longrightarrow C_n \longrightarrow \cdots \longrightarrow C_1 \longrightarrow X \longrightarrow 0$$

in $\mathcal{C}$ and cut it in two short exact sequences

$$0 \longrightarrow Y \longrightarrow C_n \longrightarrow Z \longrightarrow 0, \quad 0 \longrightarrow Z \longrightarrow C_{n-1} \longrightarrow \cdots \longrightarrow C_1 \longrightarrow X \longrightarrow 0.$$

This yields an exact sequence

$$0 \longrightarrow \mathbf{R}(Z) \longrightarrow \mathbf{R}(C_{n-1}) \longrightarrow \cdots \longrightarrow \mathbf{R}(C_1) \longrightarrow X \longrightarrow 0.$$

Also, we obtain a commutative diagram of exact sequences

$$\begin{array}{ccccccccc} 0 & \longrightarrow & Y & \longrightarrow & C'_n & \longrightarrow & \mathbf{R}(Z) & \longrightarrow & 0 \\ & & \big\downarrow{\scriptstyle \mathrm{id}} & & \big\downarrow & & \big\downarrow & & \\ 0 & \longrightarrow & Y & \longrightarrow & C_n & \longrightarrow & Z & \longrightarrow & 0, \end{array}$$

where the top sequence splits by the above step. So ξ is also represented by the exact sequence

$$0 \longrightarrow Y \longrightarrow C'_n \longrightarrow \mathbf{R}(C_{n-1}) \longrightarrow \cdots \longrightarrow \mathbf{R}(C_1) \longrightarrow X \longrightarrow 0,$$

which has a trivial class in $\mathrm{Ext}^n_{\mathcal{C}}(X, Y)$.

(iii) We may assume that $\mathrm{hd}(\mathcal{X}) := m$ and $\mathrm{hd}(\mathcal{Y}) := n$ are both finite. In view of the exact sequence (2.1) and the long exact sequence for Ext groups, it suffices to show that $\mathrm{Ext}^{m+n+2}_{\mathcal{C}}(C, C') = 0$ for all C, C' in $\mathcal{X}$ or $\mathcal{Y}$. By (ii), this holds whenever $C \in \mathcal{X}$ and $C' \in \mathcal{Y}$. Also, if $C, C' \in \mathcal{X}$, then $\mathrm{Ext}^{m+1}_{\mathcal{C}}(C, C') = 0$: indeed, every exact sequence

$$0 \longrightarrow C' \longrightarrow C_{m+1} \longrightarrow \cdots \longrightarrow C_1 \longrightarrow C \longrightarrow 0$$

is Yoneda equivalent to the exact sequence

$$0 \longrightarrow C' \longrightarrow \mathbf{R}(C_{m+1}) \longrightarrow \cdots \longrightarrow \mathbf{R}(C_1) \longrightarrow C \longrightarrow 0,$$

which in turn is equivalent to 0 by assumption. Likewise, $\mathrm{Ext}^{n+1}_{\mathcal{C}}(C, C') = 0$ for all $C, C' \in \mathcal{Y}$. So we are reduced to checking that $\mathrm{Ext}^{m+n+2}_{\mathcal{C}}(Y, X) = 0$ for all $X \in \mathcal{X}$, $Y \in \mathcal{Y}$.

For this, we adapt again the argument of Lemma 2.1. Let $\xi \in \mathrm{Ext}^{m+n+2}_{\mathcal{C}}(Y, X)$ be represented by an exact sequence

$$0 \longrightarrow X \longrightarrow C_{m+n+2} \longrightarrow \cdots \longrightarrow C_1 \longrightarrow Y \longrightarrow 0$$

in $\mathcal{C}$. This yields two exact sequences

$$0 \longrightarrow X \longrightarrow C_{m+n+2} \longrightarrow \cdots \longrightarrow C_{n+2} \longrightarrow Z \longrightarrow 0,$$

$$0 \longrightarrow Z \longrightarrow C_{n+1} \longrightarrow \cdots \longrightarrow C_1 \longrightarrow Y \longrightarrow 0.$$

As $\mathrm{Ext}^{n+1}_{\mathcal{C}}(Y, \mathbf{L}(Z)) = 0$ by the above step, the exact sequence

$$0 \longrightarrow \mathbf{R}(Z) \longrightarrow Z \longrightarrow \mathbf{L}(Z) \longrightarrow 0$$

yields a surjection $\mathrm{Ext}^{n+1}_{\mathcal{C}}(Y, \mathbf{R}(Z)) \to \mathrm{Ext}^{n+1}_{\mathcal{C}}(Y, Z)$. Thus, there exists a commutative diagram of exact sequences

$$\begin{array}{ccccccccccccc} 0 & \longrightarrow & \mathbf{R}(Z) & \longrightarrow & C'_{n+1} & \longrightarrow & \cdots & \longrightarrow & C'_1 & \longrightarrow & Y & \longrightarrow & 0 \\ & & \big\downarrow & & \big\downarrow & & & & \big\downarrow & & \big\downarrow{\scriptstyle \mathrm{id}} & & \\ 0 & \longrightarrow & Z & \longrightarrow & C_{n+1} & \longrightarrow & \cdots & \longrightarrow & C_1 & \longrightarrow & Y & \longrightarrow & 0. \end{array}$$

Also, we have an exact sequence

$$0 \longrightarrow X \longrightarrow \mathbf{R}(C_{m+n+2}) \longrightarrow \cdots \longrightarrow \mathbf{R}(C_{n+2}) \longrightarrow \mathbf{R}(Z) \longrightarrow 0,$$

with trivial class as $\mathrm{Ext}^{m+1}_{\mathcal{C}}(\mathbf{R}(Z), X) = 0$. Hence ξ is also represented by the exact sequence

$$0 \to X \to \mathbf{R}(C_{m+n+2}) \to \cdots \to \mathbf{R}(C_{n+2}) \to C'_{n+1} \to \cdots \to C'_1 \to Y \to 0,$$

which has a trivial class as well. $\square$

COROLLARY 2.4. *Assume that $\mathcal{X}$ and $\mathcal{Y}$ are semi-simple Serre subcategories of $\mathcal{C}$. Then:*

(i) *Every object of $\mathcal{X}$ is projective in $\mathcal{C}$.*
(ii) *Every object of $\mathcal{Y}$ is injective in $\mathcal{C}$.*
(iii) *$\mathcal{C}$ is hereditary.*

PROOF. (i) Let $X \in \mathcal{X}$. Then $\mathrm{Ext}^1_{\mathcal{C}}(X, Y) = 0$ for all $Y \in \mathcal{Y}$, by Lemma 2.3. Moreover, $\mathrm{Ext}^1_{\mathcal{C}}(X, X') = 0$ for all $X' \in \mathcal{X}$ by our assumption. In view of the exact sequence (2.1), it follows that $\mathrm{Ext}^1_{\mathcal{C}}(X, C) = 0$ for all $C \in \mathcal{C}$, i.e., X is projective in $\mathcal{C}$.

(ii) This is checked similarly.

(iii) This follows from Lemma 2.3 (iii).

□

2.3. The category of extensions. We still consider an abelian category $\mathcal{C}$ equipped with a torsion pair $(\mathcal{X}, \mathcal{Y})$. Let $\mathcal{E}$ be the category with objects the triples (X, Y, ξ), where $X \in \mathcal{X}$, $Y \in \mathcal{Y}$ and $\xi \in \mathrm{Ext}^1_{\mathcal{C}}(Y, X)$; the morphisms from (X, Y, ξ) to (X', Y', ξ') are the pairs of morphisms $(u : X \to X', v : Y \to Y')$ such that $u_*(\xi) = v^*(\xi')$ in $\mathrm{Ext}^1_{\mathcal{C}}(Y, X')$. We say that $\mathcal{E}$ is the *category of extensions* associated with the triple $(\mathcal{C}, \mathcal{X}, \mathcal{Y})$.

We may assign to any $C \in \mathcal{C}$, the triple

$$\mathbf{T}(C) := (\mathbf{R}(C), \mathbf{L}(C), \xi(C)),$$

where $\xi(C) \in \mathrm{Ext}^1_{\mathcal{C}}(\mathbf{L}(C), \mathbf{R}(C))$ denotes the class of the extension (2.1),

$$0 \longrightarrow \mathbf{R}(C) \longrightarrow C \longrightarrow \mathbf{L}(C) \longrightarrow 0.$$

LEMMA 2.5. *Assume that $\mathcal{X}, \mathcal{Y}$ are Serre subcategories of $\mathcal{C}$, and $\mathrm{Hom}_{\mathcal{C}}(Y, X) = 0$ for all $X \in \mathcal{X}$, $Y \in \mathcal{Y}$. Then the above assignment extends to a covariant functor $\mathbf{T} : \mathcal{C} \to \mathcal{E}$, which is an equivalence of categories.*

PROOF. Consider a morphism $f : C \to C'$ in $\mathcal{C}$. Then f lies in a unique commutative diagram of exact sequences

$$\begin{array}{ccccccccc}
0 & \longrightarrow & X_C & \xrightarrow{f_C} & C & \xrightarrow{g_C} & Y_C & \longrightarrow & 0 \\
 & & {\scriptstyle u}\downarrow & & {\scriptstyle f}\downarrow & & {\scriptstyle v}\downarrow & & \\
0 & \longrightarrow & X_{C'} & \longrightarrow & C' & \longrightarrow & Y_{C'} & \longrightarrow & 0.
\end{array}$$

Denote by $\xi \in \mathrm{Ext}^1_{\mathcal{C}}(Y_C, X_C)$, $\xi' \in \mathrm{Ext}^1_{\mathcal{C}}(Y_{C'}, X_{C'})$ the classes of the above extensions and set $u =: \mathbf{R}(f)$, $v =: \mathbf{L}(f)$. These fit into a commutative diagram with exact rows

$$\begin{array}{ccccccccc}
0 & \longrightarrow & \mathbf{R}(C) & \longrightarrow & C & \longrightarrow & \mathbf{L}(C) & \longrightarrow & 0 \\
 & & {\scriptstyle \mathbf{R}(f)}\downarrow & & \downarrow & & {\scriptstyle \mathrm{id}}\downarrow & & \\
0 & \longrightarrow & \mathbf{R}(C') & \longrightarrow & D & \longrightarrow & \mathbf{L}(C) & \longrightarrow & 0 \\
 & & {\scriptstyle \mathrm{id}}\downarrow & & \downarrow & & {\scriptstyle \mathbf{L}(f)}\downarrow & & \\
0 & \longrightarrow & \mathbf{R}(C') & \longrightarrow & C' & \longrightarrow & \mathbf{L}(C') & \longrightarrow & 0
\end{array}$$

It follows that $\mathbf{R}(f)_*(\xi) = \mathbf{L}(f)^*(\xi')$. Thus, the assignment $f \mapsto (\mathbf{R}(f), \mathbf{L}(f))$ defines the desired covariant functor $\mathbf{T}$. We now show that $\mathbf{T}$ is an equivalence of categories.

Since $(\mathcal{X}, \mathcal{Y})$ is a torsion pair, $\mathbf{T}$ is essentially surjective. We check that it is faithful. Let $C, C' \in \mathcal{C}$ and consider $f \in \mathrm{Hom}_{\mathcal{C}}(C, C')$ such that $\mathbf{R}(f) = 0 = \mathbf{L}(f)$. Then the composition $X_C \xrightarrow{f_C} C \xrightarrow{f} C'$ is zero, and hence f factors through $g : Y_C \to C'$. Moreover, the composition $Y_C \xrightarrow{g} C' \xrightarrow{g_{C'}} Y_{C'}$ is zero, and hence g factors through $h : Y_C \to X_{C'}$. By our assumption, $h = 0$; thus, $f = 0$.

Finally, we show that $\mathbf{T}$ is full. Let again $C, C' \in \mathcal{C}$ and consider $u : X_C \to X_{C'}$, $v : Y_C \to Y_{C'}$ such that $u_*(\xi) = v^*(\xi')$, where ξ (resp. ξ') denotes the class of the extension (2.1) for C (resp. C'). Since $\mathrm{Hom}_{\mathcal{C}}(Y, X) = 0$ for all $X \in \mathcal{X}$ and $Y \in \mathcal{Y}$, these extensions are uniquely determined by their classes, and in turn by C, C'. Thus, we have a commutative diagram of extensions in $\mathcal{C}$

$$\begin{array}{ccccccccc}
0 & \longrightarrow & X_C & \xrightarrow{f_C} & C & \xrightarrow{g_C} & Y_C & \longrightarrow & 0 \\
 & & \Big\downarrow{\scriptstyle u} & & \Big\downarrow & & \Big\downarrow{\scriptstyle \mathrm{id}} & & \\
0 & \longrightarrow & X_{C'} & \xrightarrow{i} & D & \xrightarrow{q} & Y_C & \longrightarrow & 0 \\
 & & \Big\downarrow{\scriptstyle \mathrm{id}} & & \Big\downarrow & & \Big\downarrow{\scriptstyle v} & & \\
0 & \longrightarrow & X_{C'} & \xrightarrow{f_{C'}} & C' & \xrightarrow{g_{C'}} & Y_{C'} & \longrightarrow & 0.
\end{array}$$

This yields a morphism $f : C \to C'$ such that $\mathbf{R}(f) = u$ and $\mathbf{L}(f) = v$. □

With the assumptions of Lemma 2.5, the subcategory $\mathcal{X}$ (resp. $\mathcal{Y}$) of $\mathcal{C}$ is identified via $\mathbf{T}$ with the full subcategory of $\mathcal{E}$ with objects the triples of the form $(X, 0, 0)$ (resp. $(0, Y, 0)$). Assuming in addition that $\mathcal{X}$ *and* $\mathcal{Y}$ *are semi-simple*, we now obtain a description of homomorphism and extension groups in $\mathcal{E}$:

PROPOSITION 2.6. *With the above assumptions, there is an exact sequence*

$$0 \longrightarrow \mathrm{Hom}_{\mathcal{E}}(Z, Z') \xrightarrow{\iota} \mathrm{Hom}_{\mathcal{X}}(X, X') \times \mathrm{Hom}_{\mathcal{Y}}(Y, Y') \longrightarrow$$

$$\xrightarrow{\varphi} \mathrm{Ext}^1_{\mathcal{C}}(Y, X') \longrightarrow \mathrm{Ext}^1_{\mathcal{E}}(Z, Z') \longrightarrow 0$$

for any $Z = (X, Y, \xi), Z' = (X', Y', \xi') \in \mathcal{E}$, *where* ι *denotes the inclusion, and* $\varphi(u, v) := u_*(\xi') - v^*(\xi)$.

PROOF. We have $\mathrm{Ker}(\iota) = 0$ and $\mathrm{Im}(\iota) = \mathrm{Ker}(\varphi)$ by the definition of the morphisms in $\mathcal{E}$. Thus, it suffices to check that $\mathrm{Coker}(\varphi) \cong \mathrm{Ext}^1_{\mathcal{E}}(Z, Z')$.

Consider the exact sequence

$$0 \longrightarrow X' \longrightarrow Z' \longrightarrow Y' \longrightarrow 0$$

in $\mathcal{E}$, with class $\xi' \in \mathrm{Ext}^1_{\mathcal{E}}(Y', X') = \mathrm{Ext}^1_{\mathcal{C}}(Y', X')$. This yields an exact sequence

$$\mathrm{Hom}_{\mathcal{E}}(Y, Y') \xrightarrow{\partial'} \mathrm{Ext}^1_{\mathcal{E}}(Y, X') \longrightarrow \mathrm{Ext}^1_{\mathcal{E}}(Y, Z') \longrightarrow \mathrm{Ext}^1_{\mathcal{E}}(Y, Y'),$$

where $\partial'(v) := v^*(\xi')$ for any $v \in \mathrm{Hom}_{\mathcal{E}}(Y, Y')$. Moreover, since $\mathcal{Y}$ is a semi-simple Serre subcategory of $\mathcal{E}$, we have $\mathrm{Hom}_{\mathcal{E}}(Y, Y') = \mathrm{Hom}_{\mathcal{Y}}(Y, Y')$ and $\mathrm{Ext}^1_{\mathcal{E}}(Y, Y') = 0$. So we obtain a natural isomorphism

$$\mathrm{Ext}^1_{\mathcal{E}}(Y, Z') \cong \mathrm{Ext}^1_{\mathcal{E}}(Y, X') / \{v^*(\xi') \mid v \in \mathrm{Hom}_{\mathcal{Y}}(Y, Y')\}. \tag{2.2}$$

Similarly, the exact sequence

$$0 \longrightarrow X \longrightarrow Z \longrightarrow Y \longrightarrow 0$$

in $\mathcal{E}$, with class $\xi \in \mathrm{Ext}^1_{\mathcal{C}}(Y, X)$, yields an exact sequence

$$\mathrm{Hom}_{\mathcal{E}}(X, Z') \xrightarrow{\partial} \mathrm{Ext}^1_{\mathcal{E}}(Y, Z') \longrightarrow \mathrm{Ext}^1_{\mathcal{E}}(Z, Z') \longrightarrow \mathrm{Ext}^1_{\mathcal{E}}(X, Z'),$$

where $\partial(u) := u_*(\xi)$ for any $u \in \mathrm{Hom}_{\mathcal{E}}(X, Z')$. Moreover, the natural map

$$\mathrm{Hom}_{\mathcal{X}}(X, X') \longrightarrow \mathrm{Hom}_{\mathcal{E}}(X, Z')$$

is an isomorphism, since $\mathrm{Hom}_{\mathcal{E}}(X, Y') = 0$. Also, $\mathrm{Ext}^1_{\mathcal{E}}(X, Z') = 0$ by Corollary 2.4. Hence we obtain a natural isomorphism

$$\mathrm{Ext}^1_{\mathcal{E}}(Z, Z') \cong \mathrm{Ext}^1_{\mathcal{E}}(Y, Z')/\{u_*(\xi) \mid u \in \mathrm{Hom}_{\mathcal{X}}(X, X')\}. \tag{2.3}$$

Putting together the isomorphisms (2.2) and (2.3) yields the desired assertion. □

2.4. Universal extensions. We still consider an abelian category $\mathcal{C}$ equipped with a torsion pair $(\mathcal{X}, \mathcal{Y})$, and make the following assumptions:

(a) $\mathcal{X}, \mathcal{Y}$ are Serre subcategories of $\mathcal{C}$.
(b) $\mathcal{X}, \mathcal{Y}$ are semi-simple.
(c) $\mathrm{Hom}_{\mathcal{C}}(Y, X) = 0$ for all $X \in \mathcal{X}$, $Y \in \mathcal{Y}$.
(d) There exists a covariant exact functor $\mathbf{F} : \mathcal{Y} \to \widetilde{\mathcal{X}}$, where $\widetilde{\mathcal{X}}$ is a semi-simple abelian category containing $\mathcal{X}$ as a Serre subcategory, and a bi-functorial isomorphism

$$\mathrm{Ext}^1_{\mathcal{C}}(Y, X) \xrightarrow{\cong} \mathrm{Hom}_{\widetilde{\mathcal{X}}}(\mathbf{F}(Y), X) \tag{2.4}$$

for all $X \in \mathcal{X}$, $Y \in \mathcal{Y}$.

REMARK 2.7. The above assumptions are satisfied by the isogeny category of algebraic groups and some natural Serre subcategories, as we will see in Subsection 3.2. Also, assumptions (a), (b) and (c) are just those of Corollary 2.4 and Proposition 2.6. Note that a weak version of (d) always holds, where we only require $\widetilde{\mathcal{X}}$ to be a category containing $\mathcal{X}$: take $\widetilde{\mathcal{X}}$ to be the opposite category of covariant functors from $\mathcal{X}$ to sets; then the functor $\mathrm{Ext}^1_{\mathcal{C}}(Y, ?)$ is an object of $\widetilde{\mathcal{X}}$ for any $Y \in \mathcal{Y}$, and the isomorphism (2.4) follows from Yoneda's lemma. But requiring $\widetilde{\mathcal{X}}$ to be abelian and semi-simple is a restrictive assumption.

Under the four above assumptions, every $C \in \mathcal{C}$ defines an extension class $\xi(C) \in \mathrm{Ext}^1_{\mathcal{C}}(Y_C, X_C)$, and in turn a morphism $\eta(C) \in \mathrm{Hom}_{\widetilde{\mathcal{X}}}(\mathbf{F}(Y_C), X_C)$. Moreover, every morphism $f : C \to C'$ in $\mathcal{C}$ induces morphisms $u : X_C \to X_{C'}$, $v : Y_C \to Y_{C'}$ such that the push-forward $u_*\xi(C) \in \mathrm{Ext}^1_{\mathcal{C}}(Y, X')$ is identified with $u \circ \eta(C) \in \mathrm{Hom}_{\widetilde{\mathcal{X}}}(\mathbf{F}(Y_C), X_{C'})$, and the pull-back $v^*\xi(C') \in \mathrm{Ext}^1_{\mathcal{C}}(Y, X')$ is identified with $\eta(C') \circ F(v) \in \mathrm{Hom}_{\widetilde{\mathcal{X}}}(\mathbf{F}(Y_C), X_{C'})$.

It follows that the category of extensions $\mathcal{E}$ (considered in Subsection 2.3) is equivalent to the category $\mathcal{F}$ with objects the triples (X, Y, η), where $X \in \mathcal{X}$, $Y \in \mathcal{Y}$ and $\eta \in \mathrm{Hom}_{\widetilde{\mathcal{X}}}(\mathbf{F}(Y), X)$; the morphisms from (X, Y, η) to (X', Y', η') are the pairs of morphisms $(u : X \to X',\ v : Y \to Y')$ such that the diagram

$$\begin{array}{ccc} \mathbf{F}(Y) & \xrightarrow{F(v)} & \mathbf{F}(Y') \\ {\scriptstyle\eta}\downarrow & & \downarrow{\scriptstyle\eta'} \\ X & \xrightarrow{u} & X' \end{array}$$

commutes. With this notation, Lemma 2.5 yields readily:

LEMMA 2.8. *The assignment* $C \mapsto (\mathbf{R}(C), \mathbf{L}(C), \eta(C))$ *extends to an equivalence of categories* $\mathcal{C} \overset{\cong}{\to} \mathcal{F}$.

Next, consider the category $\widetilde{\mathcal{F}}$ with objects the triples (X, Y, η), where $X \in \widetilde{\mathcal{X}}$, $Y \in \mathcal{Y}$ and $\eta \in \mathrm{Hom}_{\widetilde{\mathcal{X}}}(\mathbf{F}(Y), X)$; the morphisms are defined like those of $\mathcal{F}$. Then one may readily check the following:

LEMMA 2.9. *With the above notation,* $\mathcal{F}$ *is a Serre subcategory of* $\widetilde{\mathcal{F}}$. *Moreover, the triple* $(\widetilde{\mathcal{F}}, \widetilde{\mathcal{X}}, \mathcal{Y})$ *satisfies the assumptions* (a), (b), (c), *and* (d) *with the same functor* $\mathbf{F}$. *For any* $Z = (X, Y, \eta)$ *and* $Z' = (X', Y', \eta') \in \widetilde{\mathcal{F}}$, *we have an exact sequence*

$$0 \longrightarrow \mathrm{Hom}_{\widetilde{\mathcal{F}}}(Z, Z') \xrightarrow{\iota} \mathrm{Hom}_{\mathcal{X}}(X, X') \times \mathrm{Hom}_{\mathcal{Y}}(Y, Y') \longrightarrow$$

$$\xrightarrow{\psi} \mathrm{Hom}_{\widetilde{\mathcal{X}}}(\widetilde{X}, X') \longrightarrow \mathrm{Ext}^1_{\widetilde{\mathcal{F}}}(Z, Z') \longrightarrow 0,$$

where $\widetilde{X} := \mathbf{F}(Y)$ *and* $\psi(u, v) := u \circ \eta - \eta' \circ \mathbf{F}(v)$.

We now consider the covariant exact functors like in Lemma 2.3:

$$\widetilde{\mathbf{R}} : \widetilde{\mathcal{F}} \longrightarrow \widetilde{\mathcal{X}}, \quad (X, Y, \eta) \longmapsto X, \quad (u, v) \longmapsto u,$$

$$\widetilde{\mathbf{L}} : \widetilde{\mathcal{F}} \longrightarrow \mathcal{Y}, \quad (X, Y, \eta) \longmapsto Y, \quad (u, v) \longmapsto v.$$

LEMMA 2.10. *With the above notation, the assignment* $Y \mapsto (\mathbf{F}(Y), Y, \mathrm{id}_{\mathbf{F}(Y)})$ *extends to a covariant exact functor* $\mathbf{E} : \mathcal{Y} \to \widetilde{\mathcal{X}}$, *which is left adjoint to* $\widetilde{\mathbf{L}}$.

PROOF. For any morphism $v : Y_1 \to Y_2$ in $\mathcal{Y}$, the induced morphism $\mathbf{F}(v) : \mathbf{F}(Y_1) \to \mathbf{F}(Y_2)$ satisfies $(\mathbf{F}(v), v) \in \mathrm{Hom}_{\widetilde{\mathcal{F}}}(\mathbf{E}(Y_1), \mathbf{E}(Y_2))$ by the definition of the morphisms in $\widetilde{\mathcal{F}}$. We may thus set $\mathbf{E}(v) := (\mathbf{F}(v), v)$. As $\mathbf{F}$ is a covariant exact functor, so is $\mathbf{E}$. We now check the adjunction assertion: let $Y \in \mathcal{Y}$ and $(X, Y', \eta) \in \widetilde{\mathcal{F}}$. Then $\mathrm{Hom}_{\widetilde{\mathcal{F}}}(\mathbf{E}(Y), (X, Y', \eta))$ consists of the pairs $(u \in \mathrm{Hom}_{\widetilde{\mathcal{X}}}(\mathbf{F}(Y), X),\ v \in \mathrm{Hom}_{\mathcal{Y}}(Y, Y'))$ such that $u = \eta \circ F(v)$. Thus, the map induced by $\widetilde{\mathbf{L}}$,

$$\mathrm{Hom}_{\widetilde{\mathcal{F}}}(\mathbf{E}(Y), (X, Y', \eta)) \longrightarrow \mathrm{Hom}_{\mathcal{Y}}(Y, Y'), \quad (u, v) \longmapsto v$$

is an isomorphism. □

For any $Y \in \mathcal{Y}$, there is a tautological exact sequence

$$0 \longrightarrow \mathbf{F}(Y) \xrightarrow{\iota} \mathbf{E}(Y) \xrightarrow{\pi} Y \longrightarrow 0 \tag{2.5}$$

in $\widetilde{\mathcal{F}}$, which is universal in the following sense:

PROPOSITION 2.11. *Let* $Z = (X, Y, \eta) \in \widetilde{\mathcal{F}}$.

(i) *There exists a unique morphism* $\mu : \mathbf{E}(Y) \to Z$ *in* $\widetilde{\mathcal{F}}$ *such that the diagram*

$$\begin{array}{ccccccccc} 0 & \longrightarrow & \mathbf{F}(Y) & \xrightarrow{\iota} & \mathbf{E}(Y) & \xrightarrow{\pi} & Y & \longrightarrow & 0 \\ & & \eta\downarrow & & \mu\downarrow & & \mathrm{id}_Y\downarrow & & \\ 0 & \longrightarrow & X & \xrightarrow{f} & Z & \xrightarrow{g} & Y & \longrightarrow & 0 \end{array}$$

commutes, and the left square is cartesian.

(ii) *The resulting exact sequence*

$$0 \longrightarrow \mathbf{F}(Y) \xrightarrow{(\eta,\iota)} X \times \mathbf{E}(Y) \xrightarrow{f-\mu} Z \longrightarrow 0 \tag{2.6}$$

is a projective resolution of Z in $\widetilde{\mathcal{F}}$. In particular, $\widetilde{\mathcal{F}}$ has enough projectives.

(iii) *Z is projective if and only if η is a monomorphism; then $Z \cong X' \times \mathbf{E}(Y)$ for some subobject $X' \hookrightarrow X$.*

(iv) *$\mathbf{E}(Y)$ is a projective cover of Y.*

PROOF. (i) This follows from the isomorphism (2.4) together with Yoneda's lemma.

(ii) By Corollary 2.4, X and $\mathbf{F}(Y)$ are projective objects of $\widetilde{\mathcal{F}}$. Moreover, $\mathbf{E}(Y)$ is projective as well, since $\mathrm{Hom}_{\widetilde{\mathcal{F}}}(\mathbf{E}(Y), ?) \cong \mathrm{Hom}_{\mathcal{Y}}(Y, \widetilde{\mathbf{L}}(?))$, where $\widetilde{\mathbf{L}}$ is exact and $\mathcal{Y}$ is semi-simple.

(iii) If Z is projective, then of course $\mathrm{Ext}^1_{\widetilde{\mathcal{F}}}(Z, X') = 0$ for all $X' \in \widetilde{\mathcal{X}}$. In view of the projective resolution (2.6), it follows that the map

$$\mathrm{Hom}_{\widetilde{\mathcal{X}}}(X, X') \times \mathrm{Hom}_{\widetilde{\mathcal{X}}}(\mathbf{E}(Y), X') \longrightarrow \mathrm{Hom}_{\widetilde{\mathcal{X}}}(\mathbf{F}(Y), X'), \quad (u, v) \longmapsto u \circ \eta + v \circ \iota$$

is surjective. As $\mathrm{Hom}_{\widetilde{\mathcal{X}}}(\mathbf{E}(Y), X') = 0$ by Lemma 2.10, this just means that the map

$$\mathrm{Hom}_{\widetilde{\mathcal{X}}}(X, X') \longrightarrow \mathrm{Hom}_{\widetilde{\mathcal{X}}}(\mathbf{F}(Y), X'), \quad (u, v) \longmapsto u \circ \eta$$

is surjective (alternatively, this follows from the exact sequence of Lemma 2.9). Since $\widetilde{\mathcal{X}}$ is semi-simple, the pull-back map

$$\mathrm{Hom}_{\widetilde{\mathcal{X}}}(\mathbf{F}(Y), X') \longrightarrow \mathrm{Hom}_{\widetilde{\mathcal{X}}}(\mathrm{Ker}(\eta), X')$$

is surjective as well. Thus, $\mathrm{Hom}_{\widetilde{\mathcal{X}}}(\mathrm{Ker}(\eta), X') = 0$ for all X', i.e., $\mathrm{Ker}(\eta) = 0$. Using the semi-simplicity of $\widetilde{\mathcal{X}}$ again, we may choose a subobject $X' \hookrightarrow X$ such that $X = X' \oplus \mathrm{Im}(\eta)$; then the natural map $X' \times \mathbf{E}(Y) \to Z$ is an isomorphism. Conversely, $X' \times \mathbf{E}(Y)$ is projective by (ii).

(iv) This follows from (iii), since we have $\mathrm{Hom}_{\widetilde{\mathcal{F}}}(X', Y) = 0$ for any $X' \in \widetilde{\mathcal{X}}$, and $\mathrm{Hom}_{\widetilde{\mathcal{F}}}(\mathbf{E}(Y'), Y) \cong \mathrm{Hom}_{\mathcal{Y}}(Y', Y)$ for any $Y' \in \mathcal{Y}$. □

Finally, we obtain two homological characterizations of the universal objects $\mathbf{E}(Y)$, the first one being somewhat analogous to the notion of exceptional objects:

LEMMA 2.12. *The following conditions are equivalent for $Z = (X, Y, \eta) \in \widetilde{\mathcal{F}}$:*

(i) *$Z \cong \mathbf{E}(Y)$.*

(ii) *$\mathrm{End}_{\widetilde{\mathcal{F}}}(Z) \cong \mathrm{End}_{\mathcal{Y}}(Y)$ via $\widetilde{\mathbf{L}}$, and $\mathrm{Ext}^1_{\widetilde{\mathcal{F}}}(Z, Z) = 0$.*

(iii) *$\mathrm{Hom}_{\widetilde{\mathcal{F}}}(Z, X') = 0 = \mathrm{Ext}^1_{\widetilde{\mathcal{F}}}(Z, X')$ for all $X' \in \widetilde{\mathcal{X}}$.*

PROOF. (i) ⇒ (ii), (i) ⇒ (iii) This follows from Lemma 2.10 and Proposition 2.11.

(ii) ⇒ (i) In view of the exact sequence of Lemma 2.9, we may rephrase the assumption as follows: for any $v \in \mathrm{End}_{\mathcal{Y}}(Y)$, there exists a unique $u \in \mathrm{End}_{\widetilde{\mathcal{X}}}(X)$ such that $u \circ \eta = \eta \circ F(v)$; moreover, the map $\psi : (f, g) \mapsto f \circ \eta - \eta \circ F(g)$ is surjective. As a consequence, η is an epimorphism (by the uniqueness of u), and $\mathrm{Ker}(\eta)$ is stable under $F(v)$ for any $v \in \mathrm{End}_{\mathcal{Y}}(Y)$ (by the existence of u). Then $\psi(u, v)$ vanishes identically on $\mathrm{Ker}(\eta)$ for any $u \in \mathrm{End}_{\widetilde{\mathcal{X}}}(X)$, $v \in \mathrm{End}_{\mathcal{Y}}(Y)$. As ψ is surjective, this forces $\mathrm{Ker}(\eta) = 0$. Thus, η is an isomorphism, and hence the pair $(\eta : \mathbf{F}(Y) \to X, \mathrm{id} : Y \to Y)$ yields an isomorphism $\mathbf{E}(Y) \to Z$ in $\widetilde{\mathcal{F}}$.

(iii) ⇒ (i) In view of the long exact exact sequence of Lemma 2.9 again, the map

$$\psi : \mathrm{Hom}_{\widetilde{\mathcal{X}}}(X, X') \longrightarrow \mathrm{Hom}_{\widetilde{\mathcal{X}}}(\mathbf{F}(Y), X'), \quad u \longmapsto u \circ \eta$$

is an isomorphism for any $\eta \in \widetilde{\mathcal{X}}$. It follows that η is an isomorphism as well. □

2.5. Relation to module categories. Let $\mathcal{C}$ be an abelian category equipped with a torsion pair $(\mathcal{X}, \mathcal{Y})$ satisfying the assumptions (a), (b), (c), (d) of Subsection 2.4. We assume in addition that $\mathcal{C}$ is a *finite length* category, i.e., every object has a composition series. Then the semi-simple categories $\mathcal{X}$, $\mathcal{Y}$ are of finite length as well; as a consequence, each of them is equivalent to the category of all left modules of finite length over a ring, which can be constructed as follows.

Denote by I the set of isomorphism classes of simple objects of $\mathcal{X}$. Choose a representative S for each class, and let

$$D_S := \mathrm{End}_{\mathcal{X}}(S)^{\mathrm{op}};$$

then D_S is a division ring. Given $X \in \mathcal{X}$, the group $\mathrm{Hom}_{\mathcal{X}}(S, X)$ is a left D_S-vector space of finite dimension; moreover, $\mathrm{Hom}_{\mathcal{X}}(S, X) = 0$ for all but finitely many $S \in I$. Thus,

$$\mathbf{M}(X) := \bigoplus_{S \in I} \mathrm{Hom}_{\mathcal{X}}(S, X)$$

is a left module of finite length over the ring

$$R_{\mathcal{X}} := \bigoplus_{S \in I} D_S.$$

(Notice that every $R_{\mathcal{X}}$-module of finite length is semi-simple; moreover, the ring $R_{\mathcal{X}}$ is semi-simple if and only if I is finite). The assignment $X \mapsto \mathbf{M}(X)$ extends to a covariant functor

$$\mathbf{M}_{\mathcal{X}} : \mathcal{X} \xrightarrow{\cong} R_{\mathcal{X}}\text{-}\mathrm{mod}, \tag{2.7}$$

which is easily seen to be an equivalence of categories. Likewise, we have an equivalence of categories

$$\mathbf{M}_{\mathcal{Y}} : \mathcal{Y} \longrightarrow R_{\mathcal{Y}}\text{-}\mathrm{mod},$$

where $R_{\mathcal{Y}} := \bigoplus_{T \in J} D_T$.

We now make a further (and final) assumption:

(e) The equivalence (2.7) extends to an equivalence of categories

$$\mathbf{M}_{\widetilde{\mathcal{X}}} : \widetilde{\mathcal{X}} \xrightarrow{\cong} R_{\mathcal{X}}\text{-}\mathrm{Mod}^{\mathrm{ss}}. \tag{2.8}$$

The right-hand side of (2.8) is a semi-simple category containing $\mathcal{X}$ as a Serre subcategory, as required by assumption (d). When the set I is finite, assumption (e) just means that $\widetilde{\mathcal{X}} = R_{\mathcal{X}}$-Mod. For an arbitrary set I, the objects of $\widetilde{\mathcal{X}}$ are the direct sums $X = \bigoplus_{S \in I} X_S$, where each X_S is a left D_S-vector space. We say that X_S is the *isotypical component* of X of type S; its dimension (possibly infinite) is the *multiplicity* of S in X.

We now turn to the covariant exact functor $\mathbf{F} : \mathcal{Y} \to \widetilde{\mathcal{X}}$. We may identify each simple object T of $\mathcal{Y}$ with D_T, on which $R_{\mathcal{Y}}$ acts via left multiplication. Then $\mathbf{F}(D_T)$ is a semi-simple left $R_{\mathcal{X}}$-module. Also, $\mathbf{F}(D_T)$ is a right D_T-module, via the action of D_T on itself via right multiplication, which yields an isomorphism

$$D_T \xrightarrow{\cong} \mathrm{End}_{R_{\mathcal{Y}}}(D_T),$$

and the ring homomorphism $\mathrm{End}_{R_{\mathcal{Y}}}(D_T) \to \mathrm{End}_{R_{\mathcal{X}}}(\mathbf{F}(D_T))$ induced by $\mathbf{F}$. As the left and right actions of D_T on itself commute, $\mathbf{F}(D_T)$ is a D_T-$R_{\mathcal{Y}}$-bimodule, which we view as a $R_{\mathcal{X}}$-$R_{\mathcal{Y}}$-bimodule. Let

$$M := \bigoplus_{T \in J} \mathbf{F}(D_T).$$

This is again a $R_{\mathcal{X}}$-$R_{\mathcal{Y}}$-bimodule, semi-simple as a $R_{\mathcal{X}}$-module and as a $R_{\mathcal{Y}}$-module. So we may form the triangular matrix ring

$$R := \begin{pmatrix} R_{\mathcal{Y}} & 0 \\ M & R_{\mathcal{X}} \end{pmatrix}$$

as in [**ARS97**, §III.2]. More specifically, R consists of the triples (x, y, m), where $x \in R_{\mathcal{X}}$, $y \in R_{\mathcal{Y}}$, and $m \in M$; the addition and multiplication are those of the matrices $\left(\begin{smallmatrix} y & 0 \\ m & x \end{smallmatrix}\right)$, using the bimodule structure of M. Note the decomposition $R = (R_{\mathcal{X}} \oplus R_{\mathcal{Y}}) \oplus M$, where $R_{\mathcal{X}} \oplus R_{\mathcal{Y}}$ is a subring, and M is an ideal of square 0.

We say that a R-module Z is *locally finite*, if $Z = X \oplus Y$ as an $R_{\mathcal{X}} \oplus R_{\mathcal{Y}}$-module, where X is a semi-simple $R_{\mathcal{X}}$-module and Y is an $R_{\mathcal{Y}}$-module of finite length. We denote by $R\text{-}\widetilde{\mathrm{mod}}$ the full subcategory of R-Mod with objects the locally finite modules; then R-mod is a Serre subcategory of $R\text{-}\widetilde{\mathrm{mod}}$. We may now state our main homological result:

THEOREM 2.13. *With the above notation and assumptions (a), (b), (c), (d), (e), the abelian categories $\mathcal{F}$ and $\widetilde{\mathcal{F}}$ are hereditary, and $\widetilde{\mathcal{F}}$ has enough projectives. Moreover, there are compatible equivalences of categories*

$$\mathbf{M} : \mathcal{F} \xrightarrow{\cong} R\text{-}\mathrm{mod}, \quad \widetilde{\mathbf{M}} : \widetilde{\mathcal{F}} \xrightarrow{\cong} R\text{-}\widetilde{\mathrm{mod}}.$$

PROOF. The first assertion follows by combining Corollary 2.4, Lemma 2.5 and Proposition 2.11.

To show the second assertion, we can freely replace $\mathcal{X}$, $\widetilde{\mathcal{X}}$, $\mathcal{Y}$ with compatibly equivalent categories in the construction of $\mathcal{F}$, $\widetilde{\mathcal{F}}$. Thus, we may assume that $\mathcal{X} = R_{\mathcal{X}}$-mod, $\widetilde{\mathcal{X}} = R_{\mathcal{X}}\text{-}\mathrm{Mod}^{\mathrm{ss}}$ and $\mathcal{Y} = R_{\mathcal{Y}}$-mod.

The category of all left R-modules is equivalent to the category of triples (X, Y, f), where X is a $R_{\mathcal{X}}$-module, Y a $R_{\mathcal{Y}}$-module, and $f : M \otimes_{R_{\mathcal{Y}}} Y \to X$ a morphism of $R_{\mathcal{X}}$-modules. The morphisms from (X, Y, f) to (X', Y', f') are the pairs (u, v), where $u \in \mathrm{Hom}_{\mathcal{X}}(X, X')$, $v \in \mathrm{Hom}_{\mathcal{Y}}(Y, Y')$, and the following diagram commutes:

$$\begin{array}{ccc} M \otimes_{R_{\mathcal{Y}}} Y & \xrightarrow{\mathrm{id}_M \otimes v} & M \otimes_{R_{\mathcal{Y}}} Y' \\ {\scriptstyle f}\downarrow & & {\scriptstyle f'}\downarrow \\ X & \xrightarrow{u} & X'. \end{array}$$

(This result is obtained in [**ARS97**, Prop. III.2.2] for modules of finite length over an Artin algebra. The proof adapts without change to the present setting). Moreover, the full subcategory R-mod (resp. $R\text{-}\widetilde{\mathrm{mod}}$) is equivalent to the full subcategory of triples (X, Y, f), where X and Y have finite length (resp. X is semi-simple and Y has finite length).

To complete the proof, it suffices to show that the covariant exact functor $\mathbf{F} : \mathcal{Y} \to \widetilde{\mathcal{X}}$ is isomorphic to $M \otimes_{R_{\mathcal{Y}}} -$. As $\mathbf{F}$ commutes with finite direct sums, we have $\mathbf{F} = \bigoplus_{T \in J} \mathbf{F}_T$ for covariant exact functors $\mathbf{F}_T : D_T\text{-mod} \to \widetilde{\mathcal{X}}$. We now argue as in the proof of the Eilenberg-Watts theorem (see [**Ba68**, Thm. II.2.3]).

Given a left D_T-vector space V, every $v \in V$ yields a D_T-morphism $v : D_T \to V$, and hence a $R_{\mathcal{X}}$-morphism $\mathbf{F}_T(v) : \mathbf{F}_T(D_T) = \mathbf{F}(D_T) \to \mathbf{F}_T(V)$. The resulting map $V \to \mathrm{Hom}_{R_{\mathcal{X}}}(\mathbf{F}(D_T), \mathbf{F}_T(V))$ is easily checked to be a D_T-morphism. In view of the natural isomorphism

$$\mathrm{Hom}_{D_T}(V, \mathrm{Hom}_{R_{\mathcal{X}}}(\mathbf{F}(D_T), \mathbf{F}_T(V))) \cong \mathrm{Hom}_{R_{\mathcal{X}}}(\mathbf{F}(D_T) \otimes_{D_T} V, \mathbf{F}_T(V)),$$

this yields a functorial map

$$f_V : \mathbf{F}(D_T) \otimes_{D_T} V \longrightarrow \mathbf{F}_T(V).$$

When $V = D_T$, one checks that f_V is identified to the identity map of $\mathbf{F}(D_T)$; moreover, the formation of f_V commutes with finite direct sums, since so does $\mathbf{F}_T$. So f_V yields an isomorphism of functors $\mathbf{F}_T \cong \mathbf{F}(D_T) \otimes_{D_T} -$. □

REMARK 2.14. Instead of assumption (e), we may make the stronger and much simpler assumption that $\widetilde{\mathcal{X}} = \mathcal{X}$. (This holds for the isogeny category of vector extensions of abelian varieties, as we will see in §3.2.3). Then we obtain as in the proof of Theorem 3.5 that $\widetilde{\mathcal{F}} = \mathcal{F}$ *is hereditary, has enough projectives, and is equivalent to* R-mod.

Next, we obtain a separation property of the above ring R, and we describe its center $Z(R)$ as well as the center of the abelian category R-mod. We denote by Z_S (resp. Z_T) the center of the division ring D_S (resp. D_T) for any $S \in I$, $T \in J$.

PROPOSITION 2.15.

(i) *The intersection of all the left ideals of finite colength in R is zero.*
(ii) *The center $Z(R)$ consists of the triples $(x, y, 0)$, where $x = \sum_S x_S \in \bigoplus_{S \in I} Z_S$, $y = \sum_T y_T \in \bigoplus_{T \in J} Z_T$ and $x_S m = m y_T$ for all $m \in \mathbf{F}(D_T)$.*
(iii) *The center of R-mod is the completion of $Z(R)$, consisting of the pairs (x, y), where $x = (x_S) \in \prod_{S \in I} Z_S$, $y = (y_T) \in \prod_{T \in J} Z_T$ and $x_S m = m y_T$ for all $m \in \mathbf{F}(D_T)$.*

PROOF. (i) Given $S \in I$ and $T \in J$, we may form the triangular matrix ring

$$R_{S,T} := \begin{pmatrix} D_T & 0 \\ \mathbf{F}(D_T)_S & D_S \end{pmatrix},$$

where $\mathbf{F}(D_T)_S$ denotes the isotypical component of type S of the $R_{\mathcal{X}}$-module $\mathbf{F}(D_T)$. Clearly, $R_{S,T}$ is the quotient of R by a two-sided ideal $I_{S,T}$, and $\bigcap_{S,T} I_{S,T} = 0$. Thus, it suffices to show the assertion for R replaced with $R_{S,T}$.

The left D_S-vector space $\mathbf{F}(D_T)_S$ contains a family of subspaces $(M_a)_{a \in A}$ such that the dimension of each quotient $\mathbf{F}(D_T)/M_a$ is finite, and $\bigcap_{a \in A} M_a = 0$. Then $\left(\begin{smallmatrix} 0 & 0 \\ M_a & 0 \end{smallmatrix}\right)$ is a left ideal of $R_{S,T}$, as well as $\left(\begin{smallmatrix} D_T & 0 \\ 0 & 0 \end{smallmatrix}\right)$ and $\left(\begin{smallmatrix} 0 & 0 \\ 0 & D_S \end{smallmatrix}\right)$. Moreover, all these left ideals have finite colength, and their intersection is zero.

(ii) This is a direct verification.

(iii) Recall that the center of R-mod consists of the families $z = (z_N)_{N \in R\text{-mod}}$ such that $z_N \in \mathrm{End}_R(N)$ and $f \circ z_N = z_{N'} \circ f$ for any $f \in \mathrm{Hom}_R(N, N')$; in particular, z_N is central in $\mathrm{End}_R(N)$. Thus, $z_S \in Z_S$ and $z_T \in Z_T$ for all S, T. Since $N = X \oplus Y$ as a $R_{\mathcal{X}} \oplus R_{\mathcal{Y}}$-module, we see that z is uniquely determined by the families $(z_S)_{S \in I}$, $(z_T)_{T \in J}$. Moreover, we have $z_S m = m z_T$ for all $m \in \mathbf{F}(D_T)_S$, as follows e.g. from Lemma 2.9. Thus, the center of R-mod is contained in the completion of $Z(R)$. The opposite inclusion follows from (i). □

REMARK 2.16. Assume that each D_S-D_T-bimodule $\mathbf{F}(D_T)_S$ contains a family of sub-bimodules $(N_a)_{a\in A}$ such that each quotient $\mathbf{F}(D_T)_S$ has finite length as a D_S-module, and $\bigcap_{a\in A} N_a = 0$. Then R satisfies a stronger separation property, namely, its two-sided ideals of finite colength (as left modules) have zero intersection.

The above assumption obviously holds if $\mathbf{F}(D_T)_S$ has finite length as a D_S-module.It also holds if both D_S and D_T are (say) of characteristic 0 and finite-dimensional over $\mathbb{Q}$; indeed, $\mathbf{F}(D_T)_S$ is a module over $D_S \otimes_{\mathbb{Q}} D_T^{\mathrm{op}}$, and the latter is a finite-dimensional semi-simple $\mathbb{Q}$-algebra.

3. Applications to commutative algebraic groups

3.1. Some isogeny categories.

3.1.1. *Algebraic groups* [**DG70**, **SGA3**, **Br16**]. Throughout this section, we fix a ground field k, with algebraic closure $\bar{k}$ and characteristic $\mathrm{char}(k)$. An *algebraic group* G is a group scheme of finite type over k. A *subgroup* $H \subset G$ is a k-subgroup scheme; then H is a closed subscheme of G. When $\mathrm{char}(k) = 0$, every algebraic group is smooth.

Unless otherwise mentioned, all algebraic groups will be assumed *commutative.* They form the objects of an abelian category $\mathcal{C}$, with morphisms the homomorphisms of k-group schemes (see [**SGA3**, VIA, Thm. 5.4.2]). Every object in $\mathcal{C}$ is artinian (since every decreasing sequence of closed subschemes of a scheme of finite type eventually terminates), but generally not noetherian: in the multiplicative group $\mathbb{G}_m$, the subgroups of roots of unity of order ℓ^n, where ℓ is a fixed prime and n a non-negative integer, form an infinite ascending chain.

The finite group schemes form a Serre subcategory $\mathcal{F}$ of $\mathcal{C}$. The quotient category $\mathcal{C}/\mathcal{F}$ is equivalent to the localization of $\mathcal{C}$ with respect to the multiplicative system of *isogenies*, i.e., of morphisms with finite kernel and cokernel. Also, $\mathcal{C}/\mathcal{F}$ is equivalent to its full subcategory $\underline{\mathcal{C}}$ with objects the smooth connected algebraic groups (see [**Br16**, Lem. 3.1]). We say that $\underline{\mathcal{C}}$ is the *isogeny category of algebraic groups.* Every object of $\underline{\mathcal{C}}$ is artinian and noetherian, i.e., $\underline{\mathcal{C}}$ *is a finite length category* (see [**Br16**, Prop. 3.2]).

Let G be an algebraic group, with group law denoted additively. For any integer n, we have the multiplication map

$$n_G : G \longrightarrow G, \quad x \longmapsto nx.$$

We say that G is *divisible*, if n_G is an epimorphism for any $n \neq 0$. When $\mathrm{char}(k) = 0$, this is equivalent to G being connected; when $\mathrm{char}(k) > 0$, the divisible algebraic groups are the semi-abelian varieties (these will be discussed in detail in §3.2.4).

If G is divisible, then the natural map

$$\mathbb{Z} \longrightarrow \mathrm{End}_{\mathcal{C}}(G), \quad n \longmapsto n_G$$

extends to a homomorphism $\mathbb{Q} \to \mathrm{End}_{\underline{\mathcal{C}}}(G)$; in other terms, $\mathrm{End}_{\underline{\mathcal{C}}}(G)$ is a $\mathbb{Q}$-algebra. As a consequence, $\mathrm{Ext}^n_{\underline{\mathcal{C}}}(G, G')$ is a $\mathbb{Q}$-vector space for any divisible algebraic groups G, G' and any integer $n \geq 0$. By [**Br16**, Prop. 3.6]), the induced maps

$$\mathrm{Hom}_{\mathcal{C}}(G, G')_{\mathbb{Q}} \longrightarrow \mathrm{Hom}_{\underline{\mathcal{C}}}(G, G'), \quad \mathrm{Ext}^1_{\mathcal{C}}(G, G')_{\mathbb{Q}} \longrightarrow \mathrm{Ext}^1_{\underline{\mathcal{C}}}(G, G'), \tag{3.1}$$

are isomorphisms, where we set $M_{\mathbb{Q}} := M \otimes_{\mathbb{Z}} \mathbb{Q}$ for any abelian group M.

In particular, *the isogeny category* $\underline{\mathcal{C}}$ *is* $\mathbb{Q}$*-linear when* $\mathrm{char}(k) = 0$; then its objects are just the connected algebraic groups, and its morphisms are the rational multiples of morphisms in $\mathcal{C}$.

Given an extension of fields K/k and an algebraic group G over k, we obtain an algebraic group over K,

$$G_K := G \otimes_k K = G \times_{\mathrm{Spec}(k)} \mathrm{Spec}(K),$$

by extension of scalars. The assignment $G \mapsto G_K$ extends to the *base change functor*

$$\otimes_k K : \mathcal{C} = \mathcal{C}_k \longrightarrow \mathcal{C}_K,$$

which is exact and faithful. Also, G is finite if and only if G_K is finite. As a consequence, we obtain a base change functor, still denoted by

$$\otimes_k K : \underline{\mathcal{C}}_k \longrightarrow \underline{\mathcal{C}}_K,$$

and which is still exact and faithful. When K/k is purely inseparable, the above functor is an equivalence of categories (see [**Br16**, Thm. 3.11]). We say that $\underline{\mathcal{C}}_k$ is *invariant under purely inseparable field extensions.*

Thus, to study $\underline{\mathcal{C}}_k$ when $\mathrm{char}(k) = p > 0$, we may replace k with its *perfect closure*, $k_i := \bigcup_{n \geq 0} k^{1/p^n} \subset \bar{k}$, and hence assume that k is perfect. This will be very useful, since the structure of algebraic groups is much better understood over a perfect ground field (see e.g. §3.1.5).

3.1.2. *Linear algebraic groups, affine group schemes* [**DG70**, III.3]. A (possibly non-commutative) algebraic group G is called *linear* if G is isomorphic to a subgroup scheme of the general linear group GL_n for some integer $n > 0$; this is equivalent to G being affine (see e.g. [**Br15**, Prop. 3.1.1]). The smooth linear algebraic groups are the "linear algebraic groups defined over k" in the sense of [**Bo91**].

For any exact sequence

$$0 \longrightarrow G_1 \longrightarrow G \longrightarrow G_2 \longrightarrow 0$$

in $\mathcal{C}$, the group G is affine if and only if G_1 and G_2 are affine (see e.g. [**Br15**, Prop. 3.1.2]). Thus, the (commutative) linear algebraic groups form a Serre subcategory $\mathcal{L}$ of $\mathcal{C}$, which contains $\mathcal{F}$.

The property of being affine is also invariant under field extensions and isogenies, in the following sense: an algebraic group G is affine if and only if G_K is affine for some field extension K of k, if and only if H is affine for some isogeny $f : G \to H$. It follows that the quotient category $\mathcal{L}/\mathcal{F}$ is equivalent to its full subcategory $\underline{\mathcal{L}}$ with objects the smooth connected linear algebraic groups. Moreover, $\underline{\mathcal{L}}$ is invariant under purely inseparable field extensions.

The affine k-group schemes (not necessarily of finite type) form an abelian category $\widetilde{\mathcal{L}}$, containing $\mathcal{L}$ as a Serre subcategory. Moreover, every affine group scheme G is the filtered inverse limit of linear algebraic groups, quotients of G (see [**DG70**, III.3.7.4, III.3.7.5]). In fact, $\widetilde{\mathcal{L}}$ is the pro-completion of the abelian category $\mathcal{L}$, in the sense of [**DG70**, V.2.3.1].

We say that a group scheme G is *pro-finite*, if G is an inverse limit of finite group schemes; equivalently, G is affine and every algebraic quotient group of G is finite. The pro-finite group schemes form a Serre subcategory $\widetilde{\mathcal{FL}}$ of $\widetilde{\mathcal{L}}$. The quotient category $\widetilde{\mathcal{L}}/\widetilde{\mathcal{FL}}$ is the *isogeny category of affine group schemes*; it contains $\mathcal{L}/\mathcal{F}$ as a Serre subcategory.

3.1.3. *Groups of multiplicative type* [**DG70**, IV.1]. The invertible diagonal matrices form a subgroup scheme $\mathrm{D}_n \subset \mathrm{GL}_n$, which is commutative, smooth and connected; moreover, $\mathrm{D}_1 = \mathrm{GL}_1$ is isomorphic to the multiplicative group $\mathbb{G}_m$, and

$\mathrm{D}_n \cong \mathbb{G}_m^n$ (the product of n copies of $\mathbb{G}_m$). An algebraic group G is said to be *diagonalizable*, if G is isomorphic to a subgroup of D_n for some n. Also, G is called *of multiplicative type* (resp. a *torus*), if the base change $G_{\bar{k}}$ is diagonalizable (resp. isomorphic to some $\mathrm{D}_{n,\bar{k}}$). Both properties are invariant under field extensions, but not under isogenies. Also, the tori are the smooth connected algebraic groups of multiplicative type. The diagonalizable algebraic groups (resp. the algebraic groups of multiplicative type) form a Serre subcategory $\mathcal{D}$ (resp. $\mathcal{M}$) of $\mathcal{C}$.

For any diagonalizable algebraic group G, the *character group*

$$\mathbf{X}(G) := \mathrm{Hom}_{\mathcal{C}}(G, \mathbb{G}_m)$$

is a finitely generated abelian group. Moreover, the assignment $G \mapsto \mathbf{X}(G)$ extends to an anti-equivalence of categories

$$\mathbf{X} : \mathcal{D} \longrightarrow \mathbb{Z}\text{-Mod}^{\mathrm{fg}},$$

where the right-hand side denotes the category of finitely generated abelian groups (see [**DG70**, IV.1.1] for these results).

Given an algebraic group of multiplicative type G, there exists a finite Galois extension of fields K/k such that G_K is diagonalizable. Thus, G_{k_s} is diagonalizable, where k_s denotes the *separable closure* of k in $\bar{k}$. Let

$$\Gamma := \mathrm{Gal}(k_s/k) = \mathrm{Aut}(\bar{k}/k)$$

denote the *absolute Galois group* of k. Then Γ is a pro-finite topological group, the inverse limit of its finite quotients $\mathrm{Gal}(K/k)$, where K runs over the finite Galois field extensions of k. Also, Γ acts on the character group,

$$\mathbf{X}(G) := \mathrm{Hom}_{\mathcal{C}_{k_s}}(G_{k_s}, \mathbb{G}_{m,k_s}),$$

and the stabilizer of any character is an open subgroup. Thus, $\mathbf{X}(G)$ is a *discrete Galois module* in the sense of [**Se97**, §2.1]. Moreover, G is diagonalizable if and only if Γ fixes $\mathbf{X}(G)$ pointwise; then the base change map

$$\mathrm{Hom}_{\mathcal{C}_k}(G, \mathbb{G}_{m,k}) \longrightarrow \mathrm{Hom}_{\mathcal{C}_{k_s}}(G_{k_s}, \mathbb{G}_{m,k_s})$$

is an isomorphism, i.e., the two notions of character groups are compatible. Furthermore, G is a torus (resp. finite) if and only if the abelian group $\mathbf{X}(G)$ is free (resp. finite); also, note that the tori are the divisible algebraic groups of multiplicative type.

The above assignment $G \mapsto \mathbf{X}(G)$ yields an anti-equivalence of categories

$$\mathbf{X} : \mathcal{M} \longrightarrow \mathbb{Z}\Gamma\text{-Mod}^{\mathrm{fg}} \tag{3.2}$$

(*Cartier duality*), where the right-hand side denotes the category of discrete Γ-modules which are finitely generated as abelian groups. Moreover, the abelian group $\mathrm{Hom}_{\mathcal{C}}(T, T')$ is free of finite rank, for any tori T and T' (see [**DG70**, IV.1.2, IV.1.3] for these results).

Consider the full subcategory $\mathcal{FM}$ of $\mathcal{M}$ with objects the finite group schemes of multiplicative type. Then $\mathcal{FM}$ is a Serre subcategory of $\mathcal{M}$, anti-equivalent via $\mathbf{X}$ to the category of finite discrete Γ-modules. Moreover, the quotient category $\mathcal{M}/\mathcal{FM}$ is equivalent to its full subcategory $\underline{\mathcal{T}}$ with objects the tori, and we have an anti-equivalence of categories

$$\mathbf{X}_{\mathbb{Q}} : \underline{\mathcal{T}} \longrightarrow \mathbb{Q}\Gamma\text{-}\mathrm{mod}, \quad T \longmapsto \mathbf{X}(T)_{\mathbb{Q}}. \tag{3.3}$$

Here $\mathbb{Q}\Gamma$-mod denotes the category of finite-dimensional $\mathbb{Q}$-vector spaces equipped with a discrete linear action of Γ; note that $\mathbb{Q}\Gamma$-mod is semi-simple, $\mathbb{Q}$-linear and invariant under purely inseparable field extensions. In view of (3.1), this yields natural isomorphisms

$$\mathrm{Hom}_{\underline{\mathcal{T}}}(T, T') \cong \mathrm{Hom}_{\mathcal{C}}(T, T')_{\mathbb{Q}} \cong \mathrm{Hom}^{\Gamma}(\mathbf{X}(T')_{\mathbb{Q}}, \mathbf{X}(T)_{\mathbb{Q}})$$

for any tori T, T'. As a consequence, *the isogeny category $\underline{\mathcal{T}}$ is semi-simple, $\mathbb{Q}$-linear, Hom-finite, and invariant under purely inseparable field extensions.*

Next, we extend the above results to affine k-group schemes, not necessarily algebraic, by using again results of [**DG70**, IV.1.2, IV.1.3]. We say that an affine group scheme G is of multiplicative type, if so are all its algebraic quotient groups. Denote by $\widetilde{\mathcal{M}}$ the full subcategory of $\widetilde{\mathcal{L}}$ with objects the group schemes of multiplicative type; then $\widetilde{\mathcal{M}}$ is a Serre subcategory of $\widetilde{\mathcal{L}}$. Moreover, the Cartier duality (3.2) extends to an anti-equivalence of categories

$$\mathbf{X} : \widetilde{\mathcal{M}} \longrightarrow \mathbb{Z}\Gamma\text{-}\mathrm{Mod}, \tag{3.4}$$

where $\mathbb{Z}\Gamma$-Mod stands for the category of all discrete Γ-modules. Note that $\mathbb{Z}\Gamma$-Mod is an abelian category, containing $\mathbb{Z}\Gamma\text{-Mod}^{\mathrm{fg}}$ as a Serre subcategory.

Consider the full subcategory $\mathbb{Z}\Gamma\text{-Mod}^{\mathrm{tors}} \subset \mathbb{Z}\Gamma$-Mod with objects the discrete Γ-modules which are torsion as abelian groups. Then $\mathbb{Z}\Gamma\text{-Mod}^{\mathrm{tors}}$ is a Serre subcategory of $\mathbb{Z}\Gamma$-Mod, anti-equivalent via $\mathbf{X}$ to the full subcategory $\widetilde{\mathcal{FM}} \subset \widetilde{\mathcal{M}}$ with objects the pro-finite group schemes of multiplicative type.

For any $M \in \mathbb{Z}\Gamma$-Mod, the kernel and cokernel of the natural map $M \to M_{\mathbb{Q}}$ are torsion. It follows readily that the induced covariant functor $\mathbb{Z}\Gamma\text{-}\mathrm{Mod} \to \mathbb{Q}\Gamma\text{-}\mathrm{Mod}$ yields an equivalence of categories

$$(\mathbb{Z}\Gamma\text{-}\mathrm{Mod})/(\mathbb{Z}\Gamma\text{-}\mathrm{Mod}^{\mathrm{tors}}) \xrightarrow{\cong} \mathbb{Q}\Gamma\text{-}\mathrm{Mod},$$

where $\mathbb{Q}\Gamma$-Mod denotes the category of all $\mathbb{Q}$-vector spaces equipped with a discrete linear action of Γ. Thus, the category $\mathbb{Q}\Gamma$-Mod is anti-equivalent to $\widetilde{\mathcal{M}}/\widetilde{\mathcal{FM}}$. In turn, the latter category is equivalent to its full subcategory with objects the inverse limits of tori: the *isogeny category of pro-tori*, that we denote by $\underline{\widetilde{\mathcal{T}}}$. Clearly, the category $\mathbb{Q}\Gamma$-Mod is semi-simple. Thus, $\underline{\widetilde{\mathcal{T}}}$ is semi-simple as well; its simple objects are the simple tori, i.e., the tori G such that every subgroup $H \subsetneq G$ is finite.

As in Subsection 2.5, choose representatives T of the set I of isomorphism classes of simple tori, and let $D_T := \mathrm{End}_{\underline{\mathcal{T}}}(T)^{\mathrm{op}}$; then each D_T is a division ring of finite dimension over $\mathbb{Q}$. Let

$$R_{\underline{\mathcal{T}}} := \bigoplus_{T \in I} D_T,$$

then we have an equivalence of categories

$$\underline{\mathcal{T}} \xrightarrow{\cong} R_{\underline{\mathcal{T}}}\text{-}\mathrm{mod}$$

which extends to an equivalence of categories

$$\underline{\widetilde{\mathcal{T}}} \xrightarrow{\cong} R_{\underline{\mathcal{T}}}\text{-}\mathrm{Mod}^{\mathrm{ss}}.$$

Note finally that $\underline{\widetilde{\mathcal{T}}}$ is equivalent to a Serre subcategory of the isogeny category $\underline{\widetilde{\mathcal{L}}}$.

3.1.4. *Unipotent groups, structure of linear groups* [**DG70**, IV.2, IV.3]. The upper triangular matrices with all diagonal entries equal to 1 form a subgroup scheme $\mathrm{U}_n \subset \mathrm{GL}_n$, which is smooth and connected; moreover, U_1 is isomorphic to the additive group $\mathbb{G}_a$. An algebraic group G is called *unipotent* if G is isomorphic to a subgroup of U_n for some n. The (commutative) unipotent algebraic groups form a Serre subcategory $\mathcal{U}$ of $\mathcal{L}$. Also, the property of being unipotent is invariant under field extensions, in the sense of §3.1.1.

We say that an affine group scheme G is unipotent, if so are all algebraic quotients of G (this differs from the definition given in [**DG70**, IV.2.2.2], but both notions are equivalent in view of [**DG70**, IV.2.2.3]). The unipotent group schemes form a Serre subcategory $\widetilde{\mathcal{U}}$ of $\widetilde{\mathcal{L}}$, which is the pro-completion of $\mathcal{U}$ (as defined in [**DG70**, V.2.3.1]).

By [**DG70**, IV.2.2.4, IV.3.1.1], every affine group scheme G lies in a unique exact sequence

$$0 \longrightarrow M \longrightarrow G \longrightarrow U \longrightarrow 0, \tag{3.5}$$

where M is of multiplicative type and U is unipotent; moreover, $\mathrm{Hom}_{\widetilde{\mathcal{L}}}(M,U) = 0 = \mathrm{Hom}_{\widetilde{\mathcal{L}}}(U,M)$. Thus, $(\widetilde{\mathcal{M}},\widetilde{\mathcal{U}})$ (resp. $(\mathcal{M},\mathcal{U})$) is a torsion pair of Serre subcategories of $\widetilde{\mathcal{L}}$ (resp. $\mathcal{L}$), as considered in Subsection 2.2.

If the field k is perfect, then the exact sequence (3.5) has a unique splitting (see [**DG70**, IV.3.1.1]). It follows that the assignment $(M,U) \mapsto M \times U$ yields equivalences of categories

$$\widetilde{\mathcal{M}} \times \widetilde{\mathcal{U}} \xrightarrow{\cong} \widetilde{\mathcal{L}}, \quad \mathcal{M} \times \mathcal{U} \xrightarrow{\cong} \mathcal{L}.$$

In turn, this yields equivalences of isogeny categories

$$\underline{\widetilde{\mathcal{T}}} \times \underline{\widetilde{\mathcal{U}}} \xrightarrow{\cong} \underline{\widetilde{\mathcal{L}}}, \quad \underline{\mathcal{T}} \times \underline{\mathcal{U}} \xrightarrow{\cong} \underline{\mathcal{L}}. \tag{3.6}$$

In fact, *the latter equivalences hold over an arbitrary field*, as $\underline{\mathcal{T}}$, $\underline{\mathcal{U}}$ and $\underline{\mathcal{L}}$ are invariant under purely inseparable field extensions.

We now assume that $\mathrm{char}(k) = 0$. Then every unipotent algebraic group G is isomorphic to the direct sum of n copies of $\mathbb{G}_a$, where $n := \dim(G)$. In particular, G is isomorphic as a scheme to the affine space $\mathbb{A}^n$, and hence is smooth and connected. Moreover, every morphism of unipotent groups $f : G \to H$ is linear in the corresponding coordinates $x_1, \ldots, x_n$ on G. Thus, *the category $\mathcal{U}$ is equivalent to the category k-mod of finite-dimensional k-vector spaces.* This extends to an equivalence of $\widetilde{\mathcal{U}}$ to the category k-Mod of all k-vector spaces (see [**DG70**, IV.2.4.2]).

As every finite unipotent group is trivial, $\mathcal{U}$ and $\widetilde{\mathcal{U}}$ are their own isogeny categories; they are obviously semi-simple and k-linear, and $\mathcal{U}$ is Hom-finite. In view of the equivalence (3.6), it follows that *$\underline{\mathcal{L}}$ and $\underline{\widetilde{\mathcal{L}}}$ are semi-simple.*

3.1.5. *Unipotent groups in positive characteristics* [**DG70**, V.1, V.3]. Throughout this paragraph, we assume that $\mathrm{char}(k) = p > 0$; then every unipotent algebraic group G is p-torsion. The structure of these groups is much more complicated than in characteristic 0. For example, the additive group $\mathbb{G}_a$ admits many finite subgroups, e.g., the (schematic) kernel of the Frobenius endomorphism

$$F : \mathbb{G}_a \longrightarrow \mathbb{G}_a, \quad x \longmapsto x^p.$$

The ring $\mathrm{End}_{\mathcal{U}}(\mathbb{G}_a)$ is generated by k (acting by scalar multiplication) and F, with relations $Fx - x^p F = 0$ for any $x \in k$ (see [**DG70**, II.3.4.4]).

Assume in addition that k is perfect. Then the categories $\mathcal{U}$ and $\widetilde{\mathcal{U}}$ may be described in terms of modules over the *Dieudonné ring* $\mathbb{D}$ (see [**DG70**, V.1]). More specifically, $\mathbb{D}$ is a noetherian domain, generated by the ring of Witt vectors $W(k)$, the Frobenius F and the Verschiebung V; also, R is non-commutative unless $k = \mathbb{F}_p$. The left ideal $\mathbb{D}V \subset \mathbb{D}$ is two-sided, and the quotient ring $\mathbb{D}/\mathbb{D}V$ is isomorphic to $\mathrm{End}_{\mathcal{U}}(\mathbb{G}_a)$. More generally, for any positive integer n, the left ideal $\mathbb{D}V^n$ is two-sided and $\mathbb{D}/\mathbb{D}V^n \cong \mathrm{End}_{\mathcal{U}}(W_n)$, where W_n denotes the *group of Witt vectors of length* n; this is a smooth connected unipotent group of dimension n, which lies in an exact sequence

$$0 \longrightarrow W_n \longrightarrow W_{n+1} \longrightarrow \mathbb{G}_a \longrightarrow 0.$$

The $\mathrm{End}_{\mathcal{U}}(\mathbb{G}_a)$-module $\mathrm{Ext}^1_{\mathcal{U}}(\mathbb{G}_a, W_n)$ is freely generated by the class of the above extension. Moreover, the assignment

$$G \longmapsto \mathbf{M}(G) := \varinjlim \mathrm{Hom}_{\mathcal{U}}(G, W_n)$$

extends to an anti-equivalence $\mathbf{M}$ of $\widetilde{\mathcal{U}}$ with the full subcategory of $\mathbb{D}$-Mod with objects V-torsion modules. Also, G is algebraic (resp. finite) if and only if $\mathbf{M}(G)$ is finitely generated (resp. of finite length); we have $\mathbf{M}(W_n) = \mathbb{D}/\mathbb{D}V^n$ for all n. As a consequence, $\mathbf{M}$ restricts to an anti-equivalence of $\mathcal{U}$ with the full subcategory of $\mathbb{D}$-Mod with objects the finitely generated modules M which are V-torsion.

This yields a description of the isogeny categories $\underline{\mathcal{U}}, \underline{\widetilde{\mathcal{U}}}$ in terms of module categories. Let $S := \mathbb{D} \setminus \mathbb{D}V$; then we may form the left ring of fractions

$$S^{-1}\mathbb{D} =: R = R_{\underline{\mathcal{U}}}$$

by [**DG70**, V.3.6.3]. This is again a (generally non-commutative) noetherian domain; its left ideals are the two-sided RV^n in view of [**DG70**, V.3.6.11]. In particular, R has a unique maximal ideal, namely, RV; moreover, the quotient ring R/RV is isomorphic to the division ring of fractions of $\mathrm{End}_{\mathcal{U}}(\mathbb{G}_a)$. Thus, R is a discrete valuation domain (not necessarily commutative), as considered in [**KT07**]. By [**DG70**, V.3.6.7], a morphism of unipotent group schemes $f : G \to H$ is an isogeny if and only if the associated morphism $S^{-1}\mathbf{M}(f) : S^{-1}\mathbf{M}(H) \to S^{-1}\mathbf{M}(G)$ is an isomorphism. As a consequence, $S^{-1}\mathbf{M}$ yields an anti-equivalence of $\underline{\mathcal{U}}$ (resp. $\underline{\widetilde{\mathcal{U}}}$) with R-mod (resp. R-$\mathrm{Mod}^{\mathrm{tors}}$), where the latter denotes the full subcategory of R-Mod with objects the V-torsion modules.

We now show that the abelian category R-$\mathrm{Mod}^{\mathrm{tors}}$ is hereditary, and has enough projectives and a unique indecomposable projective object. Let $M \in R$-$\mathrm{Mod}^{\mathrm{tors}}$ and choose an exact sequence in R-Mod

$$0 \longrightarrow M \longrightarrow I \longrightarrow J \longrightarrow 0,$$

where I is injective in R-Mod; equivalently, the multiplication by V in I is surjective. Thus, J is injective in R-Mod as well. Let $I^{\mathrm{tors}} \subset I$ be the largest V-torsion submodule. Then we have an exact sequence in R-Mod

$$0 \longrightarrow M \longrightarrow I^{\mathrm{tors}} \longrightarrow J^{\mathrm{tors}} \longrightarrow 0.$$

Moreover, I^{tors}, J^{tors} are injective in R-Mod, and hence in R-$\mathrm{Mod}^{\mathrm{tors}}$ as well. As the injective objects of R-Mod are direct sums of copies of the division ring of fractions $K := \mathrm{Fract}(R)$ and of the quotient K/R (see e.g. [**KT07**, Thm. 6.3]), it follows that the abelian category R-$\mathrm{Mod}^{\mathrm{tors}}$ is hereditary and has enough injectives; moreover, it has a unique indecomposable injective object, namely,

$$K/R = \varinjlim RV^{-n}/R \cong \varinjlim R/RV^n$$

(the injective hull of the simple module). Also, note that R-Mod^{tors} is equipped with a duality (i.e., an involutive contravariant exact endofunctor), namely, the assignment $M \mapsto \mathrm{Hom}_R(M, K/R)$. As a consequence, we obtain an equivalence of $\widetilde{\underline{\mathcal{U}}}$ with R-Mod^{tors}, which restricts to an equivalence of $\underline{\mathcal{U}}$ with R-mod.

Thus, $\widetilde{\underline{\mathcal{U}}}$ *is hereditary and has enough projectives; its unique indecomposable projective object is* $W := \varprojlim W_n$. Also, by [**DG70**, V.3.6.11] (see also [**KT07**, Thm. 4.8]), every unipotent algebraic group is isogenous to $\bigoplus_{n \geq 1} a_n W_n$ for uniquely determined integers $a_n \geq 0$. In other terms, *every indecomposable object of* $\underline{\mathcal{U}}$ *is isomorphic to* W_n *for a unique* $n \geq 1$.

Note finally that the above structure results for $\underline{\mathcal{U}}$ extend to an arbitrary field k of characteristic p, by invariance under purely inseparable field extensions. More specifically, $\underline{\mathcal{U}}$ is equivalent to R-mod, where R denotes the ring constructed as above from the perfect closure k_i.

3.2. More isogeny categories.

3.2.1. *Abelian varieties* [**Mi86**]. An *abelian variety* is a smooth, connected algebraic group A which is proper as a k-scheme. Then A is a projective variety and a divisible commutative group scheme; its group law will be denoted additively. Like for tori, the abelian group $\mathrm{Hom}_{\mathcal{C}}(A, A')$ is free of finite rank for any abelian varieties A and A'. Moreover, we have the Poincaré complete reducibility theorem: for any abelian variety A and any abelian subvariety $B \subset A$, there exists an abelian subvariety $C \subset A$ such that the map

$$B \times C \longrightarrow A, \quad (x, y) \longmapsto x + y$$

is an isogeny.

We denote by $\mathcal{P}$ the full subcategory of $\mathcal{C}$ with objects the proper algebraic groups; then $\mathcal{P}$ is a Serre subcategory of $\mathcal{C}$, containing $\mathcal{F}$ and invariant under field extensions. Moreover, the quotient category $\mathcal{P}/\mathcal{F}$ is equivalent to its full subcategory $\underline{\mathcal{A}}$ with objects the abelian varieties.

By (3.1), we have an isomorphism

$$\mathrm{Hom}_{\mathcal{C}}(A, A')_{\mathbb{Q}} \xrightarrow{\cong} \mathrm{Hom}_{\underline{\mathcal{A}}}(A, A')$$

for any abelian varieties A, A'. Also, the base change map

$$\mathrm{Hom}_{\mathcal{C}_k}(A, A') \longrightarrow \mathrm{Hom}_{\mathcal{C}_K}(A_K, A'_K)$$

is an isomorphism for any extension of fields K/k such that k is separably closed in K (see [**Co06**, Thm. 3.19] for a modern version of this classical result of Chow).

In view of the above results, *the abelian category* $\underline{\mathcal{A}}$ *is semi-simple,* $\mathbb{Q}$*-linear, Hom-finite, and invariant under purely inseparable field extensions. Also,* $\underline{\mathcal{A}}$ *is a Serre subcategory of* $\underline{\mathcal{C}}$.

Like for tori again, $\underline{\mathcal{A}}$ is equivalent to the category of all left modules of finite length over the ring

$$R_{\underline{\mathcal{A}}} := \bigoplus_{A \in J} D_A,$$

where J denotes the set of isogeny classes of simple abelian varieties, and we set $D_A := \mathrm{End}_{\underline{\mathcal{A}}}(A)^{\mathrm{op}}$ for chosen representatives A of the classes in J. Moreover, each D_A is a division algebra of finite dimension over $\mathbb{Q}$. Such an endomorphism algebra is a classical object, considered e.g. in [**Mu08**, Chap. IV] and [**Oo88**] where it is denoted by $\mathrm{End}_k^0(A)$. The choice of a polarization of A yields an involutory anti-automorphism of A (the Rosati involution), and hence an isomorphism of D_A with its opposite algebra.

3.2.2. *General algebraic groups, quasi-compact group schemes* [**Br16**, **Pe75**, **Pe76**]. The linear algebraic groups form the building blocks for all connected algebraic groups, together with the abelian varieties. Indeed, we have *Chevalley's structure theorem*: for any connected algebraic group G, there exists an exact sequence

$$0 \longrightarrow L \longrightarrow G \longrightarrow A \longrightarrow 0,$$

where L is linear and A is an abelian variety. Moreover, there is a unique smallest such subgroup $L \subset G$, and this group is connected. If G is smooth and k is perfect, then L is smooth as well (see [**Co02**, **Br15**] for modern expositions of this classical result).

Returning to an arbitrary ground field k, it is easy to see that $\mathrm{Hom}_{\mathcal{C}}(A, L) = 0$ for any abelian variety A and any linear algebraic group L; also, the image of any morphism $L \to A$ is finite (see e.g. [**Br16**, Prop. 2.5]).

It follows that $(\underline{\mathcal{L}}, \underline{\mathcal{A}})$ is a torsion pair of Serre subcategories in $\underline{\mathcal{C}}$, and we have $\mathrm{Hom}_{\underline{\mathcal{C}}}(A, L) = 0$ for all $A \in \underline{\mathcal{A}}$, $L \in \underline{\mathcal{L}}$. Therefore, $\mathrm{Ext}^1_{\underline{\mathcal{C}}}(L, A) = 0$ for all such A, L by Lemma 2.3. In view of Chevalley's structure theorem and the vanishing of $\mathrm{Ext}^1_{\underline{\mathcal{C}}}(A', A)$ for all $A, A' \in \underline{\mathcal{A}}$, we obtain that $\mathrm{Ext}^1_{\underline{\mathcal{C}}}(G, A) = 0$ for all $G \in \underline{\mathcal{C}}$ and $A \in \underline{\mathcal{A}}$. Thus, *every abelian variety is injective in* $\underline{\mathcal{C}}$ (see [**Br16**, Thm. 5.16] for the determination of the injective objects of $\underline{\mathcal{C}}$).

By (3.6), we have $\mathrm{Ext}^1_{\underline{\mathcal{C}}}(T, U) = 0 = \mathrm{Ext}^1_{\underline{\mathcal{C}}}(U, T)$ for all $T \in \underline{\mathcal{T}}$, $U \in \underline{\mathcal{U}}$. Also, recall that $\mathrm{Ext}^1_{\underline{\mathcal{C}}}(T, A) = 0$ for all $A \in \underline{\mathcal{A}}$ and $\mathrm{Ext}^1_{\underline{\mathcal{C}}}(T, T') = 0$ for all $T' \in \underline{\mathcal{T}}$. By Chevalley's structure theorem again, it follows that $\mathrm{Ext}^1_{\underline{\mathcal{C}}}(T, G) = 0$ for all $G \in \underline{\mathcal{C}}$. Thus, *every torus is projective in* $\underline{\mathcal{C}}$. If $\mathrm{char}(k) = 0$, then $\underline{\mathcal{L}}$ is semi-simple, as seen in §3.1.4. In view of Corollary 2.4, it follows that every linear algebraic group is projective in $\underline{\mathcal{C}}$ (see [**Br16**, Thm. 5.14] for the determination of the projective objects of $\underline{\mathcal{C}}$ in arbitrary characteristics).

We now adapt the above results to the setting of *quasi-compact* group schemes. Recall that a scheme is quasi-compact if every open covering admits a finite refinement. Every affine scheme is quasi-compact, as well as every scheme of finite type (in particular, every algebraic group). Also, every connected group scheme is quasi-compact (see [**Pe75**, II.2.4, II.2.5] or [**SGA3**, VIA, Thm. 2.6.5]). The quasi-compact (commutative) group schemes form an abelian category $\widetilde{\mathcal{C}}$, containing $\mathcal{C}$ as a Serre subcategory. Moreover, every $G \in \widetilde{\mathcal{C}}$ is the limit of a filtered inverse system $((G_i)_{i \in I}, (u_{ij} : G_j \to G_i)_{i \leq j})$ such that the G_i are algebraic groups and the u_{ij} are affine morphisms (see [**Pe75**, V.3.1, V.3.6]). Also, there is a unique exact sequence in $\widetilde{\mathcal{C}}$

$$0 \longrightarrow G^0 \longrightarrow G \longrightarrow F \longrightarrow 0,$$

where G^0 is connected and F is *pro-étale* (i.e., a filtered inverse limit of finite étale group schemes); see [**Pe75**, II.2.4, V.4.1]. Finally, there is an exact sequence as in Chevalley's structure theorem

$$0 \longrightarrow H \longrightarrow G^0 \longrightarrow A \longrightarrow 0,$$

where H is an affine group scheme, and A an abelian variety (see [**Pe75**, V.4.3.1]).

Note that $\widetilde{\mathcal{C}}$ is not the pro-completion of the abelian category $\mathcal{C}$, as infinite products do not necessarily exist in $\widetilde{\mathcal{C}}$. For example, the product of infinitely many copies of a non-zero abelian variety A is not represented by a scheme (this may be checked by arguing as in [**SP16**, 91.48], with the morphism $\mathrm{SL}_2 \to \mathbb{P}^1$ replaced by the surjective smooth affine morphism $V \to A$, where V denotes the disjoint union of finitely many open affine subschemes covering A).

We define the *isogeny category of quasi-compact group schemes*, $\widetilde{\underline{\mathcal{C}}}$, as the quotient category of $\widetilde{\mathcal{C}}$ by the Serre subcategory $\widetilde{\mathcal{FC}} = \widetilde{\mathcal{FL}}$ of pro-finite group schemes. Every object of $\widetilde{\underline{\mathcal{C}}}$ is isomorphic to an extension of an abelian variety A by an affine group scheme H. Moreover, $\mathrm{Hom}_{\widetilde{\underline{\mathcal{C}}}}(A, H) = 0$ and the image of every morphism $f : H \to A$ is finite (indeed, f factors through a closed immersion $H/\mathrm{Ker}(f) \to A$ by [**Pe75**, V.3.3]). As a consequence, $(\widetilde{\underline{\mathcal{L}}}, \underline{\mathcal{A}})$ is a torsion pair of Serre subcategories of $\widetilde{\underline{\mathcal{C}}}$; moreover, $\mathrm{Hom}_{\widetilde{\underline{\mathcal{C}}}}(A, H) = 0$ for all $A \in \underline{\mathcal{A}}$, $H \in \widetilde{\underline{\mathcal{L}}}$.

Like for the category $\underline{\mathcal{C}}$, it follows that every abelian variety is projective in $\widetilde{\underline{\mathcal{C}}}$, and every pro-torus is injective; when $\mathrm{char}(k) = 0$, every affine group scheme is projective.

3.2.3. *Vector extensions of abelian varieties* [**Br16**, 5.1]. The objects of the title are the algebraic groups G obtained as extensions

$$0 \longrightarrow U \longrightarrow G \longrightarrow A \longrightarrow 0, \tag{3.7}$$

where A is an abelian variety and U is a *vector group*, i.e., $U \cong n\mathbb{G}_a$ for some n. As $\mathrm{Hom}_{\mathcal{C}}(U, A) = 0 = \mathrm{Hom}_{\mathcal{C}}(A, U)$ (see §3.1.2), the data of G and of the extension (3.9) are equivalent. Also, we have a bi-functorial isomorphism

$$\mathrm{Ext}^1_{\mathcal{C}}(A, \mathbb{G}_a) \xrightarrow{\cong} H^1(A, \mathcal{O}_A), \tag{3.8}$$

where the right-hand side is a k-vector space of dimension $\dim(A)$ (see [**Oo66**, III.17]).

If $\mathrm{char}(k) = p > 0$, then $p_U = 0$ and hence the class of the extension (3.7) is killed by p. Thus, this extension splits after pull-back by the isogeny $p_A : A \to A$.

From now on, we assume that $\mathrm{char}(k) = 0$; then the vector extensions of A are the extensions by unipotent groups. We denote by $\underline{\mathcal{V}}$ the full subcategory of $\underline{\mathcal{C}}$ with objects the vector extensions of abelian varieties. By the Chevalley structure theorem (§3.2.2) and the structure of linear algebraic groups (§3.1.4), a connected algebraic group G is an object of $\underline{\mathcal{V}}$ if and only if $\mathrm{Hom}_{\underline{\mathcal{C}}}(T, G) = 0$. As the functor $\mathrm{Hom}_{\underline{\mathcal{C}}}(T, ?)$ is exact, it follows that $\underline{\mathcal{V}}$ is a Serre subcategory of $\underline{\mathcal{C}}$. Moreover, $(\mathcal{U}, \underline{\mathcal{A}})$ is a torsion pair of Serre subcategories of $\underline{\mathcal{V}}$; they are both semi-simple in view of §§3.1.4 and 3.2.1.

By (3.1), we have an isomorphism

$$\mathrm{Ext}^1_{\mathcal{C}}(A, U)_{\mathbb{Q}} \xrightarrow{\cong} \mathrm{Ext}^1_{\underline{\mathcal{C}}}(A, U).$$

In view of (3.8), this yields bi-functorial isomorphisms

$$\mathrm{Ext}^1_{\underline{\mathcal{V}}}(A, U) \cong H^1(A, \mathcal{O}_A) \otimes_k U \cong \mathrm{Hom}_k(H^1(A, \mathcal{O}_A)^*, U), \tag{3.9}$$

where $H^1(A, \mathcal{O}_A)^*$ denotes of course the dual k-vector space. Moreover, the assignment $A \mapsto H^1(A, \mathcal{O}_A)^*$ extends to a covariant exact functor

$$\mathbf{U} : \underline{\mathcal{A}} \longrightarrow \mathcal{U},$$

as follows e.g. from [**Br16**, Cor. 5.3].

So the triple $(\underline{\mathcal{V}}, \mathcal{U}, \underline{\mathcal{A}})$ satisfies the assumptions (a), (b), (c), (d) of Subsection 2.4, with $\widetilde{\mathcal{X}} = \mathcal{U}$ and $\mathbf{F} = \mathbf{U}$. With the notation of §3.2.1, set

$$M_A := H^1(A, \mathcal{O}_A)^*$$

for any $A \in J$; then M_A is a k-D_A-bimodule.

PROPOSITION 3.1. *With the above notation, there is an equivalence of categories*

$$\mathbf{M}_{\mathcal{V}} : \underline{\mathcal{V}} \xrightarrow{\cong} R_{\underline{\mathcal{V}}}\text{-mod},$$

where $R_{\underline{\mathcal{V}}}$ stands for the triangular matrix ring

$$\begin{pmatrix} \bigoplus_{A \in J} D_A & 0 \\ \bigoplus_{A \in J} M_A & k \end{pmatrix}.$$

Moreover, the center of $\underline{\mathcal{V}}$ is $\mathbb{Q}$.

PROOF. The first assertion follows from Theorem 2.13 and Remark 2.14.

By Proposition 2.15, the center of $\underline{\mathcal{V}}$ consists of the pairs $z = (x, (y_A)_{A \in J})$ where $x \in k$, $y_A \in Z_A$ (the center of D_A) and $xm = my_A$ for all $m \in M_A$ and $A \in J$. In particular, if the simple abelian variety A satisfies $D_A = \mathbb{Q}$, then $x = y_A \in \mathbb{Q}$. As such abelian varieties exist (e.g., elliptic curves without complex multiplication), it follows that $x \in \mathbb{Q}$. Then for any $A \in J$, we obtain $y_A = x$ as $M_A \neq 0$. So $z \in \mathbb{Q}$. □

Next, recall that every abelian variety A has a universal vector extension $\mathbf{E}(A)$, by the vector group $\mathbf{U}(A)$. In view of Proposition 2.11, the projective objects of $\underline{\mathcal{V}}$ are the products of unipotent groups and universal vector extensions; moreover, every $G \in \underline{\mathcal{V}}$ has a canonical projective resolution,

$$0 \longrightarrow \mathbf{U}(A) \longrightarrow U \times \mathbf{E}(A) \longrightarrow G \longrightarrow 0, \tag{3.10}$$

where A denotes of course the abelian variety quotient of G. In particular, the abelian category $\underline{\mathcal{V}}$ is hereditary and has enough projectives. This recovers most of the results in [**Br16**, Sec. 5.1].

3.2.4. *Semi-abelian varieties* [**Br15**, 5.4]. These are the algebraic groups G obtained as extensions

$$0 \longrightarrow T \longrightarrow G \longrightarrow A \longrightarrow 0, \tag{3.11}$$

where A is an abelian variety and T is a torus. Like for vector extensions of abelian varieties, we have $\mathrm{Hom}_{\mathcal{C}}(T, A) = 0 = \mathrm{Hom}_{\mathcal{C}}(A, T)$; thus, the data of G and of the extension (3.11) are equivalent.

The Weil-Barsotti formula (see e.g. [**Oo66**, III.17, III.18]) yields a bi-functorial isomorphism

$$\mathrm{Ext}^1_{\mathcal{C}}(A, T) \xrightarrow{\cong} \mathrm{Hom}^{\Gamma}(\mathbf{X}(T), \widehat{A}(k_s)).$$

Here $\widehat{A}$ denotes the dual of A; this is an abelian variety with dimension $\dim(A)$ and with Lie algebra $H^1(A, \mathcal{O}_A)$. In view of (3.1), this yields in turn a bi-functorial isomorphism

$$\mathrm{Ext}^1_{\underline{\mathcal{C}}}(A, T) \xrightarrow{\cong} \mathrm{Hom}^{\Gamma}(\mathbf{X}(T), \widehat{A}(k_s)_{\mathbb{Q}}). \tag{3.12}$$

We denote by $\underline{\mathcal{S}}$ the full subcategory of $\underline{\mathcal{C}}$ with objects the semi-abelian varieties. By [**Br15**, Lem. 5.4.3, Cor. 5.4.6], $\underline{\mathcal{S}}$ is a Serre subcategory of $\underline{\mathcal{C}}$, invariant under purely inseparable field extensions. Moreover, $(\underline{\mathcal{T}}, \underline{\mathcal{A}})$ is a torsion pair of Serre subcategories of $\underline{\mathcal{S}}$. The assignment $A \mapsto \widehat{A}(k_s)_{\mathbb{Q}}$ extends to a contravariant exact functor $\underline{\mathcal{A}} \to \mathbb{Q}\Gamma\text{-Mod}$ in view of [**Br16**, Rem. 4.8] (see also Lemma 3.7 below). Since $\mathbb{Q}\Gamma$-Mod is anti-equivalent to $\underline{\widetilde{\mathcal{T}}}$ (§3.1.3), this yields a covariant exact functor

$$\widetilde{\mathbf{T}} : \underline{\mathcal{A}} \longrightarrow \underline{\widetilde{\mathcal{T}}}$$

together with a bi-functorial isomorphism

$$\mathrm{Ext}^1_{\underline{\mathcal{C}}}(A, T) \xrightarrow{\cong} \mathrm{Hom}_{\underline{\widetilde{\mathcal{T}}}}(\widetilde{\mathbf{T}}(A), T).$$

Thus, the triple $(\underline{\mathcal{S}}, \underline{\mathcal{T}}, \underline{\mathcal{A}})$ satisfies the assumptions (a), (b), (c), (d) of Subsection 2.4 with $\widetilde{\mathcal{X}} = \underline{\widetilde{\mathcal{T}}}$ and $\mathbf{F} = \widetilde{\mathbf{T}}$; moreover, the assumption (e) of Subsection 2.5 holds by construction. With the notation of §§3.1.3 and 3.2.1, let

$$M_{T,A} := \mathrm{Hom}^{\Gamma}(\mathbf{X}(T), \widehat{A}(k_s)_{\mathbb{Q}})$$

for all $T \in I$, $A \in J$; then $M_{T,A}$ is a D_T-D_A-bimodule. In view of Theorem 2.13, we obtain:

PROPOSITION 3.2. *There is an equivalence of categories*

$$\mathbf{M}_{\mathcal{S}} : \underline{\mathcal{S}} \xrightarrow{\cong} R_{\underline{\mathcal{S}}}\text{-mod},$$

where $R_{\underline{\mathcal{S}}}$ stands for the triangular matrix ring

$$\begin{pmatrix} \bigoplus_{A \in J} D_A & 0 \\ \bigoplus_{T \in I, A \in J} M_{T,A} & \bigoplus_{T \in I} D_T. \end{pmatrix}.$$

REMARK 3.3. If k is locally finite (i.e., the union of its finite subfields), then the abelian group $A(k)$ is torsion for any abelian variety A. It follows that $\underline{\mathcal{S}} \cong \underline{\mathcal{T}} \times \underline{\mathcal{A}}$ and $R_{\underline{\mathcal{S}}} = R_{\underline{\mathcal{T}}} \times R_{\underline{\mathcal{A}}}$. In particular, the center of $R_{\underline{\mathcal{S}}}$ is an infinite direct sum of fields.

On the other hand, if k is not locally finite, then the abelian group $A(k_s)$ has infinite rank for any non-zero abelian variety A (see [**FJ74**, Thm. 9.1]). As a consequence, A admits no universal extension in $\underline{\mathcal{S}}$. If in addition k is separably closed, then the center of $R_{\underline{\mathcal{S}}}$ is $\mathbb{Q}$, as follows by arguing as in the proof of Proposition 3.1 with $\mathbb{G}_a$ replaced by $\mathbb{G}_m$. We do not know how to determine the center of $R_{\underline{\mathcal{S}}}$ for an arbitrary (not locally finite) field k.

Next, consider the isogeny category $\underline{\widetilde{\mathcal{C}}}$ of quasi-compact group schemes, and denote by $\underline{\widetilde{\mathcal{S}}} \subset \underline{\widetilde{\mathcal{C}}}$ the full subcategory with objects the group schemes obtained as extensions of abelian varieties by pro-tori. Then $\underline{\mathcal{S}}$ is a Serre subcategory of $\underline{\widetilde{\mathcal{S}}}$; moreover, $(\underline{\widetilde{\mathcal{T}}}, \underline{\mathcal{A}})$ is a torsion pair of Serre subcategories of $\underline{\widetilde{\mathcal{S}}}$, and $\mathrm{Hom}_{\underline{\widetilde{\mathcal{S}}}}(A, T) = 0$ for any abelian variety A and any pro-torus T. Thus, $\underline{\widetilde{\mathcal{S}}}$ is equivalent to the category of extensions of abelian varieties by pro-tori (as defined in Subsection 2.3), and in turn to the category $\widetilde{R_{\underline{\mathcal{S}}}\text{-mod}}$ by Theorem 2.13.

In view of the results of Subsection 2.4, every abelian variety A has a universal extension in $\underline{\widetilde{\mathcal{S}}}$, by the pro-torus with Cartier dual $\widehat{A}(k_s)_{\mathbb{Q}}$. Moreover, the projective objects of $\underline{\widetilde{\mathcal{S}}}$ are the products of pro-tori and universal extensions; also, every $G \in \underline{\widetilde{\mathcal{S}}}$ has a canonical projective resolution, similar to (3.10). In particular, the abelian category $\underline{\widetilde{\mathcal{S}}}$ is hereditary and has enough projectives.

3.2.5. *General algebraic groups (continued).* We return to the setting of §3.2.2, and consider the isogeny category of algebraic groups, $\underline{\mathcal{C}}$, as a Serre category of the isogeny category of quasi-compact group schemes, $\underline{\widetilde{\mathcal{C}}}$. Also, recall from §§3.1.4, 3.1.5 the isogeny category of unipotent algebraic groups, $\underline{\mathcal{U}}$, a Serre subcategory of that of unipotent group schemes, $\underline{\widetilde{\mathcal{U}}}$. Likewise, we have the isogeny category of semi-abelian varieties, $\underline{\mathcal{S}}$, a Serre subcategory of the isogeny category of extensions of abelian varieties by pro-tori, $\underline{\widetilde{\mathcal{S}}}$. These are the ingredients of a structure result for $\underline{\mathcal{C}}$, $\underline{\widetilde{\mathcal{C}}}$ in positive characteristics:

PROPOSITION 3.4. *If* $\mathrm{char}(k) = p > 0$, *then the assignment* $(S, U) \mapsto S \times U$ *extends to equivalences of categories*

$$\underline{\mathcal{S}} \times \underline{\mathcal{U}} \xrightarrow{\cong} \underline{\mathcal{C}}, \quad \underline{\widetilde{\mathcal{S}}} \times \underline{\widetilde{\mathcal{U}}} \xrightarrow{\cong} \underline{\widetilde{\mathcal{C}}}.$$

PROOF. The first equivalence is obtained in [**Br16**, Prop. 5.10]. We provide an alternative proof: by Chevalley's structure theorem, every $G \in \underline{\mathcal{C}}$ lies in an exact sequence in $\underline{\mathcal{C}}$

$$0 \longrightarrow U \longrightarrow G \longrightarrow S \longrightarrow 0,$$

where $U \in \underline{\mathcal{U}}$ and $S \in \underline{\mathcal{S}}$. Moreover, we have $\mathrm{Ext}^n_{\underline{\mathcal{C}}}(U, S) = 0 = \mathrm{Ext}^n_{\underline{\mathcal{C}}}(S, U)$ for all $n \geq 0$, since the multiplication map p_S is an isomorphism in $\underline{\mathcal{C}}$, while $p^n_U = 0$ for $n \gg 0$. In particular, the above exact sequence has a unique splitting, which is functorial in U, S.

The second equivalence follows from the first one, as every quasi-compact group scheme is the inverse limit of its algebraic quotient groups. □

Next, we obtain the main result of this paper; for this, we gather some notation. Define a ring $R = R_{\underline{\mathcal{C}}}$ by

$$R = \begin{cases} R_{\underline{\mathcal{S}}} \times R_{\underline{\mathcal{U}}}, & \text{if } \mathrm{char}(k) > 0 \\ R_{\underline{\mathcal{S}}} \times_{R_{\underline{\mathcal{A}}}} R_{\underline{\mathcal{V}}}, & \text{if } \mathrm{char}(k) = 0. \end{cases}$$

More specifically,

$$R = \begin{pmatrix} \bigoplus_{A \in J} D_A & 0 \\ \bigoplus_{T \in I, A \in J} M_{T,A} & \bigoplus_{T \in I} D_T \end{pmatrix} \times (\mathbb{D} \setminus \mathbb{D}V)^{-1}\mathbb{D}$$

if $\mathrm{char}(k) > 0$, where I (resp. J) denotes the set of isogeny classes of simple tori (resp. of simple abelian varieties), $T \in I$ (resp. $A \in J$) denote representatives of their classes, $D_T := \mathrm{End}_{\mathcal{T}}(T)^{\mathrm{op}}_{\mathbb{Q}}$, $D_A := \mathrm{End}_{\mathcal{A}}(A)^{\mathrm{op}}_{\mathbb{Q}}$, and $M_{T,A} := \mathrm{Hom}^{\Gamma}(\mathbf{X}(T), \widehat{A}(k_s)_{\mathbb{Q}})$. Moreover, $\mathbb{D}$ denotes the Dieudonné ring over the perfect closure of k, and $V \in \mathbb{D}$ the Verschiebung as in §3.1.5.

If $\mathrm{char}(k) = 0$, then

$$R = \begin{pmatrix} \bigoplus_{A \in J} D_A & 0 \\ (\bigoplus_{T \in I, A \in J} M_{T,A}) \oplus (\bigoplus_{A \in J} M_A) & (\bigoplus_{T \in I} D_T) \oplus k \end{pmatrix},$$

where $M_A := H^1(A, \mathcal{O}_A)^*$. We may now state:

THEOREM 3.5. *With the above notation, the abelian categories* $\underline{\mathcal{C}}$ *and* $\underline{\widetilde{\mathcal{C}}}$ *are hereditary, and* $\underline{\widetilde{\mathcal{C}}}$ *has enough projectives. Moreover, there are compatible equivalences of categories*

$$\underline{\mathcal{C}} \xrightarrow{\cong} R\text{-}\mathrm{mod}, \quad \underline{\widetilde{\mathcal{C}}} \xrightarrow{\cong} R\text{-}\widetilde{\mathrm{mod}}.$$

PROOF. When $\operatorname{char}(k) > 0$, this follows by combining Proposition 3.4 with the structure results for $\underline{\mathcal{U}}$, $\widetilde{\underline{\mathcal{U}}}$ recalled in §3.1.5, and those for $\underline{\mathcal{S}}$, $\widetilde{\underline{\mathcal{S}}}$ obtained in Proposition 3.2. When $\operatorname{char}(k) = 0$, recall from §3.2.2 that $(\underline{\mathcal{L}}, \underline{\mathcal{A}})$ (resp. $(\widetilde{\underline{\mathcal{L}}}, \underline{\mathcal{A}})$) is a torsion pair of Serre subcategories of $\underline{\mathcal{C}}$ (resp. $\widetilde{\underline{\mathcal{C}}}$); moreover, $\underline{\mathcal{L}} \cong \underline{\mathcal{T}} \times \mathcal{U}$ and $\widetilde{\underline{\mathcal{L}}} \cong \widetilde{\underline{\mathcal{T}}} \times \widetilde{\mathcal{U}}$ by §3.1.4. In view of the bi-functorial isomorphisms (3.9) and (3.12), the assertions follow from Theorem 2.13 like in the cases of vector extensions of abelian varieties (Proposition 3.1) and of semi-abelian varieties (Proposition 3.2). □

Also, recall from Proposition 2.15 that the intersection of the left ideals of finite colength in R is zero. One may check (by using Remark 2.16) that this also holds for the two-sided ideals of finite colength as left modules.

We will show in Subsection 3.4 that the center of $\underline{\mathcal{C}}$ equals $\mathbb{Q}$ if $\operatorname{char}(k) = 0$, and contains $\mathbb{Q} \times \mathbb{Z}_p$ if $\operatorname{char}(k) = p > 0$.

3.3. Functors of points. Let G be an algebraic group. Then the group of $\bar{k}$-points, $G(\bar{k})$, is equipped with an action of the absolute Galois group Γ. We also have the subgroup of k_s-points, $G(k_s)$, which is stable under Γ. Since every $x \in G(k_s)$ lies in $G(K)$ for some finite Galois extension of fields K/k, we see that the stabilizer of x in Γ is open, i.e., $G(k_s)$ is a discrete Γ-module. Likewise, the Γ-module $G(\bar{k})$ is discrete as well.

Clearly, the assignment $G \mapsto G(\bar{k})$ extends to a covariant exact functor

$$(\bar{k}) : \mathcal{C} \longrightarrow \mathbb{Z}\Gamma\text{-}\mathrm{Mod}, \quad \mathcal{F} \longrightarrow \mathbb{Z}\Gamma\text{-}\mathrm{Mod}^{\mathrm{tors}}.$$

The assignment $G \mapsto G(k_s)$ also extends to a covariant functor

$$(k_s) : \mathcal{C} \longrightarrow \mathbb{Z}\Gamma\text{-}\mathrm{Mod}, \quad \mathcal{F} \longrightarrow \mathbb{Z}\Gamma\text{-}\mathrm{Mod}^{\mathrm{tors}}$$

which is additive and left exact. But the functor (k_s) is not exact when k is an imperfect field, as seen from the exact sequence

$$0 \longrightarrow \alpha_p \longrightarrow \mathbb{G}_a \xrightarrow{F} \mathbb{G}_a \longrightarrow 0,$$

where $p := \operatorname{char}(k)$ and F denotes the Frobenius endomorphism, $x \mapsto x^p$. Yet the functors $(\bar{k})$ and (k_s) are closely related:

LEMMA 3.6.

(i) *Let K/k be an extension of fields of characteristic $p > 0$ and assume that $K^{p^n} \subset k$. Then $p^n x \in G(k)$ for any $x \in G(K)$.*
(ii) *For any $x \in G(\bar{k})$, there exists $n = n(x) \geq 0$ such that $p^n x \in G(k_s)$.*

PROOF. (i) By assumption, we have $x \in G(k^{1/p^n})$. Consider the nth relative Frobenius morphism,

$$F^n_{G/k} : G \longrightarrow G^{(p^n)}$$

(see e.g. [**CGP15**, A.3]). Then $F^n_{G/k}(x) \in G^{(p^n)}(k)$: indeed, this holds with G replaced with any scheme of finite type, since this holds for the affine space $\mathbb{A}^n$ and the formation of the relative Frobenius morphism commutes with immersions. We also have the nth Verschiebung,

$$V^n_{G/k} : G^{(p^n)} \longrightarrow G,$$

which satisfies $V^n_{G/K} \circ F^n_{G/k} = p^n_G$ (see [**SGA3**, VIIA.4.3]). It follows that $p^n x = V^n_{G/k}(F^n_{G/k}(x))$ is in $G(k)$.

(ii) Just apply (i) to k_s instead of k, and use the fact that $\bar{k} = \bigcup_{n \geq 0} k_s^{1/p^n}$. □

As a direct consequence, we obtain:

LEMMA 3.7.

(i) *The natural map* $G(k_s)_{\mathbb{Q}} \to G(\bar{k})_{\mathbb{Q}}$ *is an isomorphism for any algebraic group* G.

(ii) *The covariant exact functor*

$$(\bar{k})_{\mathbb{Q}} : \mathcal{C} \longrightarrow \mathbb{Q}\Gamma\text{-Mod}, \quad G \longmapsto G(\bar{k})_{\mathbb{Q}} := G(\bar{k})_{\mathbb{Q}}$$

yields a covariant exact functor, also denoted by

$$(\bar{k})_{\mathbb{Q}} : \underline{\mathcal{C}} \longrightarrow \mathbb{Q}\Gamma\text{-Mod}.$$

REMARKS 3.8. (i) Assume that G is unipotent. If $\mathrm{char}(k) = p > 0$, then $G(\bar{k})$ is p-torsion and hence $G(\bar{k})_{\mathbb{Q}} = 0$. On the other hand, if $\mathrm{char}(k) = 0$, then $G \cong n\mathbb{G}_a$ and hence $G(\bar{k})_{\mathbb{Q}} \cong n\bar{k}$ as a Γ-module. Using the normal basis theorem, it follows that the multiplicity of any simple discrete Γ-module M in $G(\bar{k})_{\mathbb{Q}}$ equals $n \dim(M)$; in particular, all these multiplicities are finite.

The latter property does not extend to the case where G is a torus. For example, the multiplicity of the trivial Γ-module in $\mathbb{G}_m(\bar{k})_{\mathbb{Q}}$ is the rank of the multiplicative group k^*, which is infinite when k is not locally finite. Indeed, under that assumption, k contains either $\mathbb{Q}$ or the field of rational functions in one variable $\mathbb{F}_p(t)$; so the assertion follows from the infiniteness of prime numbers and of irreducible polynomials in $\mathbb{F}_p[t]$.

Whenn G is an abelian variety, the finiteness of multiplicities of the Γ-module $G(k)_{\mathbb{Q}}$ will be discussed in the next subsection.

(ii) For any abelian variety A and any torus T, we have a bi-functorial isomorphism

$$\mathrm{Ext}^1_{\underline{\mathcal{C}}}(A, T) \xrightarrow{\cong} \mathrm{Hom}^{\Gamma}(\mathbf{X}(T), \widehat{A}(\bar{k})_{\mathbb{Q}}),$$

as follows from (3.12) together with Lemma 3.7. Also, recall that $\widehat{A}$ is isogenous to A (see e.g. [**Mi86**]) and hence the Γ-module $\widehat{A}(\bar{k})_{\mathbb{Q}}$ is isomorphic (non-canonically) to $A(\bar{k})_{\mathbb{Q}}$.

Next, we associate an endofunctor of $\underline{\mathcal{C}}$ with any discrete Γ-module M, which is a free abelian group of finite rank. We may choose a finite Galois extension $K/k \subset k_s/k$ such that Γ acts on M via its finite quotient $\Gamma' := \mathrm{Gal}(K/k)$. For any $G \in \mathcal{C}$, consider the tensor product of commutative group functors $G_K \otimes_{\mathbb{Z}} M$. This group functor is represented by an algebraic group over K (isomorphic to the product of r copies of G_K, where r denotes the rank of M as an abelian group), equipped with an action of Γ' such that the structure map $G_K \otimes_{\mathbb{Z}} M \to \mathrm{Spec}(K)$ is equivariant. By Galois descent (see e.g. [**Co06**, Cor. 3.4]), the quotient

$$G(M) := (G_K \otimes_{\mathbb{Z}} M)/\Gamma'$$

is an algebraic group over k, equipped with a natural Γ'-equivariant isomorphism

$$G_K \otimes_{\mathbb{Z}} M \xrightarrow{\cong} G(M)_K.$$

The assignment $G \mapsto G(M)$ extends to a covariant endofunctor (M) of $\mathcal{C}$, which is exact as the base change functor $\otimes_k K$ is faithful. Moreover, one may easily check that (M) is independent of the choice of K, and hence comes with a natural Γ-equivariant isomorphism

$$G_{k_s} \otimes_{\mathbb{Z}} M \xrightarrow{\cong} G(M)_{k_s}$$

for any $G \in \mathcal{C}$. In particular, we have an isomorphism of Γ-modules

$$G(k_s) \otimes_{\mathbb{Z}} M \xrightarrow{\cong} G(M)(k_s).$$

This implies readily:

LEMMA 3.9. *With the above notation and assumptions, the endofunctor (M) of $\mathcal{C}$ yields a covariant exact endofunctor $(M)_{\mathbb{Q}}$ of $\underline{\mathcal{C}}$ that stabilizes $\underline{\mathcal{T}}$, $\underline{\mathcal{U}}$, $\underline{\mathcal{A}}$ and $\underline{\mathcal{S}}$. Moreover, $(M)_{\mathbb{Q}}$ only depends on $M_{\mathbb{Q}} \in \mathbb{Q}\Gamma$-mod, and there is a natural isomorphism of Γ-modules*

$$G(M)(k_s)_{\mathbb{Q}} \cong G(k_s) \otimes_{\mathbb{Z}} M_{\mathbb{Q}} = G(k_s)_{\mathbb{Q}} \otimes_{\mathbb{Q}} M_{\mathbb{Q}}$$

for any $G \in \mathcal{C}$.

REMARKS 3.10. (i) If G is a torus, then one readily checks that $G(M)$ is the torus with character group $\mathrm{Hom}_{\mathbb{Z}}(M, \mathbf{X}(G))$. As a consequence, the endofunctor (M) of $\underline{\mathcal{T}}$ is identified with the tensor product by the dual module M^*, under the anti-equivalence of categories of $\underline{\mathcal{T}}$ with $\mathbb{Q}\Gamma$-mod.

On the other hand, the endofunctor (M) of $\underline{\mathcal{U}}$ is just given by the assignment $G \mapsto rG$, where r denotes the rank of the free abelian group M. Indeed, we may assume that G is indecomposable and (using the invariance of $\underline{\mathcal{U}}$ under purely inseparable field extensions) that k is perfect. Then $G \cong W_n$ and hence $G(M)_{k_s} = W_{n,k_s} \otimes_{\mathbb{Z}} M \cong rW_{n,k_s}$ in $\underline{\mathcal{U}}_{k_s}$. In view of the uniqueness of the decomposition in $\underline{\mathcal{U}}$ as a direct sum of groups of Witt vectors, it follows that $G(M) \cong rW_n$ as desired.

(ii) The endofunctor (M) can be interpreted in terms of Weil restriction when M is a permutation Γ-module, i.e., M has a $\mathbb{Z}$-basis which is stable under Γ. Denote by $\Delta \subset \Gamma$ the isotropy group of some basis element, and by $K \subset k_s$ the fixed point subfield of Δ. Then K/k is a finite separable field extension, and one may check that there is a natural isomorphism $G(M) \cong \mathrm{R}_{K/k}(G_K)$ with the notation of [**CGP15**, A.5].

(iii) The assignment $G \mapsto G(M)$ is in fact a special case of a tensor product construction introduced by Milne in the setting of abelian varieties (see [**Mi72**]) and systematically studied by Mazur, Rubin and Silverberg in [**MRS07**]. More specifically, the tensor product $M \otimes_{\mathbb{Z}} G$, defined there in terms of Galois cohomology, is isomorphic to $G(M)$ in view of [**MRS07**, Thm. 1.4].

3.4. Finiteness conditions for Hom and Ext groups. Recall from §§3.1.3, 3.2.1 that the abelian categories $\underline{\mathcal{T}}$, $\underline{\mathcal{A}}$ are $\mathbb{Q}$-linear and Hom-finite. Also, recall from §3.1.4 that $\underline{\mathcal{U}} \cong \mathcal{U} \cong k$-mod is k-linear, semi-simple and Hom-finite when $\mathrm{char}(k) = 0$.

PROPOSITION 3.11.

(i) *The abelian categories $\underline{\mathcal{T}}$ and $\underline{\mathcal{A}}$ are not K-linear for any field K strictly containing $\mathbb{Q}$.*

(ii) *The abelian category $\underline{\mathcal{U}}$ is not K-linear for any field K, when* $\mathrm{char}(k) = p > 0$.

(iii) *The center of $\underline{\mathcal{C}}$ is $\mathbb{Q}$ when* $\mathrm{char}(k) = 0$.

PROOF. (i) The assertion clearly holds for $\underline{\mathcal{T}}$, as $\mathrm{End}_{\underline{\mathcal{T}}}(\mathbb{G}_m) = \mathbb{Q}$. For $\underline{\mathcal{A}}$, we replace $\mathbb{G}_m$ with appropriate elliptic curves E. Given any $t \in k$, there exists such a curve with j-invariant t (see e.g. [**Si86**, Prop. III.1.4]). If $\mathrm{char}(k) = 0$, then we choose $t \in \mathbb{Q} \setminus \mathbb{Z}$; in particular, t is not an algebraic integer. By [**Si86**, Thm. C.11.2], we have $\mathrm{End}_{\mathcal{A}}(E) = \mathbb{Z}$ and hence $\mathrm{End}_{\underline{\mathcal{A}}}(E) = \mathbb{Q}$. If $\mathrm{char}(k) = p$ and k is not algebraic over $\mathbb{F}_p$, then we may choose t transcendental over $\mathbb{F}_p$. By [**Mu08**, §22], we then have again $\mathrm{End}_{\underline{\mathcal{A}}}(E) = \mathbb{Q}$. Finally, if k is algebraic over $\mathbb{F}_p$, then every abelian variety A is defined over a finite subfield of k, and hence the associated Frobenius endomorphism lies in $\mathrm{End}_{\underline{\mathcal{A}}}(A) \setminus \mathbb{Q}$. In that case, it follows from [**Oo88**, (2.3)] that $\mathrm{End}_{\underline{\mathcal{A}}}(E)$ is an imaginary quadratic number field in which p splits, if E is ordinary. On the other hand, if E is supersingular, then $\mathrm{End}_{\underline{\mathcal{A}}}(E)$ contains no such field. It follows that the largest common subfield to all rings $\mathrm{End}_{\underline{\mathcal{A}}}(E)$ is $\mathbb{Q}$.

(ii) Assume that $\underline{\mathcal{U}}$ is K-linear for some field K. Then K is a subfield of $\mathrm{End}_{\underline{\mathcal{U}}}(\mathbb{G}_a)$, and hence $\mathrm{char}(K) = p$. More generally, for any $n \geq 1$, we have a ring homomorphism $K \to \mathrm{End}_{\underline{\mathcal{U}}}(W_n) = R/RV^n$ with the notation of §3.1.5. As these homomorphisms are compatible with the natural maps $R/RV^{n+1} = \mathrm{End}_{\underline{\mathcal{U}}}(W_{n+1}) \to \mathrm{End}_{\underline{\mathcal{U}}}(W_n) = R/RV^n$, we obtain a ring homomorphism $K \to \varprojlim R/RV^n$. Since the right-hand side has characteristic 0, this yields a contradiction.

(iii) By Proposition 2.15, it suffices to show that the center of R is $\mathbb{Q}$. Moreover, every central element of R is of the form

$$z = \begin{pmatrix} \sum_{A \in J} y_A & 0 \\ 0 & (\sum_{T \in I} x_T) + x \end{pmatrix},$$

where each y_A is central in D_A, each x_T is central in D_T, and $x \in k$; also, $x m_A = m_A y_A$ for all $m \in M_A$, and $x_T m_{T,A} = m_{T,A} y_A$ for all $m_{T,A} \in M_{T,A}$. Like in the proof of Proposition 3.1, it follows that $x \in \mathbb{Q}$ and $y_A = x$ for all A. As a consequence, $x_T = y_A$ whenever $M_{T,A} \neq 0$.

To complete the proof, it suffices to show that for any $T \in I$, there exists $A \in J$ such that $M_{T,A} \neq 0$; equivalently, the Γ-module $A(\bar{k})_{\mathbb{Q}}$ contains $\mathbf{X}(T)_{\mathbb{Q}}$. We may choose an elliptic curve E such that $E(k)$ is not torsion, i.e., $E(k)_{\mathbb{Q}} \neq 0$; then the Γ-module $E(\mathbf{X}(T))(\bar{k})$ contains $\mathbf{X}(T)$. Thus, the desired assertion holds for some simple factor A of $E(\mathbf{X}(T))$. □

REMARK 3.12. The above statement (iii) does not extend to the case where $\mathrm{char}(k) = p > 0$, since we then have $\underline{\mathcal{C}} \cong \underline{\mathcal{S}} \times \underline{\mathcal{U}}$. One may then show that the center of $\underline{\mathcal{U}}$ contains the ring of p-adic integers, $\mathbb{Z}_p = W(\mathbb{F}_p)$, with equality if and only if k is infinite. In view of Remark 3.3, it follows that the center of $\underline{\mathcal{C}}$ contains $\mathbb{Q} \times \mathbb{Z}_p$, with equality if k is separably closed.

PROPOSITION 3.13. *The abelian category $\underline{\mathcal{S}}$ is $\mathbb{Q}$-linear and Hom-finite. It is Ext-finite if and only if k satisfies the following condition:*

(MW) *The vector space $A(k)_{\mathbb{Q}}$ is finite-dimensional for any abelian variety A.*

PROOF. Recall that every semi-abelian variety is divisible. In view of (3.1), it follows that $\underline{\mathcal{S}}$ is $\mathbb{Q}$-linear. It is Hom-finite in view of the Hom-finiteness of $\underline{\mathcal{T}}$ and $\underline{\mathcal{A}}$, combined with Proposition 2.6.

By that proposition, $\underline{\mathcal{S}}$ is Ext-finite if and only if the $\mathbb{Q}$-vector space $\mathrm{Ext}^1_{\underline{\mathcal{S}}}(A, T)$ is finite-dimensional for any abelian variety A and any torus T. In view of the isomorphism (3.12) and of the anti-equivalence of categories (3.3), this amounts to the condition that the vector space $\mathrm{Hom}^{\Gamma}(M, \widehat{A}(k_s)_{\mathbb{Q}})$ be finite-dimensional for any $M \in \mathbb{Q}\Gamma\text{-mod}$. The latter condition is equivalent to (MW) by Lemma 3.9. □

REMARKS 3.14. (i) The above condition (MW) is a weak version of the Mordell-Weil theorem, which asserts that the abelian group $A(k)$ is finitely generated for any abelian variety A over a number field k.

(ii) The condition (MW) holds trivially if k is locally finite, as the abelian group $A(k)$ is torsion under that assumption.

(iii) Let K/k be a finitely generated regular extension of fields (recall that the regularity assumption means that k is algebraically closed in K, and K is separable over k). If (MW) holds for k, then it also holds for K in view of the Lang-Néron theorem (see [**Co06**] for a modern proof of this classical result). As a consequence, (MW) holds whenever k is finitely generated over a number field or a locally finite field.

One can also show that (MW) is invariant under purely transcendental extensions (not necessarily finitely generated), by using the fact that every rational map from a projective space to an abelian variety is constant.

PROPOSITION 3.15. *Assume that* $\mathrm{char}(k) = 0$. *Then the* $\mathbb{Q}$*-linear category* $\underline{\mathcal{V}}$ *(resp.* $\underline{\mathcal{C}}$*) is Hom-finite if and only if* k *is a number field. Under that assumption,* $\underline{\mathcal{V}}$ *and* $\underline{\mathcal{C}}$ *are Ext-finite as well.*

PROOF. If $\underline{\mathcal{V}}$ is Hom-finite, then $\mathrm{End}_{\underline{\mathcal{V}}}(\mathbb{G}_a)$ is finite-dimensional as a $\mathbb{Q}$-vector space. Since $\mathrm{End}_{\underline{\mathcal{V}}}(\mathbb{G}_a) = \mathrm{End}_{\mathcal{U}}(\mathbb{G}_a) = k$, this means that k is a number field.

Conversely, if k is a number field, then $\mathcal{U}$ is Hom-finite, and hence so is $\underline{\mathcal{L}}$. In view of Proposition 2.6, it follows that $\underline{\mathcal{C}}$ is Hom-finite, and hence so is $\underline{\mathcal{V}}$. By that proposition again, to prove that $\underline{\mathcal{C}}$ is Ext-finite, it suffices to check that the $\mathbb{Q}$-vector space $\mathrm{Ext}^1_{\underline{\mathcal{C}}}(A, L)$ is finite-dimensional for any abelian variety A and any connected linear algebraic group L. Since $L = U \times T$ for a unipotent group U and a torus T, this finiteness assertion follows by combining the isomorphisms (3.9), (3.12) and the Mordell-Weil theorem. □

3.5. Finiteness representation type: an example. As in §3.2.3, we consider the abelian category $\underline{\mathcal{V}}$ of vector extensions of abelian varieties over a field k of characteristic 0. Recall that $\underline{\mathcal{V}}$ is $\mathbb{Q}$-linear and hereditary, and has enough projectives; its simple objects are the additive group $\mathbb{G}_a$ and the simple abelian varieties. In particular, $\underline{\mathcal{V}}$ has infinitely many isomorphism classes of simple objects. Also, by Proposition 3.15, $\underline{\mathcal{V}}$ is Hom-finite if and only if k is a number field; then $\underline{\mathcal{V}}$ is Ext-finite as well.

We now assume that k is a number field. Choose a finite set $F = \{A_1, \ldots, A_r\}$ of simple abelian varieties, pairwise non-isogenous. Denote by $\underline{\mathcal{V}}_F$ the Serre subcategory of $\underline{\mathcal{V}}$ generated by F. More specifically, the objects of $\underline{\mathcal{V}}_F$ are the algebraic groups obtained as extensions

$$0 \longrightarrow m_0\mathbb{G}_a \longrightarrow G \longrightarrow \bigoplus_{i=1}^{r} m_i A_i \longrightarrow 0,$$

where $m_0, m_1, \ldots, m_r$ are non-negative integers. The morphisms of $\underline{\mathcal{V}}_F$ are the homomorphisms of algebraic groups.

By Theorem 2.13 and Remark 2.14, we have an equivalence of categories

$$\underline{\mathcal{V}}_F \xrightarrow{\cong} R_F\text{-mod},$$

where R_F denotes the triangular matrix ring

$$\begin{pmatrix} D_1 \oplus \cdots \oplus D_r & 0 \\ M_1 \oplus \cdots \oplus M_r & k \end{pmatrix}.$$

Here $D_i := \mathrm{End}_{\underline{\mathcal{V}}}(A_i)^{\mathrm{op}}$ is a division ring of finite dimension as a $\mathbb{Q}$-vector space, and $M_i := H^1(A_i, \mathcal{O}_{A_i})^*$ is a k-D_i-bimodule, of finite dimension as a k-vector space. Thus, R_F is a finite-dimensional $\mathbb{Q}$-algebra. Also, R_F is hereditary, since so is $\underline{\mathcal{V}}_F$.

The $\mathbb{Q}$-*species* of R_F is the directed graph Γ_F with vertices $0, 1, \ldots, r$ and edges $\varepsilon_i := (0, i)$ for $i = 1, \ldots, r$. The vertex 0 is labeled with the field k, and each vertex $i = 1, \ldots, r$ is labeled with the division ring D_i; each edge ε_i is labeled with the k-D_i-bimodule M_i. The category R_F-mod is equivalent to that of representations of the $\mathbb{Q}$-species Γ_F, as defined in [**DR76**] (see also [**Le12**]).

The *valued graph* of Γ_F is the underlying non-directed graph Δ_F, where each edge $\{0, i\}$ is labeled with the pair $(\dim_k(M_i), \dim_{D_i}(M_i))$. As all edges contain 0, we say that 0 is a *central vertex*; in particular, Δ_F is connected.

Recall that an Artin algebra is said to be *of finite representation type* if it has only finitely many isomorphism classes of indecomposable modules of finite length. In view of the main result of [**DR76**], R_F is of finite representation type if and only if Δ_F is a Dynkin diagram. By inspecting such diagrams having a central vertex, this is equivalent to Δ_F being a subgraph (containing 0 as a central vertex) of one of the following graphs:

$$\mathbf{B}_3: \quad 1 \text{ --- } 0 \overset{(1,2)}{\text{---}} 2 \qquad\qquad \mathbf{C}_3: \quad 1 \text{ --- } 0 \overset{(2,1)}{\text{---}} 2$$

$$\mathbf{D}_4: \quad \begin{array}{ccccc} 1 & \text{---} & 0 & \text{---} & 2 \\ & & | & & \\ & & 3 & & \end{array}$$

$$\mathbf{G}_2: \quad 0 \overset{(3,1)}{\text{---}} 1 \qquad\qquad \mathbf{G}_2^{\mathrm{op}}: \quad 0 \overset{(1,3)}{\text{---}} 1$$

The subgraphs obtained in this way are as follows:

$$\mathbf{A}_2: \quad 0 \text{ --- } 1 \qquad\qquad \mathbf{B}_2: \quad 0 \overset{(1,2)}{\text{---}} 1$$

$$\mathbf{C}_2: \quad 0 \overset{(2,1)}{\text{---}} 1 \qquad\qquad \mathbf{A}_3: \quad 1 \text{ --- } 0 \text{ --- } 2$$

Here all unmarked edges have value $(1, 1)$. We set

$$g_i := \dim(A_i) = \dim_k(M_i), \quad n_i := [D_i : \mathbb{Q}], \quad n := [k : \mathbb{Q}].$$

Then the label of each edge $\{0, i\}$ is $(g_i, \frac{g_i n}{n_i})$. The above list entails restrictions on these labels, and hence on the simple abelian varieties A_i and the associated division rings D_i. We will work out the consequences of these restrictions in the case where k is the field of rational numbers, which yields an especially simple result:

PROPOSITION 3.16. *When $k = \mathbb{Q}$, the algebra R_F is of finite representation type if and only if $\Delta_F = \mathbf{D}_4, \mathbf{C}_3, \mathbf{G}_2$ or a subgraph containing* 0 *as a central vertex (i.e., $\mathbf{A}_2, \mathbf{A}_3, \mathbf{C}_2$), and the abelian varieties A_i satisfy the following conditions:*

$\mathbf{D}_4$: *A_1, A_2, A_3 are elliptic curves.*

$\mathbf{C}_3$: *A_1 is an elliptic curve and A_2 is a simple abelian surface with $[D_2 : \mathbb{Q}] = 2$.*

$\mathbf{G}_2$: *A_1 is a simple abelian threefold with $[D_1 : \mathbb{Q}] = 3$.*

PROOF. Since Δ_F is a Dynkin diagram, we have $\dim_{\mathbb{Q}}(M_i) = 1$ or $\dim_{D_i}(M_i) = 1$. In the former case, we have $g_i = 1$, that is, A_i is an elliptic curve. Moreover, $D_i \hookrightarrow \mathrm{End}_{\mathbb{Q}}(M_i)$ as M_i is a k-D_i-bimodule; thus, $D_i = \mathbb{Q}$. In the latter case, we have $g_i = n_i$. The result follows from these observations via a case-by-case checking. □

REMARKS 3.17. (i) We may view the Dynkin diagrams $\mathbf{D}_4$ and $\mathbf{C}_3$ as unfoldings of $\mathbf{G}_2$. In fact, a similar picture holds for the abelian varieties under consideration: let $A := A_1 \oplus \cdots \oplus A_r$, then $A_1, \ldots, A_r$ satisfy the assertion of Proposition 3.16 if and only if $\dim(A) = 3 = \dim \mathrm{End}_{\underline{\mathcal{V}}}(A)_{\mathbb{Q}}$. In the "general" case where A is simple, this yields type $\mathbf{G}_2$; it "specializes" to types $\mathbf{C_3}$ and $\mathbf{D}_4$.

(ii) When R_F is of finite representation type, its indecomposable modules of finite length are described by the main result of [**DR76**]: the isomorphism classes of such modules correspond bijectively to the positive roots of the root system with Dynkin diagram Δ_F, by assigning with each module its dimension type (the sequence of multiplicities of the simple modules). This yields a case-by-case construction of the indecomposable objects of $\underline{\mathcal{V}}_F$. For example, in type $\mathbf{D}_4$, the indecomposable object associated with the highest root (i.e., with the sequence of multiplicities $m_0 = 2$, $m_1 = m_2 = m_3 = 1$) is the quotient of the universal vector extension $E(A_1 \oplus A_2 \oplus A_3)$ by a copy of $\mathbb{G}_a$ embedded diagonally in $U(A_1 \oplus A_2 \oplus A_3) \cong 3\mathbb{G}_a$. But we do not know any uniform construction of indecomposable objects for all types, along the lines of (i).

(iii) All the abelian varieties A_i over $\mathbb{Q}$ that occur in Proposition 3.16 satisfy the condition that $\mathrm{End}(A_i)_{\mathbb{Q}}$ is a field of dimension equal to $\dim(A_i)$. This condition defines the class of *abelian varieties of* GL_2*-type*, introduced by Ribet in [**Ri92**]; it includes all elliptic curves over $\mathbb{Q}$, and also the abelian varieties associated with certain modular forms via a construction of Shimura (see [**Sh71**, Thm. 7.14]). Assuming a conjecture of Serre on Galois representations, Ribet showed in [**Ri92**] that this construction yields all abelian varieties of GL_2-type up to isogeny.

Examples of abelian varieties of GL_2-type have been obtained by González, Guárdia and Rotger in dimension 2 (see [**GGR05**, Cor. 3.10]), and by Baran in dimension 3 (see [**Ba14**]).

(iv) Still assuming that k is a number field, the question of characterizing finite representation type makes sense, more generally, for the Serre subcategory $\underline{\mathcal{C}}_{E,F} \subset \underline{\mathcal{C}}$ generated by a finite set E of simple linear algebraic groups and a finite set F of simple abelian varieties, pairwise non-isogenous (so that $\underline{\mathcal{C}}_{\mathbb{G}_a,F} = \underline{\mathcal{V}}_F$). The abelian category $\underline{\mathcal{C}}_{E,F}$ is equivalent to $R_{E,F}$-mod, where $R_{E,F}$ is a triangular matrix algebra of finite dimension over $\mathbb{Q}$, constructed as above. The $\mathbb{Q}$-species associated with

$R_{E,F}$ is the directed graph $\Gamma_{E,F}$ with vertices $E \sqcup F$ and edges (i,j) for all $i \in E$, $j \in F$ such that $\mathrm{Ext}^1_{\underline{\mathcal{C}}}(A_j, L_i) \neq 0$; here A_j (resp. L_i) denotes the corresponding simple abelian variety (resp. linear algebraic group). In particular, if $\mathbb{G}_a \in E$ then the associated vertex is linked to all vertices in F, but some simple tori need not be linked to some simple abelian varieties. Each vertex v is labeled with the division ring D_v opposite to the endomorphism ring of the corresponding simple module, and each edge (i,j) is labeled with the D_i-D_j-bimodule $\mathrm{Ext}^1_{\underline{\mathcal{C}}}(A_j, L_i)$. Then again, the category $R_{E,F}$ is equivalent to that of representations of the $\mathbb{Q}$-species $\Gamma_{E,F}$; it is of finite representation type if and only if each connected component of the associated valued graph $\Delta_{E,F}$ is a Dynkin diagram. Note that such a diagram comes with a bipartition (by vertices in E, F).

To obtain a full characterization of finite representation type in this generality, we would need detailed information on the structure of Γ-module of $A(\bar{k})_{\mathbb{Q}}$ for any $A \in F$. But it seems that very little is known on this topic. For example, just take $E := \{\mathbb{G}_m\}$; recall that $\mathrm{Ext}^1_{\underline{\mathcal{C}}}(A, \mathbb{G}_m) \cong A(k)_{\mathbb{Q}}$ for any abelian variety A. We may thus assume that the finitely generated abelian group $A(k)$ is infinite for any $A \in F$; then $\mathbb{G}_m$ is a central vertex of $\Gamma_{\mathbb{G}_m,F}$. Arguing as in the proof of Proposition 3.16, one obtains a similar characterization of finite representation type in terms of Dynkin diagrams satisfying the following conditions:

D$_4$: $[D_i : \mathbb{Q}] = 3 = \dim A_i(k)_{\mathbb{Q}}$ for $i = 1, 2, 3$.
C$_3$: $[D_1 : \mathbb{Q}] = 1 = \dim A_1(k)_{\mathbb{Q}}$ and $[D_2 : \mathbb{Q}] = 2 = \dim A_2(k)_{\mathbb{Q}}$.
G$_2$: $[D_1 : \mathbb{Q}] = 3 = \dim A_1(k)_{\mathbb{Q}}$.

As a consequence, all simple abelian varieties A occuring in F must satisfy $[D_A : \mathbb{Q}] \leq 3$ (in particular, D_A is commutative) and $\dim A(k)_{\mathbb{Q}} = [D_A : \mathbb{Q}]$. We do not know whether such abelian varieties exist in arbitrary large dimensions.

Acknowledgments. I had the opportunity to present the results from [**Br16**] and some results from the present paper, at the Lens 2016 mini-courses and the International Conference on Representations of Algebras, Syracuse, 2016. I thank the organizers of both events for their invitation, and the participants for stimulating questions. Also, I warmly thank Claire Amiot, Brian Conrad, Stéphane Guillermou, Henning Krause, George Modoi, Idun Reiten and Gaël Rémond for very helpful discussions or e-mail exchanges.

References

[ARS97] Maurice Auslander, Idun Reiten, and Sverre O. Smalø, *Representation theory of Artin algebras*, Cambridge Studies in Advanced Mathematics, vol. 36, Cambridge University Press, Cambridge, 1997. Corrected reprint of the 1995 original. MR1476671

[Ba14] Burcu Baran, *An exceptional isomorphism between level 13 modular curves via Torelli's theorem*, Math. Res. Lett. **21** (2014), no. 5, 919–936, DOI 10.4310/MRL.2014.v21.n5.a1. MR3294556

[Ba68] Hyman Bass, *Algebraic K-theory*, W. A. Benjamin, Inc., New York-Amsterdam, 1968. MR0249491

[BR07] Apostolos Beligiannis and Idun Reiten, *Homological and homotopical aspects of torsion theories*, Mem. Amer. Math. Soc. **188** (2007), no. 883, viii+207, DOI 10.1090/memo/0883. MR2327478

[Bo91] Armand Borel, *Linear algebraic groups*, 2nd ed., Graduate Texts in Mathematics, vol. 126, Springer-Verlag, New York, 1991. MR1102012

[Br15] M. Brion, *Some structure theorems for algebraic groups*, Proc. Symp. Pure Math. **94** (2017), 53–125.

[Br16] Michel Brion, *Commutative algebraic groups up to isogeny*, Doc. Math. **22** (2017), 679–725. MR3650225

[Co02] Brian Conrad, *A modern proof of Chevalley's theorem on algebraic groups*, J. Ramanujan Math. Soc. **17** (2002), no. 1, 1–18. MR1906417

[Co06] Brian Conrad, *Chow's K/k-image and K/k-trace, and the Lang-Néron theorem*, Enseign. Math. (2) **52** (2006), no. 1-2, 37–108. MR2255529

[CGP15] Brian Conrad, Ofer Gabber, and Gopal Prasad, *Pseudo-reductive groups*, 2nd ed., New Mathematical Monographs, vol. 26, Cambridge University Press, Cambridge, 2015. MR3362817

[DR76] Vlastimil Dlab and Claus Michael Ringel, *Indecomposable representations of graphs and algebras*, Mem. Amer. Math. Soc. **6** (1976), no. 173, v+57, DOI 10.1090/memo/0173. MR0447344

[DG70] M. Demazure, P. Gabriel, *Groupes algébriques*, Masson, Paris, 1970.

[FJ74] Gerhard Frey and Moshe Jarden, *Approximation theory and the rank of abelian varieties over large algebraic fields*, Proc. London Math. Soc. (3) **28** (1974), 112–128, DOI 10.1112/plms/s3-28.1.112. MR0337997

[Ga62] Pierre Gabriel, *Des catégories abéliennes* (French), Bull. Soc. Math. France **90** (1962), 323–448. MR0232821

[Ga71] Peter Gabriel, *Indecomposable representations. II*, Symposia Mathematica, Vol. XI (Convegno di Algebra Commutativa, INDAM, Rome, 1971), Academic Press, London, 1973, pp. 81–104. MR0340377

[GGR05] Josep González, Jordi Guàrdia, and Victor Rotger, *Abelian surfaces of* GL_2*-type as Jacobians of curves*, Acta Arith. **116** (2005), no. 3, 263–287, DOI 10.4064/aa116-3-3. MR2114780

[KT07] P. A. Krylov and A. A. Tuganbaev, *Modules over discrete valuation domains. I*, J. Math. Sci. (N.Y.) **145** (2007), no. 4, 4997–5117, DOI 10.1007/s10958-007-0335-7. Algebra. MR2372456

[Le12] J. Lemay, *Valued graphs and the representation theory of Lie algebras*, Axioms **1** (2012), 111–148.

[MRS07] B. Mazur, K. Rubin, and A. Silverberg, *Twisting commutative algebraic groups*, J. Algebra **314** (2007), no. 1, 419–438, DOI 10.1016/j.jalgebra.2007.02.052. MR2331769

[Mi70] J. S. Milne, *The homological dimension of commutative group schemes over a perfect field*, J. Algebra **16** (1970), 436–441, DOI 10.1016/0021-8693(70)90017-7. MR0266929

[Mi72] J. S. Milne, *On the arithmetic of abelian varieties*, Invent. Math. **17** (1972), 177–190, DOI 10.1007/BF01425446. MR0330174

[Mi86] J. S. Milne, *Abelian varieties*, Arithmetic geometry (Storrs, Conn., 1984), Springer, New York, 1986, pp. 103–150. MR861974

[Mu08] David Mumford, *Abelian varieties*, Tata Institute of Fundamental Research Studies in Mathematics, vol. 5, Published for the Tata Institute of Fundamental Research, Bombay; by Hindustan Book Agency, New Delhi, 2008. With appendices by C. P. Ramanujam and Yuri Manin; Corrected reprint of the second (1974) edition. MR2514037

[Oo66] F. Oort, *Commutative group schemes*, Lecture Notes in Mathematics, vol. 15, Springer-Verlag, Berlin-New York, 1966. MR0213365

[Oo88] Frans Oort, *Endomorphism algebras of abelian varieties*, Algebraic geometry and commutative algebra, Vol. II, Kinokuniya, Tokyo, 1988, pp. 469–502. MR977774

[Pe75] Daniel Perrin, *Schémas en groupes quasi-compacts sur un corps* (French), Schémas en groupes quasi-compacts sur un corps et groupes henséliens, U. E. R. Math., Univ. Paris XI, Orsay, 1975, pp. 1–75. MR0409487

[Pe76] Daniel Perrin, *Approximation des schémas en groupes, quasi compacts sur un corps*, Bull. Soc. Math. France **104** (1976), no. 3, 323–335. MR0432661

[Ri92] Kenneth A. Ribet, *Abelian varieties over* $\mathbf{Q}$ *and modular forms*, Algebra and topology 1992 (Taejŏn), Korea Adv. Inst. Sci. Tech., Taejŏn, 1992, pp. 53–79. MR1212980

[Se60] Jean-Pierre Serre, *Groupes proalgébriques* (French), Inst. Hautes Études Sci. Publ. Math. **7** (1960), 67. MR0118722

[Se97] Jean-Pierre Serre, *Cohomologie galoisienne* (French), With a contribution by Jean-Louis Verdier. Lecture Notes in Mathematics, No. 5. Troisième édition, vol. 1965, Springer-Verlag, Berlin-New York, 1965. MR0201444

[SGA3] *Schémas en groupes. II: Groupes de type multiplicatif, et structure des schémas en groupes généraux* (French), Séminaire de Géométrie Algébrique du Bois Marie 1962/64 (SGA 3). Dirigé par M. Demazure et A. Grothendieck. Lecture Notes in Mathematics, Vol. 152, Springer-Verlag, Berlin-New York, 1970. MR0274459

[Sh71] Goro Shimura, *Introduction to the arithmetic theory of automorphic functions*, Publications of the Mathematical Society of Japan, No. 11. Iwanami Shoten, Publishers, Tokyo; Princeton University Press, Princeton, N.J., 1971. Kanô Memorial Lectures, No. 1. MR0314766

[Si86] Joseph H. Silverman, *The arithmetic of elliptic curves*, Graduate Texts in Mathematics, vol. 106, Springer-Verlag, New York, 1986. MR817210

[SP16] The Stack Project Authors, *Stacks Project*, `http://stacks.math.columbia.edu`, 2016.

Université Grenoble Alpes, Institut Fourier, CS 40700, 38058 Grenoble cedex 09, France

Email address: `Michel.Brion@univ-grenoble-alpes.fr`

Contemporary Mathematics
Volume **705**, 2018
http://dx.doi.org/10.1090/conm/705/14199

Noncommutative resolutions of discriminants

Ragnar-Olaf Buchweitz, Eleonore Faber, and Colin Ingalls

ABSTRACT. We give an introduction to the McKay correspondence and its connection to quotients of $\mathbb{C}^n$ by finite reflection groups. This yields a natural construction of noncommutative resolutions of the discriminants of these reflection groups. This paper is an extended version of E. F.'s talk with the same title delivered at the ICRA.

1. Introduction

This article has two objectives: first, we want to present the components of the classical McKay correspondence, which relate algebra, geometry and representation theory. On the other hand, we give a short introduction to discriminants of finite reflection groups and our construction of their noncommutative desingularizations. The details of our new results will be published elsewhere [**BFI16**].

This project grew out of the following (which is NOT how we will present our story!): Start with a commutative ring R, then a noncommutative resolution of singularities (=NCR) of R (or of $\operatorname{Spec}(R)$) is an endomorphism ring $\operatorname{End}_R M$ of a faithful module M such that the global dimension of $\operatorname{End}_R M$ is finite. Endomorphism rings of finite global dimension have got a lot of attention lately because of their connections to various parts of mathematics, such as commutative algebra, noncommutative algebraic geometry and representation theory. The problem of constructing explicit NCRs is difficult in general and one only knows some scattered examples. In particular, NCRs for non-normal rings have not been considered much in the literature, mostly only in examples where R has Krull-dimension ≤ 2 (geometrically: R is the coordinate ring of a collection of points, curves or surfaces).

In [**DFI15**] we were interested in noncommutative resolutions (=NCRs) of certain non-normal hypersurfaces, so-called free divisors, and explicitly constructed a NCR of a normal crossings divisor. After some extensive calculations we realized that there was a more conceptual way to understand this particular NCR as a quotient of the skew group ring of the polynomial ring in n variables and the reflection group $(\mu_2)^n$ by a two-sided ideal generated by an idempotent of the skew group ring. And since the normal crossings divisor is the discriminant of a real

2010 *Mathematics Subject Classification.* 14E16, 13C14, 14E15, 14A22 .

Key words and phrases. Reflection groups, hyperplane arrangements, maximal Cohen–Macaulay modules, matrix factorizations, noncommutative desingularization.

R.-O. B. was partially supported by an NSERC Discovery grant, E.F. was partially supported by an Oberwolfach Leibniz Fellowship, C.I. was partially supported by an NSERC Discovery grant.

The first author passed away on November 11, 2017.

reflection group, we had the idea to carry out the same quotient procedure for other discriminants of reflection groups. It was readily verified that indeed this construction works for any discriminant of a finite reflection group.

After some more painful calculations and serious doubts, that these quotients are endomorphism rings in general, we were finally able to also prove this for reflection groups generated by reflections of order 2, and discovered the surprising fact that the corresponding reflection arrangement is in some sense the noncommutative resolution of the discriminant of a reflection group. Moreover, our work generalizes Auslander's theorem about skew group rings of small subgroups $G \subseteq \mathrm{GL}(n, \mathbb{C})$ and thus our NCRs give rise to a McKay correspondence for reflection groups. For reflection groups $G \subseteq \mathrm{GL}(2, \mathbb{C})$ our results nicely fit in the picture of the classical McKay correspondence: we see that the reflection arrangement provides a representation generator for the torsion-free modules over the discriminant curves.

This paper is organized as follows:

CONTENTS

2. Reflection groups

Throughout the paper, we assume that $k = \mathbb{C}$ and that $G \subseteq \mathrm{GL}(V)$ is a finite subgroup acting on $V \cong \mathbb{C}^n$. Thus, $G \subseteq \mathrm{GL}(V)$ means $G \subseteq \mathrm{GL}(n, \mathbb{C})$. Then $\mathrm{Sym}_k(V) \cong k[x_1, \ldots, x_n]$ is a (graded) polynomial ring, with the standard grading $|x_i| = 1$. Assume that G acts on V, then any $g \in G$ acts also on S via $g(f(x_1, \ldots, x_n)) = f(g(x_1, \ldots, x_n))$. Note that in most of the literature (about the McKay correspondence), one considers the completion $k[[x_1, \ldots, x_n]]$ of S, but in fact, all the results we need work for S the graded polynomial ring.

Here we gather some well-known results about complex reflection groups, see e.g. [**Bou81**, **LT09**, **OT92**], we follow the notation of [**OT92**].

2.1. Basics. An element $g \in \mathrm{GL}(V)$ is called a *pseudo-reflection* if it is conjugate to a diagonal matrix $\mathsf{diag}(\zeta, 1, \ldots, 1)$ with $\zeta \neq 1$ a root of unity. If $\zeta = -1$, we call g a *(true) reflection.* A pseudo-reflection g fixes a unique hyperplane H pointwise. This hyperplane will be called the *mirror* of g. If G is generated by pseudo-reflections, it is called a *complex reflection group* (or *pseudo-reflection group*). If G is generated by true reflections, then we will call G a *true reflection group.* A finite subgroup $G \subseteq \mathrm{GL}(V)$ is called *small,* if G does not contain any pseudo-reflections.

Let now $G \subseteq \mathrm{GL}(V)$ be finite. The *invariant ring* of the action of G on V is $R = S^G = \{f \in S : g(f) = f \text{ for all } g \in G\}$. By the theorem of Hochster–Roberts [**HR74**], R is Cohen–Macaulay and it is Gorenstein if $G \subseteq \mathrm{SL}(V)$, by [**Wat74**].

If G is a pseudo-reflection group, we are even in a better situation, cf. [**Che55**], [**OT92**, Thm. 6.19]:

THEOREM 2.1 (Chevalley–Shephard–Todd).

(a) *Let $G \subseteq \mathrm{GL}(V)$ be a finite group acting on V. Then G is a pseudo-reflection group if and only if there exist homogeneous polynomials $f_1, \ldots, f_n \in R$ such that $R = k[f_1, \ldots, f_n]$.*

(b) *Moreover, S is free as an R-module, more precisely, $S \cong R \otimes_k kG$ as R-modules, where kG denotes the group ring of G.*

The polynomials f_i in the theorem are called *basic invariants* of G. They are not unique, but their degrees $\deg f_i = d_i$ are uniquely determined by the group G.

2.2. Discriminant and hyperplane arrangement. Let $G \subseteq \mathrm{GL}(V)$ be any finite subgroup. Then we denote by V/G the quotient by this action. It is an affine variety given by $V/G = \mathrm{Spec}(S^G)$, see e.g. [**CLO07**, Chapter 7]. In general V/G is singular, in particular, if G is a small subgroup of $\mathrm{GL}(V)$. If G is a pseudo-reflection group, then by Thm. 2.1, V/G is smooth and isomorphic to $V \cong k^n$. We have a natural projection:

$$\pi : V \cong \mathrm{Spec}(S) \longrightarrow V/G \cong \mathrm{Spec}(S^G) \ .$$

On the ring level, this corresponds to the natural inclusion of rings

$$S^G \cong k[f_1, \ldots, f_n] \hookrightarrow S = k[x_1, \ldots, x_n] \ .$$

Now let us define two of the central characters in our story: Let G be a finite complex reflection group. The set of mirrors of G is called the *reflection arrangement of* G, denoted by $\mathcal{A}(G)$. Note that $\mathcal{A}(G) \subseteq V$. Let $H \subseteq V$ be one of these mirrors, that is, H is fixed by a cyclic subgroup generated by a pseudo-reflection $g_H \in G$ of order $\rho_H > 1$, and let $\{l_H = 0\}$ be the defining linear equation of H. Then the reflection arrangement is given by the (reduced) equation $z := \prod_{H \subseteq \mathcal{A}(G)} l_H = 0$. The image of the reflection arrangement $\mathcal{A}(G)$ under the projection $\pi : V \to V/G$ is called the *discriminant of* G and given by a reduced polynomial Δ. Note that $\Delta \in S^G$, whereas $z \notin S^G$.

One can express the discriminant in terms of the basic invariants of G: first, one can show that

$$J = Jac(f_1, \ldots, f_n) = \det \left(\frac{\partial f_i}{\partial x_j} \right)_{i,j=1,\ldots,n} = u \prod_{H \subseteq \mathcal{A}(G)} l_H^{\rho_H - 1} \ ,$$

where $u \in k^*$. The Jacobian J is an anti-invariant for G, that is, $gJ = (\det g)^{-1} J$ for any $g \in G$. The discriminant Δ is then given by

$$\Delta = Jz = \prod_{H \subseteq \mathcal{A}(G)} l_H^{\rho_H} \ ,$$

which is an element of S^G.

In the case when G is a true reflection group, we have $\rho_H = 2$ for all H and thus $z = J$, and the computation of Δ simplifies to $\Delta = z^2$.

Let us conclude this section with some (non-)examples:

EXAMPLE 2.2. Let $G \subseteq \mathrm{SL}(V)$, $\dim V = 2$, be the cyclic group C_3 generated by $\begin{pmatrix} \zeta_3 & 0 \\ 0 & \zeta_3^{-1} \end{pmatrix}$, where ζ_3 is a third root of unity acting on $S = k[x, y]$. Then V/G is a Kleinian surface singularity, denoted by A_2. The invariants of G are $u = x^3 + y^3$, $v = xy$, $w = x^3 - y^3$ and satisfy the relation $w^2 - u^2 + 4v^3$. Thus (after a change of coordinates) $S^G \cong k[u,v,w]/(w^2 + u^2 + v^3)$. A real picture of V/G can be seen in Fig. 1.

EXAMPLE 2.3. Let $G \subseteq \mathrm{GL}(V)$, $\dim V = 2$, be a true reflection group. Then the discriminant Δ is an ADE curve. This can be seen for example by considering $\Gamma := G \cap \mathrm{SL}_2(V)$. The subgroup Γ is of index 2 in G and $G/\Gamma \cong \mu_2 = \langle \sigma \rangle$, where σ is a reflection in G. The invariant ring S^Γ gives rise to an ADE-surface singularity $X := V/\Gamma$ (cf. Thm. 4.1). The quotient $X/\langle \sigma \rangle \cong V/G$ is smooth and the branch locus of the natural projection $X \to X/\langle \sigma \rangle$, is isomorphic to an ADE curve singularity, that is the discriminant Δ of G. See [**Knö84**, Section 3] for a more thorough explanation.

In [**Ban76**] the discriminants of rank two complex reflection groups are calculated explicitly and one sees that indeed all these discriminants are curves of type ADE.

EXAMPLE 2.4. Let $G = (\mu_2)^n$ be acting on k^n by reflecting about the coordinate hyperplanes. Thus the generators of G are the diagonal matrices

$$S_i = \begin{pmatrix} 1 & 0 & 0 & 0 & 0 \\ & & & & \\ 0 & \ddots & 0 & 0 & 0 \\ 0 & 0 & -1 & 0 & 0 \\ & & & & \\ 0 & 0 & 0 & \ddots & 0 \\ 0 & 0 & 0 & 0 & 1 \end{pmatrix}.$$

The hyperplane arrangement $\mathcal{A}(G)$ is the union of the coordinate hyperplanes. This can easily be seen computationally: The basic invariants in this example are $f_i = x_i^2$, that is, $S^G = k[x_1^2, \ldots, x_n^2]$. Thus $Jac(f_1, \ldots, f_n) = 2^n x_1 \cdots x_n$ and we can take $z = J = x_1 \cdots x_n$. Since G is a true reflection group, $\Delta = z^2 = x_1^2 \cdots x_n^2 = f_1 \cdots f_n$. The coordinate ring of the discriminant is $S^G/(\Delta) = k[f_1, \ldots, f_n]/(f_1 \cdots f_n)$. In this example, the discriminant is also the union of coordinate hyperplanes, but now seen in the quotient V/G. See Fig. 1 for a picture in in the 3-dimensional case.

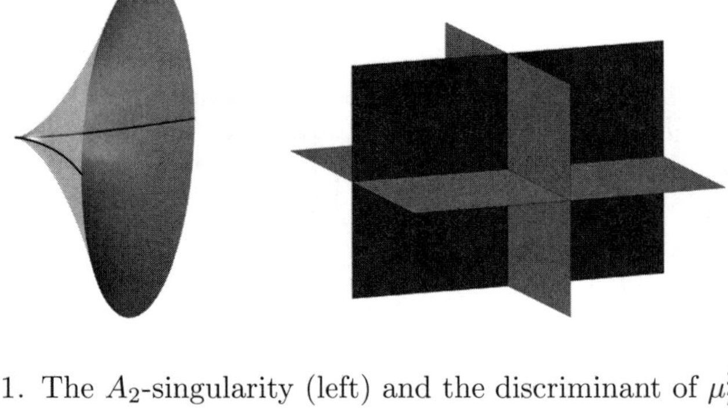

FIGURE 1. The A_2-singularity (left) and the discriminant of μ_2^3. (right).

EXAMPLE 2.5. Let $G = G(2,1,3)$, that is, the Coxeter group $B_3 := S_3 \ltimes (\mu_2)^3$ (which is also isomorphic to the direct product $S_4 \times \mu_2$) acting on k^3 by permutation and sign changes. The invariants of $S = k[x_1, x_2, x_3]$ under this group action are $u = x_1^2 + x_2^2 + x_3^2$, $v = x_1^4 + x_2^4 + x_3^4$ and $w = x_1^6 + x_2^6 + x_3^6$. Since $G(2, 1, 3)$ is a Coxeter group, it is generated by true reflections, and thus $z = J$. A calculation of the Jacobian determinant of (u, v, w) shows that

$$z = J = x_1x_2x_3(x_1^2 - x_2^2)(x_1^2 - x_3^2)(x_2^2 - x_3^2).$$

The invariant ring is $S^G = k[u, v, w]$ and an investigation of z^2 shows that it is equal to the product of w with the discriminant of the polynomial $P(t) = t^3 + ut^2 + vt + w$, see [**YS81**]. The equation for the discriminant is

$$\Delta = w(u^2v^2 - 4v^3 - 4u^3w + 18uvw - 27w^2).$$

See Fig. 2 below for pictures of the hyperplane arrangement, given by z and the discriminant Δ in $\mathbb{R}^3$.

FIGURE 2. The B_3-reflection arrangement (left) and the corresponding discriminant with equation $w(u^2v^2 - 4v^3 - 4u^3w + 18uvw - 27w^2)$ (right).

3. (Noncommutative) resolutions of singularities

Here we come to the geometric part. Let X be an affine algebraic variety, that is $X = \operatorname{Spec}(R)$, where R is a commutative finitely generated k-algebra. If R is not a regular ring, then we say that X is a *singular* algebraic variety and denote by $\operatorname{Sing} X$ its *singular locus*. A *resolution of singularities of* X is a proper birational map $\pi : \widetilde{X} \to X$ from a smooth scheme $\widetilde{X}$ to X such that π is an isomorphism on the smooth points of X.

The central result regarding resolution of singularities is

THEOREM 3.1 (Hironaka [**Hir64**]). *Let X be a scheme over a field of characteristic 0 (e.g., X an affine algebraic variety defined over $\mathbb{C}$). Then there exists a resolution of singularities $\pi : \widetilde{X} \to X$.*

One can impose some extra conditions on π, for example that $\pi^{-1}(\operatorname{Sing}(X))$ is a normal crossings divisor in $\widetilde{X}$ and π is a composition of blowups in smooth centers, or (in the case X an irreducible reduced variety over k), that X has a so-called embedded resolution, see [**Hau03**] for more on this. So existence of resolutions of singularities is (with our assumption that $k = \mathbb{C}$, that is, k of characteristic 0) not an issue. However, from an algebraic point of view, it would be desirable to have some control over $\widetilde{X}$. If X is of Krull-dimension 2, then there exists a unique *minimal resolution of singularities*, that is, a resolution Y of X such that any other resolution $\widetilde{X}$ of X factors through Y. This is false in higher dimensions, in general there does not even exist a minimal resolution.

For some Cohen–Macaulay varieties X (that is: if X is locally $\mathrm{Spec}(R)$, then R is a Cohen–Macaulay ring) there still exist preferred resolutions of singularities: namely those, that do not affect the canonical class of X. A resolution of singularities $\pi : \widetilde{X} \to X$ is called *crepant* if $\pi^*\omega_X = \omega_{\widetilde{X}}$. Crepant resolutions first appeared in the minimal model program, where the term "crepant" was coined by M. Reid to reflect the fact that the pull-back of the canonical sheaf of X does not pick up discrepancy, see e.g. [**Rei02**]. Crepant resolutions still do not always exist and are usually not isomorphic to each other, but at least their cohomologies should be the same, see [**BO02**]:

CONJECTURE 3.2 (Bondal–Orlov). *Let Y and Y^+ be two crepant resolutions of an algebraic variety X. Then there is a derived equivalence of categories*

$$D^b(\mathbf{Coh}\, Y) \simeq D^b(\mathbf{Coh}\, Y^+)\ ,$$

where $\mathbf{Coh}(-)$ *denotes the category of coherent sheaves.*

Building on work by Bridgeland–King–Reid for 3-dimensional quotient singularities [**BKR01**], Bridgeland proved [**Bri02**] the conjecture for complex 3-dimensional varieties with only terminal singularities. Inspired by their work, Van den Bergh defined a noncommutative crepant resolution (=NCCR) [**VdB04a**], also see [**VdB04b**]: if $X = \mathrm{Spec}(R)$, then the main point in their approach is to construct a (non-commutative) R-algebra such that for any crepant resolution $\pi : Y \to X$ one has $D^b(\mathbf{Coh}\, Y) \simeq D^b(\mathbf{mod}\, A)$. Here $\mathbf{mod}\, A$ denotes the category of finitely generated A-modules. The algebra A is thus a noncommutative analogue of a crepant resolution of singularities, a NCCR. Later, NCCRs were studied and constructed in various instances, see e.g., [**BLvdB10, IW14**]. The more general concept of noncommutative resolution (=NCR) was defined by Dao–Iyama–Takahashi–Vial [**DITV15**] in 2015. Unfortunately there is (so far) not a good theory about general existence and properties of NCRs, only mostly examples, see e.g. [**BIKR08, DFI15, DII$^+$16, DFI16, Leu07**]. A good introduction to NCCRs and categorical geometry with many references can be found in [**Leu12**].

But let us state the definitions:

DEFINITION 3.3. Let R be a commutative reduced noetherian ring. An R-algebra A is called a *noncommutative resolution (NCR)* of R (or of $\mathrm{Spec}(R)$) if A is of the form $\mathrm{End}_R M$ for M a finitely generated, faithful R-module and $\mathrm{gldim}(A) < \infty$. The algebra A is called a *noncommutative crepant resolution (NCCR)* of R (or of $\mathrm{Spec}(R)$) if A is a nonsingular order, that is, $\mathrm{gldim}\, A_{\mathfrak{p}} = \dim R_{\mathfrak{p}}$ for all $\mathfrak{p} \in \mathrm{Spec}(R)$ and A is a maximal Cohen–Macaulay module over R.

REMARK 3.4. In Van den Bergh's original definition, R was assumed to be a commutative normal Gorenstein domain. The reasoning behind this more general definition can be found in [**DFI15**].

In general it is not clear how to construct NCRs and whether they exist. However, there is one beautiful example, which will lead to our construction of NCRs for discriminants of reflection groups.

4. The classical McKay correspondence

Here we give brief sketches of the components of the classical McKay correspondence. Up to this date, the correspondence has been generalized in various directions, for an account and more detailed references see [**Buc12**].

4.1. Dual resolution graphs. Let $\Gamma \subseteq \mathrm{SL}(V)$ be a finite subgroup acting on $V \cong k^2$. Denote as above by $S = k[x, y]$ the symmetric algebra of V and the invariant ring $S^\Gamma = R$. Then $X := V/\Gamma = \mathrm{Spec}(R)$ is a quotient singularity. More precisely, these quotient singularities (aka *Kleinian singularities*) have been classified by Felix Klein [**Kle93**]:

THEOREM 4.1 (Klein 1884). *Let Γ and X be as above. Then X is isomorphic to an isolated surface singularity of type ADE, that is, $R \cong k[x, y, z]/(f)$:*

- A_n: $f = z^2 + y^2 + x^{n+1}$, *if Γ is a cyclic group,*
- D_n: $f = z^2 + x(y^2 + x^{n-2})$ *for $n \geq 4$, if Γ is a binary dihedral group,*
- E_6: $f = z^2 + x^3 + y^4$, *if Γ is the binary tetrahedral group,*
- E_7: $f = z^2 + x(x^2 + y^3)$, *if Γ is the binary octahedral group,*
- E_8: $f = z^2 + x^3 + y^5$, *if Γ is the binary icosahedral group.*

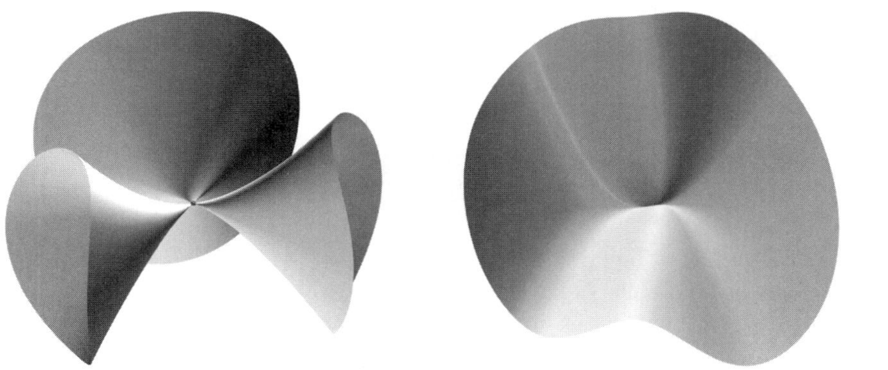

FIGURE 3. The D_4 -singularity with equation $z^2 + x(y^2 - x^2) = 0$ (left) and the E_8-singularity with equation $z^2 + x^3 + y^5 = 0$ (right).

The Kleinian singularities X are also classified by their so-called dual resolution graphs, which happen to be the corresponding ADE-Coxeter-Dynkin-diagrams (for a more detailed account of resolution of surface singularities and especially the rational double points, see [**Gre92**], [**Dur79**]): consider the minimal resolution $\pi : \widetilde{X} \to X$ of a Kleinian singularity X. The minimal resolution exists, since X is a normal surface singularity and $\widetilde{X}$ is moreover unique, see [**Lau71**]. One can for example obtain $\widetilde{X}$ by successively blowing up singular points. This process can be quite intricate, see for example the dessert in this menu [**FH10**]. The preimage of $\{0\} = \mathrm{Sing}(X)$ in $\widetilde{X}$ will be denoted by $\mathbb{E} = \bigcup_{i=1}^n \mathbb{E}_i$ and is a union of irreducible curves $\mathbb{E}_i$ on $\widetilde{X}$ (this holds for example because one can construct $\widetilde{X}$ as an embedded resolution of $X \subseteq k^3$, which implies that the preimage of $\mathrm{Sing}\, X$ must be a normal crossing divisor). Since X is a so-called rational singularity, the irreducible components $\mathbb{E}_i$ of $\mathbb{E}$ are isomorphic to $\mathbb{P}^1$'s and intersect each other at most pairwise transversally (this is to say: $\mathbb{E}$ is a normal crossings divisor on $\widetilde{X}$). Now one can form the *dual resolution graph* of X: the vertices are indexed by the $\mathbb{E}_i$, for $i = 1, \dots, n$ and there is an edge between $\mathbb{E}_i$ and $\mathbb{E}_j$ if and only if $\mathbb{E}_i \cap \mathbb{E}_j \neq \varnothing$.

Moreover, one defines the intersection matrix $E := ((\mathbb{E}_i \cdot \mathbb{E}_j))_{i,j=1,\dots,n}$. Computation shows that $(\mathbb{E}_i \cdot \mathbb{E}_i) = -2$ and $(\mathbb{E}_i \cdot \mathbb{E}_j) \in \{0, 1\}$ for $i \neq j$ and that E is symmetric and negative-definite. Moreover, there exist divisors Z' on $\widetilde{X}$ of the form $Z' = \sum_{i=1}^n m_i \mathbb{E}_i$ supported on $\mathbb{E}$ with $m_i > 0$, such that $Z' \cdot \mathbb{E}_i \leq 0$ for all i. The smallest such divisor is denoted by Z and called the *fundamental divisor of* X. One decorates the vertices of the dual resolution graph with the multiplicities m_i of Z. For an illustration of the D_4-singularity see Fig. 4.1, and the resolution with the resolution graph are below in Fig. 4.

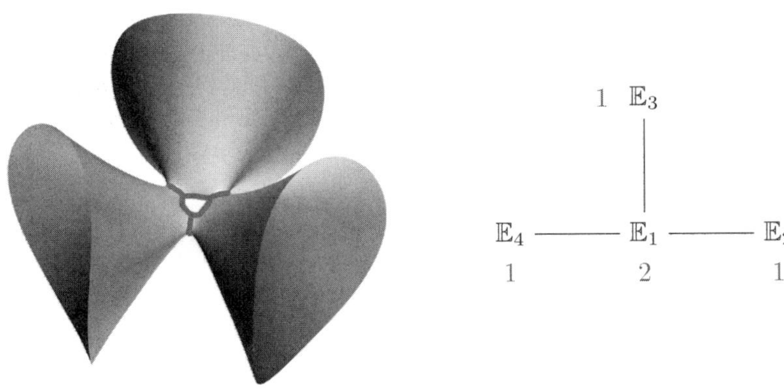

FIGURE 4. The minimal resolution of the D_4-singularity (with the exceptional curves sketched in red) and the dual resolution graph of type D_4 with the multiplicities of the fundamental divisor in red.

4.2. McKay quiver. In 1979, John McKay built the following graph out of a finite subgroup $\Gamma \subseteq \mathrm{SL}(V)$, $\dim V = 2$: the embedding of Γ in $\mathrm{SL}(V)$ defines the so-called canonical representation $c : \Gamma \hookrightarrow \mathrm{GL}(V)$ of Γ. Any finite group only has finitely many isomorphism classes of finite dimensional irreducible k-representations, given by group homomorphisms $\rho_i : \Gamma \to \mathrm{GL}(V_i)$, $i = 0, \dots, n$ for vector spaces V_i. The trivial representation will be denoted by $\rho_0 : \Gamma \to k^*$, sending any γ to 1. The *McKay quiver of* Γ consists of the vertices indexed by the V_i and there are m_{ij} arrows from V_i to V_j if and only if V_i is contained with multiplicity m_{ij} in $V_j \otimes V$.

EXAMPLE 4.2. Consider the group Γ generated by

$$\pm\begin{pmatrix} 1 & 0 \\ 0 & 1 \end{pmatrix}, \pm\begin{pmatrix} i & 0 \\ 0 & -i \end{pmatrix}, \pm\begin{pmatrix} 0 & 1 \\ -1 & 0 \end{pmatrix}, \pm\begin{pmatrix} 0 & i \\ i & 0 \end{pmatrix}.$$

This is the binary dihedral group D_4. It has five irreducible representations ρ_i, four of which are one-dimensional and one two-dimensional ρ_1, which is the canonical representation c. Using character theory, one obtains the McKay quiver of Γ (the dimensions of the irreducible representations in red):

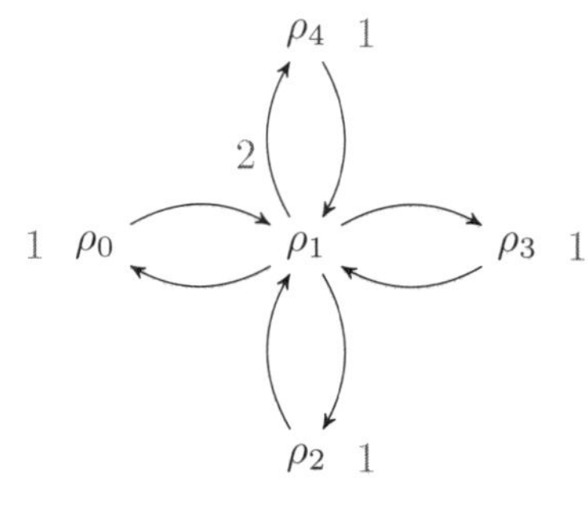

The following is due to McKay [**McK80**]:

Observation: Let Γ be as above. Then the McKay quiver of Γ is connected and contains no loops and $m_{ij} = m_{ji} \in \{0,1\}$. By calculation, one sees that the McKay quiver of Γ is the extended Coxeter–Dynkin diagram for Γ, with arrows in both directions. If one deletes the vertex corresponding to the trivial representation and collapses all arrows $V_i \rightleftarrows V_j$ to an edge, then one obtains the Coxeter–Dynkin diagram associated to Γ. If one decorates the vertices V_i with the dimensions $\dim V_i$, then one gets back the dual resolution graph of V/Γ.

Thus one obtains a $1-1$-correspondence between the irreducible components of the exceptional divisor of the minimal resolution of the quotient singularity V/Γ and the non-trivial irreducible representations of Γ! This was made precise later by [**GSV83**].

4.3. AR-quiver. But there is also an algebraic part of the correspondence: Let again $R = S^\Gamma$ for $\Gamma \subseteq \mathrm{SL}(V)$, where $V \cong k^2$. We consider reflexive modules over R. Since R is normal, M is reflexive if and only if it is maximal Cohen–Macaulay (=CM), and we write $\mathbf{CM}(R)$ for the category of CM-modules over R. By Herzog's theorem [**Her78**], there is an isomorphism of R-modules:

$$S \cong \bigoplus_{M \in \mathbf{CM}(R)} M^{a_M}, \text{ where the integer } a_M = \operatorname{rank}_R M.$$

Differently phrased: S is a *representation generator* for $\mathbf{CM}(R)$. In particular, there are only finitely many indecomposable objects in $\mathbf{CM}(R)$. One then says that R is of *finite CM-type*. By [**BGS87, Her78, Knö87**], a Gorenstein ring R is of finite CM-type if and only if it is isomorphic to a simple hypersurface singularity (if $\dim R = 2$, these are precisely the ADE-surface singularities).

Starting from the indecomposable CM-modules over R, one obtains a third quiver, the *Auslander–Reiten-quiver* (short: *AR-quiver*) of R: its vertices are given by the indecomposable CM-modules M_i, $i = 1, \dots, n$ of R and there are m_{ij} arrows from M_i to M_j if and only if in the almost split sequence $0 \to \tau M_j \to E \to M_j \to 0$ ending in M_j, M_i appears with multiplicity m_{ij} in E. Here one can also see that $\tau M_i = M_i$ for all $M_i \not\cong R$ and that the AR-quiver of R is precisely the McKay quiver of Γ. Moreover, the ranks of the indecomposable modules M_i correspond to the dimensions of the irreducible representations of Γ.

This astounding correspondence was proven by Maurice Auslander [**Aus86**] and looking at his proof more carefully, we will obtain the link to NCCRs. Therefore we need some general notation first:

Let $G \subseteq \mathrm{GL}(V)$ finite act on $S = \mathbb{S}\mathrm{ym}_k(V) = k[x_1, \dots, x_n]$ for some vector space $V \cong k^n$. Then define $A = S * G$ to be the *skew group ring* (or *twisted group ring*) of G. As an S-module, A is just $S \otimes_k kG$, but the multiplication on A is twisted by the action of G: for elements $s, s' \in S$ and $g, g' \in G$ define

$$sg \cdot s'g' := (sg(s'))(gg') ,$$

and extend by linearity. The following result is the key theorem in proving the correspondence between indecomposable CM-modules and irreducible representations. However note that the assumptions on G are more general than before: we may take any G which does not contain any pseudo-reflections and the dimension of V may be greater than 2:

THEOREM 4.3 (Auslander). *Let S be as above and assume that $G \subseteq \mathrm{GL}(V)$, $\dim V = n$, is small and set $R = S^G$. Then we have an isomorphism of algebras:*

$$A = S * G \xrightarrow{\cong} \mathrm{End}_R(S) \ , sg \mapsto (x \mapsto sg(x)) \ .$$

Moreover, $S * G$ is a CM-module over R and $\mathrm{gldim}(S * G) = n$, *and $Z(A) = R$.*

This result can be phrased in terms of NCCRs:

COROLLARY 4.4. *If $G \subseteq \mathrm{SL}(V)$, that is, R is a Gorenstein singularity, then A is an NCCR of R. If $G \subseteq \mathrm{GL}(V)$ is small and not in* $\mathrm{SL}(V)$, *then A is an NCR of R.*

PROOF. The twisted group ring $A \cong \mathrm{End}_R S$ is an endomorphism ring of a faithful CM-module over R. Since $\mathrm{gldim}\, A = n < \infty$, A is an NCR. For the crepancy in the $\mathrm{SL}(V)$ case first note that R is a normal Gorenstein domain by the theorems of Hochster–Roberts and Watanabe. Then the fact that $S * G \cong \mathrm{End}_R S$ is CM over R implies that $\mathrm{End}_R S$ is a nonsingular order, see [**VdB04a**, Lemma 4.2] and thus an NCCR. □

For $G = \Gamma \subseteq \mathrm{SL}(V)$ and $\dim V = 2$, Auslander's theorem implies a 1-1-correspondence between the indecomposable CM-modules over R, the indecomposable summands of S, the indecomposable projective modules over $S * G$, and the irreducible Γ-representations. Then McKay's observation gives the correspondence to the geometry, namely, the bijection to the irreducible components of the exceptional divisor on the minimal resolution of R.

We shortly sketch the correspondences: first, by Herzog's theorem the indecomposable objects in $\mathbf{CM}(R)$ are in bijection to the indecomposable R-summands of S. For the second part, send the indecomposable summand M of S to $\mathrm{Hom}_R(S, M)$, which is an indecomposable projective $\mathrm{End}_R(S)$-module. By Auslander's theorem, it is an indecomposable projective $S * \Gamma$-module. For the last correspondence, send a indecomposable projective $S * \Gamma$-module to $P/(x, y)P$, which is a simple module over $k\Gamma$ and hence an irreducible Γ-representation. For details about these constructions, see [**Aus86, Yos90, LW12**].

4.4. Auslander's theorem and reflections. Looking at Auslander's theorem, it is natural to ask:

QUESTION 4.5. Let $G \subseteq \mathrm{GL}(V)$, $\dim V = n$, be generated by pseudo-reflections. Is there an analogue of Auslander's theorem? In particular, is there also a noncommutative resolution hidden somewhere?

The first problem in generalizing Auslander's theorem is that if G is a pseudo-reflection group, then the map

$$A = S * G \hookrightarrow \mathrm{End}_R(S)$$

is no longer surjective. One can see this for example for $G = S_n$. Then $\mathrm{End}_R(S)$ contains the so-called Demazure operators, see e.g. [**LS92**], which are not contained in the skew group ring. However, one can show in this case that A is the intersection of $\mathrm{End}_R(S) \cap (\mathrm{End}_R(S))^\tau$ in $Q(A)$, the quotient ring of A, where τ is a particular anti-involution on $\mathrm{End}_R(S)$, for details see [**KK86**].

Another "problem" is that $R = S^G$ is no longer singular if G is generated by pseudo-reflections, by Thm. 2.1. This, however, can be remedied by looking at the discriminant $\Delta \in R$: the discriminant defines a singular hypersurface, whose singularities occur in codimension 1. More precisely: it was shown by Saito and Terao that the discriminant of a reflection arrangement $\mathcal{A}(G)$ is a so-called *free divisor* in R, that is, the Jacobian ideal of Δ is a CM-module over $R/(\Delta)$. One can also show that the reflection arrangement itself defines a free divisor in S. For more details, see [**OT92**].

The above facts already suggest our strategy to attack the problem of finding an analogue of Auslander's theorem and a noncommutative resolution for the discriminant: we will start with the twisted group ring $S*G$ and using the small group $G \cap \mathrm{SL}(V)$ we will cook up a NCR of Δ, which will also yield a 1–1-correspondence to irreducible representations of G.

5. NCRs of discriminants

For this section let us change the notation slightly: first, we will assume that G is a true reflection group, i.e., $G \subseteq \mathrm{GL}(V)$, $\dim V = n$, is generated by reflections of order 2 (some of our results also hold for pseudo-reflection groups but the main correspondence only works in this case so far). We will denote the invariant ring S^G by T. By Theorem 2.1 $T \cong k[f_1, \dots, f_n]$ for the basic invariants $f_i \in S$. Since G is a true reflection group $J = z$ and the hyperplane arrangement $\mathcal{A}(G)$ is defined by the polynomial $J = \det(\frac{\partial f_i}{\partial x_j})$. The discriminant Δ is then defined by the equation $J^2 = \Delta$, where Δ is in T. Its coordinate ring will be denoted by $T/(\Delta)$. For our construction of the NCR of Δ, we consider the small group $\Gamma = G \cap SL(V)$, with invariant ring $R = S^\Gamma$. We have an exact sequence of groups $1 \to \Gamma \to G \to H \to 1$, where the quotient is $H \cong \mu_2 = \langle \sigma \rangle$. Moreover, as above, we denote by $A = S * G$ the twisted group ring of G.

The first result deals with the problem that A is not isomorphic to $\mathrm{End}_T(S)$. We can show that A is still isomorphic to an endomorphism ring - yet over another twisted group ring:

THEOREM 5.1. *With notation introduced in the paragraph above, there is an isomorphism of rings*

$$\mathrm{End}_{R*H}(S * \Gamma) \cong S * G.$$

This result follows from generalizing the approach of H. Knörrer for curves [**Knö84**] (also using ideas from J. Bilodeau [**Bil05**]).

For simplifying notation set $B := R * H$. Now one can interpret B as the path algebra over T of a quiver Q modulo some relations. Then B has idempotents $e_{\pm} = \frac{1}{2}(1 \pm \sigma)$, where σ is the generator of H, as defined above. One crucial observation here is that $B/Be_{\pm}B$ is isomorphic to $T/(\Delta)$ as rings. Thus, using the functor

$$\mathrm{mod}(B/Be_{-}B) \xleftarrow{i^*(-)=-\otimes_B B/Be_{-}B} \mathrm{mod}(B)$$

from the standard recollement we obtain the following result:

THEOREM 5.2. *The functor i^* induces an equivalence of categories*

$$\mathbf{CM}(B)/\langle e_{-}B\rangle \simeq \mathbf{CM}(T/(\Delta)),$$

where $\langle e_{-}B\rangle$ denotes the ideal generated by the object $e_{-}B$ in the category $\mathbf{CM}(B)$.

Theorem 5.1 shows that A is in some sense too large for being a NCR of Δ. But theorem 5.2 suggests that we have to quotient A by an ideal: since $A \in \mathbf{CM}(B)$, we calculate its image in $\mathbf{CM}(B)/\langle e_{-}B\rangle$ and find that it is isomorphic to the quotient $\bar{A} = A/AeA$, where $e = \frac{1}{|G|}\sum_{g\in G} g \in A$ is the idempotent corresponding to the trivial representation of G. On the other hand, calculate the image of A under the recollement functor i^*: one gets that

$$i^*(A) = \mathrm{End}_{T/(\Delta)}(i^*(S * H)) \ .$$

Exploiting results from Stanley about the structure of R as T-module [**Sta77**], we obtain that $i^*(S * H) \cong S/(J)$, as $T/(\Delta)$-module. As a ring, $S/(J)$ is precisely the coordinate ring of the reflection arrangement $\mathcal{A}(G)$! In the last step, a result of Auslander–Platzeck–Todorov [**APT92**] about global dimension of quotients yields the following

THEOREM 5.3. *With notation as above,*

$$\bar{A} := A/AeA \cong \mathrm{End}_{T/(\Delta)}(S/(J)),$$

and $\mathrm{gldim}\,\bar{A} < \infty$. *Thus $\bar{A}$ yields a NCR of the free divisor T/Δ.*

Moreover, the indecomposable projective modules over $\bar{A}$ are in bijection with the non-trivial representations of G and also with certain CM-modules over the discriminant, namely the $T/(\Delta)$-direct summands of $S/(J)$.

Thus we not only obtain a NCR of discriminants of reflection arrangements but also a McKay correspondence for reflection groups G.

EXAMPLE 5.4. Let $\dim V = 2$ and let $G \subseteq \mathrm{GL}(V)$ be a true reflection group. Then the discriminant Δ is an ADE-curve singularity, see example 2.3. We obtain that $S/(J)$ is a generator of the category of CM-modules over $T/(\Delta)$, that is, $\mathrm{add}(S/(J)) = \mathbf{CM}(T/(\Delta))$. Moreover, the multiplicities, with which the indecomposables appear in $S/(J)$ are precisely the ranks of the corresponding irreducible representations.

As a particular example, take $G = S_3$, whose discriminant Δ is a cusp (aka A_2-singularity). For the cusp, one can show that $S/(J) \cong T/(\Delta) \oplus \mathfrak{m}^{\oplus 2}$ as $T/(\Delta)$-modules, where $\mathfrak{m}$ denotes the maximal ideal (u, v) in $T/(\Delta) = k[u, v]/(u^3 - v^2)$. Here one sees that $T/(\Delta)$ corresponds to the determinantal representation and $\mathfrak{m}$ to the canonical representation given by $G \hookrightarrow \mathrm{GL}(2, k)$.

6. Further questions

In general we are still missing a conceptual explanation for the $T/(\Delta)$-direct summands of $S/(J)$: in dimension 2, see example 5.4, any indecomposable CM-module over $T/(\Delta)$ can be obtained from $S/(J)$ (via matrix factorizations). But if $\dim V \geq 3$, that is, the Krull-dimension of $T/(\Delta)$ is greater than or equal to 2, $T/(\Delta)$ is not of finite CM-type. However, the factors of $S/(J)$ are still CM-modules and can be calculated as matrix factorizations. So far we only have few examples (the normal crossing divisor, as studied in Dao–Faber–Ingalls [**DFI15**] or the swallowtail, that is, the discriminant of S_4, where we can describe the $T/(\Delta)$-direct summands using Hovinen's classification of graded rank 1 MCM-modules [**Hov09**]).

Thus, for a geometric interpretation of the direct summands of $S/(J)$ we want to establish a similar correspondence as in [**GSV83**]. A next step would be to realize geometric resolutions as moduli spaces of isomorphism classes of representations of certain algebras as in [**CB00**].

But, in a different vein, the structure of $\bar{A}$ as an $S/(J)$-module seems to be easier to understand: $\bar{A}$ is isomorphic to the cokernel of the map φ given by left multiplication on G, i.e., the matrix of φ corresponds to the multiplication table of the group G. Since Frobenius and Dedekind, the block decomposition of this matrix is well-known. Our goal is to use this surprising discovery to learn more about the decomposition of $\bar{A}$ and hence of $S/(J)$ over the discriminant.

Moreover, looking at the quiver of $\bar{A}$, the exact form of the generating (necessarily quadratic) relations remains mysterious and will be the subject of further research.

7. Acknowledgements

We want to thank the organizers of the ICRA for making it such a pleasant and stimulating conference. In particular, we thank Graham Leuschke for his encouragement to write this proceedings article. We also thank the anonymous referee for helpful comments. We acknowledge and highly appreciate that the research presented in this paper was supported through the program "Research in Pairs" in 2015 and the program "Oberwolfach Leibniz Fellowships" in 2016 by the Mathematisches Forschungsinstitut Oberwolfach and by the Institut Mittag-Leffler (Djursholm, Sweden) in frame of the program "Representation theory" in 2015.

E.F. and C.I. want to thank R.-O.B. for the many hours we spent together working on this project—his enthusiasm, knowledge, and deep insights will be dearly missed.

References

[APT92] M. Auslander, M. I. Platzeck, and G. Todorov, *Homological theory of idempotent ideals*, Trans. Amer. Math. Soc. **332** (1992), no. 2, 667–692, DOI 10.2307/2154190. MR1052903

[Aus86] Maurice Auslander, *Rational singularities and almost split sequences*, Trans. Amer. Math. Soc. **293** (1986), no. 2, 511–531, DOI 10.2307/2000019. MR816307

[Ban76] Etsuko Bannai, *Fundamental groups of the spaces of regular orbits of the finite unitary reflection groups of dimension* 2, J. Math. Soc. Japan **28** (1976), no. 3, 447–454, DOI 10.2969/jmsj/02830447. MR0407199

[BFI16] R.-O. Buchweitz, E. Faber, and C. Ingalls. A McKay correspondence for reflection groups. 2017. arXiv:1709.04218[math.AG].

[BGS87] R.-O. Buchweitz, G.-M. Greuel, and F.-O. Schreyer, *Cohen-Macaulay modules on hypersurface singularities. II*, Invent. Math. **88** (1987), no. 1, 165–182, DOI 10.1007/BF01405096. MR877011

[BIKR08] Igor Burban, Osamu Iyama, Bernhard Keller, and Idun Reiten, *Cluster tilting for one-dimensional hypersurface singularities*, Adv. Math. **217** (2008), no. 6, 2443–2484, DOI 10.1016/j.aim.2007.10.007. MR2397457

[Bil05] Josée Bilodeau, *Auslander algebras and simple plane curve singularities*, Representations of algebras and related topics, Fields Inst. Commun., vol. 45, Amer. Math. Soc., Providence, RI, 2005, pp. 99–107. MR2146242

[BKR01] Tom Bridgeland, Alastair King, and Miles Reid, *The McKay correspondence as an equivalence of derived categories*, J. Amer. Math. Soc. **14** (2001), no. 3, 535–554, DOI 10.1090/S0894-0347-01-00368-X. MR1824990

[BLvdB10] Ragnar-Olaf Buchweitz, Graham J. Leuschke, and Michel Van den Bergh, *Non-commutative desingularization of determinantal varieties I*, Invent. Math. **182** (2010), no. 1, 47–115, DOI 10.1007/s00222-010-0258-7. MR2672281

[BO02] A. Bondal and D. Orlov, *Derived categories of coherent sheaves*, Proceedings of the International Congress of Mathematicians, Vol. II (Beijing, 2002), Higher Ed. Press, Beijing, 2002, pp. 47–56. MR1957019

[Bou81] Nicolas Bourbaki, *Éléments de mathématique* (French), Masson, Paris, 1981. Groupes et algèbres de Lie. Chapitres 4, 5 et 6. [Lie groups and Lie algebras. Chapters 4, 5 and 6]. MR647314

[Bri02] Tom Bridgeland, *Flops and derived categories*, Invent. Math. **147** (2002), no. 3, 613–632, DOI 10.1007/s002220100185. MR1893007

[Buc12] R.-O. Buchweitz. From platonic solids to preprojective algebras via the McKay correspondence. *Oberwolfach Jahresbericht*, pages 18–28, 2012.

[CB00] William Crawley-Boevey, *On the exceptional fibres of Kleinian singularities*, Amer. J. Math. **122** (2000), no. 5, 1027–1037. MR1781930

[Che55] Claude Chevalley, *Invariants of finite groups generated by reflections*, Amer. J. Math. **77** (1955), 778–782, DOI 10.2307/2372597. MR0072877

[CLO07] David Cox, John Little, and Donal O'Shea, *Ideals, varieties, and algorithms*, 3rd ed., Undergraduate Texts in Mathematics, Springer, New York, 2007. An introduction to computational algebraic geometry and commutative algebra. MR2290010

[DFI15] Hailong Dao, Eleonore Faber, and Colin Ingalls, *Noncommutative (crepant) desingularizations and the global spectrum of commutative rings*, Algebr. Represent. Theory **18** (2015), no. 3, 633–664, DOI 10.1007/s10468-014-9510-y. MR3357942

[DFI16] Brandon Doherty, Eleonore Faber, and Colin Ingalls, *Computing global dimension of endomorphism rings via ladders*, J. Algebra **458** (2016), 307–350, DOI 10.1016/j.jalgebra.2016.03.020. MR3500780

[DII+16] H. Dao, O. Iyama, S. Iyengar, R. Takahashi, M. Wemyss, and Y. Yoshino. Noncommutative resolutions using syzygies. 2016. `arXiv:1609.04842`.

[DITV15] Hailong Dao, Osamu Iyama, Ryo Takahashi, and Charles Vial, *Non-commutative resolutions and Grothendieck groups*, J. Noncommut. Geom. **9** (2015), no. 1, 21–34, DOI 10.4171/JNCG/186. MR3337953

[Dur79] Alan H. Durfee, *Fifteen characterizations of rational double points and simple critical points*, Enseign. Math. (2) **25** (1979), no. 1-2, 131–163. MR543555

[FH10] E. Faber and H. Hauser, *Today's menu: geometry and resolution of singular algebraic surfaces*, Bull. Amer. Math. Soc. (N.S.) **47** (2010), no. 3, 373–417, DOI 10.1090/S0273-0979-10-01295-4. MR2651084

[Gre92] Gert-Martin Greuel and Gerhard Pfister, *SINGULAR and applications*, Jahresber. Deutsch. Math.-Verein. **108** (2006), no. 4, 167–196. MR2287090

[GSV83] G. Gonzalez-Sprinberg and J.-L. Verdier, *Construction géométrique de la correspondance de McKay* (French), Ann. Sci. École Norm. Sup. (4) **16** (1983), no. 3, 409–449 (1984). MR740077

[Hau03] Herwig Hauser, *The Hironaka theorem on resolution of singularities (or: A proof we always wanted to understand)*, Bull. Amer. Math. Soc. (N.S.) **40** (2003), no. 3, 323–403, DOI 10.1090/S0273-0979-03-00982-0. MR1978567

[Her78] Jürgen Herzog, *Ringe mit nur endlich vielen Isomorphieklassen von maximalen, unzerlegbaren Cohen-Macaulay-Moduln* (German), Math. Ann. **233** (1978), no. 1, 21–34, DOI 10.1007/BF01351494. MR0463155

[Hir64] Heisuke Hironaka, *Resolution of singularities of an algebraic variety over a field of characteristic zero. I, II*, Ann. of Math. (2) **79** (1964), 109–203; ibid. (2) **79** (1964), 205–326. MR0199184

[Hov09] B. Hovinen. Matrix factorizations of the classical determinant. Phd thesis, University of Toronto, 2009.

[HR74] Melvin Hochster and Joel L. Roberts, *Rings of invariants of reductive groups acting on regular rings are Cohen-Macaulay*, Advances in Math. **13** (1974), 115–175, DOI 10.1016/0001-8708(74)90067-X. MR0347810

[IW14] Osamu Iyama and Michael Wemyss, *Maximal modifications and Auslander-Reiten duality for non-isolated singularities*, Invent. Math. **197** (2014), no. 3, 521–586, DOI 10.1007/s00222-013-0491-y. MR3251829

[KK86] Bertram Kostant and Shrawan Kumar, *The nil Hecke ring and cohomology of G/P for a Kac-Moody group G*, Proc. Nat. Acad. Sci. U.S.A. **83** (1986), no. 6, 1543–1545, DOI 10.1073/pnas.83.6.1543. MR831908

[Kle93] Felix Klein, *Vorlesungen über das Ikosaeder und die Auflösung der Gleichungen vom fünften Grade* (German, with German summary), Birkhäuser Verlag, Basel; B. G. Teubner, Stuttgart, 1993. Reprint of the 1884 original; Edited, with an introduction and commentary by Peter Slodowy. MR1315530

[Knö84] H. Knörrer. The Cohen–Macaulay modules over simple hypersurface singularities. 1984. Max-Planck Institut für Mathematik und Sonderforschungsbereich theoretische Physik MPI/SFB 84-51.

[Knö87] Horst Knörrer, *Cohen-Macaulay modules on hypersurface singularities. I*, Invent. Math. **88** (1987), no. 1, 153–164, DOI 10.1007/BF01405095. MR877010

[Lau71] Henry B. Laufer, *Normal two-dimensional singularities*, Princeton University Press, Princeton, N.J.; University of Tokyo Press, Tokyo, 1971. Annals of Mathematics Studies, No. 71. MR0320365

[Leu07] Graham J. Leuschke, *Endomorphism rings of finite global dimension*, Canad. J. Math. **59** (2007), no. 2, 332–342, DOI 10.4153/CJM-2007-014-1. MR2310620

[Leu12] Graham J. Leuschke, *Non-commutative crepant resolutions: scenes from categorical geometry*, Progress in commutative algebra 1, de Gruyter, Berlin, 2012, pp. 293–361. MR2932589

[LS92] Alain Lascoux and Marcel-Paul Schützenberger, *Décompositions dans l'algèbre des différences divisées* (French, with English summary), Discrete Math. **99** (1992), no. 1-3, 165–179, DOI 10.1016/0012-365X(92)90372-M. MR1158787

[LT09] Gustav I. Lehrer and Donald E. Taylor, *Unitary reflection groups*, Australian Mathematical Society Lecture Series, vol. 20, Cambridge University Press, Cambridge, 2009. MR2542964

[LW12] Graham J. Leuschke and Roger Wiegand, *Cohen-Macaulay representations*, Mathematical Surveys and Monographs, vol. 181, American Mathematical Society, Providence, RI, 2012. MR2919145

[McK80] John McKay, *Graphs, singularities, and finite groups*, The Santa Cruz Conference on Finite Groups (Univ. California, Santa Cruz, Calif., 1979), Proc. Sympos. Pure Math., vol. 37, Amer. Math. Soc., Providence, R.I., 1980, pp. 183–186. MR604577

[OT92] Peter Orlik and Hiroaki Terao, *Arrangements of hyperplanes*, Grundlehren der Mathematischen Wissenschaften [Fundamental Principles of Mathematical Sciences], vol. 300, Springer-Verlag, Berlin, 1992. MR1217488

[Rei02] Miles Reid, *La correspondance de McKay*, Astérisque **276** (2002), 53–72. Séminaire Bourbaki, Vol. 1999/2000. MR1886756

[Sta77] Richard P. Stanley, *Relative invariants of finite groups generated by pseudoreflections*, J. Algebra **49** (1977), no. 1, 134–148, DOI 10.1016/0021-8693(77)90273-3. MR0460484

[VdB04a] Michel van den Bergh, *Non-commutative crepant resolutions*, The legacy of Niels Henrik Abel, Springer, Berlin, 2004, pp. 749–770. MR2077594

[VdB04b] Michel Van den Bergh, *Three-dimensional flops and noncommutative rings*, Duke Math. J. **122** (2004), no. 3, 423–455, DOI 10.1215/S0012-7094-04-12231-6. MR2057015

[Wat74] Keiichi Watanabe, *Certain invariant subrings are Gorenstein. I, II*, Osaka J. Math. **11** (1974), 1–8; ibid. 11 (1974), 379–388. MR0354646

[Yos90] Yuji Yoshino, *Cohen-Macaulay modules over Cohen-Macaulay rings*, London Mathematical Society Lecture Note Series, vol. 146, Cambridge University Press, Cambridge, 1990. MR1079937

[YS81] Tamaki Yano and Jirō Sekiguchi, *The microlocal structure of weighted homogeneous polynomials associated with Coxeter systems. II*, Tokyo J. Math. **4** (1981), no. 1, 1–34, DOI 10.3836/tjm/1270215738. MR625118

Dept. of Computer and Mathematical Sciences, University of Toronto at Scarborough, 1265 Military Trail, Toronto, ON M1C 1A4, Canada

Department of Mathematics, University of Michigan, Ann Arbor MI, 48109, USA
Email address: emfaber@umich.edu

Department of Mathematics and Statistics, University of New Brunswick, Fredericton, NB. E3B 5A3, Canada
Email address: cingalls@unb.ca

Contemporary Mathematics
Volume **705**, 2018
http://dx.doi.org/10.1090/conm/705/14198

Polyhedral models for tensor product multiplicities

Jiarui Fei

ABSTRACT. These are expanded notes from the lecture given at the 17th International Conference on Representations of Algebras (ICRA XVII) held in Syracuse in August 2016. We discuss how to compute the tensor product multiplicities for representations of simply-connected simple algebraic groups by counting lattice points in polytopes. We focus on the explicit construction of a new family of polyhedral models.

Introduction

Let G be the simply-connected simple algebraic group over $k = \mathbb{C}$. Recall that G is uniquely determined up to isomorphism by its Dynkin diagram. All finite-dimensional representations of G form a semisimple tensor category. So given two irreducible representations $L(\mu)$ and $L(\nu)$ of highest weights μ and ν, we have a direct sum decomposition of their tensor product

$$L(\mu) \otimes L(\nu) = \bigoplus_{\lambda} c_{\mu\,\nu}^{\lambda} L(\lambda).$$

To determine the multiplicities $c_{\mu\,\nu}^{\lambda}$ is an important problem in Lie theory.

The first idea is that instead of working with each individual multiplicity, we construct a multigraded algebra A^G for each fixed G such that each multiplicity is counted by the dimension of certain graded piece. The construction involves the invariant algebra $k[G]^{U^-}$ and its dual $k[G]^U$, where (U^-, U) is an opposite pair of maximal unipotent subgroups of G. Consider the action of $G \times G$ on $k[G]$ by the left and right translations:

$$(g_1, g_2)\varphi(g) = \varphi(g_1^{-1} g g_2) \text{ for } \varphi \in k[G].$$

Then the invariant algebra $k[G]^{U^-}$ for the left translation of U^- as a G-module is a direct sum of all irreducible representations of G [**21**]. Similar decomposition holds for $k[G]^U$ where U acts by the right translation.

$$k[G]^{U^-} = \bigoplus_{\mu} L(\mu), \quad k[G]^U = \bigoplus_{\lambda} L(\lambda)^\vee.$$

Let A be the triple tensor algebra

$$A := k[G]^{U^-} \otimes k[G]^{U^-} \otimes k[G]^U = \bigoplus_{\mu,\nu,\lambda} c_{\mu\,\nu}^{\lambda} L(\lambda) \otimes L(\lambda)^\vee.$$

2010 *Mathematics Subject Classification.* 16G20, 13F60; Secondary 13A50, 52B20.

To extract the multiplicity $c_{\mu\,\nu}^{\lambda}$ we take the G-invariant subalgebra A^G of A. This algebra is multigraded by triples of dominant weights of G. The dimension of its (μ, ν, λ)-degree component is exactly $c_{\mu\,\nu}^{\lambda}$.

Next we show that the algebra A^G is in fact an *upper cluster algebra* [**5**] of some ice quiver Δ if G is simply laced. It happens frequently (but not always) that a basis of an upper algebra is parametrized by lattice points in some polyhedral cone [**11**]–[**14**]. A natural choice for lattices points is all possible g-*vectors* appearing in the upper cluster algebra. Roughly speaking, if an element z in an upper cluster algebra has a well-defined g-vector g with respect to some *cluster* $\mathbf{x}$, then the Laurent monomial $\mathbf{x}^{\mathsf{g}}$ is the "leading" term in the Laurent expansion of z. However, it is very hard to characterize them without introducing the representation theory of *quivers with potentials* [**9, 10**]. After assigning a suitable potential for Δ, we can look at the so-called μ-*supported* g-vectors. It turns out that they are all possible ones and they all lie in a (full-dimensional) polyhedral cone G_Δ in $\mathbb{R}^{\Delta_0}$. Here, Δ_0 is the set of vertices in Δ.

To explicitly describe this cone, we consider the hyperplane presentation of G_Δ

$$\{\mathsf{g} \in \mathbb{R}^{\Delta_0} \mid \mathsf{g}H \geq 0\}. \tag{0.1}$$

The crucial idea introduced in [**11**] is that the matrix H can be determined by certain stability conditions. More concretely, the columns of H are given by dimension vectors of all subrepresentations of so-called *boundary representations*. The boundary representations are in bijection with the *frozen* vertices of Δ. So the problem boils down to explicitly describe those boundary representations. To achieve this, we need more structure on the ice quiver Δ. It turns out that it is a so-called iARt quiver Δ_Q^2, where Q is any quiver whose underlying graph is the same as that of G. We coin this name because Δ_Q^2 is exactly the ice Auslander-Reiten quiver (with translation) for the category of projective presentations in $\operatorname{Rep}(Q)$. Finally the tensor product multiplicity $c_{\mu\,\nu}^{\lambda}$ is counted by the lattice points in some fibre polytope $\mathsf{G}_\Delta(\mu, \nu, \lambda)$, which is cut from G_Δ by hyperplanes $\mathsf{g}\boldsymbol{\sigma} = (\mu, \nu, \lambda)$. The triple-weight matrix $\boldsymbol{\sigma}$ can be directly read off from the iARt quiver.

In this lecture note, we will mainly focus on the construction of the iARt quivers and their boundary representations following [**14**]. We shall expand some examples there. We also include a large example, namely the exceptional group of type E_8.

Finding polyhedral models for the tensor product multiplicities has a long history dating back to 1970's (eg. [**3, 4, 20, 24**]). One highlight is Knutson-Tao's solution of the saturation conjecture for GL_n using their hive model [**22**]. We refer the readers to [**14**, Introduction and Section 7] for a historical account and comparison. A geometric realization of the algebra A^G was considered by Fock and Goncharov in [**16**]. A weaker version of our result in type A was already obtained there in the context of their cluster varieties.

Outline of the Note. In Section 1 we recall the definition of graded upper cluster algebras. In Section 2 we recall the Auslander-Reiten theory of presentations and define the iARt quivers. In Section 3 we review the generic cluster character in the setting of quivers of potentials. In Section 4 we construct the iARt QPs and compute their g-vector cones. Then we state the main result – Theorem 4.10. In Section 5 we summarize what is known for the non-simply laced cases.

Notations and Conventions. All algebras and modules/representations are finite-dimensional over the base field k. Our vectors are exclusively row vectors. All modules are right modules. Arrows are composed from left to right, i.e., ab is the path $\cdot \xrightarrow{a} \cdot \xrightarrow{b} \cdot$. Unless otherwise stated, unadorned Hom and $\otimes$ are all over the base field k, and the superscript $*$ is the trivial dual for vector spaces. For direct sum of n copies of M, we write nM instead of the traditional $M^{\oplus n}$.

1. Graded Upper Cluster Algebras

In this note, we will not consider the upper cluster algebras in full generalities as introduced in [**5**]. We only concern about those *of geometric type* and over a field k. To define such an upper cluster algebra one needs to specify a *seed* $(\Delta, \mathbf{x})$ in some ambient field $\mathcal{F} \supset k$. Here, Δ is an ice quiver with no loops or oriented 2-cycles and the *extended cluster* $\mathbf{x} = \{x_1, x_2, \ldots, x_q\}$ is a collection of algebraically independent (over k) elements of $\mathcal{F}$ attaching to each vertex of Δ.

An *ice quiver* $\Delta = (\Delta_0, \Delta_1)$ is a quiver, where some vertices in Δ_0 are designated as *mutable* while the rest are *frozen*. We usually label the quiver such that the first p vertices are mutable. Although we allow arrows between frozen vertices, those arrows are irrelevant to define an upper cluster algebra. The only information we need is the submatrix B_Δ of first p rows of the B-*matrix* $B(\Delta)$ given by

$$b_{u,v} = |\text{arrows } u \to v| - |\text{arrows } v \to u|.$$

The elements of $\mathbf{x}$ associated with the mutable vertices are called *cluster variables*; they form a *cluster*. The elements associated with the frozen vertices are called *frozen variables*, or *coefficient variables*.

DEFINITION 1.1. Let u be a mutable vertex of Δ. The *quiver mutation* μ_u transforms Δ into the new quiver $\Delta' = \mu_u(\Delta)$ via a sequence of three steps.

(1) For each pair of arrows $v \to u \to w$, introduce a new arrow $v \to w$ (unless both v and w are frozen, in which case do nothing);
(2) Reverse the direction of all arrows incident to u;
(3) Remove all oriented 2-cycles.

DEFINITION 1.2. A *seed mutation* μ_u at a (mutable) vertex u transforms $(\Delta, \mathbf{x})$ into the seed $(\Delta', \mathbf{x}') = \mu_u(\Delta, \mathbf{x})$ defined as follows. The new quiver is $\Delta' = \mu_u(\Delta)$. The new extended cluster is $\mathbf{x}' = \mathbf{x} \cup \{x_u'\} \setminus \{x_u\}$ where the new cluster variable x_u' replacing x_u is determined by the *exchange relation*

$$x_u\, x_u' = \prod_{v \to u} x_v + \prod_{u \to w} x_w.$$

We note that the mutated seed $(\Delta', \mathbf{x}')$ contains the same frozen variables as the original seed $(\Delta, \mathbf{x})$. It is easy to check that one can recover $(\Delta, \mathbf{x})$ from $(\Delta', \mathbf{x}')$ by performing a seed mutation again at u. Two seeds $(\Delta, \mathbf{x})$ and $(\Delta^\dagger, \mathbf{x}^\dagger)$ that can be obtained from each other by a sequence of mutations are called *mutation-equivalent*, denoted by $(\Delta, \mathbf{x}) \sim (\Delta^\dagger, \mathbf{x}^\dagger)$.

Let $\mathcal{L}(\mathbf{x})$ be the Laurent polynomial algebra in $\mathbf{x}$ over the base field k. If Δ is an ice quiver, we denote by $\mathcal{L}_\Delta(\mathbf{x})$ the Laurent polynomial in $\mathbf{x}$ which is polynomial in frozen variables, that is, $\mathcal{L}_\Delta(\mathbf{x}) := k\left[x_1^{\pm 1}, \ldots, x_p^{\pm 1}, x_{p+1}, \ldots x_q\right]$.

DEFINITION 1.3. The *upper cluster algebra* with seed $(\Delta, \mathbf{x})$ is

$$\overline{\mathcal{C}}(\Delta, \mathbf{x}) := \bigcap_{(\Delta^\dagger, \mathbf{x}^\dagger) \sim (\Delta, \mathbf{x})} \mathcal{L}_\Delta(\mathbf{x}^\dagger).$$

Note that the definition of $\overline{\mathcal{C}}(\Delta, \mathbf{x})$ depends only, up to a natural isomorphism, on the mutation equivalence class of the ice quiver Δ (in fact only on B_Δ). So we may drop $\mathbf{x}$ and simply write $\overline{\mathcal{C}}(\Delta)$ or $\overline{\mathcal{C}}(B_\Delta)$.

REMARK 1.4. Our definition of an upper cluster algebra is slightly different from the original one in [**5**], where $\mathcal{L}_\Delta(\mathbf{x}^\dagger)$ is replaced by $\mathcal{L}(\mathbf{x}^\dagger)$. The Laurent Phenomenon [**5, 17**] says that an upper cluster algebra contains all cluster and frozen variables. In particular, any cluster algebra is contained its upper cluster algebra. The containment can be strict in general.

DEFINITION 1.5. A *weight configuration* $\boldsymbol{\sigma}$ of $\mathbb{Z}^n$ on an ice quiver Δ is an assignment for each vertex v of Δ a (weight) vector $\boldsymbol{\sigma}(v) \in \mathbb{Z}^n$ such that for each mutable vertex u, we have that

$$\sum_{v \to u} \boldsymbol{\sigma}(v) = \sum_{u \to w} \boldsymbol{\sigma}(w). \tag{1.1}$$

The *mutation* μ_u also transforms $\boldsymbol{\sigma}$ into a weight configuration $\boldsymbol{\sigma}'$ on the mutated quiver $\mu_u(\Delta)$ defined as

$$\boldsymbol{\sigma}'(v) = \begin{cases} \sum\limits_{u \to w} \boldsymbol{\sigma}(w) - \boldsymbol{\sigma}(u) & \text{if } v = u, \\ \boldsymbol{\sigma}(v) & \text{if } v \neq u. \end{cases}$$

By slight abuse of notation, we shall view $\boldsymbol{\sigma}$ as a matrix whose v-th row is the weight vector $\boldsymbol{\sigma}(v)$. In this matrix notation, the condition (1.1) is equivalent to that $B_\Delta \boldsymbol{\sigma}$ is a zero matrix.

Given a weight configuration $(\Delta; \boldsymbol{\sigma})$, we can assign a multidegree (or weight) to the upper cluster algebra $\overline{\mathcal{C}}(\Delta, \mathbf{x})$ by setting $\deg(x_v) = \boldsymbol{\sigma}(v)$ for $v \in \Delta_0$. Then the mutation preserves multihomogeneousity. We say that this upper cluster algebra is $\boldsymbol{\sigma}$-graded, and denoted by $\overline{\mathcal{C}}(\Delta, \mathbf{x}; \boldsymbol{\sigma})$.

2. Auslander-Reiten theory of Presentations

2.1. A Brief Review of AR-theory. The Auslander-Reiten theory were developed originally for (finite-dimensional) module categories of Artin algebras, but without much difficulty almost all can be generalized to k-linear, Hom-finite, and Krull-Schmidt categories [**25**].

Let k be an algebraically closed field, and $\mathcal{A}$ be a k-linear, Hom-finite, and Krull-Schmidt category. For any two objects $M, N \in \mathcal{A}$, let $\operatorname{rad}_{\mathcal{A}}(M, N)$ be the space of all non-invertible morphisms from M to N. We define $\operatorname{rad}^2_{\mathcal{A}}(M, N)$ to consist of all morphisms of form gf, where $f \in \operatorname{rad}_{\mathcal{A}}(M, L)$ and $g \in \operatorname{rad}_{\mathcal{A}}(L, N)$ for some $L \in \mathcal{A}$. We denote $\operatorname{Irr}_{\mathcal{A}}(M, N) := \operatorname{rad}_{\mathcal{A}}(M, N) / \operatorname{rad}^2_{\mathcal{A}}(M, N)$. Let $\operatorname{ind}(\mathcal{A})$ be the full subcategory of all indecomposable objects in $\mathcal{A}$.

DEFINITION 2.1. For $M, N \in \operatorname{ind}(\mathcal{A})$, an *irreducible morphism* $f : M \to N$ is an element in $\operatorname{Irr}_{\mathcal{A}}(M, N)$.

Let $M = \bigoplus_{i=1}^t m_i M_i$ be an object in $\mathcal{A}$ with M_i indecomposable and pairwise non-isomorphic. For $f \in \operatorname{Hom}_{\mathcal{A}}(L, M)$ with L indecomposable, we can write f as

$f = (f_1, \dots, f_t)$ where $f_i = (f_{i,1}, \dots, f_{i,m_i}) : L \to m_i M_i$. Then f is called *left minimal almost split* if the residual classes of $f_{i,j}$'s in $\mathrm{Irr}_{\mathcal{A}}(L, M_i)$ form a k-basis for all i, and if there is an indecomposable $M' \in \mathcal{A}$ such that $\mathrm{Irr}_{\mathcal{A}}(L, M') \neq 0$, then $M' \cong M_i$ for some i. Similarly we can define *right minimal almost split* for $g : M \to N$.

An exact sequence $0 \to L \to M \to N \to 0$ in $\mathcal{A}$ is called *almost split* if $L \to M$ is left minimal almost split, and $M \to N$ is right minimal almost split. In this case, L is called the *translation* of N, denoted by τN.

The definition for left (right) minimal almost split is not the original one (see [**2**, Definition IV.1.1]), but it follows from [**2**, Proposition IV.4.2].

If $\mathcal{A}$ is the module category of some finite dimensional k-algebra A, then the translation τ is defined for every indecomposable non-projective A-module. The classical AR-theory [**1**, Theorem V.1.15] says that it is given by the trivial dual of *Auslander's transpose functor* (see [**1**, IV.1]).

2.2. Presentations. Let $A := kQ/I$ be a path algebra with relations over k. We assume that all relations are admissible and A is finite dimensional. Let $C^2A := \mathrm{Ch}_2(\mathrm{proj}\text{-}A)$ be the category of projective presentations. To be more precise, the objects in C^2A are 2-term complexes $P_+ \xrightarrow{f} P_-$ in $\mathrm{proj}\text{-}A$ (with P_+ and P_- in some fixed degrees). The morphisms are commutative diagrams. Note that this category is Krull-Schmidt, but not abelian. For any $\beta \in \mathbb{Z}_{\geqslant 0}^{Q_0}$ we denote $\bigoplus_{i\in Q_0} \beta(i)P_i$ by $P(\beta)$, where P_i is the indecomposable projective module corresponding to the i-th vertex of Q. If $P_\pm = P(\beta_\pm)$, then the *weight vector* $(\mathsf{f}_-, \mathsf{f}_+)$ of f is (β_-, β_+). The *reduced weight vector* f is the difference $\mathsf{f}_+ - \mathsf{f}_-$.

Presentations of forms $0 \to P$, $P \to 0$, and $P \xrightarrow{\mathrm{Id}} P$ are called *negative, positive*, and *neutral*. They are also denoted by O_P^-, O_P^+ and Id_P respectively. If $P = P_i$, then they are called i-th negative, positive, and neutral presentation, and denoted by O_i^-, O_i^+ and Id_i respectively. For a representation M of A, we write $f(M) : P_+(M) \to P_-(M)$ for its minimal presentation.

LEMMA 2.2. [**7**] *Any presentation f decomposes as $f = f_+ \oplus f_{\mathrm{Id}} \oplus f'$, where f_+ is positive, f_{Id} is neutral, and f' is the minimal presentation of* $\mathrm{Coker}(f)$.

COROLLARY 2.3. *An indecomposable presentation is one of the following four kinds. They are i-th negative, positive, neutral presentations, and minimal presentations of indecomposable non-projective representations of A.*

There are some trivial relations for the Hom-functors between the two categories C^2A and $\mathrm{mod}\,A$. For example, we have for any $P_+ \xrightarrow{f} P_- \in C^2A$ and $P \in \mathrm{mod}\,A$,

(1) $\mathrm{Hom}_{C^2A}(\mathrm{O}_P^-, f) \cong \mathrm{Hom}_A(P, P_-)$;
(2) $\mathrm{Hom}_{C^2A}(f, \mathrm{O}_P^+) \cong \mathrm{Hom}_A(P_+, P)$;
(3) $\mathrm{Hom}_{C^2A}(\mathrm{Id}_P, f) \cong \mathrm{Hom}_A(P, P_+)$;
(4) $\mathrm{Hom}_{C^2A}(f, \mathrm{Id}_P) \cong \mathrm{Hom}_A(P_-, P)$.

The next easy lemma explains a fundamental relation in terms of the exact sequence.

LEMMA 2.4. *We have the following exact sequence*

$$\mathrm{Hom}_{C^2A}(-, \mathrm{Id}_{P_+}) \to \mathrm{Hom}_{C^2A}(-, f) \to \mathrm{Hom}_A(\mathrm{Coker}\,-, \mathrm{Coker}\,f) \to 0.$$

Moreover, if f is injective, then the leftmost map is injective as well.

Based on these we can fully understand the relationship between almost split sequences in C^2A and $\operatorname{mod} A$ when $A = kQ$ is a path algebra. Readers can find the detailed statements in [**14**, Lemmas 2.8 and 2.9]. The most important consequence is that we can extend the classical Auslander-Reiten translation τ from projective-free representations to the negative-and-neutral-free presentations as in [**7**]. For $f = f(M)$, we define $\tau f := f(\tau M)$. For each $i \in Q_0$ we define $\tau(\mathrm{O}_i^+) := f(I_i)$. This allows us to construct the AR quiver for the category $C^2Q := C^2kQ$.

2.3. iARt Quivers. We slightly upgrade the classical Auslander-Reiten quiver by adding the translation arrows. The following definition is basically taken from [**1**, VII.1]. Let $\mathcal{A}$ be a category as in Section 2.1.

DEFINITION 2.5 (ARt quiver). The ARt valued quiver $\Delta(\mathcal{A})$ of $\mathcal{A}$ is defined as follows:

(1) The vertex of $\Delta(\mathcal{A})$ are the isomorphism classes of objects in $\operatorname{ind} \mathcal{A}$.
(2) There are $\dim_k \operatorname{Irr}_{\mathcal{A}}(M, N)$ *morphism arrows* from M to N.
(3) There is a *translation arrow* from N to τN if τN is defined.

A vertex u in an ARt quiver is called *transitive* if the translation and its inverse are both defined at u.

DEFINITION 2.6 (iARt quivers).

- The iARt quiver $\Delta_{\mathcal{A}}$ is obtained from the ARt quiver $\Delta(\mathcal{A})$ by freezing all vertices whose translations are not defined.
- The iARt quiver $\Delta_{\mathcal{A}}^2$ is obtained from the ARt quiver $\Delta(C^2\mathcal{A})$ by freezing all non-transitive vertices.

When $\mathcal{A}$ is the module category of a finite-dimensional algebra, the frozen vertices of $\Delta_{\mathcal{A}}$ are precisely indecomposable projective modules. However, it is not clear which vertices are frozen in $\Delta_{\mathcal{A}}^2$. If the algebra is hereditary, then they are precisely the negative, positive and neutral presentations.

In particular, if we take $\mathcal{A} := \operatorname{Rep}(Q)$ for some quiver Q without oriented cycles, we get two ARt quivers $\Delta(Q) := \Delta(\operatorname{Rep}(Q))$ and $\Delta(C^2Q)$. We denote the corresponding iARt quivers by Δ_Q and Δ_Q^2. The ice quiver Δ_Q was already considered in the cluster algebra theory [**5, 19**].

PROPOSITION 2.7. *The iARt quiver Δ_Q is a full subquiver of Δ_Q^2 if we identify vertices of Δ_Q with vertices of Δ_Q^2 via minimal presentations. Moreover, Δ_Q^2 can be obtained from Δ_Q as follows.*

(1) *We add $|Q_0|$ frozen vertices corresponding to O_i^+ and another $|Q_0|$ frozen vertices corresponding to Id_i.*
(2) *For each arrow $i \to j$ in Q, we draw morphism arrows $\mathrm{O}_j^+ \to \mathrm{O}_i^+$, and $f(I_i) \to \mathrm{O}_j^+$. We add translation arrows from O_i^+ to $f(I_i)$.*
(3) *We draw morphism arrows $f(S_i) \to \mathrm{Id}_i$, and $\mathrm{Id}_i \to \tau^{-1} f(S_i)$.*

From now on, we let Q be a Dynkin quiver. In this case, any indecomposable presentation f is uniquely determined by its weight vector. So we are able to label the vertices of Δ_Q^2 by the weight vectors of corresponding presentations. We will use the "exponential form" as a shorthand. For example, a vector $(3, 1, 0, 0, -2, 0, -1)$ is written as $(5^27, 1^32)$. Moreover, the translation arrows will be drawn as dashed arrows.

EXAMPLE 2.8. Let Q be the equioriented A_n-quiver. Then the iARt quiver Δ_Q^2 is the ice hive quiver Δ_n constructed in [**11**] up to some arrows between frozen vertices. Here is a picture for type A_4.

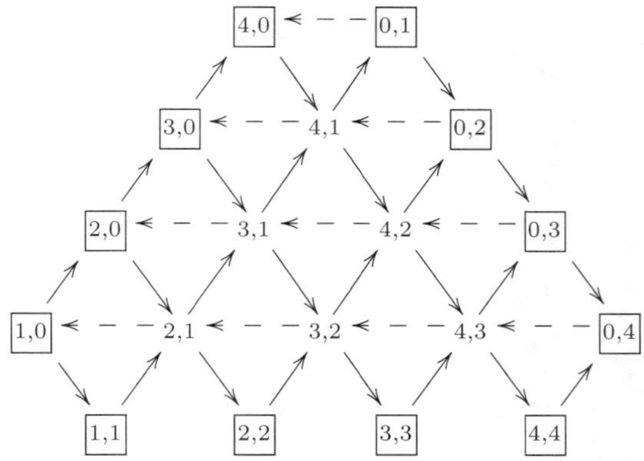

EXAMPLE 2.9 (The iARt quiver Δ_Q^2 for Q a D_4-quiver).

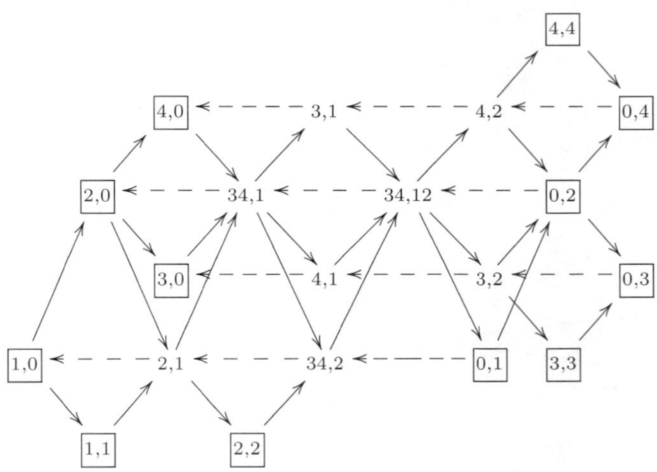

Readers can find a few other iARt quivers in [**14**, Appendix A]. The iARt quiver Δ_Q^2 for a Q of type E_8 is displayed in the next page.

Since Q is of finite representation type, each non-neutral $f \in \operatorname{ind}(C^2Q)$ is translated from a unique indecomposable positive presentation, that is, $f = \tau^t(\mathrm{O}_i^+)$ for some $i \in Q_0$ and $t \in \mathbb{Z}_{\geqslant 0}$. Now for each $f \in \operatorname{ind}(C^2Q)$, we assign a triple-weight vector as follows.

DEFINITION 2.10. If f is translated from O_i^+, then the *triple weights* $\widetilde{\mathsf{f}} \in \mathbb{Z}^{3|Q_0|}$ of f is given by $(\mathsf{e}(f), \mathsf{f}_-, \mathsf{f}_+)$ where $\mathsf{e}(f) := \mathsf{e}_i$ is the unit vector supported at i. If $f = \mathrm{Id}_i$, then we set $\widetilde{\mathsf{f}} := (0, \mathsf{e}_i, \mathsf{e}_i)$.

We can easily check the following using the properties of an AR-quiver.

LEMMA 2.11. *The assignment* $\boldsymbol{\sigma}_Q^2 : \widetilde{\mathsf{f}} \mapsto f$ *defines a weight configuration for the iARt quiver* Δ_Q^2.

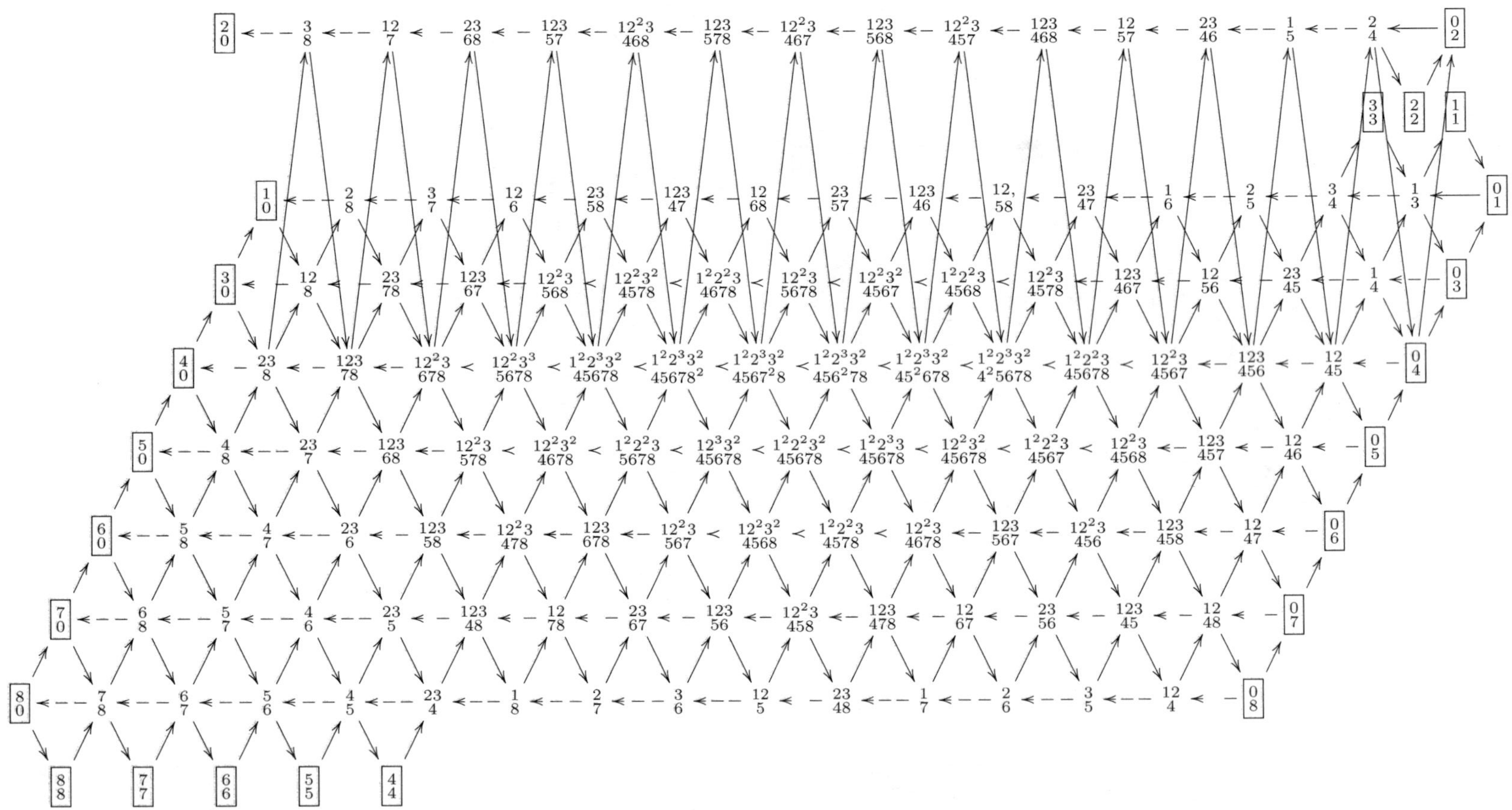

FIGURE 1. The iARt quiver Δ_Q^2 for a Q of type E_8.

3. Cluster Character from Quiver with Potential

3.1. The Quiver with Potential model. The mutation of quivers with potentials is invented in [**9**, **10**] to model the cluster algebras. Following [**9**], we define a potential W on an ice quiver Δ as a (possibly infinite) linear combination of oriented cycles in Δ. More precisely, a *potential* is an element of the *trace space* $\operatorname{Tr}(\widehat{k\Delta}) := \widehat{k\Delta}/[\widehat{k\Delta}, \widehat{k\Delta}]$, where $\widehat{k\Delta}$ is the completion of the path algebra $k\Delta$ and $[\widehat{k\Delta}, \widehat{k\Delta}]$ is the closure of the commutator subspace of $\widehat{k\Delta}$. The pair (Δ, W) is an *ice quiver with potential*, or IQP for short. For each arrow $a \in \Delta_1$, the *cyclic derivative* ∂_a on $\widehat{k\Delta}$ is defined to be the linear extension of

$$\partial_a(a_1 \cdots a_d) = \sum_{k=1}^{d} a^*(a_k) a_{k+1} \cdots a_d a_1 \cdots a_{k-1}.$$

For each potential W, its *Jacobian ideal* ∂W is the (closed two-sided) ideal in $\widehat{k\Delta}$ generated by all $\partial_a W$. The *Jacobian algebra* $J(\Delta, W)$ is the quotient algebra $\widehat{k\Delta}/\partial W$. If W is polynomial and $J(\Delta, W)$ is finite-dimensional, then the completion is unnecessary to define $J(\Delta, W)$. This is the case throughout this note.

EXAMPLE 3.1. Consider the following quiver with potential $W = a_1 b_1 c - a_2 b_2 c$.

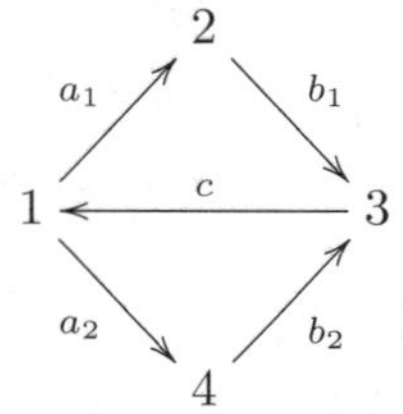

Then the Jacobian ideal ∂W is generated by $b_1 c$, ca_1, $b_2 c$, ca_2, and $a_1 b_1 - a_2 b_2$.

The key notion introduced in [**9**, **10**] is the *mutation* of quivers with potentials and their decorated representations. For an ice quiver with *nondegenerate potential* (see [**9**]), the mutation in certain sense "lifts" the mutation in Definition 1.1. However, we do not explicitly need these notions here. The fact we need is that any rigid potential is nondegenerate. We recall the definition of rigidity.

DEFINITION 3.2. A potential W is called *rigid* on a quiver Δ if every potential on Δ is cyclically equivalent to an element in the Jacobian ideal ∂W. Such a QP (Δ, W) is also called *rigid*.

It is clear that the potential in Example 3.1 is rigid.

3.2. The Generic Cluster Character.

DEFINITION 3.3. To any $\mathsf{g} \in \mathbb{Z}^{\Delta_0}$ we associate the *reduced* presentation space

$$\operatorname{PHom}_J(\mathsf{g}) := \operatorname{Hom}_J(P([\mathsf{g}]_+), P([-\mathsf{g}]_+)),$$

where $[\mathsf{g}]_+$ is the vector satisfying $[\mathsf{g}]_+(u) = \max(\mathsf{g}(u), 0)$. We denote by $\operatorname{Coker}(\mathsf{g})$ the cokernel of a general presentation in $\operatorname{PHom}_J(\mathsf{g})$.

Reader should be aware that $\operatorname{Coker}(\mathsf{g})$ is just a notation rather than a specific representation. If we write $M = \operatorname{Coker}(\mathsf{g})$, this simply means that we take a presentation general enough (according to context) in $\operatorname{PHom}_J(\mathsf{g})$, then let M to be its cokernel.

DEFINITION 3.4. A vector $\mathsf{g} \in \mathbb{Z}^{\Delta_0}$ is called *μ-supported* if $\operatorname{Coker}(\mathsf{g})$ is μ-supported. Let $G(\Delta, W)$ be the set of all μ-supported g-vectors in $\mathbb{Z}^{\Delta_0}$.

This definition certainly makes sense due to the semi-continuity of the rank function.

For a vector $\mathsf{g} \in \mathbb{Z}^{\Delta_0}$, we write $\mathbf{x}^{\mathsf{g}}$ for the monomial $\prod_{v\in\Delta_0} x_v^{\mathsf{g}(v)}$. We set $y_u = \mathbf{x}^{-b_u}$ where b_u is the u-th row of the matrix $B(\Delta)$, and let $\mathbf{y} = \{y_u\}$.

DEFINITION 3.5 ([**26**]). We define the *generic character* $C_W : G(\Delta, W) \to \mathbb{Z}(\mathbf{x})$ by

$$C_W(\mathsf{g}) = \mathbf{x}^{\mathsf{g}} \sum_{\mathsf{e}} \chi\big(\operatorname{Gr}^{\mathsf{e}}(\operatorname{Coker}(\mathsf{g}))\big)\mathbf{y}^{\mathsf{e}}, \tag{3.1}$$

where $\operatorname{Gr}^{\mathsf{e}}(M)$ is the variety parameterizing e-dimensional quotient representations of M, and $\chi(-)$ denotes the topological Euler-characteristic.

REMARK 3.6. This definition also makes sense because $\chi\big(\operatorname{Gr}^{\mathsf{e}}(\operatorname{Coker}(f))\big)$ is constant for f in some dense subset of $\operatorname{PHom}_J(\mathsf{g})$. We also observe that y_v is irrelevant in the definition for any frozen vertex v because g is μ-supported. Finally, we note that if $(\Delta, \mathbf{x})$ is $\boldsymbol{\sigma}$-graded, then $C_W(\mathsf{g})$ is multihomogeneous of degree $\mathsf{g}\boldsymbol{\sigma}$ because the variables in $\mathbf{y}$ have zero degrees.

THEOREM 3.7 ([**11**, Corollary 5.14], *cf.* [**26**, Theorem 1.1]). *Suppose that IQP (Δ, W) is non-degenerate and B_Δ has full rank. The generic character C_W maps $G(\Delta, W)$ (bijectively) to a set of linearly independent elements in $\overline{\mathcal{C}}(\Delta)$ containing all cluster monomials.*

DEFINITION 3.8. We say that an IQP (Δ, W) *models* an algebra $\mathcal{C}$ if the generic cluster character maps $G(\Delta, W)$ (bijectively) onto a basis of $\mathcal{C}$. If $\mathcal{C}$ is the upper cluster algebra $\overline{\mathcal{C}}(\Delta)$, then we simply say that (Δ, W) is a *cluster model.* Such a model is called *polyhedral* if $G(\Delta, W)$ is given by lattice points in some polyhedron.

Many polyhedral cluster models were found, including the IQPs introduced in [**11**–**13**], and the ones in the next section.

4. iARt QPs

4.1. The iARt QP (Δ_Q^2, W_Q^2). We keep assuming Q is a Dynkin quiver. A *translation triangle* in an iARt quiver is an oriented cycle of the form

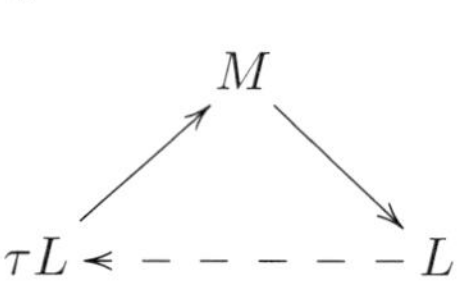

For each iARt quiver Δ_Q^2, we define the potential W_Q^2 as an alternating sum of all translation triangles. We make this more precise as follows. We can also label each non-neutral $f = \tau^{-t}\,\mathrm{O}_i^-$ by the pair $(i,t)^* = (i^*(f), t^*(f))$. The arrows of Δ_Q^2 are thus classified into three classes

Type A arrows	$(i,t)^* \to (j,t+1)^*,\ f(S_i) \to \mathrm{Id}_i;$
Type B arrows	$(i,t)^* \to (j,t)^*,\ \mathrm{Id}_i \to \tau^{-1} f(S_i);$
Type C arrows	$(i,t)^* \dashrightarrow (i,t-1)^*.$

Let a (resp. b and c) denote the sum of all type A (B and C) arrows. The potential W_Q^2 is defined as $acb - abc$. Thus the Jacobian ideal is generated by the elements

$$(ac - ca)e_u,\ (cb - bc)e_u,\ (ba - ab)e_u.$$

LEMMA 4.1. *The IQP* (Δ_Q^2, W_Q^2) *is rigid and* $J(\Delta_Q^2, W_Q^2)$ *is finite-dimensional.*

Let $f : P_+ \to P_-$ be a presentation in $\operatorname{ind}(C^2Q)$. We denote by $\boldsymbol{I}_f$ the indecomposable injective representation of $J := J(\Delta_Q^2, W_Q^2)$ corresponding to the vertex f. For $j \in Q_0$, let $j^* = i^*(\mathrm{O}_j^+) \in Q_0$. It is well-known that $j \mapsto j^*$ is a (possibly trivial) involution, which does not depend on the orientation of Q.

LEMMA 4.2. *We have the following for the injective modules in* $J(\Delta_Q^2, W_Q^2)$

(1) $\boldsymbol{I}_{\mathrm{O}_i^-}(f) \cong \operatorname{Hom}_Q(P_{i^*(f)}, P_i)^*$;
(2) $\boldsymbol{I}_{\mathrm{O}_i^+}(f) \cong \operatorname{Hom}_Q(P_+, P_i)^*$;
(3) $\boldsymbol{I}_{\mathrm{Id}_i}(f) \cong \operatorname{Hom}_Q(P_-, P_i)^*$.

4.2. The Cone $\mathsf{G}_{\Delta_Q^2}$. We consider the following set of *boundary representations* T_v

$$0 \to T_v \to \boldsymbol{I}_v \to \bigoplus_{i \to j} \boldsymbol{I}_{\mathrm{O}_j^-} \qquad \text{for } v = \mathrm{O}_i^-,$$

$$0 \to T_v \to \boldsymbol{I}_v \to \bigoplus_{i \to j} \boldsymbol{I}_{\mathrm{O}_j^+} \qquad \text{for } v = \mathrm{O}_i^+,$$

$$0 \to T_v \to \boldsymbol{I}_v \to \bigoplus_{i \to j} \boldsymbol{I}_{\mathrm{Id}_j} \qquad \text{for } v = \mathrm{Id}_i.$$

The map from $\boldsymbol{I}_{\mathrm{O}_i^\pm}$ to $\boldsymbol{I}_{\mathrm{O}_j^\pm}$ is given by the morphism arrow $\mathrm{O}_j^\pm \to \mathrm{O}_i^\pm$. For the map from $\boldsymbol{I}_{\mathrm{Id}_i}$ to $\boldsymbol{I}_{\mathrm{Id}_j}$, let us recall from Lemma 4.2(3) that $\operatorname{Hom}(\boldsymbol{I}_{\mathrm{Id}_i}, \boldsymbol{I}_{\mathrm{Id}_j}) \cong \boldsymbol{P}_{\mathrm{Id}_j}(\mathrm{Id}_i) \cong \operatorname{Hom}_Q(P_j, P_i)$. We take the map to be the irreducible map in $\operatorname{Hom}_Q(P_j, P_i)$.

For any map between projective modules $f : P(\beta_1) \to P(\beta_2)$, the *$i$-th top restriction* of f is the induced map $\operatorname{top} P(\beta_1(i)) \to \operatorname{top} P(\beta_2(i))$.

THEOREM 4.3. *We have the following description for the modules* T_v.

(1) *The module* $T_{\mathrm{O}_i^-}$ *is the indecomposable module supported on all vertices translated from* $\mathrm{O}_{i^*}^+$ *with dimension vector* $(1, 1, \ldots, 1)$;
(2) *The defining linear map* $f \to g$ *in* $T_{\mathrm{O}_i^+}$ *is given by the i-th top restriction of* φ_+;
(3) *The defining linear map* $f \to g$ *in* T_{Id_i} *is given by the i-th top restriction of* φ_-,

where $\varphi = (\varphi_+, \varphi_-)$ *is the irreducible morphism from* f *to* g:

$$\begin{array}{ccc} P_+ & \xrightarrow{f} & P_- \\ {\scriptstyle \varphi_+}\downarrow & & \downarrow{\scriptstyle \varphi_-} \\ R_+ & \xrightarrow{g} & R_- \end{array}$$

In particular, the dimension vector θ_v of T_v is given by

$$\begin{aligned} \theta_v(f) &= \mathsf{e}(f)(i^*) && \textit{for } v = \mathrm{O}_i^-, \\ \theta_v(f) &= \mathsf{f}_+(i) && \textit{for } v = \mathrm{O}_i^+, \\ \theta_v(f) &= \mathsf{f}_-(i) && \textit{for } v = \mathrm{Id}_i. \end{aligned}$$

DEFINITION 4.4. We define a cone $\mathsf{G}_{\Delta_Q^2} \subset \mathbb{R}^{(\Delta_Q^2)_0}$ by $\mathsf{g}(\underline{\dim} S) \geq 0$ for all nontrivial subrepresentations $S \subset T_v$ and all frozen v.

So all these dimension vectors constitute the H-matrix in (0.1).

THEOREM 4.5. *The set of lattice points $\mathsf{G}_{\Delta_Q^2} \cap \mathbb{Z}^{(\Delta_Q^2)_0}$ is exactly $G(\Delta_Q^2, W_Q^2)$.*

EXAMPLE 4.6 (Example 2.8 continued). It is obvious that the boundary modules are all like $T_{\mathrm{O}_i^-}$'s. It was shown in [**11**] that the cone $\mathsf{G}_{\Delta_Q^2}$ is unimodularly equivalent to Knutson-Tao's hive cone [**22**].

EXAMPLE 4.7 (Example 2.9 continued). In this case, it is almost trivial to list all subrepresentations for all T_v. For example, there are 7 nontrivial subrepresentations for O_1^+:

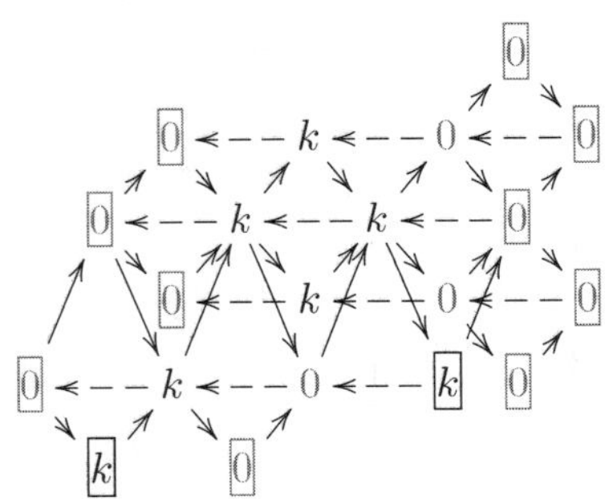

Readers can easily find that there are 3 nontrivial subrepresentations for all O_i^-, and $7, 6, 1, 1$ nontrivial subrepresentations for $\mathrm{O}_1^+, \mathrm{O}_2^+, \mathrm{O}_3^+, \mathrm{O}_4^+$ respectively, and $1, 2, 7, 7$ nontrivial subrepresentations for $\mathrm{Id}_1, \mathrm{Id}_2, \mathrm{Id}_3, \mathrm{Id}_4$ respectively. All their dimension vectors are needed to define $\mathsf{G}_{\Delta_Q^2}$, so there are 44 inequalities. This example can be easily generalized to Q of type D_n with a similar orientation.

EXAMPLE 4.8. For Q of type E_8 (see Figure 1), it is impossible to write down all defining conditions (ie., the H-matrix) for $\mathsf{G}_{\Delta_Q^2}$ by hand. For example, the nontrivial subrepresentations of T_{Id_2} have 147249 different dimension vectors. We found all of them using the algorithm outlined in [**14**, Appendix B.3]. Readers can download the full H-matrices of all exceptional types from author's web page [**15**].

DEFINITION 4.9. For any vector $\sigma \in \mathbb{Z}^{3|Q_0|}$, we define the convex polytope $\mathsf{G}_{\Delta_Q^2}(\sigma)$ as $\mathsf{G}_{\Delta_Q^2}$ cut out by the hyperplane sections $\mathbf{g}\boldsymbol{\sigma}_Q^2 = \sigma$.

4.3. Main Theorem. Let G be a simply-connected simple complex algebraic group of type A, D, E. Let Q be a *quiver* of G, that is, the underlying graph of Q is the Dynkin diagram of G. Recall that the weight lattice of G has a $\mathbb{Z}$-basis $\{\varpi\}_{i \in Q_0}$ of fundamental weights. We can thus identify a dominant weight $\sum_i \lambda(i)\varpi_i$ by a non-negative vector $\lambda \in \mathbb{Z}^{Q_0}$. In particular, we shall identify a triple of dominant weights (μ, ν, λ) with a non-negative vector in $\mathbb{Z}^{3|Q_0|}$. We are now able to state the main theorem. We recall the invariant algebra A^G and the multiplicity $c_{\mu,\nu}^{\lambda}$ defined in the introduction.

THEOREM 4.10. *Let G be a simply-connected simple complex algebraic group of type A, D, E, and Q be any quiver of G. Then the algebra A^G is isomorphic to the graded upper cluster algebra $\overline{\mathcal{C}}(\Delta_Q^2; \boldsymbol{\sigma}_Q^2)$. Moreover, (Δ_Q^2, W_Q^2) is a cluster model. In particular, $c_{\mu\,\nu}^{\lambda}$ is counted by lattice points in $\mathsf{G}_{\Delta_Q^2}(\mu, \nu, \lambda)$.*

REMARK 4.11. If Q' is another quiver of G, then Q and Q' are related by a sequence of BGP-reflections. We can show that Δ_Q^2 and $\Delta_{Q'}^2$ are mutation-equivalent so that they define the same upper cluster algebra, which is predicted by the theorem.

It is known that the upper cluster algebra $\overline{\mathcal{C}}(\Delta_Q^2)$ strictly contains the corresponding cluster algebra in all types except A_n (see [**14**, Example 9.3] for an example of type D_4). Whether they are equal is still open in type A.

We expect the theorem can be extended to non-simply laced cases (see next section).

5. Remarks on Non-simply Laced Cases

To generalize the above work to the non-simply laced cases (type B, C, F, G), we can replace the ordinary quivers with the valued ones.

DEFINITION 5.1. A *valued quiver* is a triple $Q = (Q_0, Q_1, C)$ where

(1) Q_0 is a set of vertices, usually labelled by natural numbers $1, 2, \ldots, n$;
(2) Q_1 is a set of arrows, which is a subset of $Q_0 \times Q_0$;
(3) $C = \{(c_{i,j}, c_{j,i}) \in \mathbb{N} \times \mathbb{N} \mid (i,j) \in Q_1\}$ is called the *valuation* of Q.

It is called *symmetrizable* if there is $d = \{d_i \in \mathbb{N} \mid i \in Q_0\}$ such that $d_i c_{i,j} = c_{j,i} d_j$ for every $(i,j) \in Q_1$.

To draw a valued quiver (Q_0, Q_1, C), we first draw its *ordinary quiver* (Q_0, Q_1), then put valuations above its arrows, eg., $i \xrightarrow{(c_{i,j}, c_{j,i})} j$. We will omit the valuation if (i,j) is trivially valued, i.e., $c_{i,j} = c_{j,i} = 1$.

Readers can find a short introduction on the representation theory of valued quivers in [**14**, Section 1.1]. A standard reference is Dlab and Ringel's article [**6**]. The (upper) cluster algebras are also defined for symmetrizable valued quivers [**17**].

All Section 2 can be immediately generalized to valued quivers (see [**14**, Section 2, 3] for detail). The iARt quiver Δ_Q^2 for a valued quiver Q will also be a valued quiver. In particular, we need to modify the second part of the recipe in Proposition 2.7: For each $i \xrightarrow{(a,b)} j$ in Q, we draw morphism arrows $\mathrm{O}_j^+ \xrightarrow{(a,b)} \mathrm{O}_i^+$, and $f(I_i) \xrightarrow{(b,a)} \mathrm{O}_j^+$. We add translation arrows from O_i^+ to $f(I_i)$ (with trivial valuation).

EXAMPLE 5.2. Here is an iARt quiver for G of type G_2.

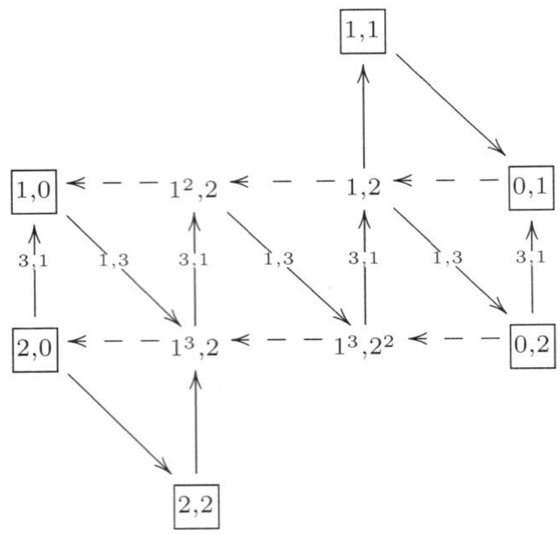

At this stage we do not have enough machinery to deal with the valued analogue of QP (eg., [**23**]). Although we can still define an analogous Jacobian algebra, a cluster character is still missing in the valued setting. If we mimic Definition 3.5, it is not clear how to show the image actually lands in the desired upper cluster algebra.

However, we observe that the cone $\mathsf{G}_{\Delta_Q^2}$ can be defined without introducing the Jacobian algebra. Without the Jacobian algebra we are unable to define the module T_v by injective presentations, but it still makes perfect sense to define T_v via Theorem 4.3. Once T_v's are defined, we define the cone $\mathsf{G}_{\Delta_Q^2}$ by Definition 4.4.

EXAMPLE 5.3. The most complicated T_v for Q of type G_2 is T_{Id_1}. Its dimension vector is

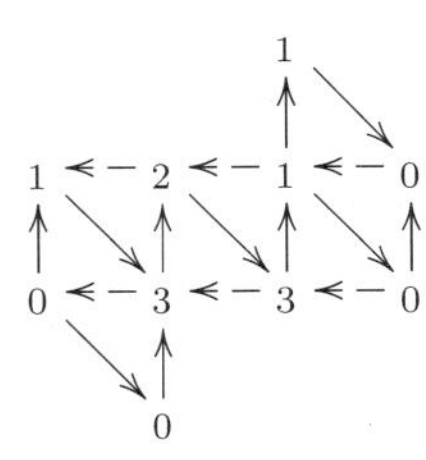

and its nontrivial subrepresentations have 13 distinct dimension vectors. However, we only need 5 of them to define $\mathsf{G}_{\Delta_Q^2}$ because some are consequences of others. Since the full H-matrix is rather small, we display it here. The rightmost column indicates the vertices corresponding to the rows. The H-matrices for other exceptional types can be downloaded from author's web page.

$$H = \begin{pmatrix} 1 & 0 & 0 & 0 & 0 & 0 & 0 & 1 & 1 & 1 & 1 & 0 & 1 & 1 & 1 & 1 & 0 & 0 \\ 1 & 1 & 0 & 0 & 0 & 0 & 0 & 1 & 1 & 0 & 0 & 0 & 2 & 2 & 0 & 0 & 0 & 0 \\ 0 & 0 & 0 & 1 & 0 & 0 & 0 & 2 & 2 & 2 & 0 & 0 & 3 & 3 & 3 & 0 & 0 & 0 \\ 0 & 0 & 0 & 1 & 1 & 0 & 0 & 1 & 0 & 0 & 0 & 0 & 3 & 0 & 0 & 0 & 0 & 0 \\ 1 & 1 & 1 & 0 & 0 & 0 & 0 & 0 & 0 & 0 & 0 & 0 & 0 & 0 & 0 & 0 & 0 & 0 \\ 0 & 0 & 0 & 1 & 1 & 1 & 0 & 0 & 0 & 0 & 0 & 0 & 0 & 0 & 0 & 0 & 0 & 0 \\ 0 & 0 & 0 & 0 & 0 & 0 & 1 & 0 & 0 & 0 & 0 & 0 & 0 & 0 & 0 & 0 & 0 & 0 \\ 0 & 0 & 0 & 0 & 0 & 0 & 0 & 1 & 1 & 1 & 1 & 1 & 0 & 0 & 0 & 0 & 0 & 0 \\ 0 & 0 & 0 & 0 & 0 & 0 & 0 & 0 & 0 & 0 & 0 & 0 & 1 & 1 & 1 & 1 & 1 & 0 \\ 0 & 0 & 0 & 0 & 0 & 0 & 0 & 0 & 0 & 0 & 0 & 0 & 0 & 0 & 0 & 0 & 0 & 1 \end{pmatrix} \qquad \begin{matrix} 1,2 \\ 1^2,2 \\ 1^3,2^2 \\ 1^3,2 \\ \boxed{1,0} \\ \boxed{2,0} \\ \boxed{0,1} \\ \boxed{0,2} \\ \boxed{1,1} \end{matrix}$$

We can show that $\overline{\mathcal{C}}(\Delta_Q^2; \boldsymbol{\sigma}_Q^2)$ is always a graded subalgebra of A^G no matter if G is simply laced. The conjecture is that

CONJECTURE 5.4. The first and last statements of Theorem 4.10 are still true even if G is not simply laced.

Acknowledgment

The author would like to thank the ICRA committee and the NCTS (National Center for Theoretical Sciences) for the financial support.

References

[1] Maurice Auslander, Idun Reiten, and Sverre O. Smalø, *Representation theory of Artin algebras*, Cambridge Studies in Advanced Mathematics, vol. 36, Cambridge University Press, Cambridge, 1997. Corrected reprint of the 1995 original. MR1476671

[2] I. Assem, D. Simson, A. Skowroński, *Elements of the Representation Theory of Associative Algebras,* London Mathematical Society Student Texts 65, Cambridge University Press, 2006.

[3] A. D. Berenstein and A. V. Zelevinsky, *Triple multiplicities for* $\mathrm{sl}(r+1)$ *and the spectrum of the exterior algebra of the adjoint representation*, J. Algebraic Combin. **1** (1992), no. 1, 7–22, DOI 10.1023/A:1022429213282. MR1162639

[4] Arkady Berenstein and Andrei Zelevinsky, *Tensor product multiplicities, canonical bases and totally positive varieties*, Invent. Math. **143** (2001), no. 1, 77–128, DOI 10.1007/s002220000102. MR1802793

[5] Arkady Berenstein, Sergey Fomin, and Andrei Zelevinsky, *Cluster algebras. III. Upper bounds and double Bruhat cells*, Duke Math. J. **126** (2005), no. 1, 1–52, DOI 10.1215/S0012-7094-04-12611-9. MR2110627

[6] Vlastimil Dlab and Claus Michael Ringel, *Indecomposable representations of graphs and algebras*, Mem. Amer. Math. Soc. **6** (1976), no. 173, v+57, DOI 10.1090/memo/0173. MR0447344

[7] Harm Derksen and Jiarui Fei, *General presentations of algebras*, Adv. Math. **278** (2015), 210–237, DOI 10.1016/j.aim.2015.03.012. MR3341790

[8] Harm Derksen and Jerzy Weyman, *Semi-invariants of quivers and saturation for Littlewood-Richardson coefficients*, J. Amer. Math. Soc. **13** (2000), no. 3, 467–479, DOI 10.1090/S0894-0347-00-00331-3. MR1758750

[9] Harm Derksen, Jerzy Weyman, and Andrei Zelevinsky, *Quivers with potentials and their representations. I. Mutations*, Selecta Math. (N.S.) **14** (2008), no. 1, 59–119, DOI 10.1007/s00029-008-0057-9. MR2480710

[10] Harm Derksen, Jerzy Weyman, and Andrei Zelevinsky, *Quivers with potentials and their representations II: applications to cluster algebras*, J. Amer. Math. Soc. **23** (2010), no. 3, 749–790, DOI 10.1090/S0894-0347-10-00662-4. MR2629987

[11] J. Fei, *Cluster algebras and semi-invariant rings I. Triple flags,* Proc. Lond. Math. Soc. (2017), no. 1, 1–32. MR3669931

[12] Jiarui Fei, *Cluster algebras and semi-invariant rings II: projections*, Math. Z. **285** (2017), no. 3-4, 939–966, DOI 10.1007/s00209-016-1733-7. MR3623736

[13] Jiarui Fei, *Cluster algebras, invariant theory, and Kronecker coefficients I*, Adv. Math. **310** (2017), 1064–1112, DOI 10.1016/j.aim.2017.02.016. MR3620703

[14] ———, *Tensor product multiplicities via upper cluster algebras,* arXiv:1603.02521.

[15] ———, Author's web page.

[16] Vladimir Fock and Alexander Goncharov, *Moduli spaces of local systems and higher Teichmüller theory*, Publ. Math. Inst. Hautes Études Sci. **103** (2006), 1–211, DOI 10.1007/s10240-006-0039-4. MR2233852

[17] Sergey Fomin and Andrei Zelevinsky, *Cluster algebras. I. Foundations*, J. Amer. Math. Soc. **15** (2002), no. 2, 497–529, DOI 10.1090/S0894-0347-01-00385-X. MR1887642

[18] Sergey Fomin and Andrei Zelevinsky, *Cluster algebras. IV. Coefficients*, Compos. Math. **143** (2007), no. 1, 112–164, DOI 10.1112/S0010437X06002521. MR2295199

[19] Christof Geiss, Bernard Leclerc, and Jan Schröer, *Auslander algebras and initial seeds for cluster algebras*, J. Lond. Math. Soc. (2) **75** (2007), no. 3, 718–740, DOI 10.1112/jlms/jdm017. MR2352732

[20] Alexander Goncharov and Linhui Shen, *Geometry of canonical bases and mirror symmetry*, Invent. Math. **202** (2015), no. 2, 487–633, DOI 10.1007/s00222-014-0568-2. MR3418241

[21] Roe Goodman and Nolan R. Wallach, *Symmetry, representations, and invariants*, Graduate Texts in Mathematics, vol. 255, Springer, Dordrecht, 2009. MR2522486

[22] Allen Knutson and Terence Tao, *The honeycomb model of* $\mathrm{GL}_n(\mathbf{C})$ *tensor products. I. Proof of the saturation conjecture*, J. Amer. Math. Soc. **12** (1999), no. 4, 1055–1090, DOI 10.1090/S0894-0347-99-00299-4. MR1671451

[23] Daniel Labardini-Fragoso and Andrei Zelevinsky, *Strongly primitive species with potentials I: mutations*, Bol. Soc. Mat. Mex. (3) **22** (2016), no. 1, 47–115, DOI 10.1007/s40590-015-0063-9. MR3473752

[24] Peter Littelmann, *A Littlewood-Richardson rule for symmetrizable Kac-Moody algebras*, Invent. Math. **116** (1994), no. 1-3, 329–346, DOI 10.1007/BF01231564. MR1253196

[25] Shiping Liu, *Auslander-Reiten theory in a Krull-Schmidt category*, São Paulo J. Math. Sci. **4** (2010), no. 3, 425–472, DOI 10.11606/issn.2316-9028.v4i3p425-472. MR2856194

[26] Pierre-Guy Plamondon, *Generic bases for cluster algebras from the cluster category*, Int. Math. Res. Not. IMRN **10** (2013), 2368–2420, DOI 10.1093/imrn/rns102. MR3061943

National Center for Theoretical Sciences, Taipei 10617, Taiwan

Current address: School of Mathematical Sciences, Shanghai Jiao Tong University, Shanghai 200240, China

Email address: jiarui@sjtu.edu.cn

Contemporary Mathematics
Volume **705**, 2018
http://dx.doi.org/10.1090/conm/705/14197

Special multiserial algebras, Brauer configuration algebras and more: A survey

Edward L. Green and Sibylle Schroll

ABSTRACT. We survey results on multiserial algebras, special multiserial algebras and Brauer configuration algebras. A structural property of modules over a special multiserial algebra is presented. Almost gentle algebras are introduced and we describe some results related to this class of algebras. We also report on the structure of radical cubed zero symmetric algebras.

1. Introduction

In this paper we present a brief survey of the results of [**15**–**18**]. These papers represent an attempt to generalize biserial algebras, special biserial algebras, and Brauer graph algebras. We show that the generalizations satisfy properties that are analogous to the properties of the algebras they generalize. Biserial, special biserial, and Brauer graph algebras have been extensively studied, see, for example, [**1**, **2**, **5**, **7**, **10**, **11**, **13**, **14**, **19**, **20**, **23**, **25**, **27**, **28**, **31**, **32**, **34**] and also the survey articles [**29**, **30**] and the references therein.

The generalizations of these algebras are the multiserial algebras [**24**, **33**], the special multiserial algebras [**18**, **33**] and the Brauer configuration algebras [**15**]. These algebras are typically wild, but as our results indicate, they seem to be worthwhile algebras to study. For proofs of the results presented in this survey, examples, and further details, the reader is referred to [**15**–**18**].

After presenting the definitions of multiserial algebras and special multiserial algebras, we define Brauer configuration algebras and algebras defined by cycles. We then look at how these algebras are related to each other in Theorems 2.2, 5.1, and 5.2.

Like biserial algebras, gentle algebras have received a great deal of attention, see for example [**3**, **5**, **6**, **21**, **28**]. In [**16**] we define almost gentle algebras, a natural generalization of gentle algebras, and show that trivial extensions of such algebras by the dual of the algebra are Brauer configuration algebras, Theorem 7.1.

The most surprising result concerns modules over a special multiserial algebra. We define multserial modules and show that, even though special multiserial algebras tend to be wild, every finitely generated module over a special multiserial algebra is a multiserial module, Theorem 8.1.

2010 *Mathematics Subject Classification.* 16S37, 14M99,16W60.

This work was supported through the UK Engineering and Physical Sciences Research Council grantEP/K026364/1.

There have recently been substantial advances in the classification of tame symmetric radical cubed zero algebras [**4**, **12**, **31**]. We show that over an algebraically closed field, every indecomposable symmetric radical cubed zero algebra is a Brauer configuration algebra, Theorem 9.1.

Throughout this paper, K denotes a field, Q a finite quiver with vertex set Q_0, and arrow set Q_1, and KQ the path algebra. We let J be the ideal in KQ generated by the arrows of Q. We assume, for simplicity, that every algebra occurring in the survey is of the form KQ/I with $J^N \subseteq I \subseteq J^2$, for some $N \geq 3$ and $\mathrm{rad}^2(KQ/I) \neq 0$. We also assume that the algebras are indecomposable as rings.

2. Multiserial and special multiserial algebras

We begin with a brief discussion of biserial algebras and special biserial algebras. Recall that a K-algebra Λ is *biserial* if for every indecomposable projective left or right module P, there are uniserial left or right modules U and V, such that $\mathrm{rad}(P) = U + V$ and $U \cap V$ is either zero or simple.

A K-algebra Λ is a *special biserial* algebra if the following two properties hold:

- S1 For each arrow $a \in Q_1$, there is at most one arrow $b \in Q_1$ such that $ab \notin I$, and there is at most one arrow $c \in Q_1$ such that $ca \notin I$.
- S2 For each vertex $v \in Q_0$, there are at most 2 arrows starting at v and at most 2 arrows ending at v.

A third type of algebra, that is closely related, is a Brauer graph algebra which is constructed from a Brauer graph. A *Brauer graph* is a 4-tuple $\Gamma = (\Gamma_0, \Gamma_1, \mu, \mathfrak{o})$ where (Γ_0, Γ_1) is a finite (undirected) graph where loops and multiple edges are permitted, and $\mu \colon \Gamma_0 \to \mathbb{Z}_{\geq 1}$ is a function, called the *multiplicity function*. To describe $\mathfrak{o}$, called the *orientation* of Γ, we need a few more definitions. A vertex $v \in \Gamma_0$ is a *truncated vertex* if v is an endpoint of exactly one edge and $\mu(v) = 1$. All other vertices are called *nontruncated vertices*. The orientation $\mathfrak{o}$ is a choice, for each nontruncated vertex v, of a cyclic ordering of the edges having v as an endpoint. Since the construction of a Brauer graph algebra is a special case of the construction of a Brauer configuration algebra, we postpone the construction until later in the paper. It should be noted that Brauer graphs and Brauer graph algebras generalize Brauer trees and Brauer tree algebras, which appear in the study of group rings of finite and tame representation type [**8**, **9**, **22**]. We have the following connections.

THEOREM 2.1. [**27**, **28**, **32**] *Let $\Lambda = KQ/I$ with $\mathrm{rad}^2(\Lambda) \neq 0$. If Λ is a special biserial algebra, then Λ is biserial. Furthermore, if K is algebraically closed then Λ is a symmetric special biserial algebra if and only if Λ is a Brauer graph algebra.*

One of the most important features of special biserial algebras is that their representation theory has a combinatorial framework in terms of strings and bands where the finitely generated indecomposable modules are given by string and band modules [**34**]. Thus special biserial algebras are tame [**34**]. The algebras we now define are usually of wild representation type.

Let Λ be a K-algebra and let M be a finitely generated Λ-module. We say that M is *left multiserial* (respectively *right multiserial*) if there is a positive integer n and left (respectively right) uniserial modules $U_1, \ldots, U_n$, such that $\mathrm{rad}(M) = \sum_{i=1}^n U_i$ and, if $i \neq j, U_i \cap U_j$ is either zero or simple. We say that Λ is *multiserial* if every indecomposable projective left or right module is left or right multiserial [**15**], see also [**24, 33**].

We say $\Lambda = KQ/I$ is *special multiserial* if condition (S1) holds, that is if:

For each arrow $a \in Q_1$, there is at most one arrow $b \in Q_1$ such that $ab \notin I$, and there is at most one arrow $c \in Q_1$ such that $ca \notin I$.

Since this is one of the defining properties of a special biserial algebra, a special biserial algebra is a special multiserial algebra. The next result generalizes part of Theorem 2.1.

THEOREM 2.2. [**15**] *A special multserial algebra is a multiserial algebra.*

3. Algebras defined by cycles

Before turning our attention to Brauer configurations and Brauer configuration algebras, we introduce another new class of algebras, *algebras defined by cycles* [**18**].

Let Q be a quiver. A *simple cycle* at a vertex v is a cycle $C = a_1a_2\cdots a_n$ in Q where the a_i are arrows, a_1 starts at v, a_n ends at v and there are no repeated arrows in C, that is $a_i \neq a_j$ for all $i \neq j$. The cyclic permutations of C are $a_1 \dots a_n$, $a_2a_3\cdots a_na_1, \dots, a_na_1a_2\cdots a_{n-1}$. A *defining set of cycles* is a set $\mathcal{S}$ of simple cycles of length at least 1, such that

D1 if $C \in \mathcal{S}$, then every cyclic permutation of C is in $\mathcal{S}$,
D2 every arrows occurs in some $C \in \mathcal{S}$, and
D3 if an arrow occurs in two cycles in $\mathcal{S}$, then the cycles are a cyclic permutation of each other.

Note that some of the cycles can be loops. If $\mathcal{S}$ is a defining set of cycles, then a *multiplicity function for* $\mathcal{S}$ is a function $\nu\colon \mathcal{S} \to \mathbf{Z}_{\geq 1}$ such that if $C, C' \in \mathcal{S}$ and C is a cyclic permutation of C', then $\nu(C) = \nu(C')$. Furthermore, we require that, if $C \in \mathcal{S}$ is a loop, then $\nu(C) > 1$. We call $(\mathcal{S}, \nu)$ a *defining pair.*

If $(\mathcal{S}, \nu)$ is a defining pair, then define $I_{(\mathcal{S},\nu)}$ to be the ideal in KQ generated by all elements of the following 3 types:

Type 1 $C^{\nu(C)} - C'^{\nu(C')}$, if C, C' are cycles in $\mathcal{S}$ at some vertex v,
Type 2 $C^{\nu(C)}a_1$, if $C = a_1a_2\cdots a_m$,
Type 3 ab, where $a, b \in Q_1$ and ab is not a subpath of C^2, for all $C \in \mathcal{S}$ such that C is not a loop.

DEFINITION 3.1. Let Q be a quiver, $(\mathcal{S}, \nu)$ a defining pair on Q and $I_{(\mathcal{S},\nu)}$ the ideal defined by $(\mathcal{S}, \nu)$. Then we call $\Lambda = KQ/I_{(\mathcal{S},\nu)}$ the *algebra defined by* $(\mathcal{S}, \nu)$ and we say Λ is *defined by cycles.*

4. Brauer configurations and Brauer configuration algebras

We now turn to Brauer configurations. A Brauer configuration is a 4-tuple of the form $\Gamma = (\Gamma_0, \Gamma_1, \mu, \mathfrak{o})$. The elements of the finite set Γ_0 are called *vertices* and the elements of Γ_1 are called *polygons.* If $V \in \Gamma_1$, then V is a finite multiset of vertices, that is, the elements of V are vertices, possibly with repetitions. The function $\mu\colon \Gamma_0 \to \mathbf{Z}_{\geq 1}$ is called the *multiplicity function.* A vertex $\alpha \in \Gamma_0$ is called *truncated* if $\mu(\alpha) = 1$, the vertex α occurs exactly once in one polygon V where $V = \{\alpha, \beta\}$ and

(the number of total occurences of β in polygons counting repetitions)$\cdot(\mu(\beta)) \geq 2$.

If a vertex is not truncated it is *nontruncated.* The *orientation* $\mathfrak{o}$ is a choice, for each nontruncated vertex $\alpha \in \Gamma_0$, of a cyclic ordering of the polygons that contain α, counting repetitions.

In [**15**] we called the above "reduced" Brauer configurations. There are some technical restrictions imposed by the assumption that the associated Brauer configuration algebra, (to be defined), is indecomposable. These restrictions can be found in [**15**].

An example will help to clarify the above definition. Suppose $\Gamma = (\Gamma_0, \Gamma_1, \mu, \mathfrak{o})$ where $\Gamma_0 = \{1,2,3,4\}, \Gamma_1 = \{V_1, V_2, V_3\}$ where $V_1 = \{1,2,3\}, V_2 = \{1,1,3\}$, $V_3 = \{2,3\}$ and $V_4 = \{2,4\}$. Let $\mu(i) = 1$, for $i = 1, \ldots, 4$. The only truncated vertex is 4. The orientation $\mathfrak{o}$ is a choice for each nontruncated vertex of an ordering of the polygons that contain the vertex. For example, for 1, the ordering might be $V_1 < V_2 < V_3 < V_2 (< V_1)$ with the last inequality because it is a cyclic ordering. Any cyclic permutation, for example, $V_2 < V_3 < V_2 < V_1 (< V_2)$ is considered the same ordering. On the other hand, the ordering at 1 could also have been chosen to be $V_1 < V_2 < V_2 < V_3 (< V_1)$ which yields a different configuration.

A Brauer configuration is a generalization of a Brauer graph since every Brauer graph is a Brauer configuration with the restriction that every polygon is a set with 2 vertices.

There are "realizations" of a Brauer configuration obtained by attaching actual polygons to the elements in Γ_1 with the vertices of the polygon labelled by the appropriate elements of Γ_0. Then one identifies vertices of the actual polygons that have the same label. Another way to realize a Brauer configuration is as a hypergraph where the set of vertices is given by Γ_0 and where Γ_1 corresponds to the set of hyperedges [**16**].

We now define the Brauer configuration algebra Λ_Γ associated to a Brauer configuration $\Gamma = (\Gamma_0, \Gamma_1, \mu, \mathfrak{o})$. The quiver Q_Γ of Λ_Γ has vertex set Γ_1. If $\Gamma_1 = \{V_1, \ldots, V_n\}$, we write $(Q_\Gamma)_0 = \{v_1, \ldots, v_n\}$ to distinguish between the polygons in Γ and the vertices in Q_Γ. For every nontruncated vertex $\alpha \in \Gamma_0$, there is a cyclically ordered sequence of polygons containing α given by the orientation $\mathfrak{o}$. For each nontruncated $\alpha \in \Gamma_0$, we view the ordered sequence of polygons as a simple cycle C_α in Q_Γ. If $\alpha, \beta \in \Gamma_0$ are nontruncated with $\alpha \neq \beta$, then we require that C_α and C_β have no arrows in common. Note that if α is in exactly one polygon V and $\mu(\alpha) > 1$, then C_α is a loop at v in Q_Γ.

The arrow set of Q_Γ are exactly the arrows occuring in the C_α, for α a nontruncated vertex in Γ. Set $\mathcal{S} = \{C_\alpha \mid \alpha \text{ a nontruncated vertex in } \Gamma\}$. Define $\nu\colon \mathcal{S} \to \mathbb{Z}_{\geq 1}$ by $\nu(C_\alpha) = \mu(\alpha)$, where, on the right hand side, μ is the multiplicity function of Γ. It is easy to see that $(\mathcal{S}, \nu)$ is a defining pair.

The Brauer configuration algebra Λ_Γ associated to Γ is the algebra $KQ_\Gamma/I_{(\mathcal{S},\nu)}$ defined by $(\mathcal{S}, \nu)$. This description of Λ_Γ is slightly different than the original description given in [**15**] but both yield isomorphic algebras.

5. Connection results

The following results connect all the algebras.

THEOREM 5.1. [**17,18**] *Let K be an algebraically closed field and let $\Lambda = KQ/I$ with* $\operatorname{rad}(\Lambda)^2 \neq 0$, *be a symmetric K-algebra. Then the following statements are equivalent.*

(1) Λ *is a symmetric special multiserial algebra.*
(2) Λ *is defined by cycles.*
(3) Λ *is a Brauer configuration algebra.*

Another result generalizing a result about special biserial algebras [**34**] is the following.

THEOREM 5.2. [**18**] *Let K be a field. Every special multiserial K-algebra is a quotient of a symmetric special multiserial K-algebra.*

6. Examples

EXAMPLE 6.1. Let Q be the quiver

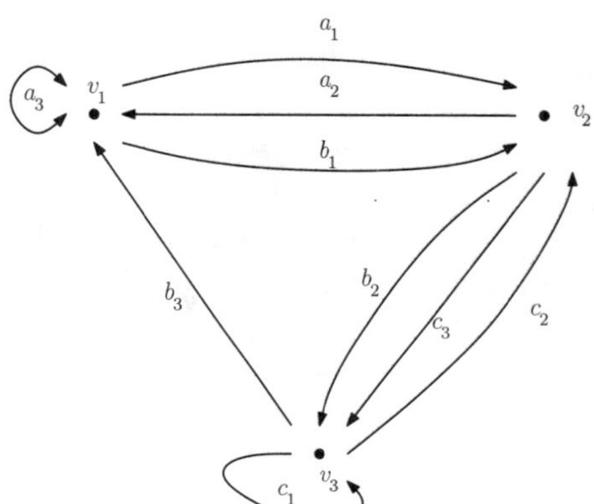

Let $\mathcal{S}$ consist of the following cycles and their cyclic permutations:

$$a_1a_2a_3, b_1b_2b_3, c_1c_2c_3$$

and define $\nu\colon \mathcal{S} \to \mathbb{Z}_{\geq 1}$ by $\nu \equiv 1$. We see that $(\mathcal{S}, \nu)$ is a defining pair. Let Λ be defined by $(\mathcal{S}, \nu)$. In particular, a generating set for the ideal of relations for Λ is given in Section 3. By Theorem 5.1, Λ is a Brauer configuration algebra and a special multiserial algebra. The Brauer configuration is $\Gamma = (\Gamma_0, \Gamma_1, \mu, \mathfrak{o})$ where $\Gamma_0 = \{1, 2, 3\}$, $\Gamma_1 = \{V_1, V_2, V_3\}$ with $V_1 = \{1, 1, 2\}, V_2 = \{2, 3, 3\}$, and $V_3 = \{1, 2, 3\}$. The multiplication function μ is identically 1 and the orientation is given below.

A realization of the Brauer configuration is

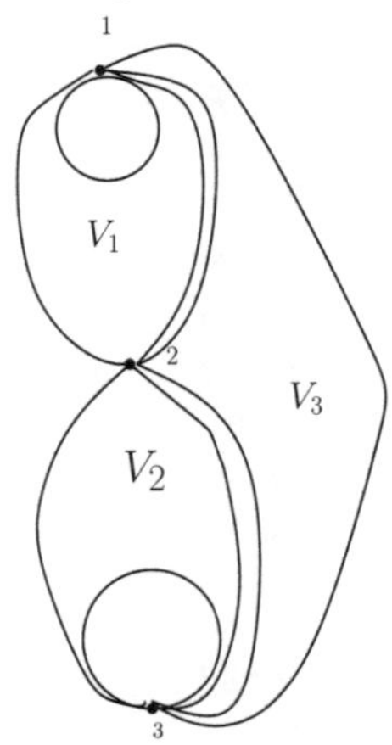

with clockwise orientation at each vertex.

EXAMPLE 6.2. Let the quiver be

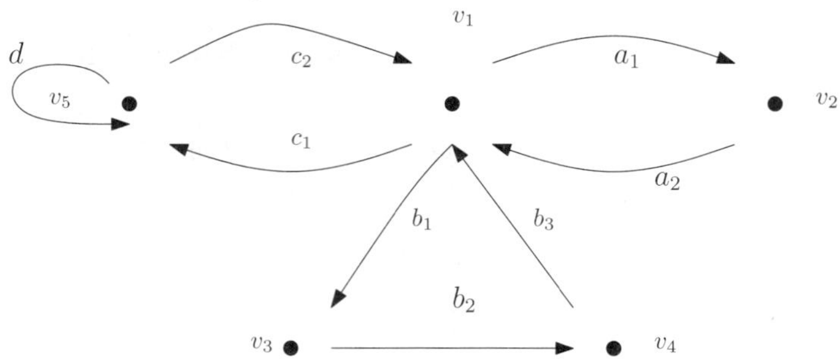

Let $\mathcal{S}$ be the set of simple cycles consisting of all the permutations of the cycles $a_1a_2, b_1b_2b_3, c_1c_2, d$ with $\nu : \mathcal{S} \to \mathbb{Z}_{\geq 1}$ defined by $\nu(a_1a_2) = 1 = \nu(b_1b_2b_3) = \nu(c_1c_2)$, and $\nu(d) = 2$. We see that $(\mathcal{S}, \nu)$ is a defining pair. Take Λ to be the algebra defined by $(\mathcal{S}, \nu)$.

One can check that Λ is isomorphic to the Brauer configuration algebra associated to the Brauer configuration $\Gamma = (\Gamma_0, \Gamma_1, \mu, \mathfrak{o})$ with $\Gamma_0 = \{1, 2, 3, 4, 5, 6, 7\}, \Gamma_1 = \{V_1, V_2, V_3, V_4, V_5\}$ with $V_1 = \{1, 2, 3\}, V_2 = \{1, 4\}, V_3 = \{2, 5\}, V_4 = \{2, 6\}, V_5 = \{3, 7\}$. The multiplicity function is 1 for all vertices except for vertex 7 where $\mu(7) = 2$. We see that vertices 4,5,6 are the truncated vertices. The orientation is given by the clockwise ordering of the polygons at each nontruncated vertex in the realization of the Brauer configuration below.

A realization of the Brauer configuration is:

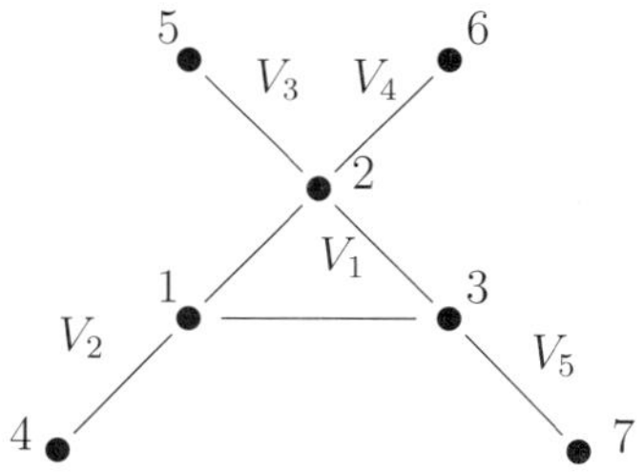

7. Almost gentle algebras

Recall that an algebra $\Lambda = KQ/I$ is *gentle* if it is special biserial, that is if (S1) and (S2) hold, and if additionally S2^{op} holds, that is for any arrow a there exists at most one arrow b such that $ab \in I$ and at most one arrow c such that $ca \in I$ and if I is generated by paths of length 2.

An obvious way to generalize the notion of a gentle algebras is: An algebra $\Lambda = KQ/I$ is *almost gentle* if it is special multiserial and if I can be generated by paths of length 2.

A characterisation of gentle algebras is through the trivial extension by their minimal injective co-generator. That is, an algebra A is gentle if and only if the trivial extension $T(A) = A \ltimes \text{Hom}_k(A, k)$ is special biserial [**26**].

We have the following results.

THEOREM 7.1. [**16**] *If A is an almost gentle algebra then $T(A)$ is a symmetric special multiserial algebra and hence it is isomorphic to a Brauer configuration* *algebra. If $T(A)$ is isomorphic to the Brauer configuration algebra associated to the Brauer configuration $\Gamma = (\Gamma_0, \Gamma_1, \mu, \mathfrak{o})$, then $\mu \equiv 1$.*

There is a converse. Let $\Lambda = KQ/I$ be a symmetric special multiserial algebra. Then by Theorem 5.1, there is a defining pair $(\mathcal{S}, \nu)$ such that Λ is defined by $(\mathcal{S}, \nu)$. Assume that $\nu \equiv 1$. We say a set of arrows D is an *admissible cut* if D consists of one arrow for each permutation class of cycles in $\mathcal{S}$. Let $\widehat{Q}$ be the quiver obtained from Q by removing the arrows in D. Let $\widehat{I}$ be the ideal $I \cap K\widehat{Q}$.

THEOREM 7.2. [**16**] *Let $\Lambda = KQ/I$ be a symmetric special multiserial algebra viewed as an algebra defined by cycles with defining pair $(\mathcal{S}, \nu)$ where $\nu \equiv 1$. Keeping the notation of the previous paragraph, let D be an admissible cut in Q. Then $K\widehat{Q}/\widehat{I}$ is an almost gentle algebra.*

If A is an almost gentle algebra, then there is an admissible cut in $T(A)$ such that the construction above yields an algebra isomorphic to A. Moreover, if Λ is a symmetric special multiserial algebra (with $\nu \equiv 1$) and D an admissible cut, then $T(K\widehat{Q}/\widehat{I})$ is isomorphic to Λ. The proofs of these results are found in [**16**].

8. Representations of multiserial algebras

As mentioned earlier, special multiserial algebras are typically wild. Thus, it is surprising that there is a structural result about all finitely generated modules over such an algebra. Recall that a left (respectively, right) finitely generated module is multiserial if it is left multiserial (respectively, right multiserial), that is if $\text{rad}(M)$ as a left (resp., right) module is a sum of uniserial modules that intersect pairwise in the zero module or in a simple module.

THEOREM 8.1. *Let A be a special multiserial K-algebra, and M an indecomposable finitely generated left (respectively, right) A-module. Then M is multiserial.*

9. Radical cubed zero

In this section we present some results about symmetric algebras with the condition that their Jacobson radical cubed is zero. The proof of the next result requires an algebraically closed field.

THEOREM 9.1. [**17**] *Let K be an algebraically closed field and let $A = KQ/I$ be a finite dimensional basic indecomposable K-algebra. Suppose that A is symmetric and that $\text{rad}^3(A) = 0$ but $\text{rad}^2(A) \neq 0$. Then A is isomorphic to a Brauer configuration algebra.*

It is well-known that there are equivalence relations on symmetric matrices with non-negative integer entries, as well as on finite undirected graphs, such that there is a one-to-one correspondence between the respective equivalence classes, see [**15**] for details. To each equivalence class one can associate a radical cubed zero symmetric algebra. Now consider the radical cubed zero Brauer configuration algebras with Brauer configuration $\Gamma = (\Gamma_0, \Gamma_1, \mu, \mathfrak{o})$ that satisfy the following 2 properties:

(1) there are no repeated vertices occurring in any polygon in Γ_1,
(2) if $\alpha \in \Gamma_0$ then either α is in exactly one polygon and $\mu(\alpha) = 2$ or α is in exactly two polygons and $\mu(\alpha) = 1$.

We show that there is an equivalence relation on radical cubed zero Brauer configuration algebras with properties (1) and (2) such that the equivalence classes are in one-to-one correspondence with the radical cubed zero symmetric algebras associated to the equivalence classes of finite graphs [**15**].

References

[1] Takuma Aihara, *Derived equivalences between symmetric special biserial algebras*, J. Pure Appl. Algebra **219** (2015), no. 5, 1800–1825, DOI 10.1016/j.jpaa.2014.07.012. MR3299707

[2] Adachi, Takahide; Aihara, Takuma; Chan, Aaron. Tilting Brauer graph algebras I: Classification of two-term tilting complexes, arXiv:1504.04827.

[3] Ibrahim Assem, Thomas Brüstle, Gabrielle Charbonneau-Jodoin, and Pierre-Guy Plamondon, *Gentle algebras arising from surface triangulations*, Algebra Number Theory **4** (2010), no. 2, 201–229, DOI 10.2140/ant.2010.4.201. MR2592019

[4] David J. Benson, *Resolutions over symmetric algebras with radical cube zero*, J. Algebra **320** (2008), no. 1, 48–56, DOI 10.1016/j.jalgebra.2008.02.033. MR2417976

[5] Rafał Bocian and Andrzej Skowroński, *Symmetric special biserial algebras of Euclidean type*, Colloq. Math. **96** (2003), no. 1, 121–148, DOI 10.4064/cm96-1-11. MR2013714

[6] Andrew T. Carroll and Calin Chindris, *On the invariant theory for acyclic gentle algebras*, Trans. Amer. Math. Soc. **367** (2015), no. 5, 3481–3508, DOI 10.1090/S0002-9947-2014-06191-6. MR3314814

[7] Roberto Vila-Freyer and William Crawley-Boevey, *The structure of biserial algebras*, J. London Math. Soc. (2) **57** (1998), no. 1, 41–54, DOI 10.1112/S0024610798005821. MR1624781

[8] E. C. Dade, *Blocks with cyclic defect groups*, Ann. of Math. (2) **84** (1966), 20–48, DOI 10.2307/1970529. MR0200355

[9] P. W. Donovan, *Dihedral defect groups*, J. Algebra **56** (1979), no. 1, 184–206, DOI 10.1016/0021-8693(79)90332-6. MR527164

[10] Duffield, Drew. Auslander-Reiten components of Brauer graph algebras, arXiv:1509.02478.

[11] Karin Erdmann and Andrzej Skowroński, *On Auslander-Reiten components of blocks and self-injective biserial algebras*, Trans. Amer. Math. Soc. **330** (1992), no. 1, 165–189, DOI 10.2307/2154159. MR1144759

[12] Karin Erdmann and Øyvind Solberg, *Radical cube zero weakly symmetric algebras and support varieties*, J. Pure Appl. Algebra **215** (2011), no. 2, 185–200, DOI 10.1016/j.jpaa.2010.04.012. MR2720683

[13] Kent R. Fuller, *Biserial rings*, Ring theory (Proc. Conf., Univ. Waterloo, Waterloo, 1978), Lecture Notes in Math., vol. 734, Springer, Berlin, 1979, pp. 64–90. MR548124

[14] J. A. Green, *Walking around the Brauer Tree*, J. Austral. Math. Soc. **17** (1974), 197–213. Collection of articles dedicated to the memory of Hanna Neumann, VI. MR0349830

[15] Edward L. Green and Sibylle Schroll, *Brauer configuration algebras: a generalization of Brauer graph algebras*, Bull. Sci. Math. **141** (2017), no. 6, 539–572. MR3698159

[16] Green. Edward L.; Schroll, Sibylle. Almost gentle algebras and their trivial extensions, arXiv:1603.03587.

[17] Edward L. Green and Sibylle Schroll, *Multiserial and special multiserial algebras and their representations*, Adv. Math. **302** (2016), 1111–1136, DOI 10.1016/j.aim.2016.07.006. MR3545948

[18] Edward L. Green and Sibylle Schroll, *Special multiserial algebras are quotients of symmetric special multiserial algebras*, J. Algebra **473** (2017), 397–405, DOI 10.1016/j.jalgebra.2016.10.033. MR3591156

[19] Edward L. Green, Sibylle Schroll, and Nicole Snashall, *Group actions and coverings of Brauer graph algebras*, Glasg. Math. J. **56** (2014), no. 2, 439–464, DOI 10.1017/S0017089513000372. MR3187910

[20] Edward Green, Sibylle Schroll, Nicole Snashall, and Rachel Taillefer, *The Ext algebra of a Brauer graph algebra*, J. Noncommut. Geom. **11** (2017), no. 2, 537–579, DOI 10.4171/JNCG/11-2-4. MR3669112

[21] Ch. Geiß and I. Reiten, *Gentle algebras are Gorenstein*, Representations of algebras and related topics, Fields Inst. Commun., vol. 45, Amer. Math. Soc., Providence, RI, 2005, pp. 129–133. MR2146244

[22] G. J. Janusz, *Indecomposable modules for finite groups*, Ann. of Math. (2) **89** (1969), 209–241, DOI 10.2307/1970666. MR0244307

[23] Michael Kauer, *Derived equivalence of graph algebras*, Trends in the representation theory of finite-dimensional algebras (Seattle, WA, 1997), Contemp. Math., vol. 229, Amer. Math. Soc., Providence, RI, 1998, pp. 201–213, DOI 10.1090/conm/229/03319. MR1676221

[24] V. V. Kirichenko and Yu. V. Yaremenko, *Multiserial rings* (Russian, with English and Ukrainian summaries), Ukraïn. Mat. Zh. **48** (1996), no. 9, 1223–1235, DOI 10.1007/BF02595360; English transl., Ukrainian Math. J. **48** (1996), no. 9, 1388–1401 (1997). MR1429608

[25] Robert J. Marsh and Sibylle Schroll, *The geometry of Brauer graph algebras and cluster mutations*, J. Algebra **419** (2014), 141–166, DOI 10.1016/j.jalgebra.2014.08.002. MR3253283

[26] Zygmunt Pogorzały and Andrzej Skowroński, *Self-injective biserial standard algebras*, J. Algebra **138** (1991), no. 2, 491–504, DOI 10.1016/0021-8693(91)90183-9. MR1102821

[27] K. W. Roggenkamp, *Biserial algebras and graphs*, Algebras and modules, II (Geiranger, 1996), CMS Conf. Proc., vol. 24, Amer. Math. Soc., Providence, RI, 1998, pp. 481–496. MR1648648

[28] Sibylle Schroll, *Trivial extensions of gentle algebras and Brauer graph algebras*, J. Algebra **444** (2015), 183–200, DOI 10.1016/j.jalgebra.2015.07.037. MR3406174

[29] Schroll, Sibylle. Brauer graph algebras, a survey, arXiv:1612.00061.

[30] Schröer, Jan. Biserial / Special Biserial /String / Gentle Algebras, preprint, http://www.math.uni-bonn.de/people/schroer/fd-atlas-files/FD-BiserialAlgebras.pdf

[31] Andrzej Skowroński, *Selfinjective algebras: finite and tame type*, Trends in representation theory of algebras and related topics, Contemp. Math., vol. 406, Amer. Math. Soc., Providence, RI, 2006, pp. 169–238, DOI 10.1090/conm/406/07658. MR2258046

[32] Andrzej Skowroński and Josef Waschbüsch, *Representation-finite biserial algebras*, J. Reine Angew. Math. **345** (1983), 172–181. MR717892

[33] Hans-Joachim von Höhne and Josef Waschbüsch, *Die struktur n-reihiger Algebren* (German), Comm. Algebra **12** (1984), no. 9-10, 1187–1206, DOI 10.1080/00927878408823049. MR738544

[34] Burkhard Wald and Josef Waschbüsch, *Tame biserial algebras*, J. Algebra **95** (1985), no. 2, 480–500, DOI 10.1016/0021-8693(85)90119-X. MR801283

Edward L. Green, Department of Mathematics, Virginia Tech, Blacksburg, VA 24061 USA

Email address: green@math.vt.edu

Sibylle Schroll, Department of Mathematics, University of Leicester, University Road, Leicester LE1 7RH, UK

Email address: schroll@le.ac.uk

Contemporary Mathematics
Volume **705**, 2018
http://dx.doi.org/10.1090/conm/705/14191

Nakayama-type phenomena in higher Auslander–Reiten theory

Gustavo Jasso and Julian Külshammer

Abstract. This paper surveys recent contructions in higher Auslander–Reiten theory. We focus on those which, due to their combinatorial properties, can be regarded as higher dimensional analogues of path algebras of linearly oriented type $\mathbb{A}$ quivers. These include higher dimensional analogues of Nakayama algebras, of the mesh category of type $\mathbb{Z}\mathbb{A}_\infty$ and the tubes, and of the triangulated category generated by an m-spherical object. For $m = 2$, the latter category can be regarded as the higher cluster category of type $\mathbb{A}_\infty$ whose cluster-tilting combinatorics are controlled by the triangulations of the cylic apeirotope.

1. Introduction

The category of finite dimensional representations of the quiver

$$\mathbb{A}_\infty^\infty\colon\ \dots \longrightarrow -1 \longrightarrow 0 \longrightarrow 1 \longrightarrow \dots$$

is well understood. The indecomposable representations are the interval modules and its Auslander–Reiten quiver is of type $\mathbb{Z}\mathbb{A}_\infty$. The category of nilpotent finite dimensional representations of the cylic quiver $\tilde{\mathbb{A}}_{n-1}$ with n vertices, obtained as the quotient of $\mathbb{A}_\infty^\infty$ by identifyng to vertices if they are equal modulo n, has a similar description. In this case its Auslander–Reiten quiver is rank n tube. Note that both the mesh category of type $\mathbb{Z}\mathbb{A}_\infty$ and the tubes are frequently occurring shapes of Auslander–Reiten components. For our purposes, it is important to note that the categories of representations of the Nakayama algebras embedd into those of the above quivers and hence admit a nice combinatorial description.

In this survey we introduce higher dimensional analogues of the quivers $\mathbb{A}_\infty^\infty$ and $\tilde{\mathbb{A}}_{n-1}$ from the viewpoint of Iyama's higher Auslander–Reiten theory [**Iya07b**, **Iya07a**]. This allows us to construct higher dimensional analogues of the Nakayama algebras, whose higher Auslander–Reiten theory is controlled by higher dimensional analogues of the mesh category of type $\mathbb{Z}\mathbb{A}_\infty$ and the tubes [**JK16a**]. To complement loc. cit., we give a few examples to illustrate how the homological behaviour of the higher Nakayama algebras differs from that of their classical versions.

We also construct certain Calabi–Yau triangulated categories which can be regarded as higher dimensional analogues of the triangulated categories generated

2010 *Mathematics Subject Classification.* 16G70; Secondary 16G20.

Part of this research was carried out during a visit of the second author at MPIM Bonn. The authors would like to thank the anonymous referee for helpful comments.

by spherical objects of dimension greater than or equal to 2. In particular, in analogy to the case of a 2-dimensional spherical object, we obtain a higher version of the cluster category of type $\mathbb{A}_\infty$ introduced by Holm and Jørgensen [**HJ12**]. In this case, the cluster-tilting combinatorics of this category is controlled by triangulations of the cyclic apeirotope (a higher dimensional analog of the ∞-gon).

Let us briefly desribe the structure of this article. Section 2 recalls the Auslander correspondence and Iyama's generalisation of it obtained by introducing cluster-tilting subcategories. We also include a short introduction to higher Auslander–Reiten theory. In Section 3 we recall the construction of the higher Nakayama algebras in terms of the higher dimensional analogues of the quivers $\mathbb{A}_\infty^\infty$ and $\tilde{\mathbb{A}}_{n-1}$. In Section 4 we give examples of higher Nakayama algebras which have different homological properties than those of their classical versions. Finally, in Section 5 we introduce the higher dimensional analogues of the triangulated categories generated by spherical objects and of the cluster category of type $\mathbb{A}_\infty$.

CONVENTIONS 1.1. We fix a positive integer d as well as a field $\Bbbk$. We denote the duality of the category of finite dimensional $\Bbbk$-vector spaces by $D := \mathsf{Hom}_\Bbbk(-, \Bbbk)$. By algebra we mean associative unital $\Bbbk$-algebra. All modules we consider are finite dimensional left modules. Two algebras Λ and Λ' are *Morita equivalent* if their categories of modules are equivalent. Let Λ be a finite dimensional algebra. We denote by $\mathsf{ind}\,\Lambda$ a complete set of representatives of the isomorphisms classes of finite dimensional indecomposable Λ-modules. Let M be a Λ-module (or, more generally, an object in some additive category). We denote by $\mathsf{add}\,M$ the full subcategory of $\mathsf{mod}\,\Lambda$ consisting of all Λ-modules which are direct summands of M^n for some positive integer n.

2. Preliminaries

In this section we recall the basics of higher Auslander–Reiten theory, namely Auslander–Iyama correspondence and the basic theory of d-representation-finite d-hereditary algebras. The reader is referred to [**Iya07a, Iya11, Iya08, IO11, IJ16, JK16c**] for further reading.

2.1. The Auslander–Iyama correspondence. For the purpose of motivation we first recall classical Auslander correspondence. This establishes a relationship between representation-finiteness, a *representation-theoretic* property, and being an Auslander algebra, a *homological* property. Moreover, the minimal projective resolutions of simple modules of projective dimension 2 over Auslander algebras provided the first evidence for the existence of almost-split sequences. Thus, Auslander correspondence can also be considered as one of the seminal results in Auslander–Reiten theory.

On the representation-theoretic side let Λ be a finite dimensional algebra. Recall that Λ is *representation-finite* if there are only finitely many indecomposable Λ-modules up to isomorphism. An equivalent way of saying this is that there exists a finite dimensional Λ-module M such that every indecomposable Λ-module is a direct summand of M. Such a Λ-module M is called a *representation generator of* $\mathsf{mod}\,\Lambda$.

On the homological side, let Γ be a finite dimensional algebra and

$$0 \to \Gamma \to I^0 \to I^1 \to \cdots \to I^d \to \cdots$$

a minimal injective coresolution of Γ. We remind the reader that the *dominant dimension of* Γ is defined as

$$\operatorname{domdim}\Gamma := 1 + \sup\left\{ i \in \mathbb{Z} \,\middle|\, I^i \text{ is projective} \right\}.$$

A finite dimensional algebra Γ is an *Auslander algebra* if

$$\operatorname{gldim}\Gamma \leq 2 \leq \operatorname{domdim}\Gamma.$$

Auslander algebras were introduced by the eponymous author in [**Aus71**] where the following theorem is also proven.

THEOREM 2.1 (Auslander correspondence). *There is a one-to-one correspondence between Morita equivalence classes of representation-finite algebras and Morita equivalence classes of Auslander algebras. The correspondence associates to a representation-finite algebra Λ the algebra $\operatorname{End}_\Lambda(M)^{\mathrm{op}}$ for a representation generator M of $\operatorname{mod}\Lambda$. In the reverse direction, it associates to an Auslander algebra Γ the algebra $\operatorname{End}_\Gamma(I^0)^{\mathrm{op}}$ where I^0 is the injective hull of the regular representation.*

Auslander correspondence is in fact a particular instance of Morita–Tachikawa correspondence. This correspondence shifts the focus on the 'representation-theoretic' side away from the algebra Λ to the category $\operatorname{add} M$. In Auslander correspondence knowledge of Λ is (up to Morita equivalence) of course equivalent to knowledge of $\operatorname{add} M$ for M a representation generator.

Recall that a Λ-module M is a *generator-cogenerator of* $\operatorname{mod}\Lambda$ if every indecomposable projective and every indecomposable injective Λ-module is a direct summand of M. In particular, a representation generator is an example of a generator-cogenerator. A *Morita–Tachikawa pair* is a pair (Λ, M) consisting of a finite dimensional algebra Λ and generator-cogenerator M of $\operatorname{mod}\Lambda$. Two Morita–Tachikawa pairs (Λ, M) and (Λ', M') are *equivalent* if $\operatorname{add} M$ and $\operatorname{add} M'$ are equivalent categories. Note that this implies that the algebras Λ and Λ' are Morita equivalent: Indeed, a module P is a projective object in $\operatorname{add} M$ (in the sense that every epimorphism to it splits) if and only if it is a projective Λ-module. It is easy to see that no other module can be projective as M is a generator. For the converse let P be a projective Λ-module and $f\colon M \to P$ be an epimorphism. Let g be the composition of the projection $P \to \operatorname{Coker} f$ and the inclusion $\operatorname{Coker} f \to I$ into its injective hull. Then, $gf = 0$ implies that $g = 0$ as f is an epimorphism. In particular $\operatorname{Coker} f = 0$. It follows that $\operatorname{add}\Lambda \cong \operatorname{add}\Lambda'$ as both are defined by categorical properties in $\operatorname{add} M \cong \operatorname{add} M'$. Thus, Λ and Λ' are Morita equivalent.

THEOREM 2.2 (Morita–Tachikawa correspondence). *There is a one-to-one correspondence between equivalence classes of Morita–Tachikawa pairs and Morita equivalence classes of finite dimensional algebras Γ such that $\operatorname{domdim}\Gamma \geq 2$. The correspondence associates to a Morita–Tachikawa pair (Λ, M) the algebra $\operatorname{End}_\Lambda(M)^{\mathrm{op}}$ and, in the other direction, associates to an algebra Γ the Morita–Tachikawa pair $(\operatorname{End}_\Gamma(I^0)^{\mathrm{op}}, I^0)$ for I^0 the injective hull of Γ as a left Γ-module.*

Almost thirty years later, Iyama realised that Auslander correspondence, together with it the most fundamental aspects of Auslander–Reiten theory, could be vastly generalised. In some sense, Iyama's approach was inverse to that of Auslander. Indeed, the class of Auslander algebras admits an "obvious" generalisation: a finite dimensional algebra Γ is a *d-Auslander algebra* if the inequalities

$$\operatorname{gldim}\Gamma \leq d+1 \leq \operatorname{domdim}\Gamma$$

are satisfied.

In view of the Morita–Tachikawa correspondence, the modules which are in correspondence with d-Auslander algebras must be particular kinds of generator-cogenerators. These are the d-cluster-tilting modules, which we now define.

DEFINITION 2.3 ([**Iya07a**,**Iya11**,**IO11**]). Let Λ be a finite dimensional algebra. A Λ-module M is *d-cluster-tilting* if

$$\begin{aligned}\operatorname{add} M &= \left\{ X \in \operatorname{mod}\Lambda \,\middle|\, \forall i \in \{1,\dots,d-1\} : \operatorname{Ext}^i_\Lambda(X,M)=0 \right\}\\ &= \left\{ Y \in \operatorname{mod}\Lambda \,\middle|\, \forall i \in \{1,\dots,d-1\} : \operatorname{Ext}^i_\Lambda(M,Y)=0 \right\}.\end{aligned}$$

An algebra Λ is called *weakly d-representation-finite* if there exists a d-cluster-tilting module.

It is easy to see that a 1-cluster-tilting Λ-module is precisely a representation generator of $\operatorname{mod}\Lambda$, and that every d-cluster-tilting Λ-module is a generator-cogenerator.

A *d-Auslander–Iyama pair* is a pair (Λ, M) where Λ is a finite dimensional algebra and M is a d-cluster-tilting module. Two d-Auslander–Iyama pairs (Λ, M) and (Λ', M') are *equivalent* if $\operatorname{add} M$ and $\operatorname{add} M'$ are equivalent. The following theorem appeared originally as [**Iya07a**, Theorem 0.2].

THEOREM 2.4 (d-dimensional Auslander–Iyama correspondence). *There is a one-to-one correspondence between equivalence classes of d-Auslander–Iyama pairs and Morita equivalence classes of d-Auslander algebras. The correspondence associates to a d-Auslander–Iyama pair* (Λ, M) *the d-Auslander algebra* $\operatorname{End}_\Lambda(M)^{\mathrm{op}}$, *and in the inverse direction to a d-Auslander algebra* Γ *the pair* $(\operatorname{End}_\Gamma(I^0)^{\mathrm{op}}, I^0)$ *where* I^0 *is the injective hull of the regular representation of* Γ.

Recently, there have been further generalisations of this correspondence (and thus instances of Morita–Tachikawa correspondence) by Iyama–Solberg [**IS16**] and Marczinzik [**Mar17**] replacing $\operatorname{gldim}\Gamma$ by $\operatorname{injdim}_\Gamma \Gamma$ and $\operatorname{findim}\Gamma$, respectively.

2.2. Higher Auslander–Reiten theory. Following the steps in the development of Auslander–Reiten theory, the d-dimensional Auslander–Iyama correspondence served as the starting point for a $(d+1)$-dimensional Auslander–Reiten theory. The minimal projective resolutions of simple modules of projective dimension $d+1$ over d-Auslander algebras give rise to the prototypical d-almost-split sequences. The only caveat is that, in order to obtain analogous properties to those in classical Auslander–Reiten theory, one needs to restrict to modules in a d-cluster-tilting subcategory, a generalisation of the category $\operatorname{add} M$ for a d-cluster-tilting module M.

DEFINITION 2.5 ([**Iya07a**, (2.4)], [**Iya07b**, Definition 2.2], [**Jas16**, Definition 3.14]). Let $\mathcal{A}$ be an abelian category. A generating-cogenerating functorially finite (full) subcategory $\mathcal{C}$ of $\mathcal{A}$ is a *d-cluster-tilting subcategory*[1] if

$$\begin{aligned}\mathcal{C} &= \{\, X \in \mathcal{A} \mid \forall i \in \{1, \dots, d-1\} : \mathsf{Ext}^i_{\mathcal{A}}(X, \mathcal{C}) = 0 \,\} \\ &= \{\, Y \in \mathcal{A} \mid \forall i \in \{1, \dots, d-1\} : \mathsf{Ext}^i_{\mathcal{A}}(\mathcal{C}, Y) = 0 \,\}.\end{aligned}$$

An object $C \in \mathcal{A}$ is a *d-cluster-tilting object* if $\mathsf{add}\, C$ is a d-cluster-tilting subcategory of $\mathcal{A}$.

The authors are not aware of any examples of abelian categories $\mathcal{A}$ having a subcategory $\mathcal{C}$ which satisfies the assumptions of a d-cluster-tilting subcategory except for being generating-cogenerating. However, this assumption is needed in the proofs of even the basic statements of the theory in [**Jas16**]. If $\mathcal{A}$ has enough projectives and enough injectives, this condition is automatic as $\mathcal{C}$ always contains all the projective and all the injective objects of $\mathcal{A}$.

Let $\mathcal{C}$ be a d-cluster-tilting subcategory of an abelian category $\mathcal{A}$. Then, a *d-almost split sequence* is an exact sequence

$$0 \to C_0 \to C_1 \to \cdots \to C_d \to C_{d+1} \to 0$$

with $C_i \in \mathcal{C}$ for all i and C_0 indecomposable such that every non-split epimorphism $C' \to C_{d+1}$ for some $C' \in \mathcal{C}$ factors through the map $C_d \to C_{d+1}$; moreover, one requires that for each $i \in \{0, 1, \dots, d\}$ the morphisms $C_i \to C_{i+1}$ lies in the Jacobson radical of $\mathsf{mod}\, \Lambda$. In this case, $C_0 \cong \tau_d(C_{d+1})$ where

$$\tau_d(M) := \tau\Omega^{d-1}(M)$$

is the d-dimensional replacement of the classical Auslander–Reiten translation τ. Here, Ω denotes Heller's syzygy functor.

In 'classical' representation theory, representation-finite hereditary algebras were among the first algebras whose Auslander–Reiten theory was systematically investigated, leading to fruitful results, see for example [**ASS06, SS07a, SS07b**]. Modelled on this, weakly d-representation finite algebras of small global dimension were among the first to be investigated.

DEFINITION 2.6. A finite dimensional algebra Λ is *d-representation-finite d-hereditary* if it is weakly d-representation-finite and $\mathsf{gldim}\, \Lambda \leq d$.

As an exercise, the reader can verify that d is the smallest possible global dimension for a non-semisimple weakly d-representation-finite algebra. Thus, a d-representation-finite d-hereditary algebra has global dimension either 0 or d. Much of the research in higher Auslander–Reiten theory has focused on the class of d-representation-finite d-hereditary algebras, see for example [**IO11, HI11a, HI11b, OT12, IO13**]. In the beginning of Section 3 we recall Iyama's construction of the higher Auslander algebras of type $\mathbb{A}$, which are a particularly important class of d-representation-finite d-hereditary algebras.

There is an analogous result to the fact that for a representation-finite hereditary algebra all indecomposable modules can be constructed from the indecomposable injectives by iterative application of the Auslander–Reiten translation.

[1] Note that d-cluster-tilting subcategories where originally called "maximal $(d-1)$-orthogonal subcategories" in [**Iya07a**].

THEOREM 2.7 ([**Iya11**, Proposition 1.3(b)]). *Let Λ be a d-representation-finite d-hereditary algebra with n simple modules $S_1, \dots, S_n$ up to isomorphism. Then, the Λ-module*

$$M = \bigoplus_{i=1}^{n} \bigoplus_{j \geq 0} \tau_d^j I_i,$$

where I_i denotes the indecomposable injective Λ-module with socle S_i, is the (up to isomorphism) unique basic d-cluster-tilting Λ-module.

Another nice feature of d-representation-finite d-hereditary algebras is that their derived categories also have a d-cluster-tilting subcategory. Note that d-cluster-tilting subcategories of triangulated categories can be defined in an obvious way (omitting the condition of being generating-cogenerating) with the obvious generalisation of almost split triangles to almost split d-angles.

For a finite dimensional algebra Λ, set

$$\mathcal{U}(\Lambda) := \operatorname{add} \{ \nu_d^i \Lambda \mid i \in \mathbb{Z} \},$$

where $\nu = D\Lambda \otimes_\Lambda^{\mathbb{L}} -$ is the derived Nakayama functor and $\nu_d := \nu[-d]$. Thus, $\mathcal{U}(\Lambda)$ is a full subcategory of the bounded derived category $D^b(\operatorname{mod} \Lambda)$.

THEOREM 2.8. *Let Λ be a d-representation-finite d-hereditary algebra. Then, $\mathcal{U}(\Lambda)$ is a d-cluster-tilting subcategory of $D^b(\operatorname{mod} \Lambda)$.*

Note that an analogous construction for more general weakly d-representation-finite algebra is not known and a naive extension is even not possible, see the authors' note [**JK16b**].

3. Higher Nakayama algebras

In this section we recall the construction of higher Nakayama algebras from the authors' article [**JK16a**]. For this, we introduce a certain universal category whose importance in higher Auslander–Reiten theory is already mentioned by Iyama in [**Iya11**, Introduction].

Every partially ordered $(P, \leq)$ has an associated $\Bbbk$-category $A(P)$ called the *incidence $\Bbbk$-category of the poset P*, whose objects are the elements of P and with morphisms

$$A(P)(x,y) := \begin{cases} \Bbbk f_{yx} & \text{if } x \leq y, \\ 0 & \text{otherwise,} \end{cases}$$

where f_{yx} denotes a basis vector of a corresponding one-dimensional space. The composition in $A(P)$ induced by $f_{zy} \circ f_{yx} := f_{zx}$. If the set P is finite, this is just the incidence algebra of the poset $(P, \leq)$.

DEFINITION 3.1. Let $\leq$ be the product order on $\mathbb{Z}^d$, i.e. $\alpha \leq \beta$ if and only if $\alpha_i \leq \beta_i$ for all $i \in \{1, \dots, d\}$. Let $A^{(d)} := A(\mathbb{Z}^d, \leq)$.

The $\Bbbk$-category $A^{(d)}$ can be explicitly described as follows. Let $Q^{(d)}$ be the quiver with vertex set $\mathbb{Z}^d$ and arrows

$$a_i = a_i(\alpha) \colon \alpha \to \alpha + e_i$$

where $i \in \{1, \dots, d\}$. Let I be the two-sided ideal of the path category $\Bbbk Q^{(d)}$ of $Q^{(d)}$ generated by all commutativity relations

$$\{ a_i a_j - a_j a_i \mid 1 \leq i < j \leq d \}.$$

With these conventions, $A^{(d)}$ is isomorphic to the quotient category $\Bbbk Q^{(d)}/I$. From both descriptions it is clear that $A^{(1)}$ is isomorphic to the path category of $\mathbb{A}_\infty^\infty$ with linear orientation.

Note that $A^{(d)}$ is naturally a graded $\Bbbk$-category with respect to path length. It is elementary to verify that $A^{(d)}$ is the universal covering of the polynomial ring in d variables in the sense of [**Gre83**] (see also [**MVdlP83**]), that it is of global dimension d and a Koszul category in the sense of [**MOS09**].

In the sequel we heavily use a certain automorphism φ_d on $A^{(d)}$ (and idempotent quotients thereof) defined by

$$\varphi_d(\mu_1, \dots, \mu_d) = (\mu_1 - 1, \dots, \mu_d - 1),$$
$$\varphi_d(a_i(\mu_1, \dots, \mu_d)) = a_i(\mu_1 - 1, \dots, \mu_d - 1).$$

3.1. The universal d-Nakayama category. In this subsection, we introduce another category, a certain idempotent quotient of $A^{(d)}$, which serves as the universal higher Nakayama category, much like the path category of $\mathbb{A}_\infty^\infty$ serves as the universal Nakayama category. Set

$$\mathbf{os}^{(d)} := \left\{ (\mu_1, \dots, \mu_d) \in \mathbb{Z}^d \,\middle|\, \mu_1 \geq \mu_2 \geq \cdots \geq \mu_d \right\}.$$

DEFINITION 3.2. For $d \geq 2$, the idempotent quotient $A_\infty^{(d)}$ of $A^{(d)}$ by the objects not in $\mathbf{os}^{(d)}$ is called the *universal d-Nakayama category.*

Note that $A_\infty^{(d)}$ is different from the incidence category of $\mathbf{os}^{(d)}$ as there are certain zero relations at the 'boundary' of $A_\infty^{(d)}$. Furthermore observe that $A_\infty^{(2)}$ is the mesh category of a $\mathbb{Z}\mathbb{A}_\infty$-component, one of the most typical Auslander–Reiten components in 'classical' Auslander–Reiten theory. This is why we chose to give them the similar name $A_\infty^{(d)}$ as we see them as higher analogues of this category.

DEFINITION 3.3. Let $\lambda \in \mathbf{os}^{(d+1)}$. The *interval module* M_λ is the $A_\infty^{(d)}$-module with composition factors

$$[M_\lambda \colon S_\kappa] = \begin{cases} \Bbbk & \text{if } \kappa \in [(\lambda_2, \dots, \lambda_{d+1}), (\lambda_1, \dots, \lambda_d)], \\ 0 & \text{else,} \end{cases}$$

where $[(\lambda_2, \dots, \lambda_{d+1}), (\lambda_1, \dots, \lambda_d)]$ denotes the interval in the partial order on $\mathbf{os}^{(d)}$. The maps on the arrows a_i are identity maps between different copies of $\Bbbk$ and 0 elsewhere.

The category $A_\infty^{(d)}$ satisfies analogous properties to being d-representation-finite in the infinite setting:

THEOREM 3.4 ([**JK16a**, Theorem 2.3.5]). *The abelian category* $\operatorname{mod} A_\infty^{(d)}$ *of finite dimensional $A_\infty^{(d)}$-modules is of global dimension d and has a d-cluster-tilting subcategory consisting of the additive hull $\mathcal{M}_\infty^{(d)}$ of all interval modules M_λ for $\lambda \in$* $\mathbf{os}^{(d+1)}$.

In the derived situation,

$$\mathcal{U}_\infty^{(d)} := \operatorname{add}\left\{ M_\lambda[di] \,\middle|\, \lambda \in \mathbf{os}^{(d+1)}, i \in \mathbb{Z} \right\}$$

is a d-cluster-tilting subcategory of $D^b(\operatorname{mod} A_\infty^{(d)})$.

3.2. Higher Nakayama algebras. In this section, we survey on the results of the authors' paper [**JK16a**] on the construction of higher analogues of Nakayama algebras.

DEFINITION 3.5. A finite dimensional algebra Λ is called a *Nakayama algebra* (or *uniserial*) if every indecomposable projective and every indecomposable injective module is uniserial.

It is well-known that (over an algebraically closed field), the connected Nakayama algebras are precisely the quotients by admissible ideals of the path algebras of the linearly oriented Dynkin quiver

$$\mathbb{A}_n\colon\ 0 \longrightarrow 1 \longrightarrow 2 \longrightarrow \cdots \longrightarrow n-1,$$

and the linearly oriented extended Dynkin quiver

$\widetilde{\mathbb{A}}_{n-1}\colon$

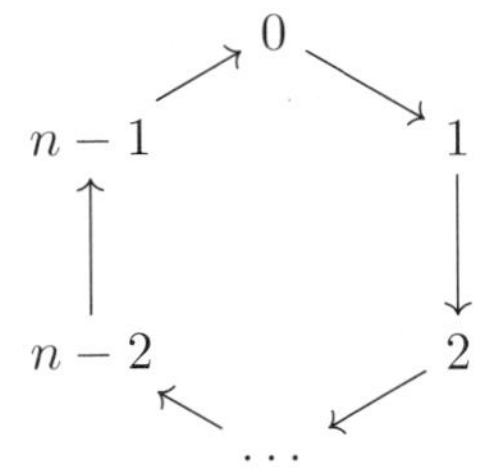

Note that the path algebra of $\mathbb{A}_n$ is an idempotent truncation of $A_\infty^{(1)}$ while the path algebra of $\widetilde{\mathbb{A}}_{n-1}$ is obtained from $A_\infty^{(1)}$ by dividing out by the action of φ_1^n. There are analogues of the path algebras of $\mathbb{A}_n$ and $\widetilde{\mathbb{A}}_{n-1}$, respectively. The former was already introduced in [**Iya11**] and studied extensively in [**OT12**]. For $n=1$, the latter can be seen as a higher preprojective algebra of type $\mathbb{A}_\infty$. We start by introducing the higher analogues of the path algebra of $\mathbb{A}_n$, in [**IO11**] called higher Auslander algebras of type $\mathbb{A}$.

DEFINITION 3.6. Let $\mathbf{os}_n^{(d)} := \left\{ (\mu_1,\dots,\mu_d)\in \mathbf{os}^{(d)} \,\middle|\, n-1\geq \mu_1, \mu_d\geq 0 \right\}$. The *$d$-Auslander algebra $A_n^{(d)}$ of type $\mathbb{A}_n$* is defined to be the (full) subcategory of $A_\infty^{(d)}$ on the vertices in $\mathbf{os}_n^{(d)}$.

THEOREM 3.7 ([**Iya11**, Theorem 1.18]). *Let $d\geq 1$. Then, $A_n^{(d)}$ is a d-representation-finite d-hereditary algebra. Moreover, $A_n^{(d+1)} \cong \mathsf{End}_{A_n^{(d)}}(M_n^{(d)})$ where $M_n^{(d)}$ is the unique basic d-cluster-tilting $A_n^{(d)}$-module. The indecomposable direct summands of $M_n^{(d)}$ are precisely the interval modules M_λ for $\lambda\in\mathbf{os}_n^{(d+1)}$.*

The following theorem of Iyama characterises the higher Auslander algebras of type $\mathbb{A}$ as the only finite dimensional algebras which are d-representation finite d-hereditary and $(d-1)$-Auslander algebras at the same time. Thus, in particular, these are the only d-representation-finite d-hereditary algebra which can be constructed inductively as in Theorem 3.7.

THEOREM 3.8 ([**Iya11**, Theorem 1.19]). *Let $d\geq 2$ and Λ a finite dimensional algebra. Then, Λ is d-representation-finite d-hereditary as well as a $(d-1)$-Auslander algebra if and only if it is Morita equivalent to $A_n^{(d)}$ where n is the number of simple Λ-modules in the d-cluster-tilting subcategory of* $\mathsf{mod}\,\Lambda$.

For constructing the higher analogue of the path algebra of $\tilde{\mathbb{A}}$ note that $A_\infty^{(d)}$ is closed under φ_d. Recall that φ_d was defined to be the quiver automorphism satisfying

$$\varphi_d(\mu_1,\dots,\mu_d) = (\mu_1 - 1,\dots,\mu_d - 1),$$
$$\varphi_d(a_i(\mu_1,\dots,\mu_d)) = a_i(\mu_1 - 1,\dots,\mu_d - 1).$$

Define the higher analogue of the path algebra of $\tilde{\mathbb{A}}_{n-1}$ by $\tilde{A}_{n-1}^{(d)} := A_\infty^{(d)}/\varphi_d^n$. It is easy to see that the images of the $A_\infty^{(d)}$-modules M_λ and M_μ under the pushdown functor are isomorphic if and only if $\lambda \equiv \mu \mod n\mathbb{Z}(1,1,\dots,1)$. As the quiver of this category contains oriented cycles, we consider the category of finite dimensional nilpotent modules instead of the category of finite dimensional modules.

THEOREM 3.9 ([**JK16a**, Theorem 2.3.4]). *The category* $\mathsf{nil}\,\tilde{A}_{n-1}^{(d)}$ *has a d-cluster-tilting subcategory* $\mathcal{T}_n^{(d)}$ *whose objects are given by the images of the interval modules* M_λ *where* $\lambda \in \mathbf{os}_n^{(d+1)}$. *Moreover,* $\mathcal{T}_n^{(d)} \cong \mathsf{add}\,\tilde{A}_{n-1}^{(d+1)}$.

The name $\mathcal{T}_n^{(d)}$ is chosen because for $d = 1$, $\mathcal{T}_n^{(d)}$ is the tube of rank n from 'classical' Auslander–Reiten theory.

After having defined the analogues of the path algebras of $\mathbb{A}_n$ and $\tilde{\mathbb{A}}_{n-1}$, the next step is to find analogues for admissible relations for Nakayama algebras. In contrast to classical Nakayama algebras, which arise as quotients by admissible ideals of the algebras $A_n^{(1)}$ and $\tilde{A}_{n-1}^{(1)}$, the higher Nakayama algebras arise as idempotent quotients from $A_n^{(d)}$ and $\tilde{A}_{n-1}^{(d)}$, respectively.

Recall that a convenient combinatorial way to label Nakayama algebras is by means of the vector

$$(\mathsf{dim}\,P_0, \mathsf{dim}\,P_1, \dots, \mathsf{dim}\,P_{n-1})$$

where P_i is the indecomposable projective to the vertex i. This vector is called the *Kupisch series* corresponding to the Nakayama algebra. Given a Kupisch series $\underline{\ell} = (\ell_0,\dots,\ell_{n-1})$, the corresponding set of relations one needs to divide out by is given by the paths $\{\, i - \ell_i \rightsquigarrow i \mid \ell_{i-1} \geq \ell_i \,\}$. Note that the interval modules $M_{(\mu_1,\mu_2)}$ for the Nakayama algebra with Kupisch series $\underline{\ell}$ satisfy $\mu_1 - \mu_2 + 1 \leq \ell_{\mu_1}$ as $\mu_1 - \mu_2 + 1$ gives the length of the interval module $M_{(\mu_1,\mu_2)}$ and μ_1 gives its top and every module is a quotient of the corresponding indecomposable projective module with the same top.

Analogously, in the higher setting for $d \geq 2$, the higher Nakayama algebra $A_{\underline{\ell}}^{(d)}$ (of type $\mathbb{A}$, i.e. for $\ell_0 = 1$) associated to the Kupisch series $\underline{\ell}$ is defined to be the idempotent quotient of $A_n^{(d)}$ by the idempotents not in

$$\mathbf{os}_{\underline{\ell}}^{(d)} := \left\{ (\mu_1,\dots,\mu_d) \in \mathbf{os}_n^{(d)} \,\middle|\, \mu_1 - \mu_d + 1 \leq \ell_{\mu_1} \right\}.$$

For the Nakayama algebras of type $\tilde{\mathbb{A}}_{n-1}$ given the Kupisch series ℓ one defines an infinite series $\tilde{\ell}$ by periodically extending ℓ. One then defines

$$\mathbf{os}_{\tilde{\underline{\ell}}}^{(d)} := \left\{ (\mu_1,\dots,\mu_d) \in \mathbf{os}^{(d)} \,\middle|\, \mu_1 - \mu_d + 1 \leq \tilde{\ell}_{\mu_1} \right\}.$$

Note that by construction, $\mathbf{os}_{\tilde{\underline{\ell}}}^{(d)}$ is stable under φ_d^n. Define $\tilde{A}_{\underline{\ell}}^{(d)}$ to be the idempotent quotient of $\tilde{A}_{n-1}^{(d)}$ by the idempotents not corresponding to the ones in $\mathbf{os}_{\tilde{\underline{\ell}}}^{(d)}$.

THEOREM 3.10. *For every Kupisch series $\underline{\ell}$ and every $d \geq 1$, the algebra $A_{\underline{\ell}}^{(d)}$ (respectively $\tilde{A}_{\underline{\ell}}^{(d)}$ in type $\tilde{\mathbb{A}}$) is weakly d-representation-finite. The modules in the corresponding d-cluster-tilting module are the interval modules M_λ with $\lambda \in \mathbf{os}_{\underline{\ell}}^{(d+1)}$ (respectively $\lambda \in \mathbf{os}_{\tilde{\underline{\ell}}}^{(d+1)}$ in type $\tilde{\mathbb{A}}$).*

Among all weakly d-representation-finite algebras, the higher Nakayama algebras belong to a more special class of algebras, since their cluster-tilting modules M satisfy the stronger condtion

$$\mathsf{Ext}^i(M, M) \neq 0 \implies i \in d\mathbb{Z}.$$

One of their special features is that exact sequences of length d in the cluster-tilting subcategory give rise to long exact Ext^d-sequences, see [**IJ16**]. Such algebras are called *$d\mathbb{Z}$-representation-finite*. They can also be characterised by the fact that the d-cluster-tilting module is closed under Ω^d.

4. Obstructions to an alternative definition of higher Nakayama algebras

The reader may ask whether there is a characterisation of higher Nakayama algebras similar to projectives and injectives being uniserial. The main problem in answering this question is that it is not clear what the analogue of a filtration is when there are only higher Ext-groups, but no Ext^1. In this section we discuss a few more problems with such definition and with statements involving the simples in a cluster-tilting subcategory.

4.1. Simple modules and the Ext^d-quiver of the higher Nakayama algebras. We start by characterising simples only in terms of the cluster tilting subcategory.

DEFINITION 4.1. Let $\mathcal{A}$ be an additive category and $S \in \mathcal{A}$ a non-zero object. We say that S is *Schur simple* if every non-zero morphism $X \to S$ is an epimorphism and every non-zero morphism $S \to Y$ is a monomorphism.

It turns out that, with this definition, Schur simples of a cluster tilting subcategory are precisely the simple modules.

LEMMA 4.2. *Let $\mathcal{A}$ be an abelian category and $\mathcal{M} \subseteq \mathcal{A}$ a functorially finite generating-cogenerating subcategory. Then, an object $S \in \mathcal{M}$ is Schur simple if and only if it is a simple object in $\mathcal{A}$.*

PROOF. Let $S \in \mathcal{M}$ be simple in $\mathcal{A}$. Then, of course, it is Schur simple. For the other direction we show that a Schur simple object in $\mathcal{M}$ is in fact Schur simple in $\mathcal{A}$, and hence simple. For this, let $S \in \mathcal{M}$ be Schur simple and $X \in \mathcal{A}$ be arbitrary with a morphism $f: S \to X$. Since $\mathcal{M}$ is covariantly finite, there exists a left $\mathcal{M}$-approximation $g: X \to M$ for some $M \in \mathcal{M}$. Since S is Schur simple in $\mathcal{M}$, the composition gf is a monomorphism or zero. Since the left approximation g is a monomorphism since $\mathcal{M}$ is cogenerating, it follows that f is a monomorphism or zero. Dually, one shows that every non-zero morphism $X \to S$ in $\mathcal{A}$ is an epimorphism. It follows that S is Schur simple in $\mathcal{A}$ which is equivalent to being simple. □

DEFINITION 4.3. Let $\mathcal{A}$ be an abelian category and $\mathcal{M} \subseteq \mathcal{A}$ a d-cluster-tilting subcategory. The Ext^d-*quiver* of $\mathcal{M}$ is the quiver which has as vertices the isomorphism classes of Schur simples in $\mathcal{M}$ and the number of arrows $[S] \to [S']$ is given by $\dim_{\Bbbk} \mathsf{Ext}^d_{\mathcal{A}}(S, S')$.

The Ext^d-quivers of the higher Nakayama algebras are precisely what one would expect, namely equal to the Ext^1-quivers of the corresponding (classical) Nakayama algebras.

PROPOSITION 4.4. *Let $\underline{\ell}$ be a Kupisch series. Then, the the Ext^d-quiver of $A^{(d)}_{\underline{\ell}}$ is $\mathbb{A}_n$ in case $\ell_0 = 1$. while the Ext^d-quiver of $\tilde{A}^{(d)}_{\underline{\ell}}$ is $\widetilde{\mathbb{A}}_{n-1}$ in case $\ell_0 \neq 1$.*

PROOF. This follows immediately from the results of [**JK16a**]. □

We now give an example showing that, in general the Ext^d-quiver for d-cluster-tilting subcategories of module categories of finite dimensional algebras is ill-behaved.

EXAMPLE 4.5. Let A be the preprojective algebra of type $\mathbb{A}_3$ which is also the higher Nakayama algebra $\tilde{A}^{(2)}_{(3)}$, that is, A is the preprojective algebra of $\mathbb{A}_3$, i.e. the path algebra of the quiver

$$3 \underset{y_2}{\overset{x_2}{\rightleftarrows}} 2 \underset{y_1}{\overset{x_1}{\rightleftarrows}} 1$$

bound by the relations $y_1 x_1 = 0$, $x_2 y_2 = 0$ and $x_1 y_1 = y_2 x_2$. The reader can verify that

$$M = A \oplus 1 \oplus \begin{smallmatrix} 2 \\ 1 \end{smallmatrix} \oplus \begin{smallmatrix} 1 \\ 2 \end{smallmatrix} \quad \text{and} \quad N = A \oplus \begin{smallmatrix} 2 \\ 1 \end{smallmatrix} \oplus \begin{smallmatrix} 2 \\ 3 \end{smallmatrix} \oplus \begin{smallmatrix} 2 \\ 13 \end{smallmatrix}$$

are 2-cluster-tilting A-modules, M being the one given by interval modules. For the convenience of the reader, the Auslander–Reiten quiver of $\mathsf{mod}\, A$ is given below, where indecomposable A-modules are described by their radical filtration and the indecomposable direct summands of M and N are indicated with rectangles and circles respectively.

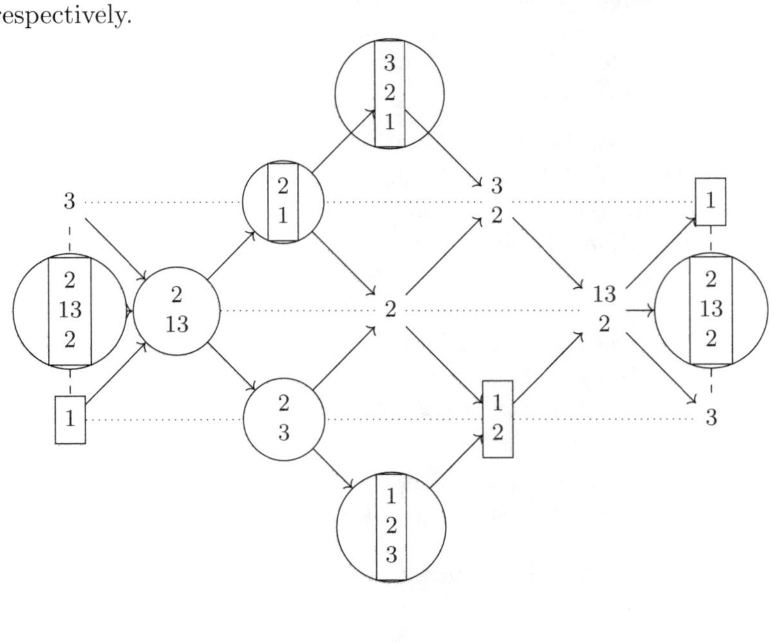

The Ext^2-quiver of $\mathsf{add}\, M$ has one vertex and one arrow and the Ext^2-quiver of $\mathsf{add}\, N$ is empty since N has no simple summands. Thus, this example shows that in general

- the Ext^d-quiver of $\mathcal{M}$ depends on $\mathcal{M}$ and not only on A, and that
- the Ext^d-quiver of $\mathcal{M}$ may be empty.

In particular, the Ext^d-quiver of $\mathcal{M}$ in general does not determine $\mathcal{M}$ nor A.

4.2. Global dimension of the higher Nakayama algebras. A further example shows that even for the higher Nakayama algebras, where the simple modules in many ways behave like simple modules in the classical sense, e.g. in the sense that they all lie on a unique τ_d-orbit (see [**JK16a**]), in some respect they are different. The subsequent example shows that the global dimension of one of the Nakayama algebras is not attained by a projective dimension of a simple in the cluster-tilting subcategory as well as the global dimension of $A_{\underline{\ell}}^{(d)}$ is in general not equal to $d \cdot \mathsf{gldim}\, A_{\underline{\ell}}^{(1)}$. Moreover, the example also shows that, in general,

$$\dim \mathsf{Ext}^2_{A_{\underline{\ell}}^{(1)}}(S_i, S_j) \neq \dim \mathsf{Ext}^{2d}_{A_{\underline{\ell}}^{(d)}}(S_{(i,i,\cdots,i)}, S_{(j,j,\cdots,j)}).$$

This suggest that there is no straightforward way to generalise relations in $A_{\underline{\ell}}^{(1)}$, which are given by the $\mathsf{Ext}^2_{A_{\underline{\ell}}^{(1)}}$ between simples, to higher values of d.

EXAMPLE 4.6. Let $\underline{\ell} = (1, 2, 2, 3, 3, 4, 3)$. The following figure shows the module category of $A_{\underline{\ell}}^{(1)}$. The numbers on the vertices show the projective dimensions of the corresponding indecomposable modules.

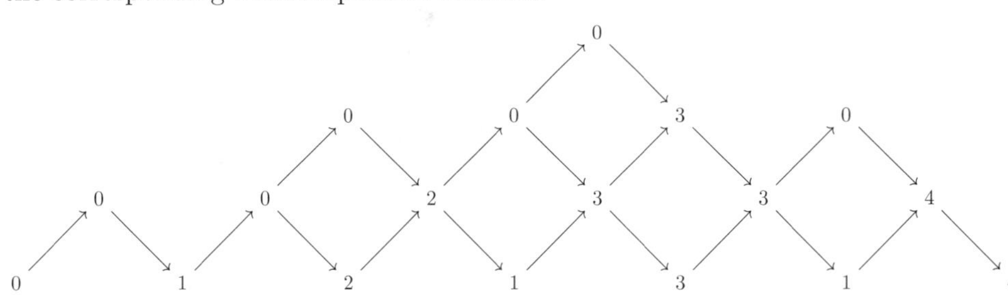

In particular, the global dimension is 4 and is attained by the simple module S_6. The following figure shows the 2-cluster-tilting module of $A_{\underline{\ell}}^{(2)}$ whose indecomposable direct summands are the interval modules. The numbers on the vertices show the projective dimensions divided by $d = 2$ of the corresponding indecomposable modules.

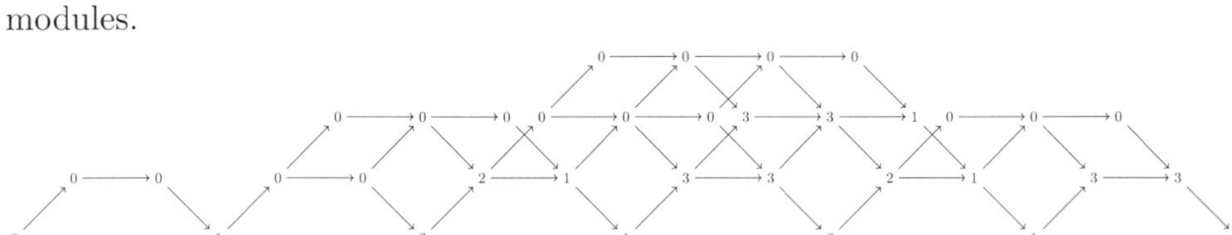

Note that the global dimension of the algebra is 6, but is not attained by a simple module in the cluster tilting subcategory. Also note that

$$\mathsf{gldim}\, A_{\underline{\ell}}^{(2)} = 6 \neq 2 \cdot 4 = 2 \cdot \mathsf{gldim}\, A_{\underline{\ell}}^{(1)}.$$

Finally, note that

$$0 \neq \dim \mathsf{Ext}^2_{A^{(1)}_{\underline{\ell}}}(S_6, S_3) \neq \dim \mathsf{Ext}^4_{A^{(2)}_{\underline{\ell}}}(S_{(6,6)}, S_{(3,3)}) = 0$$

where the rightmost vanishing condition can be shown by noting that the fifth term in a minimal projective resolution of $S_{(6,6)}$ does not have a summand $S_{(3,3)}$ in its top.

It is not even true that finiteness of global dimension is preserved when passing to the higher analogue[2]. To show this, consider the same example as before but with the last and the first vertex identified, i.e. the higher Nakayama algebra of type $\tilde{\mathbb{A}}_5$ with Kupisch series $(2,3,3,4,3,2)$. Then, $\mathsf{projdim}\, S_0 = \infty$ for $\tilde{A}^{(1)}_{\underline{\ell}}$, but $\mathsf{gldim}\, \tilde{A}^{(2)}_{\underline{\ell}} = 6 < \infty$.

4.3. Dominant dimension of the higher Nakayama algebras. Concerning the dominant dimension, the authors proved in [**JK16a**] that $\mathsf{domdim}\, A \geq d$ for every d-Nakayama algebra A. It is however also not true that $\mathsf{domdim}\, \tilde{A}^{(d)}_{\underline{\ell}} = d \cdot \mathsf{domdim}\, \tilde{A}^{(1)}_{\underline{\ell}}$ as the example of the higher Nakayama algebra $\tilde{A}^{(d)}_{(3,4,4)}$ shows. For $d = 1$ its dominant dimension is 4 as $P_1 = I_1$ and $P_2 = I_2$ are projective-injective and the non-injective module P_0 has an injective coresolution

$$0 \to P_0 \to I_1 \to I_2 \to I_2 \to I_0 \to 0.$$

On the other hand, $\tilde{A}^{(2)}_{(3,4,4)}$ has 11 indecomposable projective of which 8 are projective-injective. We label them by their label as interval modules for the category $A^{(d)}_{\infty}$. The projective-injectives are $P_{(1,0)} = I_{(3,1)}$, $P_{(2,1)} = I_{(4,2)}$, $P_{(2,2)} = I_{(4,3)}$, $P_{(1,1)} = I_{(4,1)}$, $P_{(4,2)} = I_{(2,1)}$, $P_{(2,0)} = I_{(3,2)}$, $P_{(4,3)} = I_{(2,2)}$, and $P_{(4,1)} = I_{(1,1)}$. The non-injective modules $P_{(0,0)}$, $P_{(3,1)}$, and $P_{(3,2)}$ have the following injective coresolutions, respectively.

$$0 \to P_{(3,1)} \to I_{(1,1)} \to I_{(4,1)} \to I_{(4,2)} \to I_{(4,3)} \to I_{(2,2)} \to I_{(3,2)} \to I_{(0,0)} \to 0$$
$$0 \to P_{(3,2)} \to I_{(2,1)} \to I_{(4,1)} \to I_{(1,0)} \to I_{(2,0)} \to I_{(2,2)} \to I_{(3,2)} \to I_{(0,0)} \to 0$$
$$0 \to P_{(0,0)} \to I_{(3,1)} \to I_{(4,1)} \to I_{(1,1)} \to I_{(2,1)} \to I_{(2,2)} \to I_{(4,3)} \to I_{(2,0)} \to 0$$

Hence, $\mathsf{domdim}\, \tilde{A}^{(2)}_{(3,4,4)} = 6 \neq 2 \cdot 4 = 2 \cdot \mathsf{domdim}\, \tilde{A}^{(1)}_{(3,4,4)}$.

5. Cluster categories of type $\mathbb{A}_n$ and $\mathbb{A}_\infty$

In this section, we survey on the relationship between cluster tilting subcategories and cluster categories, especially the connection to triangulations of polygons and generalisations thereof. Already in the first paper on cluster algebras [**FZ02**], Fomin and Zelevinsky studied an example of a cluster algebra of type $\mathbb{A}_n$, the coordinate ring of the affine cone over the Grassmannian of 2-dimensional subspaces of an $(n+3)$-dimensional (complex) vector space. The clusters of the cluster algebra are in bijection with triangulations of the polygon with n-vertices. The flip, that is the operation on triangulations replacing one diagonal by the unique other one yielding a triangulation, corresponds to mutating a cluster variable. Cluster algebras were later categorified in [**BMR**$^+$**06**], and the authors of that paper realised that clusters of the cluster algebra of type $\mathbb{A}_n$ correspond to cluster-tilting objects in

[2]The authors want to thank Yiping Chen for posing this question.

the corresponding cluster category of type $\mathbb{A}_n$, which is a certain 2-Calabi–Yau category. As certain tilting objects in $D^b(\operatorname{mod} A_n^{(1)})$ provide cluster-tilting objects in the cluster category of type $\mathbb{A}_n$, the concept of cluster-tilting objects can be seen as an extension of the concept of tilting objects. There have been two generalisations of this picture. Staying in the '2-dimensional' situation, the situation has been generalised to $n = \infty$ by Holm and Jørgensen in [**HJ12**] categorifying triangulations of the apeirogon, in that paper called infinitygon. The corresponding cluster category also arises as the perfect derived category of the cochain differential graded algebra associated to the 2-sphere. The other generalisation was to the higher setting, replacing the algebra $A_n^{(1)}$ by the higher Auslander algebras $A_n^{(d)}$. This generalisation is due to Oppermann and Thomas in [**OT12**] and they prove that a corresponding cluster category categorifies the triangulations of the cyclic polytope in dimension $2d$. In a forthcoming paper [**JK17**], the authors will unify the two generalisations. In this section, we survey on the results of [**HJ12, OT12, JK17**]. In the first subsection, we introduce higher analogues of the cluster categories of type $\mathbb{A}_n$. The second subsection gives the correspondence between tilting modules for $A_n^{(d)}$ and triangulations of a cyclic polytope, following [**OT12**]. The third subsection is concerned with the results of [**OT12**] and [**JK17**] providing a bijection between the cluster-tilting subcategories of a cluster category and triangulations of a cyclic polytope and a corresponding infinite version for the cyclic apeirotope.

5.1. Categorical construction. We start this section by recalling the definitions of a Calabi–Yau triangulated category and of a spherical object.

DEFINITION 5.1. Let $\mathcal{D}$ be a $\Bbbk$-linear Hom-finite triangulated category.

(i) A functor $\mathbb{S}\colon \mathcal{D} \to \mathcal{D}$ is called a *Serre functor* if there is a bifunctorial isomorphism
$$D\operatorname{Hom}_{\mathcal{D}}(X,Y) \cong \operatorname{Hom}_{\mathcal{D}}(Y,\mathbb{S}X).$$
(ii) The category $\mathcal{D}$ is said to be *δ-Calabi–Yau* if $[\delta]$, the δ-th power of the shift functor, is a Serre functor.
(iii) Let $\mathcal{D}$ be a δ-Calabi–Yau triangulated category. An object $X \in \mathcal{D}$ is called *δ-spherical* if
$$\operatorname{Hom}_{\mathcal{D}}(X,X[i]) = \begin{cases} \Bbbk & \text{if } i = 0,\delta, \\ 0 & \text{else.} \end{cases}$$

The thick subcategory generated by a δ-spherical object, i.e. the smallest triangulated category containing the δ-spherical object and being closed under direct summands is independent of the specific δ-spherical object (especially of the ambient triangulated category), see e.g. [**KYZ09**, Theorem 2.1]. This category arises in topology as follows: Let $m \geq 2$ be an integer. Let $X = S^m$ be the m-sphere. Let $C^*(X,\Bbbk)$ be the cochain differential graded algebra of X with coefficients in $\Bbbk$ (with respect to the cup product. Let $D^c(C^*(X,\Bbbk))$ be the perfect derived category of $C^*(X,\Bbbk)$, i.e. the subcategory spanned by all compact objects in the derived category of the differential graded algebra $C^*(X,\Bbbk)$. The following result is partially due to Jørgensen and partially folklore.

THEOREM 5.2 (cf. [**Jør04**]). *The category $D^c(C^*(X,\Bbbk))$ is an m-Calabi–Yau category. In particular, it has almost split triangles. Its Auslander–Reiten quiver* *consists of* $(m-1)$ *components of type* $\mathbb{Z}\mathbb{A}_\infty$. *Furthermore,* $D^c(C^*(X,\Bbbk))$ *is classically generated by an* m*-spherical object* S*, i.e.* $D^c(C^*(X,\Bbbk))$ *itself is the smallest triangulated subcategory of* $D^c(C^*(X,\Bbbk))$ *containing* S.

There are several other incarnations of this category throughout mathematics. Almost by definition, $D^c(C^*(X,\Bbbk)) \cong D^c(\Bbbk[\varepsilon]/(\varepsilon^2))$ where $\Bbbk[\varepsilon]/(\varepsilon^2)$ is regarded as a differential graded algebra with trivial differential and ε in degree m. Using Koszul duality, it is also equivalent to the category $D^f(\Bbbk[T])$, the derived category of complexes with finite dimensional (total) cohomology over the differential graded algebra $\Bbbk[T]$ with trivial differential and T in degree $-m+1$.

In representation theory, more precisely in the theory of cluster categories, the category is better known as the m-cluster category of type $\mathbb{A}_\infty$ and can be defined as $D^b(\operatorname{mod} A_\infty^{(1)})/\mathbb{S}[-m]$ where $\mathbb{S}$ denotes the Serre functor of $D^b(\operatorname{mod} A_\infty^{(1)})$ which exists because $D^b(\operatorname{mod} A_\infty^{(1)})$ has almost-split triangles.

Our goal in [**JK17**] was to obtain similar results starting from $A_\infty^{(d)}$ instead of $A_\infty^{(1)}$. As explained in the introduction to this section, the construction of a higher cluster category of type $\mathbb{A}_n$ was achieved by Oppermann–Thomas in [**OT12**]. Our proof follows a similar strategy but additional technical complications arise. The first step, similar to [**OT12**] is to realise that $D^b(\operatorname{mod} A_\infty^{(d)})/\mathbb{S}[-md]$ is not triangulated. But, thanks to a construction by Keller [**Kel05**], it is possible to embed it into its triangulated hull which we denote by $\mathcal{C}_{\infty,m}^{(d)}$. The technically difficult part is to prove that this category is in fact Hom-finite. This is achieved by establishing a derived equivalence $D^b(\operatorname{mod} A_\infty^{(d)}) \cong D^b(\operatorname{mod} Z_\infty^{(d)})$ where $Z_\infty^{(d)}$ is a locally bounded category. Recall that a subcategory $\mathcal{O} \subseteq \mathcal{D}$ of a triangulated category is called *weakly* $d\mathbb{Z}$*-cluster-tilting* if it satisfies the analogous Ext-vanishing property to the case of an abelian category, but is not necessarily functorially finite (nor generating-cogenerating) and furthermore it is closed under shift by $[d]$.

Then, a straightforward generalisation of the methods in [**Ami09**, **Ami11**, **Guo11**] shows the following theorem:

THEOREM 5.3 ([**OT12**, Theorems 5.14 and 5.25] for finite n and $m = 2$, [**JK17**]). *Let* n *be finite or infinity. Let* $A_n^{(d)}$ *be the algebra (or category) defined in Section 3. Let* $\mathcal{D}$ *be the category* $D^b(\operatorname{mod} A_n^{(d)})$. *Let* $\mathbb{S}$ *be the corresponding Serre functor on* $\mathcal{D}$. *The triangulated hull* $\mathcal{C}_{n,m}^{(d)}$ *of* $\mathcal{D}/\mathbb{S}[-md]$ *is triangulated* Hom*-finite* md*-Calabi–Yau.*

Let $\mathcal{U}$ *be the subcategory* $\mathcal{U}(A_n^{(d)})$ *of* $D^b(\operatorname{mod} A_n^{(d)})$. *The subcategory* $\mathcal{O}_{n,m}^{(d)} := \mathcal{U}/\mathbb{S}[-md]$ *is weakly* $d\mathbb{Z}$*-cluster-tilting in* $\mathcal{C}_{\infty,m}^{(d)}$. *For finite* n *the 'weakly' can be omitted. For infinite* n*, the category* $\mathcal{O}_{\infty,m}^{(d)}$ *is classically generated by an* md*-spherical object* S*, i.e. closing the subcategory* S *under shifts and* $(d+2)$*-angles gives the whole category* $\mathcal{O}_{\infty,m}^{(d)}$.

An open question is still whether it is possible to get rid of the 'weakly' in the theorem in case $n = \infty$. The problem is that it is not clear whether in this case $D^b(\operatorname{mod} A_\infty^{(d)})/\mathbb{S}[-md]$ is functorially finite in its triangulated hull which would be a sufficient condition.

Another open question is whether for $n = \infty$ there is a natural construction associating these categories to certain topological objects which generalises the case of $d = 1$.

5.2. Tilting modules for $A_n^{(d)}$. Recall that there is a bijection between the isomorphism classes of indecomposable basic tilting $A_n^{(1)}$-modules and the triangulations of the regular polygon with $n+2$ vertices; moreover, this bijection is compatible with tilting mutation on one side and flip of triangulations on the other side [**BK04, OT12**].

The higher dimensional analogues of regular polygons are widely considered to be the cyclic polytopes [**DLRS10**, Section 6.1]. In order to define them let $\{\,(t,t^2,\dots,t^\delta)\,|\,t\in\mathbb{R}\,\}$ be the *moment curve.* Choose p points on it. Then, a *cyclic polytope* is the convex hull $C(p,\delta)$ of such points. A *triangulation* of a cyclic polytope is a subdivision of it into δ-simplices whose vertices are also vertices of $C(p,\delta)$. By a result of Dey [**Dey93**] the analogy to triangulations of regular polygons is particularly strong if $\delta=2d$ is even. In this case, a triangulation is determined by the d-dimensional internal simplices, more precisely:

THEOREM 5.4 (cf. [**OT12**, Theorem 2.4]). *Specifying a triangulation of $C(p,2d)$ is equivalent to giving a collection of $\binom{p-d-1}{d}$ non-intersecting d-simplices which do not lie on a lower boundary facet with vertices in the p points.*

For $d=1$, every maximal collection of 1-simplices has $p-2$ elements. For $d>2$ this is in general no longer true, see [**OT12**].

There is an involutive operation on the set of triangulations of $C(p,\delta)$ called "bistellar flip". On the corresponding collection of non-intersecting d-simplices it can be described by removing a d-simplex and replacing it by the unique other d-simplex giving rise to a triangulation. When $\delta=2$, this reduces to the usual flip of triangulations of planar polygons, see *loc. cit.* for further details.

The corresponding notion on the categorical side is tilting mutation: Given two tilting modules T and T' contained in $\mathsf{add}\,M_n^{(d)}$ which only differ in indecomposable direct summands X and X', T is called the mutation of T' in X' and vice versa. In this case, there exists either an exact sequence

$$0\to X\to T_d\to\cdots\to T_1\to X'\to 0$$

with $T_i\in\mathsf{add}\,\overline{T}$ where $\overline{T}$ is a direct complement to X in T (resp. to X' in T') or an exact sequence in the other direction.

THEOREM 5.5 ([**OT12**, Thm. 1.1]). *There is a bijection*

$$\{\textit{internal } d\textit{-simplices of } C(n+2d,2d)\}$$
$$\updownarrow$$
$$\left\{\textit{indecomposable non-projective-injective summands of } M_n^{(d)}\right\}$$

which in turn induces a bijection

$$\{\textit{triangulations of } C(n+2d,2d)\}$$
$$\updownarrow$$
$$\left\{\textit{basic tilting modules for } A_n^{(d)} \textit{ contained in } M_n^{(d)}\right\}.$$

Moreover, this bijection is compatible with tilting mutation on one side and bistellar flip of triangulations on the other side.

The second bijection follows from the fact that under the first bijection two d-simplices intersect if and only if the corresponding indecomposable modules have non-trivial Ext^d in some direction. Note that because of the property of $M_n^{(d)}$ being cluster-tilting and the fact that $\mathsf{gldim}\, A_n^{(d)} = d$, the space Ext^d is the only possible extension not to vanish. This theorem gives a first categorification of the triangulations of a cyclic polytope. In the forthcoming subsection, we give a second one using cluster tilting subcategories of a cluster category instead of tilting modules in a module category.

5.3. Triangulations of cyclic polytopes and apeirotopes. In this subsection, we describe the combinatorial model for the cluster-tilting subcategories of the category $\mathcal{O}_{n,2}^{(d)}$ for finite or infinite n, introduced in Subsection 5.1, in more detail. This has been done for finite n by Oppermann–Thomas [**OT12**] and for $n = \infty$ by the authors in [**JK17**] building on earlier work [**FZ02, BMR$^+$06, HJ12**]. Following [**OT12**], we need to replace functorially finiteness in the definition of cluster-tilting by a stronger condition in the category $\mathcal{O}_{n,2}^{(d)}$:

DEFINITION 5.6. An object $T \in \mathcal{O}_{n,2}^{(d)}$ is called *cluster-tilting* if

(i) $\mathsf{Hom}_{\mathcal{O}_{n,2}^{(d)}}(T, T[d]) = 0$, and

(ii) Any $X \in \mathcal{O}_{n,2}^{(d)}$ occurs in a $(d+2)$-angle

$$X[-d] \to T_d \to T_{d-1} \to \cdots \to T_1 \to T_0 \to X$$

with $T_i \in \mathsf{add}\, T$.

With this definition, for finite n, Oppermann and Thomas were able to prove that cluster-tilting subcategories of $\mathcal{O}_{n,2}^{(d)}$ are precisely the cluster-tilting subcategories of $\mathcal{C}_{n,2}^{(d)}$ contained in $\mathcal{O}_{n,2}^{(d)}$. For infinite n the problem is again that it is not clear whether $\mathcal{O}_{\infty,2}^{(d)}$ is functorially finite in $\mathcal{C}_{\infty,2}^{(d)}$. The only ingridient missing before stating Oppermann–Thomas' categorification of triangulations of the cyclic polytope by cluster-tilting subcategories of the higher cluster category of type $\mathbb{A}$ is the notion of mutation of cluster-tilting objects.

DEFINITION 5.7. Let X and Y be indecomposable objects of $\mathcal{O}_{n,2}^{(d)}$ and $\overline{T}$ be such that $X \oplus \overline{T}$ and $Y \oplus \overline{T}$ are cluster-tilting objects. Then $X \oplus \overline{T}$ is called a mutation of $Y \oplus \overline{T}$ at Y and vice versa.

In this case, there are *exchange $(d+2)$-angles*

$$X \to E_d \to \cdots \to E_1 \to Y$$

and

$$Y \to F_1 \to \cdots \to F_d \to X$$

with $E_i, F_j \in \mathsf{add}\, \overline{T}$.

THEOREM 5.8 ([**OT12**, Thm. 1.2]). *There is a bijection*

$$\{\textit{internal } d\textit{-simplices of } C(n+2d+1, 2d)\}$$
$$\updownarrow$$
$$\left\{\textit{indecomposable objects in } \mathcal{O}_{n,2}^{(d)}\right\}$$

which in turn induces a bijection

$$\{\textit{triangulations of } C(n+2d+1,2d)\}$$
$$\updownarrow$$
$$\left\{\textit{basic } 2d\textit{-cluster-tilting objects in } \mathcal{O}^{(d)}_{n,2}\right\}.$$

The bistellar flip on the left hand side corresponds to the mutation of cluster-tilting objects on the right hand side.

For our generalisation to the infinite situation, we consider the *cyclic apeirotope*, i.e. the convex hull $C(\infty,2d)$ of $\{\,(t,t^2,\dots,t^{2d})\,|\,t\in\mathbb{Z}\,\}$. An *Ind-finite triangulation* is a collection of internal d-simplices such that for all $I\subseteq\mathbb{Z}$ finite, there exists an interval $[a,b]$ containing I such that the arcs with end points in $\{\,(t,t^2,\dots,t^{2d})\,|\,t\in\mathbb{Z}\cap[a,b]\,\}$ form the internal d-simplices of a triangulation.

THEOREM 5.9. *There is a bijection*

$$\{\textit{internal } d\textit{-simplices of } C(\infty,2d)\}$$
$$\updownarrow$$
$$\left\{\textit{indecomposable objects in } \mathcal{O}^{(d)}_{\infty,2}\right\}$$

which in turn induces a bijection

$$\{\textit{Ind-finite triangulations of } C(\infty,2d)\}$$
$$\updownarrow$$
$$\left\{\textit{basic } 2d\textit{-cluster-tilting subcategories in } \mathcal{O}^{(d)}_{\infty,2}\right\}.$$

The bistellar flip on the left hand side again corresponds to the mutation of d-cluster-tilting objects on the right hand side.

References

[Ami09] Claire Amiot, *Cluster categories for algebras of global dimension 2 and quivers with potential* (English, with English and French summaries), Ann. Inst. Fourier (Grenoble) **59** (2009), no. 6, 2525–2590. MR2640929

[Ami11] Claire Amiot, *On generalized cluster categories*, Representations of algebras and related topics, EMS Ser. Congr. Rep., Eur. Math. Soc., Zürich, 2011, pp. 1–53, DOI 10.4171/101-1/1. MR2931894

[ASS06] Ibrahim Assem, Daniel Simson, and Andrzej Skowroński, *Elements of the representation theory of associative algebras. Vol. 1*, London Mathematical Society Student Texts, vol. 65, Cambridge University Press, Cambridge, 2006. Techniques of representation theory. MR2197389

[Aus71] Maurice Auslander, *Representation dimension of Artin algebras*, Lecture Notes, Queen Mary College, London, 1971.

[BK04] Aslak Bakke Buan and Henning Krause, *Tilting and cotilting for quivers and type* $\tilde{A}_n$, J. Pure Appl. Algebra **190** (2004), no. 1-3, 1–21, DOI 10.1016/j.jpaa.2003.11.004. MR2043318

[BMR+06] Aslak Bakke Buan, Robert Marsh, Markus Reineke, Idun Reiten, and Gordana Todorov, *Tilting theory and cluster combinatorics*, Adv. Math. **204** (2006), no. 2, 572–618, DOI 10.1016/j.aim.2005.06.003. MR2249625

[Dey93] Tamal Krishna Dey, *On counting triangulations in d dimensions*, Comput. Geom. **3** (1993), no. 6, 315–325, DOI 10.1016/0925-7721(93)90005-Q. MR1259613

[DLRS10] Jesús A. De Loera, Jörg Rambau, and Francisco Santos, *Triangulations*, Algorithms and Computation in Mathematics, vol. 25, Springer-Verlag, Berlin, 2010. Structures for algorithms and applications. MR2743368

[FZ02] Sergey Fomin and Andrei Zelevinsky, *Cluster algebras. I. Foundations*, J. Amer. Math. Soc. **15** (2002), no. 2, 497–529, DOI 10.1090/S0894-0347-01-00385-X. MR1887642

[Gre83] Edward L. Green, *Graphs with relations, coverings and group-graded algebras*, Trans. Amer. Math. Soc. **279** (1983), no. 1, 297–310, DOI 10.2307/1999386. MR704617

[Guo11] Lingyan Guo, *Cluster tilting objects in generalized higher cluster categories*, J. Pure Appl. Algebra **215** (2011), no. 9, 2055–2071, DOI 10.1016/j.jpaa.2010.11.015. MR2786597

[HI11a] Martin Herschend and Osamu Iyama, *n-representation-finite algebras and twisted fractionally Calabi-Yau algebras*, Bull. Lond. Math. Soc. **43** (2011), no. 3, 449–466, DOI 10.1112/blms/bdq101. MR2820136

[HI11b] Martin Herschend and Osamu Iyama, *Selfinjective quivers with potential and 2-representation-finite algebras*, Compos. Math. **147** (2011), no. 6, 1885–1920, DOI 10.1112/S0010437X11005367. MR2862066

[HJ12] Thorsten Holm and Peter Jørgensen, *On a cluster category of infinite Dynkin type, and the relation to triangulations of the infinity-gon*, Math. Z. **270** (2012), no. 1-2, 277–295, DOI 10.1007/s00209-010-0797-z. MR2875834

[IJ16] Osamu Iyama and Gustavo Jasso, *Higher Auslander correspondence for dualizing R-varieties*, Algebr. Represent. Theory **20** (2017), no. 2, 335–354, DOI 10.1007/s10468-016-9645-0. MR3638352

[IO11] Osamu Iyama and Steffen Oppermann, *n-representation-finite algebras and n-APR tilting*, Trans. Amer. Math. Soc. **363** (2011), no. 12, 6575–6614, DOI 10.1090/S0002-9947-2011-05312-2. MR2833569

[IO13] Osamu Iyama and Steffen Oppermann, *Stable categories of higher preprojective algebras*, Adv. Math. **244** (2013), 23–68, DOI 10.1016/j.aim.2013.03.013. MR3077865

[IS16] Osamu Iyama and Øyvind Solberg, *Auslander-Gorenstein algebras and precluster tilting*, preprint, 2016.

[Iya07a] Osamu Iyama, *Auslander correspondence*, Adv. Math. **210** (2007), no. 1, 51–82, DOI 10.1016/j.aim.2006.06.003. MR2298820

[Iya07b] Osamu Iyama, *Higher-dimensional Auslander-Reiten theory on maximal orthogonal subcategories*, Adv. Math. **210** (2007), no. 1, 22–50, DOI 10.1016/j.aim.2006.06.002. MR2298819

[Iya08] Osamu Iyama, *Auslander-Reiten theory revisited*, Trends in representation theory of algebras and related topics, EMS Ser. Congr. Rep., Eur. Math. Soc., Zürich, 2008, pp. 349–397, DOI 10.4171/062-1/8. MR2484730

[Iya11] Osamu Iyama, *Cluster tilting for higher Auslander algebras*, Adv. Math. **226** (2011), no. 1, 1–61, DOI 10.1016/j.aim.2010.03.004. MR2735750

[Jas16] Gustavo Jasso, *n-abelian and n-exact categories*, Math. Z. **283** (2016), no. 3-4, 703–759, DOI 10.1007/s00209-016-1619-8. MR3519980

[JK16a] Gustavo Jasso and Julian Külshammer, *Higher Nakayama algebras I: Construction*, preprint, arXiv: 1604.03500, 2016.

[JK16b] ———, *The naive approach for constructing the derived category of a d-abelian category fails*, note, arXiv: 1604.03473, 2016.

[JK16c] Gustavo Jasso and Sondre Kvamme, *An introduction to higher Auslander–Reiten theory*, preprint, 2016.

[JK17] Gustavo Jasso and Julian Külshammer, *Spherical objects and cluster-tilting*, in preparation, 2017.

[Jør04] Peter Jørgensen, *Auslander-Reiten theory over topological spaces*, Comment. Math. Helv. **79** (2004), no. 1, 160–182, DOI 10.1007/s00014-001-0795-4. MR2031704

[Kel05] Bernhard Keller, *On triangulated orbit categories*, Doc. Math. **10** (2005), 551–581. MR2184464

[KYZ09] Bernhard Keller, Dong Yang, and Guodong Zhou, *The Hall algebra of a spherical object*, J. Lond. Math. Soc. (2) **80** (2009), no. 3, 771–784, DOI 10.1112/jlms/jdp054. MR2559128

[Mar17] Rene Marczinzik, *Finitistic Auslander algebras*, preprint, 2017.

[MOS09] Volodymyr Mazorchuk, Serge Ovsienko, and Catharina Stroppel, *Quadratic duals, Koszul dual functors, and applications*, Trans. Amer. Math. Soc. **361** (2009), no. 3, 1129–1172, DOI 10.1090/S0002-9947-08-04539-X. MR2457393

[MVdlP83] R. Martínez-Villa and J. A. de la Peña, *The universal cover of a quiver with relations*, J. Pure Appl. Algebra **30** (1983), no. 3, 277–292, DOI 10.1016/0022-4049(83)90062-2. MR724038

[OT12] Steffen Oppermann and Hugh Thomas, *Higher-dimensional cluster combinatorics and representation theory*, J. Eur. Math. Soc. (JEMS) **14** (2012), no. 6, 1679–1737, DOI 10.4171/JEMS/345. MR2984586

[SS07a] Daniel Simson and Andrzej Skowroński, *Elements of the representation theory of associative algebras. Vol. 2*, London Mathematical Society Student Texts, vol. 71, Cambridge University Press, Cambridge, 2007. Tubes and concealed algebras of Euclidean type. MR2360503

[SS07b] Daniel Simson and Andrzej Skowroński, *Elements of the representation theory of associative algebras. Vol. 3*, London Mathematical Society Student Texts, vol. 72, Cambridge University Press, Cambridge, 2007. Representation-infinite tilted algebras. MR2382332

Mathematisches Institut, Universität Bonn, Endenicher Allee 60, D-53115 Bonn, GERMANY

Email address: `gjasso@math.uni-bonn.de`
URL: `https://gustavo.jasso.info`

Institut für Algebra und Zahlentheorie, Universität Stuttgart, Pfaffenwaldring 57, 70569 Stuttgart, GERMANY

Email address: `julian.kuelshammer@mathematik.uni-stuttgart.de`
URL: `http://www.iaz.uni-stuttgart.de/LstAlgZahl/Kuelshammer/`

Contemporary Mathematics
Volume **705**, 2018
http://dx.doi.org/10.1090/conm/705/14196

K-polynomials of type A quiver orbit closures and lacing diagrams

Ryan Kinser

Abstract. This article contains an overview of the author's joint work with Allen Knutson and Jenna Rajchgot on K-polynomials of orbit closures for type A quivers. It is written to an audience interested in interactions between representations of algebras, algebraic geometry, and commutative algebra. A few open problems resulting from the work are also explained.

1. Background and context

We denote a *quiver* by $Q = (Q_0, Q_1, t, h)$, where Q_0 is the vertex set, Q_1 the arrow set, and $t, h\colon Q_1 \to Q_0$ give the tail and head of an arrow $ta \xrightarrow{a} ha$. We fix a field $\Bbbk$, and will proceed with the assumption that $\Bbbk$ is algebraically closed in order to simplify the geometric language, although this is completely inessential to our main result since all schemes involved are defined over $\mathbb{Z}$.

Given a quiver Q and *dimension vector* $\mathbf{d}\colon Q_0 \to \mathbb{Z}_{\geq 0}$, we study the *representation variety*

$$\mathtt{rep}_Q(\mathbf{d}) = \prod_{a \in Q_1} \mathrm{Mat}(\mathbf{d}(ta), \mathbf{d}(ha)), \tag{1.1}$$

where $\mathrm{Mat}(m, n)$ denotes the variety of matrices with m rows, n columns, and entries in $\Bbbk$. We consider the right action of the *base change group*

$$GL(\mathbf{d}) = \prod_{z \in Q_0} GL(\mathbf{d}(z)) \tag{1.2}$$

on $\mathtt{rep}_Q(\mathbf{d})$ given by

$$M \cdot g = (g_{ta}^{-1} M_a g_{ha})_{a \in Q_1}, \tag{1.3}$$

where $g = (g_z)_{z \in Q_0} \in GL(\mathbf{d})$ and $M = (M_a)_{a \in Q_1} \in \mathtt{rep}_Q(\mathbf{d})$. A *representation* of Q is a collection of (finite-dimensional) $\Bbbk$-vector spaces $(V_z)_{z \in Q_0}$ assigned to the vertices of Q, along with a collection of $\Bbbk$-linear maps $(\varphi_a\colon V_{ta} \to V_{ha})_{a \in Q_1}$ assigned to the arrows. Thus, the points of $\mathtt{rep}_Q(\mathbf{d})$ are in bijection with representations of Q along with a fixed basis at each vertex. For algebraic context, we mention

2010 *Mathematics Subject Classification.* 14M12, 05E15, 14C17, 19E08, 16G20.

that there is a natural definition of a *morphism* between two representations which yields a category $\mathtt{rep}(Q)$ of all representations of Q. This category is abelian and in fact equivalent to the category of right modules over the *path algebra* $\Bbbk Q$. Then, at least when Q has no oriented cycles so that $\Bbbk Q$ is finite dimensional, the dimension vector of a representation is equivalent to its class in the Grothendieck group of $\mathtt{rep}(Q)$. We refer the interested reader to standard references such as [**Sch14**, **ASS06**, **ARS97**] for further details.

Then simply from the definitions, orbits in $\mathtt{rep}_Q(\mathbf{d})$ under $GL(\mathbf{d})$ are in bijection with isomorphism classes of representations of dimension vector $\mathbf{d}$; for a representation M of Q, we denote by $\mathcal{O}_M$ the orbit of M in $\mathtt{rep}_Q(\mathbf{d})$, and by $\overline{\mathcal{O}_M}$ the closure of this orbit. Orbit closures in $\mathtt{rep}_Q(\mathbf{d})$ have remarkable connections with the representation theory of Q and related objects. Here we highlight a few connections, and refer the reader to surveys such as [**Bon98**, **Zwa11**, **HZ14**] for detailed treatments of the connections to representation theory.

Commutative algebra. Orbit closures in $\mathtt{rep}_Q(\mathbf{d})$ come with a natural set of polynomials vanishing on them obtained from projective resolutions of the indecomposable representations of Q. These polynomials are minors of certain matrices whose entries are the natural coordinate functions on $\mathtt{rep}_Q(\mathbf{d})$, possibly repeated, and 0s (see [**RZ13**, §4] or [**KR15**, §3]). So ideals generated by these minors can be seen as generalizations of determinantal ideals. From this perspective it is then natural to ask when these ideals are primary, prime, normal, Cohen-Macaulay, etc. There are some surprisingly general results, such as the fact that the ideals obtained in this way are always primary [**Bon96**] when Q is a quiver of Dynkin type A, D, or E (generalized to all representation-finite algebras by Zwara [**Zwa99**]).

Lie theory. Each Dynkin quiver Q determines a finite-dimensional, simple complex Lie algebra $\mathfrak{g}_Q$ (which is actually independent of the orientation of Q) and thus a universal enveloping algebra $U(\mathfrak{g}_Q)$. Ringel's work constructing the upper half $U(\mathfrak{n})$ of this algebra as a Hall algebra [**Rin90**] was geometrically realized by Lusztig [**Lus90**, **Lus91**] as a convolution algebra of constructible functions on $\mathtt{rep}_Q(\mathbf{d})$ which are constant on the orbit closures (there was also an unpublished manuscript by Schofield on the subject around this time). More recently, Geiss, Leclerc, and Schröer have made strides towards generalizing some of the above mentioned work to arbitrary symmetrizable Kac-Moody Lie algebras (not necessarily simply laced) [**GLS16**]. Their work uses convolution algebras of constructible functions which are constant on orbit closures in representation schemes of certain Iwanaga-Gorenstien algebras of dimension 1.

Representations of algebras. We will just mention here two kinds of results relating the representation theory of a quiver to the geometry of orbit closures in its representation varieties, and refer the interested reader to the surveys cited above for more. The *degeneration order* on representations of Q (of the same dimension vector) is defined by $M \leq_{deg} N$ if and only if $\overline{\mathcal{O}_M} \supseteq \overline{\mathcal{O}_N}$. It turns out that this order is closely connected to algebraic properties of M and N. The nicest results are for Q of Dynkin or extended Dynkin type, where for example $M \leq_{deg} N$ if and only if $\dim \operatorname{Hom}_Q(M, X) \leq \dim \operatorname{Hom}_Q(N, X)$ for all indecomposable representations X [**Bon95**]. The latter condition is called the *Hom order* and typically denoted simply by $M \leq N$; this was further proven to be equivalent to a related *Ext order* for extended Dynkin quivers (and more generally, tame concealed algebras) by Zwara [**Zwa98**].

Another remarkable characterization due to Zwara [**Zwa00**], building on work of Riedtmann [**Rie86**], is that $M \leq_{deg} N$ exactly when there exists another representation Z (of unknown dimension) and an exact sequence of the form

$$0 \to N \to M \oplus Z \to Z \to 0,$$

and that in this case the degeneration can be realized by a rational curve $\mathbb{A}^1 \to \overline{\mathcal{O}_M}$.

Algebraic geometry. The work described in this article is most directly inspired by the literature on degeneracy loci. To start with the simplest case, given a nonsingular algebraic variety X and a map between vector bundles $\phi\colon E \to F$ on X, let $\phi_x\colon E_x \to F_x$ denote the induced map on fibers over $x \in X$. Then for $r \in \mathbb{Z}_{\geq 0}$, we may consider the *degenerarcy locus* $\Omega_r = \{x \in X \mid \operatorname{rank} \phi_x \leq r\}$, which is a closed subvariety of X since it is defined by the vanishing of minors in local coordinates around each point. Then it turns out that, when ϕ is sufficiently general, the fundamental class of Ω_r in the Chow or cohomology ring of X has a universal expression as a Schur function evaluated at the Chern roots of E and F (the Giambelli-Thom-Porteous determinantal formula). A brief history tracing this formula back to its geometric and algebraic roots in the 1800s can be found in the Introduction of Fulton's paper [**Ful92**]; this paper established many ideas essential to our work.

The connection with quivers originated with Buch and Fulton [**Ful99, BF99**], generalizing to sequences of vector bundle maps. In quiver language, viewing $\phi\colon E \to F$ in the setup above as a "representation of the quiver A_2 by vector bundles on X", their work replaces A_2 by an arbitrary *equioriented* type A quiver (all arrows pointing in the same direction). It is natural from there to seek generalizations to other quivers, although the strongest results are to be expected for Dynkin quivers. Buch elevated the formulas to the level of K-theory in [**Buc02a**]; see also [**Buc05a, FR02, BFR05, KMS06, Rim, All14, Rim14**] for other important contributions. Much of the state of the art can be found in Buch's article [**Buc08**], which focuses on Dynkin quivers.

The aim of this article is to give a somewhat self-contained overview of the "K-theoretic component formula" proven in the author's joint work with Allen Knutson and Jenna Rajchgot [**KKR**]. Readers interested in the interface between representation theory of algebras and algebraic geometry may find this article to be a more accessible introduction to the result, as it contains a little more commentary and references to the representation theory side than the original paper. It also contains a single running example illustrating most of the key ideas. In particular, we highlight the role of lacing diagrams in the equivariant geometry of orbit closures in $\mathtt{rep}_Q(\mathbf{d})$, since these should be intuitive to anyone familiar with representations of type A quivers. We shall also pose some open questions with the hope of motivating further work on the topic.

2. Lacing diagrams

Assume Q is a quiver of Dynkin type A for this section. Lacing diagrams were introduced by Abeasis and del Fra in [**ADF85**] as a tool to combinatorially characterize the degeneration order on orbits in $\mathtt{rep}_Q(\mathbf{d})$ (which they call the "geometrical ordering"). Knutson, Miller, and Shimozono introduced a refinement of Abeasis and del Fra's diagrams [**KMS06**] for equioriented type A quivers, realizing that allowing the laces to cross enables us to keep track of subtle combinatorial information relevant to the equivariant geometry of orbit closures in $\mathtt{rep}_Q(\mathbf{d})$. Buch and Rimányi utilized lacing diagrams to study equivariant geometry of $\mathtt{rep}_Q(\mathbf{d})$ for arbitrarily oriented Q of type A in [**BR07**].

A *lacing diagram* of dimension vector $\mathbf{d}$ for Q consists of:

(1) for each vertex $z \in Q_0$, a column of $\mathbf{d}(z)$ dots;
(2) for each arrow $a \in Q_1$, a set of arrows from dots in column ta to dots in column ha, such that no dot has more than one incoming or outgoing arrow.

We will develop a running example throughout this article using the following quiver.

(2.1) $Q =$ 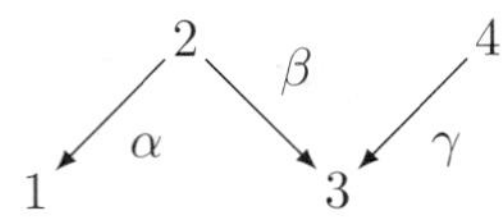

Three lacing diagrams of dimension vector $\mathbf{d} = (\mathbf{d}(1), \mathbf{d}(2), \mathbf{d}(3), \mathbf{d}(4)) = (1, 3, 2, 1)$ for Q are seen in Figure 1.

Equivalently, one may simply define a lacing diagram as a sequence $\mathbf{w} = (w_a)_{a \in Q_1}$ where each w_a is a $\mathbf{d}(ta) \times \mathbf{d}(ha)$ *partial permutation matrix*, meaning its entries are all 0 or 1, with at most one 1 in each row and each column. Note that this definition clearly identifies a lacing diagram $\mathbf{w}$ as a specific point of $\mathtt{rep}_Q(\mathbf{d})$. For details of our conventions, which may differ from other authors for technical reasons, see [**KKR**, §2.8]. The matrix representation of the leftmost lacing diagram in Figure 1 is

$$\mathbf{w} = (w_\alpha, w_\beta, w_\gamma) = \left(\begin{bmatrix} 1 \\ 0 \\ 0 \end{bmatrix}, \begin{bmatrix} 1 & 0 \\ 0 & 1 \\ 0 & 0 \end{bmatrix}, \begin{bmatrix} 0 & 1 \end{bmatrix} \right). \tag{2.2}$$

Note that the diagrammatic presentation of a lacing diagram makes it very easy to recognize the orbit of the point in $\mathtt{rep}$: this is equivalent to knowing its direct sum decomposition into indecomposables, which are just the individual laces. For example, one clearly sees that all three lacing diagrams above lie in the same orbit because they have the same number of laces connecting any two columns. They are also more convenient for performing certain combinatorial manipulations described below. On the other hand, the partial permutation matrix viewpoint provides a more natural interface with combinatorial commutative algebra.

REMARK 2.1. One possible way to generalize lacing diagrams to arbitrary quivers would be to consider *tree modules*, or more precisely, the *coefficient quivers* of tree modules [**Rin98**, §2]. Roughly speaking, a tree module is an indecomposable

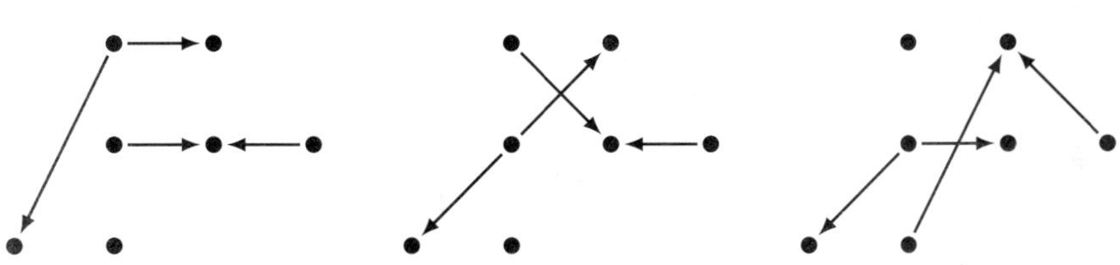

FIGURE 1. Three lacing diagrams for Q.

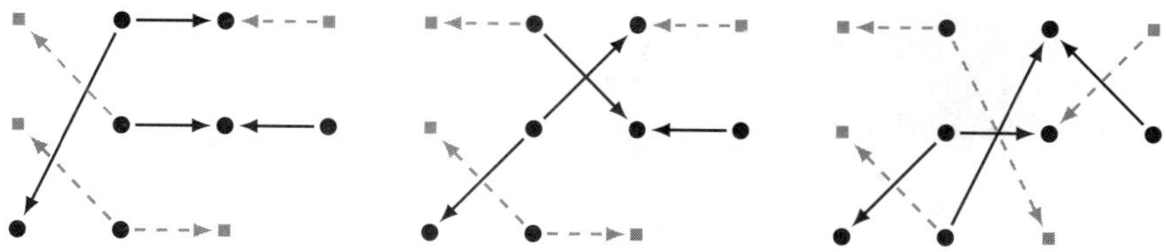

FIGURE 2. Three completed lacing diagrams for Q.

quiver representation whose structure can be completely encoded by another quiver whose underlying graph is a tree. Ringel has shown that any indecomposable quiver representation with no self extensions is a tree module, so in particular any indecomposable representation of a Dynkin quiver is a tree module, and therefore an arbitrary representation of a Dynkin quiver can be presented as a disjoint union of trees. For type A quivers, these specialize exactly to lacing diagrams. Tree modules have been studied extensively by Ringel, Weist and others, see for example [**Kra91**, **Kin10**, **Wei10**, **Wei12**, **Kin13**, **Rin13b**, **Rin13a**].

Although the example shows that an orbit may be represented by many different lacing diagrams, we will see now that there are certain distinguished lacing diagrams. Each lacing diagram has hidden "virtual" laces which must be revealed in order to read off the combinatorial information encoded in the diagram. To do this, we need to fix another convention: let $\{1, 2, \ldots, n\}$ denote the vertex set of Q, such that there is an arrow between i and j (in either direction) if and only $|i - j| = 1$. Arrows of the form $i \xrightarrow{a} i + 1$ will be called rightward arrows, and the others leftward arrows. We complete a lace diagram $\mathbf{w} = (w_a)$ to its *extended lacing diagram* $(c_a(w_a))$ by extending each partial permutation matrix to a full permutation matrix according to the following convention: if a is rightward pointing, then $c_a(w_a)$ is the unique permutation of minimal size and Coxeter length such that w_a lies in the northwest corner of $c_a(w_a)$, and for a leftward pointing, $c_a(w_a)$ is the same but containing w_a in the southeast corner instead. The length $|\mathbf{w}|$ of $\mathbf{w}$ is defined as

$$|\mathbf{w}| = \sum_{a \in Q_1} \ell(c_a(w_a)),$$

where the function ℓ gives the Coxeter length of a permutation with respect to the standard generating set of adjacent transpositions. A *minimal lacing diagram* is one whose length is minimal among all lacing diagrams in $\mathcal{O}_{\mathbf{w}}$. The completion is encoded in a visual diagram in a natural way by adding "virtual" dots and laces (red squares and dashed lines in our examples).

The completions of the lacing diagrams in Figure 1 are shown in Figure 2. The first two are minimal, with 2 crossings each, but the last is not since it lies in the same orbit but the completion has 5 crossings.

The following proposition gives a first taste of the connections between the combinatorics of lacing diagrams, equivariant geometry of $\mathtt{rep}_Q(\mathbf{d})$, and representation theory of Q. It follows from [**BR07**, Cor. 2] and the Artin-Voigt formula (see, for example, [**Rin80**, Lemma 2.3]).

PROPOSITION 2.2. *Let* $\mathbf{w}$ *be a minimal lacing diagram in* $\mathtt{rep}_Q(\mathbf{d})$. *Then* $|\mathbf{w}|$ *is equal to both the codimension of* $\overline{\mathcal{O}_{\mathbf{w}}}$ *in* $\mathtt{rep}_Q(\mathbf{d})$, *and* $\dim_{\Bbbk} \operatorname{Ext}^1_Q(\mathbf{w}, \mathbf{w})$.

All minimal lacing diagrams in a given orbit are related by a sequence of moves of the following form, where both dots in the middle column and at least one dot in each outer column must not be virtual.

(2.3) 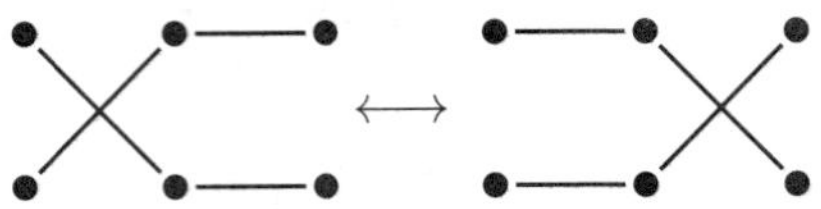

The remaining minimal lacing diagrams for the particular orbit in the running example are in Figure 3, so there are 5 total.

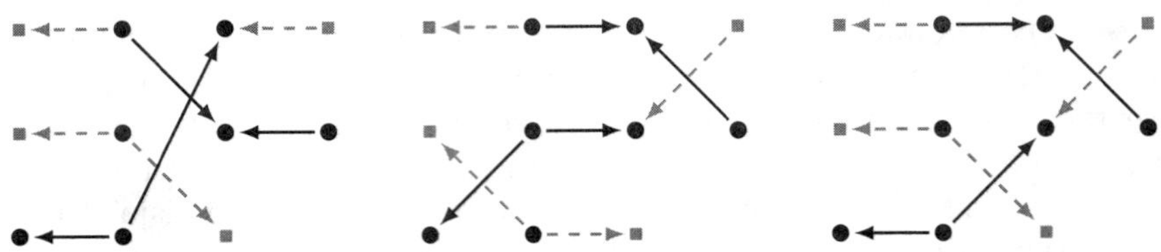

FIGURE 3. The remaining minimal lacing diagrams for $\mathcal{O}_{\mathbf{w}}$

There are also *K-theoretic transformations* of lacing diagrams

(2.4) 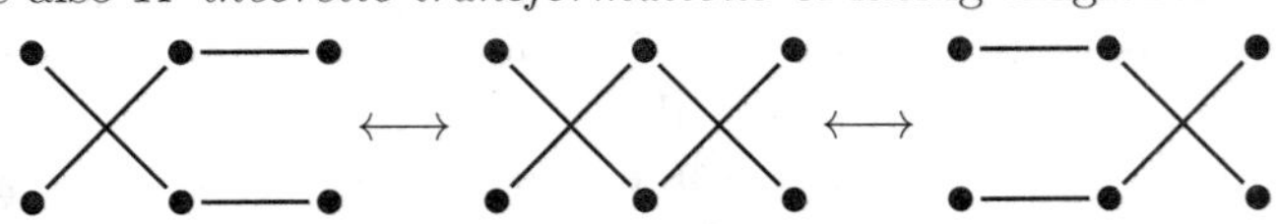

with the same condition on the dots, and in addition the two middle dots should be consecutive in their column. Suppose $\mathbf{w}$ is a minimal lacing diagram. A lacing diagram $\mathbf{w}'$ is said to be a *K-theoretic lacing diagram* for the orbit $\mathcal{O}_{\mathbf{w}}$ if $\mathbf{w}'$ can be obtained from $\mathbf{w}$ by a sequence of K-theoretic transformations of diagrams. The K-theoretic lacing diagrams for the running example are shown in Figures 4 and 5. Note that last lacing diagram in Figure 2 lies in $\mathcal{O}_{\mathbf{w}}$ but is not a K-theoretic lacing diagram for that orbit.

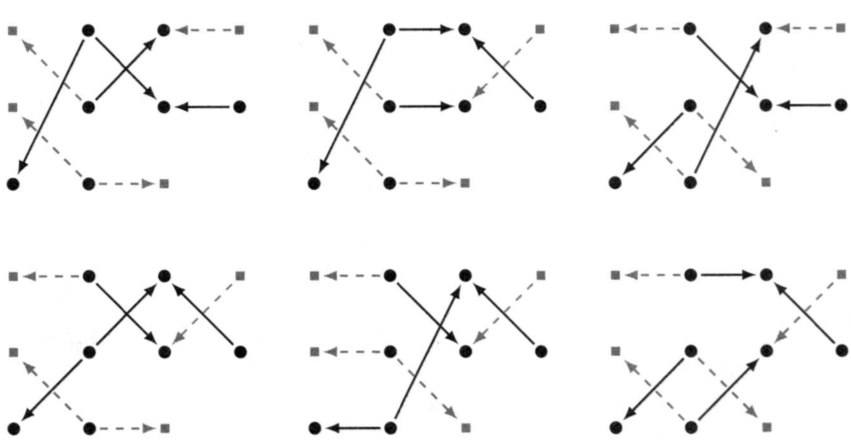

FIGURE 4. 3 crossing K-theoretic diagrams for $\mathcal{O}_{\mathbf{w}}$

We see in the above example that it is not necessarily true that a K-theoretic lacing diagram for $\mathcal{O}_{\mathbf{w}}$ will lie in the same orbit as $\mathbf{w}$. We will see below that non-minimal K-theoretic diagrams for $\overline{\mathcal{O}_{\mathbf{w}}}$ correspond to higher degree terms in

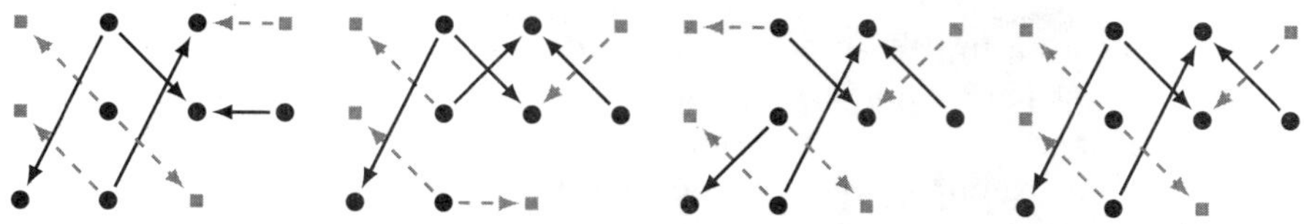

FIGURE 5. 4 or 5 crossing K-theoretic diagrams for $\mathcal{O}_{\mathbf{w}}$

the K-polynomial of $\overline{\mathcal{O}_{\mathbf{w}}}$, arising from an inclusion-exclusion process on K-classes of irreducible components of a partial Gröbner degeneration of $\overline{\mathcal{O}_{\mathbf{w}}}$. A purely representation theoretic meaning of K-theoretic diagrams is not clear.

3. K-polynomials of quiver orbit closures

We start this section by recalling the definition of K-polynomials in the case of closed subvarieties of an affine space. Fix and action of the algebraic torus $T = (\mathbb{k}^\times)^n$ on affine space $\mathbb{A}^N$ and let $X \subseteq \mathbb{A}^N$ be a T-stable closed subvariety. This induces a $\mathbb{Z}^n$-grading on the coordinate ring

$$\mathbb{k}[X] = \bigoplus_{\mathbf{e}\in\mathbb{Z}^n} \mathbb{k}[X]_{\mathbf{e}}, \tag{3.1}$$

where the homogeneous pieces are the weight spaces (see, for example, [**KKR**, §2.6] for a detailed account). We assume for the remainder of the discussion that this grading is *positive*, meaning that the degree $\mathbf{0}$ piece consists only of the coefficient field $\mathbb{k}$ (which will always be the case in our application to quiver representations below). This grading gives rise to a Hilbert series, defined below, which lives in the additive group $\prod_{\mathbf{e}\in\mathbb{Z}^n} \mathbb{Z}\mathbf{a}^{\mathbf{e}}$, where $\mathbf{a} = (a_1, \dots, a_n)$ is an alphabet, $\mathbf{e} = (e_1, \dots, e_n) \in \mathbb{Z}^n$, and $\mathbf{a}^{\mathbf{e}} = a_1^{e_1} \cdots a_n^{e_n}$ is a monomial.

The (multigraded) *Hilbert series* of X is defined as

$$H(X; \mathbf{a}) = \sum_{\mathbf{e}\in\mathbb{Z}^n} (\dim_{\mathbb{k}} \mathbb{k}[X]_{\mathbf{e}})\, \mathbf{a}^{\mathbf{e}}. \tag{3.2}$$

This important invariant has a drawback which we want to circumvent: suppose that $I \subseteq \mathbb{k}[\mathbb{A}^N]$ is the defining ideal of X, and that we extend scalars to $I \otimes_{\mathbb{k}} \mathbb{k}[\mathbb{A}^{N'}] \subseteq \mathbb{k}[\mathbb{A}^{N+N'}]$ to work in a larger ambient space (e.g., take the ideal generated by minors in a matrix of variables, then consider the ideal generated by the same minors in a larger matrix). This will change the Hilbert series. On the other hand, it can be shown that the ratio

$$\mathcal{K}(X; \mathbf{a}) := \frac{H(X; \mathbf{a})}{H(\mathbb{A}^N; \mathbf{a})} \tag{3.3}$$

remains unchanged by extending scalars as above. This expression is actually a Laurent polynomial in the variables $\mathbf{a}$, known as the *K-polynomial of* $\overline{\mathcal{O}_M}$.

For the remainder of this article, the varieties we work with will always come equipped with a natural torus action (which the reader will be reminded of), and thus it is unambiguous to omit the variable set $\mathbf{a}$ from the notation. We will follow this practice in order to significantly simplify the notation.

For an arbitrary quiver Q and dimension vector $\mathbf{d}$, we consider the (maximal) torus $T \subseteq GL(\mathbf{d})$ consisting of all collections of diagonal matrices. Then $\mathtt{rep}_Q(\mathbf{d})$ and all $\overline{\mathcal{O}_M}$ in it inherit an action of T from this inclusion, and thus the coordinate ring $\Bbbk[\overline{\mathcal{O}_M}]$ has a natural $\mathbb{Z}^d$-grading, where $d = \sum_{z \in Q_0} \mathbf{d}(z)$. Following the convention of the last paragraph, we may simply denote the K-polynomial of an orbit closure with respect to this torus action by $\mathcal{K}(\overline{\mathcal{O}_M})$.

In the running example, we make the identification of $\mathtt{rep}_Q(\mathbf{d})$ with the product of matrix spaces whose general element is shown in (3.4). The row and column labels illustrate the alphabets associated to the tori acting by row and column scaling. Thus, the degree of a coordinate function picking out a matrix entry is its row label minus its column label; for example the degree of the coordinate function picking out the entry b_3 is $u_2 - s_1$.

$$\mathtt{rep}_Q(\mathbf{d}) = \left\{ \left(\begin{array}{cc} & v_1 \\ u_1 & c_1 \\ u_2 & c_2 \\ u_3 & c_3 \end{array}, \begin{array}{ccc} & s_1 & s_2 \\ u_1 & b_1 & b_2 \\ u_2 & b_3 & b_4 \\ u_3 & b_5 & b_6 \end{array}, \begin{array}{ccc} & s_1 & s_2 \\ t_1 & a_1 & a_2 \end{array} \right) \right\}. \tag{3.4}$$

The orbit closure of the lacing diagrams of our running example (Figure 1) is defined by the rank condition

$$\operatorname{rank} \begin{bmatrix} 0 & a_1 & a_2 \\ c_1 & b_1 & b_2 \\ c_2 & b_3 & b_4 \\ c_3 & b_5 & b_6 \end{bmatrix} \leq 2, \tag{3.5}$$

so that the ideal of this orbit closure is generated by the 3×3-minors of the matrix in (3.5).

REMARK 3.1. The invariant $\mathcal{K}(\overline{\mathcal{O}_M})$ has several equivalent formulations, whose relations are carefully explained in [**Buc08**, §§3,4]. For example, it represents an element of the ring of virtual rational representations of $GL(\mathbf{d})$. This ring in turn can be identified with the Grothendieck group of the category of $GL(\mathbf{d})$-equivariant coherent sheaves on $\mathtt{rep}_Q(\mathbf{d})$, known as the $GL(\mathbf{d})$-equivariant K-homology of $\mathtt{rep}_Q(\mathbf{d})$, which is isomorphic to the $GL(\mathbf{d})$-equivariant K-cohomology ring of $\mathtt{rep}_Q(\mathbf{d})$ since this variety is nonsingular.

The building blocks of our formulas for K-polynomials of quiver orbit closures are the double Grothendieck polynomials $\mathfrak{G}_w(\mathbf{a}; \mathbf{b})$ of Lascoux and Schützenberger. Instead of the original recursive definition (see for example [**LS82**, **FL94**]), we will introduce them below as K-polynomials of certain closed subvarieties of matrix spaces, since this is how they naturally arise in our work and significantly simplifies the overview. The connection between this formulation and the original definition is explained in [**KKR**, §2.7], following from results in [**Buc02a**] or [**KM05**].

Given a space of matrices $X = \operatorname{Mat}(m, n)$, consider the natural multiplication action of $B_- \times B_+$ on X where $B_- \subseteq GL(m)$ denotes the group of invertible lower triangular matrices, and $B_+ \subseteq GL(n)$ the group of invertible upper triangular matrices. Each $B_- \times B_+$-orbit on X has a unique partial permutation matrix; denote by X_w the closure of the orbit containing the partial permutation matrix w, which is called a *matrix Schubert variety*. The equations defining X_w as a closed subscheme are collections of minors corresponding to imposing upper bounds on the ranks of all northwest justified submatrices in the space X. A more detailed introduction to matrix Schubert varieties and their properties can be found in [**MS05**, Ch. 15].

Each matrix Schubert variety $X_w \subseteq X$ carries an action of the subgroup of diagonal matrices

$$(\Bbbk^\times)^m \times (\Bbbk^\times)^n \subseteq B_- \times B_+ \tag{3.6}$$

where the factor $(\Bbbk^\times)^m$ acts by scaling rows, and the factor $(\Bbbk^\times)^n$ acts by scaling columns. The coordinate ring $\Bbbk[X_w]$ thus inherits a $\mathbb{Z}^{m+n}$ grading, and we identify this grading group with the free abelian group on the alphabet $(\mathbf{a}, \mathbf{b}) := (a_1, \ldots, a_m, b_1, \ldots, b_n)$. Then we will take

$$\mathfrak{G}_w(\mathbf{a}; \mathbf{b}) = \mathcal{K}(X_w; \mathbf{a}, \mathbf{b}) \tag{3.7}$$

to be the *double Grothendieck polynomial* indexed by w. As mentioned above, we will omit the variables and simply denote it by $\mathfrak{G}_w$ below. In addition to the original definition mentioned above, there are many other combinatorial formulas for Grothendieck polynomials [**FK94**, **Len99**, **Len00**, **BR04**, **Buc02b**].

To connect with our running example, consider the leftmost lacing diagram in Figure 3; written in matrix form we find the middle matrix to be

$$w_b = \begin{matrix} & s_1 & s_2 \\ u_1 & \\ u_2 & \\ u_3 & \end{matrix}\!\!\!\!\!\!\!\!\begin{matrix} \\ \begin{bmatrix} 0 & 1 \\ 0 & 0 \\ 1 & 0 \end{bmatrix} \end{matrix}. \tag{3.8}$$

The corresponding Grothendieck polynomial can be calculated (for example, using Lascoux-Schützenberger's recursive definition or one of the other combinatorial formulas cited above) to be

$$\mathfrak{G}_{w_b} = \left(1 - \frac{u_1}{s_1}\right)\left(1 - \frac{u_2}{s_1}\right). \tag{3.9}$$

Restricting our attention to type A quivers now, we need to introduce some "opposite" versions of the above concepts to deal with arbitrary orientation. For each partial permutation matrix w, the *opposite matrix Schubert variety* $X^w \subseteq X$ is the closure of the $B_+ \times B_-$-orbit containing w, where $B_+ \subseteq GL(m)$ denotes the group of invertible upper triangular matrices, and $B_- \subseteq GL(n)$ the group of invertible lower triangular matrices. Likewise, we get an *opposite Grothendieck polynomial*

$$\mathfrak{G}^w(\mathbf{a}; \mathbf{b}) = \mathcal{K}(X^w; \mathbf{a}, \mathbf{b}), \tag{3.10}$$

denoted $\mathfrak{G}^w$ for short. If w is a permutation matrix, then $\mathfrak{G}^w$ is just the standard double Grothendieck polynomial for the $180°$ rotation of w, with the orders of the individual input alphabets $\mathbf{a}, \mathbf{b}$ reversed.

Then a lacing diagram $\mathbf{w} = (w_a)_{a \in Q_1}$ determines a product of matrix Schubert varieties and opposite matrix Schubert varieties

$$X_{\mathbf{w}} = \prod_{\xrightarrow{a} \in Q_1} X_{w_a} \times \prod_{\xleftarrow{a} \in Q_1} X^{w_a} \subseteq \mathtt{rep}_Q(\mathbf{d}) \tag{3.11}$$

where the first product is over rightward arrows of Q and the second product over leftward arrows of Q. This is a T-stable subvariety and the K-polynomial of $X_{\mathbf{w}}$ is simply the following product of Grothendieck polynomials (and their opposites):

$$\mathfrak{G}_{\mathbf{w}} = \left(\prod_{\xrightarrow{a} \in Q_1} \mathfrak{G}_{w_a} \right) \left(\prod_{\xleftarrow{a} \in Q_1} \mathfrak{G}^{w_a} \right). \tag{3.12}$$

These are the building blocks of our main result below. For example, consider the unique 5-crossing K-theoretic lace diagram for the running example, which is the rightmost entry of Figure 5. Its matrix representation is

$$\mathbf{w} = (w_a, w_b, w_c) = \left(\begin{array}{c} \\ u_1 \\ u_2 \\ u_3 \end{array} \begin{array}{c} v_1 \\ \begin{bmatrix} 1 \\ 0 \\ 0 \end{bmatrix} \end{array}, \begin{array}{c} \\ u_1 \\ u_2 \\ u_3 \end{array} \begin{array}{c} s_1 \quad s_2 \\ \begin{bmatrix} 0 & 1 \\ 0 & 0 \\ 1 & 0 \end{bmatrix} \end{array}, \begin{array}{c} \\ t_1 \end{array} \begin{array}{c} s_1 \quad s_2 \\ \begin{bmatrix} 1 & 0 \end{bmatrix} \end{array} \right) \tag{3.13}$$

and the corresponding product of Grothendieck polynomials is

$$\mathfrak{G}_{\mathbf{w}} = \mathfrak{G}^{w_c} \mathfrak{G}_{w_b} \mathfrak{G}^{w_a} = \left(1 - \frac{u_2}{v_1}\right)\left(1 - \frac{u_3}{v_1}\right)\left(1 - \frac{u_1}{s_1}\right)\left(1 - \frac{u_2}{s_1}\right)\left(1 - \frac{t_1}{s_2}\right). \tag{3.14}$$

4. The component formula

At this point, we have enough background in place to state our capstone formula from [**KKR**] for K-polynomials of type A orbit closures. This formula is called the "component formula" since it generalizes a formula of the same name from [**KMS06**], where it was proven in the case of equioriented type A quivers. We remark that the name comes from the geometry technique of the proof outlined below; the name "lace formula" would suit it equally well. A detailed account of the motivations for this formula and its relation to existing literature can be found in Sections 1.1 and 1.3 of [**KKR**], respectively. Let us at least mention here though that it was conjectured by Buch and Rimányi in [**BR07**], where they proved the cohomological version using the *interpolation method* of Fehér and Rimányi [**FR04, BFR05**].

THEOREM 4.1 (Theorems 4.37 and 5.20 of [**KKR**]). *Let Q be an arbitrary quiver of Dynkin type A, let $\mathbf{d}$ be a dimension vector for Q, and $\overline{\mathcal{O}_M} \subseteq \mathtt{rep}_Q(\mathbf{d})$ an orbit closure. Then the K-polynomial of $\overline{\mathcal{O}_M}$ is given by the formula*

$$\mathcal{K}(\overline{\mathcal{O}_M}) = \sum_{\mathbf{w}} (-1)^{|\mathbf{w}| - \operatorname{codim} \overline{\mathcal{O}_M}} \mathfrak{G}_{\mathbf{w}} \tag{4.1}$$

where the sum is over K-theoretic lacing diagrams for $\overline{\mathcal{O}_M}$.

So in our running example we would sum 15 polynomials of the form (3.14) in 12 variables, indexed by the 15 K-theoretic lacing diagrams shown in the figures of Section 2.

We will now give a bird's eye view of the proof of Theorem 4.1. The first main idea is to reduce the problem to type A quivers of a specific orientation, namely the bipartite (i.e., sink-source) orientation. Given Q of type A, one simply inserts a "backwards" arrow in the middle of each length two path to get an associated bipartite type A quiver $\widetilde{Q}$. Then there is a dimension vector $\widetilde{\mathbf{d}}$ for $\widetilde{Q}$ such that the equivariant geometry of $\mathtt{rep}_Q(\mathbf{d})$ can be relatively easily reduced to that of $\mathtt{rep}_{\widetilde{Q}}(\widetilde{\mathbf{d}})$.

It was pointed out to us by Jorge Vitória that the path algebras of Q and $\widetilde{Q}$ are related by a universal localization in the sense of [**Sch85**, §4]. In contrast, several prominent results preceding ours on the geometry of orbit closures for arbitrary type A quivers [**BZ01, RZ13**] were reduced to the equioriented case using Zwara's work on Hom-controlled functors [**Zwa02**] and sophisticated representation theoretical arguments. The results in that case had been proven in work of Lakshmibai and Magyar [**LM98**] using the connections with Schubert varieties mentioned below.

Reducing to the case of bipartite Q allows us to make liberal use of the bipartite Zelevinsky map from the author's previous work with Rajchgot [**KR15**] throughout all parts of the proof. This map embeds $\mathtt{rep}_Q(\mathbf{d})$ in a partial flag variety $GL(d)/P$, such that orbit closures in $\mathtt{rep}_Q(\mathbf{d})$ are identified with open subschemes of Schubert varieties. This map generalizes results of Zelevinsky [**Zel85**] and Lakshmibai-Magyar [**LM98**] on the case of equioriented type A quivers, and in particular yields defining equations of the prime ideals of the orbit closures in that case. While passing back and forth between $\mathtt{rep}_Q(\mathbf{d})$ and $GL(d)/P$ via this map is ubiquitous throughout the proof, it is primarily for technical purposes, and thus will not be discussed further.

The second main idea can be explained in a purely geometric way, though a significant amount of combinatorics is necessary for the proof. We construct a "simultaneous flat degeneration" of the action of $GL(\mathbf{d})$ on $\overline{\mathcal{O}_M}$ to an action of $B_+ \times_T B_-$ on a scheme $\overline{\mathcal{O}_M}(0)$. Here, $B_+ \times_T B_-$ denotes a certain subgroup of $GL(\mathbf{d}) \times GL(\mathbf{d})$ consisting of various upper and lower triangular matrices, determined by the orientation of Q, and the degeneration $\overline{\mathcal{O}_M}(0)$ is a union of a certain (potentially nonreduced) subschemes whose underlying varieties are various $X_{\mathbf{w}} \subseteq \mathtt{rep}_Q(\mathbf{d})$, as defined in (3.11) above. The key point of this stage of our proof is that

$$\overline{\mathcal{O}_M}(0) = \bigcup_{\mathbf{w} \text{ minimal}} X_{\mathbf{w}} \tag{4.2}$$

(i.e., the degeneration is reduced and its irreducible components are precisely the $B_+ \times_T B_-$-orbit closures of the minimal lace diagrams for M). Because of flatness, the K-polynomial of $\overline{\mathcal{O}_M}$ is equal to the K-polynomial of $\overline{\mathcal{O}_M}(0)$.

We remark that this is already enough to prove the cohomological component formula of [**BR07**], since this invariant only requires knowledge of the irreducible components of a scheme. On the other hand, the K-polynomial contains deeper information depending on how the irreducible components intersect. A simple but instructive comparison can be found in [**Buc05b**, Example 1]. Thus, roughly speaking, it remains to understand the configuration of the irreducible components of the degeneration with respect to one another.

The third and final main idea is a computation of the Möbius functions of certain posets in terms of lacing diagram combinatorics. To be more precise, a theorem of Knutson [**Knu**] allows us to compute the K-polynomial of $\overline{\mathcal{O}_M}(0)$ by computing the values of the Möbius function of the poset $\mathcal{M}$ of $B_+ \times_T B_-$-orbit closures contained in $\overline{\mathcal{O}_M}(0)$, with respect to containment order. The goal is to show that the Möbius function is nonzero on an orbit closure $X_{\mathbf{w}}$ precisely when $\mathbf{w}$ is a K-theoretic lacing diagram for $\mathcal{O}_M$, and in this case the value is ± 1, alternating with the number of crossings of $\mathbf{w}$. In Figure 6 we see the Hasse diagram of the poset of K-theoretic lacing diagrams for $\mathcal{O}_M$ in our running example, ordered by containment of their corresponding $B_+ \times_T B_-$-orbit closures, with each element

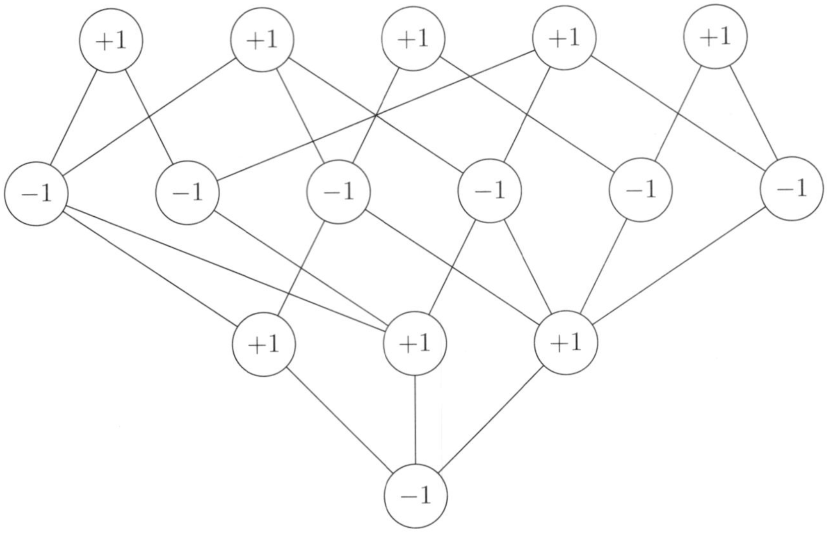

FIGURE 6. The Hasse diagram of the poset of K-theoretic diagrams for $\mathcal{O}_{\mathbf{w}}$ in the running example, with each element labeled by the value the Möbius function takes there

labeled by the value of the Möbius function. Starting at the top left and reading each row to the right, then proceeding down to the next row, the ordering is the same as the order in which they are displayed in the figures of Section 2.

This Möbius function is difficult to understand directly, so we embed the incidence algebra of $\mathcal{M}$ in the incidence algebra of the *pipe complex* Δ studied by Woo and Yong [**WY12**] (which is an example of the subword complexes studied by Knutson and Miller [**KM04**]). It is easier to compute the Möbius function of Δ because this poset is a simplicial complex which is homeomorphic to a ball. Then the primary challenge of this step is to relate the combinatorics of the pipe complex to those of the K-theoretic lacing diagrams in order to obtain the desired results on the Möbius function of $\mathcal{M}$, thus completing the proof.

5. Open problems

Finally, we collect some ideas for future directions to be pursued. The reader is also referred to Zwara's survey [**Zwa11**] for a wealth of interesting problems and questions about the geometry of orbit closures in $\mathtt{rep}_Q(\mathbf{d})$ in more generality.

Formulas for K-polynomials. For an arbitrary quiver Q without oriented cycles, Buch has given a general shape of formulas for K-polynomials of $GL(\mathbf{d})$-invariant closed subvarieties $\Omega \subseteq \mathtt{rep}_Q(\mathbf{d})$ (e.g., orbit closures) in [**Buc08**]. He showed that each such K-polynomial can be written as a certain sum of products of *stable double Grothendieck polynomials* indexed by partitions. His expression has a uniqueness property that leads to well-defined *quiver coefficients* $c_\mu(\Omega) \in \mathbb{Z}$ indexed by sequences of partitions $\mu = (\mu_i)_{i \in Q_0}$. His conjecture on the properties of these numbers is only fully proven for equioriented type Q quivers, and quivers of type A_3. Given the formal similarity of his formula to our component formula, it would be interesting to better understand their relation.

PROBLEM 5.1. Use the component formula (4.1) to prove Buch's Conjecture 1.1 of [**Buc08**] for all type A quivers.

Embeddings of representation varieties. Bobiński and Zwara have shown in [**BZ02**] that the smooth equivalence classes of singularities which appear in type A quiver orbit closures (varying over all type A quivers Q and $\mathbf{d}$) are precisely those which appear in orbit closures in flag varieties $GL(n)/B$ (Schubert varieties). It has been reported by András Lőrincz that his work on b-functions of quiver semi-invariants [**Lőra, Lőrb**] implies that there exist singularities in type D quiver orbit closures which are not smoothly equivalent to any singularity of any type A quiver orbit closure. Thus, we should look beyond the realm of flag varieties $GL(n)/B$ if we want to generalize the type A Zelevinsky map and relate more general Dynkin quiver orbit closures to the wider algebraic geometry literature. To retain as many of the desirable properties of the type A Zelevinsky map as possible, *spherical varieties* make reasonable candidates for target varieties; that is, varieties X such that:

(1) X is normal (or even smooth);
(2) X comes equipped with the action of a connected, reductive algebraic group G;
(3) X has finitely many B-orbits, where $B \subset G$ is a Borel subgroup.

These are well studied so that we may hope to transport their properties to $\mathtt{rep}_Q(\mathbf{d})$ (see the survey [**Per14**]).

PROBLEM 5.2. Let Q be a quiver of Dynkin type D or E. Find a collection of varieties $X(\mathbf{d})$ along with maps $\zeta_{\mathbf{d}}\colon \mathtt{rep}_Q(\mathbf{d}) \to X(\mathbf{d})$ such that:

(1) each $X(\mathbf{d})$ is spherical with respect to action of a reductive group $G(\mathbf{d})$, say with Borel subgroup $B(\mathbf{d})$;
(2) each $\zeta_{\mathbf{d}}$ is equivariant with respect to an embedding of algebraic groups $GL(\mathbf{d}) \to G(\mathbf{d})$;
(3) the restriction of $\zeta_{\mathbf{d}}$ to an orbit closure gives an isomorphism between $\overline{\mathcal{O}_M}$ and the intersection of a $B(\mathbf{d})$-orbit closure with the image of $\zeta_{\mathbf{d}}$, and these intersections differ from $B(\mathbf{d})$-orbit closures in $X(\mathbf{d})$ by a smooth factor.

Joint work in progress with Rajchgot, inspired by results of Bobiński-Zwara and Brion [**BZ02, Bri03**], proposes a solution to this problem for Q bipartite of type D and each $X(\mathbf{d})$ a *double Grassmannian* $Gr_a(V) \times Gr_b(V)$ with $G(\mathbf{d}) = GL(V)$ acting diagonally (a, b, V all depend on $\mathbf{d}$).

Ideals of orbits closures. The proofs in our paper [**KKR**] produce a *Gröbner basis* for the prime ideal of $\overline{\mathcal{O}_M}$. Gröbner bases could be useful for attacking certain open problems, such as showing that a proposed generating set for the ideal of $\overline{\mathcal{O}_M}$ does indeed define a prime ideal, or for combinatorial cohomology formulas for $\overline{\mathcal{O}_M}$.

PROBLEM 5.3. Let Q be a Dynkin quiver and $\overline{\mathcal{O}_M}$ an orbit closure in some $\mathtt{rep}_Q(\mathbf{d})$. Determine a Gröbner basis for the prime ideal defining $\overline{\mathcal{O}_M}$.

References

[ADF85] S. Abeasis and A. Del Fra, *Degenerations for the representations of a quiver of type $\mathcal{A}_m$*, J. Algebra **93** (1985), no. 2, 376–412, DOI 10.1016/0021-8693(85)90166-8. MR786760

[All14] Justin Allman, *Grothendieck classes of quiver cycles as iterated residues*, Michigan Math. J. **63** (2014), no. 4, 865–888, DOI 10.1307/mmj/1417799229. MR3286674

[ARS97] Maurice Auslander, Idun Reiten, and Sverre O. Smalø, *Representation theory of Artin algebras*, Cambridge Studies in Advanced Mathematics, vol. 36, Cambridge University Press, Cambridge, 1997. Corrected reprint of the 1995 original. MR1476671

[ASS06] Ibrahim Assem, Daniel Simson, and Andrzej Skowroński, *Elements of the representation theory of associative algebras. Vol. 1*, London Mathematical Society Student Texts, vol. 65, Cambridge University Press, Cambridge, 2006. Techniques of representation theory. MR2197389

[BF99] Anders Skovsted Buch and William Fulton, *Chern class formulas for quiver varieties*, Invent. Math. **135** (1999), no. 3, 665–687, DOI 10.1007/s002220050297. MR1669280

[BFR05] Anders S. Buch, László M. Fehér, and Richárd Rimányi, *Positivity of quiver coefficients through Thom polynomials*, Adv. Math. **197** (2005), no. 1, 306–320, DOI 10.1016/j.aim.2004.10.019. MR2166185

[Bon95] Klaus Bongartz, *Degenerations for representations of tame quivers*, Ann. Sci. École Norm. Sup. (4) **28** (1995), no. 5, 647–668. MR1341664

[Bon96] Klaus Bongartz, *On degenerations and extensions of finite-dimensional modules*, Adv. Math. **121** (1996), no. 2, 245–287, DOI 10.1006/aima.1996.0053. MR1402728

[Bon98] Klaus Bongartz, *Some geometric aspects of representation theory*, Algebras and modules, I (Trondheim, 1996), CMS Conf. Proc., vol. 23, Amer. Math. Soc., Providence, RI, 1998, pp. 1–27. MR1648601

[BR04] Anders S. Buch and Richárd Rimányi, *Specializations of Grothendieck polynomials* (English, with English and French summaries), C. R. Math. Acad. Sci. Paris **339** (2004), no. 1, 1–4, DOI 10.1016/j.crma.2004.04.015. MR2075223

[BR07] Anders Skovsted Buch and Richárd Rimányi, *A formula for non-equioriented quiver orbits of type A*, J. Algebraic Geom. **16** (2007), no. 3, 531–546, DOI 10.1090/S1056-3911-07-00441-9. MR2306279

[Bri03] Michel Brion, *Multiplicity-free subvarieties of flag varieties*, Commutative algebra (Grenoble/Lyon, 2001), Contemp. Math., vol. 331, Amer. Math. Soc., Providence, RI, 2003, pp. 13–23, DOI 10.1090/conm/331/05900. MR2011763

[Buc02a] Anders Skovsted Buch, *Grothendieck classes of quiver varieties*, Duke Math. J. **115** (2002), no. 1, 75–103, DOI 10.1215/S0012-7094-02-11513-0. MR1932326

[Buc02b] Anders Skovsted Buch, *A Littlewood-Richardson rule for the K-theory of Grassmannians*, Acta Math. **189** (2002), no. 1, 37–78, DOI 10.1007/BF02392644. MR1946917

[Buc05a] Anders Skovsted Buch, *Alternating signs of quiver coefficients*, J. Amer. Math. Soc. **18** (2005), no. 1, 217–237, DOI 10.1090/S0894-0347-04-00473-4. MR2114821

[Buc05b] Anders Skovsted Buch, *Combinatorial K-theory*, Topics in cohomological studies of algebraic varieties, Trends Math., Birkhäuser, Basel, 2005, pp. 87–103, DOI 10.1007/3-7643-7342-3_3. MR2143073

[Buc08] Anders Skovsted Buch, *Quiver coefficients of Dynkin type*, Michigan Math. J. **57** (2008), 93–120, DOI 10.1307/mmj/1220879399. Special volume in honor of Melvin Hochster. MR2492443

[BZ01] Grzegorz Bobiński and Grzegorz Zwara, *Normality of orbit closures for Dynkin quivers of type* $\mathbb{A}_n$, Manuscripta Math. **105** (2001), no. 1, 103–109, DOI 10.1007/PL00005871. MR1885816

[BZ02] Grzegorz Bobiński and Grzegorz Zwara, *Schubert varieties and representations of Dynkin quivers*, Colloq. Math. **94** (2002), no. 2, 285–309, DOI 10.4064/cm94-2-10. MR1967381

[FK94] Sergey Fomin and Anatol N. Kirillov, *Grothendieck polynomials and the Yang-Baxter equation*, Formal power series and algebraic combinatorics/Séries formelles et combinatoire algébrique, DIMACS, Piscataway, NJ, sd, pp. 183–189. MR2307216

[FL94] William Fulton and Alain Lascoux, *A Pieri formula in the Grothendieck ring of a flag bundle*, Duke Math. J. **76** (1994), no. 3, 711–729, DOI 10.1215/S0012-7094-94-07627-8. MR1309327

[FR02] László Fehér and Richárd Rimányi, *Classes of degeneracy loci for quivers: the Thom polynomial point of view*, Duke Math. J. **114** (2002), no. 2, 193–213, DOI 10.1215/S0012-7094-02-11421-5. MR1920187

[FR04] László M. Fehér and Richárd Rimányi, *Calculation of Thom polynomials and other cohomological obstructions for group actions*, Real and complex singularities, Contemp. Math., vol. 354, Amer. Math. Soc., Providence, RI, 2004, pp. 69–93, DOI 10.1090/conm/354/06475. MR2087805

[Ful92] William Fulton, *Flags, Schubert polynomials, degeneracy loci, and determinantal formulas*, Duke Math. J. **65** (1992), no. 3, 381–420, DOI 10.1215/S0012-7094-92-06516-1. MR1154177

[Ful99] William Fulton, *Universal Schubert polynomials*, Duke Math. J. **96** (1999), no. 3, 575–594, DOI 10.1215/S0012-7094-99-09618-7. MR1671215

[GLS16] Christof Geiß, Bernard Leclerc, and Jan Schröer, *Quivers with relations for symmetrizable Cartan matrices III: Convolution algebras*, Represent. Theory **20** (2016), 375–413, DOI 10.1090/ert/487. MR3555157

[HZ14] B. Huisgen-Zimmermann, *Fine and coarse moduli spaces in the representation theory of finite dimensional algebras*, Expository lectures on representation theory, Contemp. Math., vol. 607, Amer. Math. Soc., Providence, RI, 2014, pp. 1–34, DOI 10.1090/conm/607/12086. MR3204864

[Kin10] Ryan Kinser, *Rank functions on rooted tree quivers*, Duke Math. J. **152** (2010), no. 1, 27–92, DOI 10.1215/00127094-2010-006. MR2643056

[Kin13] Ryan Kinser, *Tree modules and counting polynomials*, Algebr. Represent. Theory **16** (2013), no. 5, 1333–1347, DOI 10.1007/s10468-012-9359-x. MR3102957

[KKR] Ryan Kinser, Allen Knutson, and Jenna Rajchgot, *Three combinatorial formulas for type A quiver polynomials and K-polynomials*, `arxiv:1503.05880`.

[KM04] Allen Knutson and Ezra Miller, *Subword complexes in Coxeter groups*, Adv. Math. **184** (2004), no. 1, 161–176, DOI 10.1016/S0001-8708(03)00142-7. MR2047852

[KM05] Allen Knutson and Ezra Miller, *Gröbner geometry of Schubert polynomials*, Ann. of Math. (2) **161** (2005), no. 3, 1245–1318, DOI 10.4007/annals.2005.161.1245. MR2180402

[KMS06] Allen Knutson, Ezra Miller, and Mark Shimozono, *Four positive formulae for type A quiver polynomials*, Invent. Math. **166** (2006), no. 2, 229–325, DOI 10.1007/s00222-006-0505-0. MR2249801

[Knu] Allen Knutson, *Frobenius splitting and Möbius inversion*, `arxiv:0902.1930v1`.

[KR15] Ryan Kinser and Jenna Rajchgot, *Type A quiver loci and Schubert varieties*, J. Commut. Algebra **7** (2015), no. 2, 265–301, DOI 10.1216/JCA-2015-7-2-265. MR3370487

[Kra91] Henning Krause, *Maps between tree and band modules*, J. Algebra **137** (1991), no. 1, 186–194, DOI 10.1016/0021-8693(91)90088-P. MR1090218

[Len99] Cristian Lenart, *Noncommutative Schubert calculus and Grothendieck polynomials*, Adv. Math. **143** (1999), no. 1, 159–183, DOI 10.1006/aima.1998.1795. MR1680646

[Len00] Cristian Lenart, *Combinatorial aspects of the K-theory of Grassmannians*, Ann. Comb. **4** (2000), no. 1, 67–82, DOI 10.1007/PL00001276. MR1763950

[LM98] V. Lakshmibai and Peter Magyar, *Degeneracy schemes, quiver schemes, and Schubert varieties*, Internat. Math. Res. Notices **12** (1998), 627–640, DOI 10.1155/S1073792898000397. MR1635873

[Lőra] András Cristian Lőrincz, *The b-functions of semi-invariants of quivers*, J. Algebra **482** (2017), 346–363, DOI 10.1016/j.jalgebra.2017.03.028. MR3646294

[Lőrb] ———, *Singularities of zero sets of semi-invariants for quivers*, `arxiv:1509.04170`.

[LS82] Alain Lascoux and Marcel-Paul Schützenberger, *Structure de Hopf de l'anneau de cohomologie et de l'anneau de Grothendieck d'une variété de drapeaux* (French, with English summary), C. R. Acad. Sci. Paris Sér. I Math. **295** (1982), no. 11, 629–633. MR686357

[Lus90] G. Lusztig, *Canonical bases arising from quantized enveloping algebras*, J. Amer. Math. Soc. **3** (1990), no. 2, 447–498, DOI 10.2307/1990961. MR1035415

[Lus91] G. Lusztig, *Quivers, perverse sheaves, and quantized enveloping algebras*, J. Amer. Math. Soc. **4** (1991), no. 2, 365–421, DOI 10.2307/2939279. MR1088333

[MS05] Ezra Miller and Bernd Sturmfels, *Combinatorial commutative algebra*, Graduate Texts in Mathematics, vol. 227, Springer-Verlag, New York, 2005. MR2110098

[Per14] Nicolas Perrin, *On the geometry of spherical varieties*, Transform. Groups **19** (2014), no. 1, 171–223, DOI 10.1007/s00031-014-9254-0. MR3177371

[Rie86] Christine Riedtmann, *Degenerations for representations of quivers with relations*, Ann. Sci. École Norm. Sup. (4) **19** (1986), no. 2, 275–301. MR868301

[Rim] Richárd Rimányi, *On the cohomological Hall algebra of Dynkin quivers*, arxiv:1303.3399.

[Rim14] R. Rimányi, *Quiver polynomials in iterated residue form*, J. Algebraic Combin. **40** (2014), no. 2, 527–542, DOI 10.1007/s10801-013-0497-y. MR3239295

[Rin80] Claus Michael Ringel, *The rational invariants of the tame quivers*, Invent. Math. **58** (1980), no. 3, 217–239, DOI 10.1007/BF01390253. MR571574

[Rin90] Claus Michael Ringel, *Hall algebras and quantum groups*, Invent. Math. **101** (1990), no. 3, 583–591, DOI 10.1007/BF01231516. MR1062796

[Rin98] Claus Michael Ringel, *Exceptional modules are tree modules*, Proceedings of the Sixth Conference of the International Linear Algebra Society (Chemnitz, 1996), Linear Algebra Appl. **275/276** (1998), 471–493, DOI 10.1016/S0024-3795(97)10046-5. MR1628405

[Rin13a] Claus Michael Ringel, *Distinguished bases of exceptional modules*, Algebras, quivers and representations, Abel Symp., vol. 8, Springer, Heidelberg, 2013, pp. 253–274, DOI 10.1007/978-3-642-39485-0_11. MR3183888

[Rin13b] Claus Michael Ringel, *Indecomposable representations of the Kronecker quivers*, Proc. Amer. Math. Soc. **141** (2013), no. 1, 115–121, DOI 10.1090/S0002-9939-2012-11296-1. MR2988715

[RZ13] Christine Riedtmann and Grzegorz Zwara, *Orbit closures and rank schemes*, Comment. Math. Helv. **88** (2013), no. 1, 55–84, DOI 10.4171/CMH/278. MR3008913

[Sch85] A. H. Schofield, *Representation of rings over skew fields*, London Mathematical Society Lecture Note Series, vol. 92, Cambridge University Press, Cambridge, 1985. MR800853

[Sch14] Ralf Schiffler, *Quiver representations*, CMS Books in Mathematics/Ouvrages de Mathématiques de la SMC, Springer, Cham, 2014. MR3308668

[Wei10] Thorsten Weist, *Tree modules of the generalised Kronecker quiver*, J. Algebra **323** (2010), no. 4, 1107–1138, DOI 10.1016/j.jalgebra.2009.11.033. MR2578596

[Wei12] Thorsten Weist, *Tree modules*, Bull. Lond. Math. Soc. **44** (2012), no. 5, 882–898, DOI 10.1112/blms/bds019. MR2975149

[WY12] Alexander Woo and Alexander Yong, *A Gröbner basis for Kazhdan-Lusztig ideals*, Amer. J. Math. **134** (2012), no. 4, 1089–1137, DOI 10.1353/ajm.2012.0031. MR2956258

[Zel85] A. V. Zelevinskiĭ, *Two remarks on graded nilpotent classes* (Russian), Uspekhi Mat. Nauk **40** (1985), no. 1(241), 199–200. MR783619

[Zwa98] Grzegorz Zwara, *Degenerations for representations of extended Dynkin quivers*, Comment. Math. Helv. **73** (1998), no. 1, 71–88, DOI 10.1007/s000140050046. MR1610587

[Zwa99] Grzegorz Zwara, *Degenerations for modules over representation-finite algebras*, Proc. Amer. Math. Soc. **127** (1999), no. 5, 1313–1322, DOI 10.1090/S0002-9939-99-04714-0. MR1476404

[Zwa00] Grzegorz Zwara, *Degenerations of finite-dimensional modules are given by extensions*, Compositio Math. **121** (2000), no. 2, 205–218, DOI 10.1023/A:1001778532124. MR1757882

[Zwa02] G. Zwara, *Smooth morphisms of module schemes*, Proc. London Math. Soc. (3) **84** (2002), no. 3, 539–558, DOI 10.1112/S0024611502013382. MR1888422

[Zwa11] Grzegorz Zwara, *Singularities of orbit closures in module varieties*, Representations of algebras and related topics, EMS Ser. Congr. Rep., Eur. Math. Soc., Zürich, 2011, pp. 661–725, DOI 10.4171/101-1/13. MR2931906

University of Iowa, Department of Mathematics, Iowa City, IA, USA
Email address: ryan-kinser@uiowa.edu

Contemporary Mathematics
Volume **705**, 2018
http://dx.doi.org/10.1090/conm/705/14195

Krull–Gabriel dimension and the Ziegler spectrum

Rosanna Laking

Abstract. We provide an introduction to the the Krull-Gabriel dimension of a ring, as well as many related ideas. In particular, we outline how Krull-Gabriel dimension relates to the Cantor-Bendixson rank of the Ziegler spectrum and with the Jacobson radical of the module category. We also include a list of examples of rings and categories where the Krull-Gabriel dimension has been calculated.

In this survey we describe the notion of the Krull–Gabriel dimension $\mathrm{KG}(R)$ of a ring R (or more generally of a small preadditive category) and how it relates to notions of purity and the Ziegler spectrum of a module category.

The dimension was first defined in this form by Geigle [**12**] as a variation of the Krull dimension of an abelian category $\mathcal{A}$ defined by Gabriel in [**10**] (this has subsequently been referred to as the Gabriel dimension of $\mathcal{A}$). When R is a finite-dimensional algebra, the Krull–Gabriel dimension of R is zero if and only if R is of finite representation type [**2**]. In this context, it seems reasonable to consider the Krull–Gabriel dimension as a measure of how far R is from finite representation type, however the precise connection between representation type and Krull–Gabriel dimension is far from well-understood.

In order to define $\mathrm{KG}(R)$, we consider a transfinite filtration of the category $\mathcal{F} = \mathrm{F}_{\mathrm{fp}}(R)$ of finitely presented functors from mod-R to the category Ab of abelian groups. The initial part of the filtration is given by the sequence

$$0 = \mathcal{F}_{-1} \subseteq \mathcal{F}_0 \subseteq \mathcal{F}_1 \subseteq \mathcal{F}_2 \cdots \subseteq \mathcal{F}_n \subseteq \cdots$$

of Serre subcategories $\mathcal{F}_n$ of $\mathcal{F}$ consisting of the objects that become finite length in $\mathcal{F}/\mathcal{F}_{n-1}$ for $n \geq 0$. By taking unions at limit ordinals, this process may be continued transfinitely in the obvious way. We define $\mathrm{KG}(R) := \min\{\alpha \mid \mathcal{F}_\alpha = \mathcal{F}\}$ if such an ordinal exists and $\mathrm{KG}(R) = \infty$ otherwise.

There is a close relationship between the Serre localisations of $\mathrm{F}_{\mathrm{fp}}(R)$ and the Ziegler spectrum $\mathrm{Zg}(R\text{-Mod})$ of R-Mod. The Ziegler spectrum is a topological space with its underlying set given by isomorphism classes of indecomposable pure-injective objects in R-Mod. Both the closed subsets of $\mathrm{Zg}(R\text{-Mod})$ and the Serre localisations of $\mathrm{F}_{\mathrm{fp}}(R)$ are parametrised by the hereditary torsion pairs of finite type

2010 *Mathematics Subject Classification.* Primary 16G99, 18E15 ; Secondary 16G60, 18E30.

The author was supported by the DFG SFB / Transregio 45 during the period that this article was written.

in the category $\mathrm{F}(R)$ of all functors from mod-R to Ab. In particular, if the Krull–Gabriel dimension of R is defined then it coincides with the Cantor–Bendixson rank of $\mathrm{Zg}(R\text{-Mod})$.

Throughout this paper we will consider modules over a skeletally small preadditive category $\mathcal{R}$. One motivation for this level of generality is that it encompasses the theory of purity in a compactly generated triangulated category. See Example 1.14 for more details. References will often be given for the case where $\mathcal{R}$ is a ring as well as the more general case.

The paper is organised as follows. In the first section we outline notions of purity in categories of $\mathcal{R}$-modules. There are many equivalent ways of giving these definitions but we favour the definitions given in terms of the embedding of $\mathcal{R}$-Mod into $\mathrm{F}(\mathcal{R})$. Similarly, we follow the definition of the Ziegler spectrum with an explanation of how the topology can be seen in terms of localisations of $\mathrm{F}(\mathcal{R})$. In the second section we define the Krull–Gabriel dimension of $\mathcal{R}$ and explain the connections between this dimension and the Cantor–Bendixson rank of $\mathrm{Zg}(\mathcal{R}\text{-Mod})$. Finally, in Section 3 we give some examples where the Krull–Gabriel dimension of $\mathcal{R}$ has been calculated. We first give examples where $\mathcal{R}$ is a ring: serial rings, Dedekind domains, tame hereditary algebras, string algebras and canonical algebras. We also give examples of compactly generated triangulated categories where the dimension has been calculated; these include derived and homotopy categories of derived-discrete algebras, and derived categories of hereditary algebras.

1. Purity in categories of modules

Throughout this survey, let $\mathcal{R}$ denote a skeletally small preadditive category. In this section we introduce pure-exact sequences in the category of $\mathcal{R}$-modules. We then discuss the theory surrounding the pure-injective $\mathcal{R}$-modules i.e. the injective objects relative to the pure-exact structure. In particular, we consider the Ziegler spectrum of the module category.

Definition 1.1. A **left $\mathcal{R}$-module** is a covariant (additive) functor $M\colon \mathcal{R} \to \mathrm{Ab}$ where Ab denotes the category of abelian groups. A left $\mathcal{R}$-module M is **finitely presented** if there exists an exact sequence

$$\bigoplus_{i=1}^{n} \mathrm{Hom}_{\mathcal{R}}(r_i, -) \to \bigoplus_{j=1}^{m} \mathrm{Hom}_{\mathcal{R}}(s_i, -) \to M \to 0$$

for objects $r_i, s_j \in \mathcal{R}$. We will denote the category of left $\mathcal{R}$-modules by $\mathcal{R}$-Mod and the full subcategory of finitely presented left $\mathcal{R}$-modules by $\mathcal{R}$-mod.

A **(finitely presented) right $\mathcal{R}$-module** is a (finitely presented) left $\mathcal{R}^{\mathrm{op}}$-module. We will denote the category of right $\mathcal{R}$-modules by Mod-$\mathcal{R}$ and the full subcategory of finitely presented right $\mathcal{R}$-modules by mod-$\mathcal{R}$.

Note that if $\mathcal{R}$ has a single object r, then $\mathrm{End}_{\mathcal{R}}(r)$ is a unital ring and the above are equivalent to the usual definitions.

Notation 1.2. The category mod-$\mathcal{R}$ is a small preadditive category and we will often consider the category of left (mod-$\mathcal{R}$)-modules. To avoid cumbersome notation we will fix the following notation for these particular module categories:

$$\mathrm{F}(\mathcal{R}) := (\text{mod-}\mathcal{R})\text{-Mod} \qquad \mathrm{C}(\mathcal{R}) := \text{Mod-}(\mathcal{R}\text{-mod})$$

$$\mathrm{F}_{\mathrm{fp}}(\mathcal{R}) := (\text{mod-}\mathcal{R})\text{-mod} \qquad \mathrm{C}_{\mathrm{fp}}(\mathcal{R}) := \text{mod-}(\mathcal{R}\text{-mod})$$

Since it will always be clear from the context which category we are working in, we will denote representable functors $\mathrm{Hom}_{\mathcal{R}}(r,-)$ and $\mathrm{Hom}_{\mathcal{R}}(-,r)$ by $(r,-)$ and $(-,r)$ respectively.

Let M be an object in Mod-$\mathcal{R}$. We will also use the notation $(M,-)$ and $(-,M)$ for the representable functors $\mathrm{Hom}_{\text{Mod-}\mathcal{R}}(M,-)$ and $\mathrm{Hom}_{\text{Mod-}\mathcal{R}}(-,M)$ restricted to mod-$\mathcal{R}$.

1.1. The pure-exact structure. Next we introduce the pure-exact structure on $\mathcal{R}$-Mod. There are many equivalent definitions of a pure-exact sequence; we place an emphasis on those given by embedding $\mathcal{R}$-Mod into a category of functors since this is in line with the perspective taken in the subsequent sections.

DEFINITION 1.3 (See, for example, [**18**, Lemma 1.2.13]). Let $0 \to X \xrightarrow{f} Y \xrightarrow{g} Z \to 0$ be a short exact sequence in $\mathcal{R}$-Mod. If the following equivalent statements are satisfied, then the sequence is called a **pure-exact sequence**, the morphism f is called a **pure monomorphism** and g is called a **pure epimorphism**.

(1) The sequence
$$0 \to - \otimes_{\mathcal{R}} X \to - \otimes_{\mathcal{R}} Y \to - \otimes_{\mathcal{R}} Z \to 0$$
is exact in $\mathrm{F}(\mathcal{R})$.

(2) The sequence
$$0 \to (-,X) \to (-,Y) \to (-,Z) \to 0$$
is exact in $\mathrm{C}(\mathcal{R})$.

(3) The sequence $0 \to X \xrightarrow{f} Y \xrightarrow{g} Z \to 0$ is isomorphic to a filtered colimit of split exact sequences.

The first two conditions in Definition 1.3 allude to the following fully faithful embeddings of $\mathcal{R}$-Mod into larger module categories.

The (restricted) tensor embedding: Consider the functor $\Phi\colon \mathcal{R}\text{-Mod} \to \mathrm{F}(\mathcal{R})$ defined by $M \mapsto - \otimes_{\mathcal{R}} M$ and $f \mapsto - \otimes_{\mathcal{R}} f$ for all $\mathcal{R}$-modules M and morphisms f. The functor Φ has the following properties.

- Φ induces an equivalence between $\mathcal{R}$-Mod and the full subcategory of **fp-injective objects** in $\mathrm{F}(\mathcal{R})$ i.e. those functors F such that $\mathrm{Ext}^1(X,F) = 0$ for all X in $\mathrm{F}_{\mathrm{fp}}(\mathcal{R})$.
- Φ takes pure-exact sequences to exact sequences.
- A functor F in $\mathrm{F}(\mathcal{R})$ is right exact if and only if $F \cong \Phi(M)$ for some left $\mathcal{R}$-module M.

The proof of these assertions can be found in [**18**, Theorem B.16] for when $\mathcal{R}$ is a ring. Note that the argument given in [**7**, Theorem 3.3, Remark 3.3(2)] extends this result to the general case since there is a duality $\mathrm{F}_{\mathrm{fp}}(\mathcal{R}) \to \mathrm{F}_{\mathrm{fp}}(\mathcal{R}^{\mathrm{op}})$ for any $\mathcal{R}$, see [**33**, Theorem 10.3.4].

The (restricted) Yoneda embedding: Consider the functor $\Psi\colon \mathcal{R}\text{-Mod} \to \mathrm{C}(\mathcal{R})$ defined by $M \to (-,M)$ and $f \to (-,f)$ for all $\mathcal{R}$-modules M and all morphisms f. Then Ψ has the following properties.

- Ψ induces an equivalence between $\mathcal{R}$-Mod and the full subcategory of **flat objects** in $\mathrm{C}(\mathcal{R})$ i.e. those F such that $F \otimes -$ is exact.
- Ψ takes pure-exact sequences to exact sequences.

- A functor F in $\mathrm{C}(\mathcal{R})$ is left exact if and only if $F \cong \Psi(M)$ for some left $\mathcal{R}$-module M.

The proof of these assertions can be found in [**18**, Theorem B.11] for when $\mathcal{R}$ is a ring. For the general case, see [**7**, Theorem 1.4(2)].

We can therefore see the pure-exact structure on $\mathcal{R}$-Mod as a reflection of the exact structure on the full subcategories of fp-injective objects in $\mathrm{F}(\mathcal{R})$ or equally the exact structure on the full subcategory of flat objects in $\mathrm{C}(\mathcal{R})$.

DEFINITION 1.4. A module M in $\mathcal{R}$-Mod is called **pure-injective** if M is injective with respect to pure monomorphisms. That is, for every pure monomorphism $f\colon A \to B$ and morphism $g\colon A \to M$, there exists a morphism $h\colon B \to M$ such that the following diagram commutes.

$$\begin{array}{ccccc} 0 & \longrightarrow & A & \xrightarrow[\text{pure}]{f} & B \\ & & {\scriptstyle g}\downarrow & \swarrow {\scriptstyle h} & \\ & & M & & \end{array}$$

The definition of a **pure-projective** module is given dually.

The following theorem was originally proved in [**15**] for the case where $\mathcal{R}$ is a ring. The general case follows by combining [**7**, Lemma 3.5.1, Remark 3.3] with [**33**, Theorem 10.3.4].

THEOREM 1.5. *The functor* $\Phi\colon \mathcal{R}\text{-Mod} \to \mathrm{F}(\mathcal{R})$ *induces an equivalence*

$$\Phi\colon \mathrm{P.inj}(\mathcal{R}\text{-Mod}) \xrightarrow{\sim} \mathrm{Inj}(\mathrm{F}(\mathcal{R}))$$

where $\mathrm{P.inj}(\mathcal{R}\text{-Mod})$ *denotes the full subcategory of pure-injective modules in* $\mathcal{R}$-Mod *and* $\mathrm{Inj}(\mathrm{F}(\mathcal{R}))$ *denotes the full subcategory of injective functors in* $\mathrm{F}(\mathcal{R})$.

Similarly, the contravariant Yoneda embedding induces an equivalence between the pure-projective modules in $\mathcal{R}$-Mod and the projective functors in $\mathrm{C}(\mathcal{R})$ (see [**18**, Theorem B.11]).

1.2. The Ziegler spectrum. The set of isomorphism classes of indecomposable pure-injective modules carries with it a natural topology. We introduce this space here and describe how it connects to localisations of the functor category $\mathrm{F}(\mathcal{R})$.

DEFINITION 1.6. Define the **Ziegler spectrum** $\mathrm{Zg}(\mathcal{R}\text{-Mod})$ to be the topological space with points and basic open sets given by the following data:

Points: The isomorphism classes of indecomposable pure-injective modules in $\mathcal{R}$-Mod.

Topology: The following sets define a basis of open sets for the topology on $\mathrm{Zg}(\mathcal{R}\text{-Mod})$:

$$(F) := \{M \in \mathrm{Zg}(\mathcal{R}\text{-Mod}) \mid (F, - \otimes_{\mathcal{R}} M) \neq 0\}$$

where F ranges over functors in $\mathrm{F}_{\mathrm{fp}}(\mathcal{R})$ and $(F, - \otimes_{\mathcal{R}} M)$ denotes the group of natural transformations from F to $- \otimes_{\mathcal{R}} M$.

The topology on $\mathrm{Zg}(\mathcal{R}\text{-Mod})$ was originally defined in [**45**] in terms of pairs of pp-formulas (ϕ/ψ) and encodes much of the model theoretic information in $\mathcal{R}$-Mod. For an account of how the model theoretic definition of the Ziegler spectrum connects with the perspective given here see [**33**].

DEFINITION 1.7 ([**8**, §2.3]). Let $\mathcal{X}$ be a full subcategory of R-Mod. Then $\mathcal{X}$ is called a **definable subcategory** if the following equivalent statements are satisfied.

(1) There exists a set $\mathcal{S}$ of functors in $\mathrm{F}_{\mathrm{fp}}(\mathcal{R})$ such that

$$\mathcal{X} = \{M \in \mathcal{R}\text{-Mod} \mid (F, - \otimes_{\mathcal{R}} M) = 0 \text{ for all } F \in \mathcal{S}\}.$$

(2) The subcategory $\mathcal{X}$ is closed under direct products, direct limits and pure-submodules i.e. submodules where the canonical inclusion is a pure monomorphism.

There is a bijection between definable subcategories of $\mathcal{R}$-Mod and the closed subsets of $\mathrm{Zg}(\mathcal{R}\text{-Mod})$ given by $\mathcal{X} \mapsto \mathcal{X} \cap \mathrm{Zg}(\mathcal{R}\text{-Mod})$ and $\mathbf{X} \mapsto \langle \mathbf{X} \rangle$ where $\langle \mathbf{X} \rangle$ is the smallest definable subcategory containing $\mathbf{X}$ [**45**].

1.3. The Ziegler spectrum via localisations. Since the closed subsets of $\mathrm{Zg}(\mathcal{R}\text{-Mod})$ are parametrised by the Serre subcategories of $\mathrm{F}_{\mathrm{fp}}(\mathcal{R})$, it is natural to consider localisations of this category. In fact, there is also a bijective correspondence between the Serre subcategories and hereditary torsion pairs of finite type in $\mathrm{F}(\mathcal{R})$. In this section we consider the Ziegler spectrum from the perspective of localisations of $\mathrm{F}(\mathcal{R})$ at hereditary torsion classes of finite type.

Let $\mathcal{B}$ be a class of objects in $\mathrm{F}(\mathcal{R})$, then we will use the notation $\mathcal{B}^{\perp}$ to denote the class $\{F \in \mathrm{F}(\mathcal{R}) \mid (G, F) = 0 \text{ for all } G \in \mathcal{B}\}$ and the notation ${}^{\perp}\mathcal{B}$ to denote the class $\{G \in \mathrm{F}(\mathcal{R}) \mid (G, F) = 0 \text{ for all } F \in \mathcal{B}\}$.

DEFINITION 1.8. Let $(\mathcal{T}, \mathcal{F})$ be a pair of subclasses of the objects of $\mathrm{F}(\mathcal{R})$. Then we say that $(\mathcal{T}, \mathcal{F})$ is a **torsion pair** if the following equivalent conditions are satisfied.

(1) $\mathcal{F} = \mathcal{T}^{\perp}$ and $\mathcal{T}$ is closed under quotient objects, extensions and arbitrary direct sums.
(2) $\mathcal{T} = {}^{\perp}\mathcal{F}$ and $\mathcal{F}$ is closed under subobjects, extensions and arbitrary products.

We call $\mathcal{T}$ a **torsion class** and $\mathcal{F}$ a **torsion-free class**. A torsion pair is **hereditary** if $\mathcal{T}$ is closed under subobjects, or equivalently, if $\mathcal{F}$ is closed under injective envelopes. For any $F \in \mathrm{F}(\mathcal{R})$, let $t(F) := \sum\{G \in \mathcal{T} \mid G \text{ is a subobject of } F\}$. This induces a functor $t\colon \mathrm{F}(\mathcal{R}) \to \mathrm{F}(\mathcal{R})$ called the **torsion functor** of $(\mathcal{T}, \mathcal{F})$. If t commutes with direct limits, then $(\mathcal{T}, \mathcal{F})$ is said to be of **finite type**. Let $\mathcal{E}$ be a class of objects in $\mathrm{F}(\mathcal{R})$. Then $({}^{\perp}\mathcal{E}, ({}^{\perp}\mathcal{E})^{\perp})$ is a torsion pair and is **cogenerated by** $\mathcal{E}$.

If $\mathcal{T}$ is a torsion class of a hereditary torsion pair, then $\mathcal{T}$ is a **Serre subcategory** of $\mathrm{F}(\mathcal{R})$. That is, for any short exact sequence $0 \to X \to Y \to Z \to 0$ in $\mathrm{F}(\mathcal{R})$, we have that $X, Z \in \mathcal{T}$ if and only if $Y \in \mathcal{T}$.

THEOREM 1.9 ([**19**, Corollary 4.3]). *Let $\mathcal{T}$ be a class of objects in $\mathrm{F}(\mathcal{R})$. Then the following statements are equivalent.*

(1) *The pair $(\mathcal{T}, \mathcal{T}^{\perp})$ is a hereditary torsion pair of finite type.*
(2) *There exists a closed subset $\mathbf{X}$ of $\mathrm{Zg}(\mathcal{R}\text{-Mod})$ such that $(\mathcal{T}, \mathcal{T}^{\perp})$ is the torsion pair cogenerated by the set $\mathcal{E} := \{- \otimes_{\mathcal{R}} M \mid M \in \mathbf{X}\}$.*

There is a bijective correspondence between hereditary torsion pairs of finite type in $\mathrm{F}(\mathcal{R})$ and Serre subcategories of $\mathrm{F}_{\mathrm{fp}}(\mathcal{R})$ given by the following mutually inverse bijections:

$$(\mathcal{T},\mathcal{F}) \mapsto \mathcal{T} \cap \mathrm{F}_{\mathrm{fp}}(\mathcal{R}) \quad \text{and} \quad \mathcal{S} \mapsto (\varinjlim \mathcal{S}, (\varinjlim \mathcal{S})^{\perp})$$

where $\varinjlim(\mathcal{S})$ is the class of objects obtained by taking all direct limits of objects in $\mathcal{S}$. This can be found in [**19**, Corollary 2.10].

DEFINITION 1.10. Let $\mathcal{A}$ be an abelian category. For any Serre subcategory $\mathcal{S}$ of $\mathcal{A}$, we define the **quotient category** $\mathcal{A}/\mathcal{S}$ to be the category with the same objects as $\mathcal{A}$ and morphisms given by

$$\mathrm{Hom}_{\mathcal{A}/\mathcal{S}}(X,Y) := \varinjlim \mathrm{Hom}_{\mathcal{A}}(X',Y/Y')$$

where X' and Y' range over subobjects of X and Y respectively such that $X/X', Y' \in \mathcal{S}$. For each quotient category, there is a **localisation functor** $q\colon \mathcal{A} \to \mathcal{A}/\mathcal{S}$ defined on objects to be $q(X) = X$. For more details see, for example, [**29**].

THEOREM 1.11 ([**19**, Theorem 2.6], [**10**, III.4, Corolaire 2]). *Let $(\mathcal{T},\mathcal{F})$ be a hereditary torsion pair of finite type in $\mathrm{F}(\mathcal{R})$ and let $\mathcal{S} = \mathcal{T} \cap \mathrm{F}_{\mathrm{fp}}(\mathcal{R})$ be the corresponding Serre subcategory of $\mathrm{F}_{\mathrm{fp}}(\mathcal{R})$. Then the following statements hold.*

(1) *There is an equivalence of categories*

$$\mathrm{F}_{\mathrm{fp}}(\mathcal{R})/\mathcal{S} \xrightarrow{\sim} \mathrm{fp}(\mathrm{F}(\mathcal{R})/\mathcal{T})$$

where $\mathrm{fp}(\mathrm{F}(\mathcal{R})/\mathcal{T})$ is the full subcategory of finitely presented objects in $\mathrm{F}(\mathcal{R})/\mathcal{T}$.

(2) *The functor $q\colon \mathrm{F}(\mathcal{R}) \to \mathrm{F}(\mathcal{R})/\mathcal{T}$ has a right adjoint (i.e. $\mathcal{T}$ is a **localising subcategory**) denoted $s\colon \mathrm{F}(\mathcal{R})/\mathcal{T} \to \mathrm{F}(\mathcal{R})$.*

(3) *The (indecomposable) injective objects in $\mathrm{F}(\mathcal{R})/\mathcal{T}$ are exactly those that are isomorphic to $q(E)$ for some (indecomposable) injective object E in $\mathcal{F}$.*

(4) *For any injective object E in $\mathcal{F}$ we have $E \cong s \circ q(E)$.*

REMARK 1.12. Combining the previous theorem with Theorems 1.9 and 1.5, we have that the Ziegler spectrum $\mathrm{Zg}(\mathcal{R}\text{-Mod})$ is homeomorphic to the topological space with points given by the set $\mathcal{I}(\mathcal{R})$ isomorphism classes of indecomposable injective objects in $\mathrm{F}(\mathcal{R})$ and the closed sets given by the sets of indecomposable injective objects in $\mathrm{F}(\mathcal{R})/\mathcal{T}$ for hereditary torsion pairs $(\mathcal{T},\mathcal{F})$ of finite type (or, equivalently by the sets $\mathcal{F} \cap \mathcal{I}(\mathcal{R})$).

Given a closed set $\mathbf{X}$ in $\mathrm{Zg}(\mathcal{R}\text{-Mod})$, the full subcategory of $\mathrm{F}_{\mathrm{fp}}(\mathcal{R})$ with objects $\mathrm{ann}(\mathbf{X}) := \{F \in \mathrm{F}_{\mathrm{fp}}(\mathcal{R}) \mid (F, - \otimes_{\mathcal{R}} M) = 0 \text{ for all } M \in \mathbf{X}\}$ is a Serre subcategory. The compact open sets $(F) \cap \mathbf{X}$ of the relative topology on $\mathbf{X}$ are parametrised by the objects F in $\mathrm{F}_{\mathrm{fp}}(\mathcal{R})/\mathrm{ann}(\mathbf{X})$.

EXAMPLE 1.13. **Finite-dimensional algebras.** Let R be a finite-dimensional algebra. Then $\mathrm{Zg}(R\text{-Mod})$ is a compact topological space (in fact, this is the case whenever R is a ring). Every indecomposable module in R-mod is pure-injective and hence a point of $\mathrm{Zg}(R\text{-Mod})$ and they enjoy some particular nice properties. For example, every indecomposable module M in R-mod, the set $\{M\}$ is open and closed. Moreover, the open set $U := R\text{-mod} \cap \mathrm{Zg}(R\text{-Mod})$ contains all of the isolated points in $\mathrm{Zg}(R\text{-Mod})$ and is dense in $\mathrm{Zg}(R\text{-Mod})$. That is, for every non-empty open set V, the set $U \cap V$ is non-empty. It follows directly that $\mathrm{Zg}(R\text{-Mod})$ is finite if and only if R is of finite representation type.

For some classes of finite-dimensional algebras, the Ziegler spectrum has been explicitly described. Examples include, tame hereditary algebras [**30**, **39**], domestic string algebras [**25**, **37**], and canonical algebras of tubular type (the topology has been described but the algebraic structure of some of the points is not known) [**16**], [**14**].

EXAMPLE 1.14. **Compactly generated triangulated categories.** Let $\mathcal{C}$ be a compactly generated triangulated category and let $\mathcal{C}^c$ be the full subcategory of compact objects in $\mathcal{C}$. For definitions and more details on such categories see [**28**].

The theory of purity and the Ziegler spectrum of compactly generated triangulated categories is well-developed (see [**4**, **11**, **21**, **22**]). Let $\mathrm{Zg}(\mathcal{C})$ denote the Ziegler spectrum of $\mathcal{C}$ as it is defined in [**21**]. Then this topological space is homeomorphic to a closed subset of $\mathrm{Zg}(\text{Mod-}\mathcal{C}^c)$ and all the definitions surrounding the theory of purity in $\mathcal{C}$ coincide with those given in the preceding sections. This is explicitly proved in [**1**]; we give a brief account here.

Recall that a right $\mathcal{C}^c$-module N is fp-injective if $\mathrm{Ext}^1(M, N) = 0$ for all finitely presented right $\mathcal{C}^c$-modules M. The full subcategory $\mathcal{D}$ of fp-injective modules in $\text{Mod-}\mathcal{C}^c$ is a definable subcategory and so $\mathbf{Z} := \mathcal{D} \cap \mathrm{Zg}(\text{Mod-}\mathcal{C}^c)$ is a closed subset of the $\mathrm{Zg}(\text{Mod-}\mathcal{C}^c)$. In fact, $\mathbf{Z} = \{(-, N) \in \text{Mod-}\mathcal{C}^c \mid N \in \mathrm{Zg}(\mathcal{C})\}$ and these are exactly the indecomposable injective objects of $\text{Mod-}\mathcal{C}^c$ [**22**, Corollary 1.9]. Although the following result is not explicitly stated, much of the preliminary work needed for this result is contained in [**22**].

THEOREM 1.15 ([**1**, Theorem 1.9]). *The closed subset* $\mathbf{Z}$ *of* $\mathrm{Zg}(\text{Mod-}\mathcal{C}^c)$ *with the relative topology is homeomorphic to* $\mathrm{Zg}(\mathcal{C})$. *Moreover there is an equivalence*

$$\mathrm{F_{fp}}(\mathcal{C}^c)/\mathrm{ann}(\mathbf{Z}) \xrightarrow{\sim} (\text{mod-}\mathcal{C}^c)^{\mathrm{op}}$$

yielding the following description of the compact open sets of $\mathbf{Z}$*:*

$$(M) := \{(-, N) \in \mathcal{Z} \mid (M, (-, N)) \neq 0\}$$

where M *is a module in* $\text{mod-}\mathcal{C}^c$.

There are some examples of compactly generated triangulated categories where the Ziegler spectrum is known. For example, the homotopy category $\mathrm{K}(\mathrm{Proj}\Lambda)$ of projective modules over a derived discrete algebra [**1**]; the derived category $\mathrm{D}(R\text{-Mod})$ where R is a right hereditary ring [**11**, Theorem 8.1]; and the derived category $\mathrm{D}(R\text{-Mod})$ where R is Von Neumann regular [**11**, Theorem 8.5].

2. The Krull–Gabriel dimension of $\mathcal{R}$

In this section we define the Krull–Gabriel dimension of $\mathcal{R}$. We begin by defining the more general notion of the Krull–Gabriel dimension of an abelian category $\mathcal{A}$ and then consider the special case where $\mathcal{A} = \mathrm{F_{fp}}(\mathcal{R})$; we refer to this dimension as the Krull–Gabriel dimension of $\mathcal{R}$. We then discuss the connections between the Krull–Gabriel analysis of $\mathrm{F_{fp}}(\mathcal{R})$ and the Ziegler spectrum of $\mathcal{R}$-Mod.

2.1. The Krull–Gabriel dimension of an abelian category. Let $\mathcal{A}$ be an abelian category. The following filtration of $\mathcal{A}$ was introduced by Gabriel in [**10**, IV.1]. Let $\mathcal{A}_0$ be the Serre subcategory of $\mathcal{A}$ generated (as a Serre subcategory) by the simple objects in $\mathcal{A}$. For each ordinal α, we define a Serre subcategory $\mathcal{A}_\alpha$ via the following transfinite induction. For $\beta < \alpha$, let $q_\beta\colon \mathcal{A} \to \mathcal{A}/\mathcal{A}_\beta$ denote the corresponding localisation functor.

- If $\alpha = \beta + 1$, then define $\mathcal{A}_\alpha$ to be the Serre subcategory of $\mathcal{A}$ generated by objects X such that $q_\beta(X)$ is zero or simple in $\mathcal{A}/\mathcal{A}_\beta$.
- If α is a limit ordinal, then define $\mathcal{A}_\alpha := \bigcup_{\beta<\alpha} \mathcal{A}_\beta$.

DEFINITION 2.1. Let $\mathcal{A}$ be an abelian category. If $\mathcal{A} = \mathcal{A}_\alpha$ for some ordinal α, then we define the **Krull–Gabriel dimension** $\mathrm{KG}(\mathcal{A})$ to be α. If no such ordinal exists then the **Krull–Gabriel dimension of $\mathcal{A}$ is undefined** and this is denoted $\mathrm{KG}(\mathcal{A}) = \infty$.

Note that Gabriel and many subsequent authors refers to this as the Krull dimension of $\mathcal{A}$. The name Krull–Gabriel dimension was adopted later.

EXAMPLE 2.2. In [**3**, Theorem 3.9, Theorem 4.3], Baer shows that if $\mathcal{P}$ (respectively $\mathcal{R}$) is the preprojective (respectively regular) component of the Auslander-Reiten quiver of R-mod where R is a hereditary connected Artin algebra of infinite representation type, then $\mathrm{KG}(\mathcal{P}\text{-mod})$ is undefined if and only if $\mathrm{KG}(\mathcal{R}\text{-mod})$ is undefined if and only if R is wild.

DEFINITION 2.3. Let a be an object in an abelian category $\mathcal{A}$. The **Krull–Gabriel dimension** $\mathrm{KG}(a)$ **of** a is defined to be the greatest ordinal such that $q_\alpha(a) \neq 0$. If no such ordinal exists then the Krull–Gabriel dimension of a is undefined and this is denoted $\mathrm{KG}(a) = \infty$.

Let $L(a)$ denote the modular lattice consisting of subobjects of a. Let $\sim$ denote the congruence relation on $L(a)$ generated by the relations: $b \sim c$ whenever $b < c$ and, for all $d \in L(a)$ such that $b < d < c$, either $b = d$ or $c = d$. Let $L'(a)$ denote the modular lattice consisting of the $\sim$-congruence classes in $L(a)$.

Then let $L_0(a) := L'(a)$ and for each ordinal α we define a modular lattice $L_\alpha(a)$ as follows.

- If $\alpha = \beta + 1$, then define $L_\alpha(a) := L'_\beta(a)$.
- If α is a limit ordinal, then define $L_\alpha(a) := \varinjlim_{\beta<\alpha} L_\beta(a)$.

The following lemma follows directly from the definitions.

LEMMA 2.4. *Let a be an object in an abelian category $\mathcal{A}$. Then, for every ordinal α, we have $L_\alpha(a) \cong L(q_\alpha(a))$.*

This allows us to understand an important distinction between abelian categories $\mathcal{A}$ with defined Krull–Gabriel dimension and those with undefined Krull–Gabriel dimension.

DEFINITION 2.5. A **dense chain** in a lattice L is a non-empty sublattice C such that, for all ordered pairs $x < y$ with $x, y \in C$, there exists some $z \in C$ such that $x < z < y$.

PROPOSITION 2.6 (See, for example, [**23**, Lemma B.8]). *For an abelian category $\mathcal{A}$, the following conditions are equivalent.*

(1) *There exists an object a in $\mathcal{A}$ such that $L(a)$ contains a dense chain.*
(2) *There exists an object a such that* $\mathrm{KG}(a) = \infty$ *if and only if* $\mathrm{KG}(\mathcal{A}) = \infty$.

2.2. The Krull–Gabriel dimension of $\mathcal{R}$. The Krull–Gabriel dimension of $\mathcal{R}$, denoted $\mathrm{KG}(\mathcal{R})$, is defined to be the Krull–Gabriel dimension of the category $\mathrm{F}_{\mathrm{fp}}(\mathcal{R})$. The dimension was first studied in [**12**]. Let F be a functor in $\mathrm{F}_{\mathrm{fp}}(\mathcal{R})$. By definition there exists an exact sequence of functors

$$(B,-) \stackrel{(f,-)}{\longrightarrow} (A,-) \longrightarrow F \longrightarrow 0$$

where $f\colon A \to B$ is a morphism in mod-$\mathcal{R}$. That is, we have $F \cong (A,-)/\mathrm{im}(f,-)$. Moreover, it is well-known that the finitely presented subfunctors of $(A,-)$ are of the form $\mathrm{im}(g,-)$ for some $g\colon A \to C$ (see, for example, [**33**, Lemma 10.2.2]). It follows that subfunctors of F are of the form $\mathrm{im}(g,-)/\mathrm{im}(f,-)$ where $f = hg$ for some $h\colon C \to B$. For a careful treatment of this approach see, for example, [**34**].

With this description of subfunctors of $F \in \mathrm{F}_{\mathrm{fp}}(\mathcal{R})$ in mind, it is clear that the next result is a direct application of Proposition 2.6.

THEOREM 2.7 ([**31**]). *Let $\mathcal{R}$ be a skeletally small preadditive category. The Krull–Gabriel dimension of $\mathcal{R}$ is undefined if and only if there exist $n \geq 1$ and a collection of finitely presented n-pointed modules $\{(M_i, \underline{m}^{(i)}) \mid i \in \mathbb{Q}_{\geq 0},\ \underline{m}^{(i)} = (m_1^{(i)}, ..., m_n^{(i)}) \in M_i^n\}$ together with a collection of morphisms $\{f_{ij}\colon M_j \to M_i \mid i < j \in \mathbb{Q}_{\geq 0}\}$ such that the following conditions are satisfied for each $i < j$ and $1 \leq k \leq n$.*

(1) *The morphism $f_{ij}\colon M_j \to M_i$ is such that $f_{ij}(m_k^{(j)}) = m_k^{(i)}$.*
(2) *There is no $g\colon M_i \to M_j$ such that $g(m_k^{(i)}) = m_k^{(j)}$.*

2.3. Krull–Gabriel dimension and the Ziegler spectrum. Since every simple object in $\mathrm{F}(\mathcal{R})$ has an indecomposable injective hull, we have a direct connection between the first stage of the Krull–Gabriel filtration and $\mathrm{Zg}(\mathcal{R}\text{-Mod})$ via the equivalence induced by the tensor embedding (see Theorem 1.5). The following proposition tells us that this connection also exists for the latter stages.

PROPOSITION 2.8. *Let $\mathcal{R}$ be a small preadditive category. Consider a closed subset $\mathbf{X}$ in $\mathrm{Zg}(\mathcal{R}\text{-Mod})$ with $\mathcal{T}_{\mathbf{X}}$ the torsion class in $\mathrm{F}(\mathcal{R})$ associated to $\mathbf{X}$ under the correspondence in Theorem 1.9. Let S be a functor in $\mathrm{F}_{\mathrm{fp}}(\mathcal{R})$ such that $q_{\mathbf{X}}(S)$ is simple in $\mathrm{F}(\mathcal{R})/\mathcal{T}_X$ where $q_{\mathbf{X}}\colon \mathrm{F}(\mathcal{R}) \to \mathrm{F}(\mathcal{R})/\mathcal{T}_{\mathbf{X}}$ is the corresponding localisation functor. Then the following statements hold.*

(1) *The injective envelope $E(q_{\mathbf{X}}(S))$ of $q_{\mathbf{X}}(S)$ in $\mathrm{F}(\mathcal{R})/\mathcal{T}_{\mathbf{X}}$ is indecomposable.*
(2) *There is a pure-injective $\mathcal{R}$-module $N \in \mathrm{Zg}(\mathcal{R}\text{-Mod})$ such that $E(q_{\mathbf{X}}(S))$ is isomorphic to $q_{\mathbf{X}}(- \otimes_{\mathcal{R}} N)$ and $(S) \cap \mathbf{X} = \{N\}$.*

PROOF. It is clear that $E(q_{\mathbf{X}}(S))$ is indecomposable because $q_{\mathbf{X}}(S)$ is uniform i.e. any pair of non-zero subobjects of $q_{\mathbf{X}}(S)$ have non-zero intersection (see [**43**, Proposition 2.8]). By the uniqueness of injective envelopes, as well as Theorem 1.5 and the last part of Theorem 1.11, there is a unique $N \in \mathrm{Zg}(\mathcal{R}\text{-Mod})$ such that $E(q_{\mathbf{X}}(S)) = q_{\mathbf{X}}(- \otimes_{\mathcal{R}} N)$. By Remark 1.12, N is isolated in $\mathbf{X}$ by the open set corresponding to S. □

DEFINITION 2.9. If all isolated points in a closed set $\mathbf{X}$ of $\mathrm{Zg}(\mathcal{R}\text{-Mod})$ arise as injective envelopes of finitely presented simple functors in $\mathrm{F}(\mathcal{R})/\mathcal{T}_{\mathbf{X}}$, then we say that **$\mathbf{X}$ satisfies the isolation condition**.

Note that Jensen and Lenzing prove the following theorem in the case where $\mathcal{R}$ is a ring but the argument works equally well in the general case.

THEOREM 2.10 ([**18**, Proposition 8.52, Theorem 8.53]). *If the Krull–Gabriel dimension of* $\mathcal{R}$ *is defined, then every pure-injective module has an indecomposable direct summand and, for every ordinal* α*, the closed subset* $\mathbf{X}_\alpha$ *of* Zg($\mathcal{R}$-Mod) *corresponding to* $\mathrm{F}_{\mathrm{fp}}(\mathcal{R})/\mathrm{F}_{\mathrm{fp}}(\mathcal{R})_\alpha$ *satisfies the isolation condition.*

REMARK 2.11. If the isolation condition holds for Zg($\mathcal{R}$-Mod) (e.g. if KG($\mathcal{R}$) is defined), then the Krull–Gabriel dimension of $\mathcal{R}$ coincides with the Cantor–Bendixson rank of Zg($\mathcal{R}$-Mod) which is defined as follows.

For a topological space $\mathbf{Z}$ we define $\mathbf{Z}'$ to be the closed subset of $\mathbf{Z}$ consisting of non-isolated points. Setting $\mathbf{Z}_0 = \mathbf{Z}$, we define $\mathbf{Z}_{\alpha+1} := \mathbf{Z}'_\alpha$ for each ordinal α and let $\mathbf{Z}_\lambda := \bigcap_{\alpha<\lambda} \mathbf{Z}_\alpha$ for each limit ordinal λ. The **Cantor–Bendixson rank** of $\mathbf{Z}$ is defined to be the least α such that $\mathbf{Z}_{\alpha+1} = \emptyset$. If no such ordinal exists then the Cantor–Bendixson rank of $\mathbf{Z}$ is said to be undefined.

Thus a description of the open neighbourhoods in Zg($\mathcal{R}$-Mod) gives rise to a description of the functors in $\mathrm{F}_{\mathrm{fp}}(\mathcal{R})/\mathrm{F}_{\mathrm{fp}}(\mathcal{R})_\alpha$ for each ordinal α. This was the strategy employed by the authors of [**25**]; for a description of the simple functors see [**24**].

There are no known examples of rings R where the isolation condition does not hold for Zg(R-Mod). In other words, it is an open question as to whether the Krull–Gabriel dimension of R and the Cantor–Bendixson rank of Zg(R-Mod) always coincide.

REMARK 2.12. Let R be a ring. In [**45**], Ziegler shows that if every pure-injective module in R-Mod has an indecomposable direct summand or if R is countable then Zg(R-Mod) satisfies the isolation condition.

REMARK 2.13. As Jensen and Lenzing point out in [**18**, Remark 8.54], if the Krull–Gabriel dimension of $\mathcal{R}$ is defined, then the classification of indecomposable pure-injective modules amounts to the classification of simple functors in $\mathrm{F}(\mathcal{R})_{\alpha+1}/\mathrm{F}(\mathcal{R})_\alpha$ for each ordinal $\alpha < \mathrm{KG}(\mathcal{R})$.

This procedure is demonstrated in [**18**] for Dedekind domains and the Kronecker algebra. This is also the method employed by the authors of [**1**] in order to classify the indecomposable pure-injective objects in K(ProjΛ) where Λ is a derived-discrete algebra.

3. Examples

In this final section we give some examples where the Krull–Gabriel dimension of $\mathcal{R}$ has been calculated.

3.1. Serial rings. Let R be a unital ring. Then R is a **serial ring** if it is serial when considered as both a left and a right R-module. Equivalently there exists a collection $e_1, \ldots e_n$ of pairwise orthogonal idempotents in R such that $e_1 + \ldots e_n = 1$ and the modules e_iR and Re_i are **uniserial** for each $1 \leq i \leq n$, i.e. the lattice of submodules for each module is a chain.

In [**36**], Puninski proves the following about the Krull–Gabriel dimension of R when R is a serial ring. For the definition of Krull dimension see, for example, [**27**, Chapter 6].

THEOREM 3.1. *Let R be a serial ring. Then the following statements hold.*

(1) *The Krull–Gabriel dimension* $\mathrm{KG}(R)$ *is defined if and only if the Krull dimension of R is defined.*
(2) *If α is the Krull dimension of R, then* $\mathrm{KG}(R) \leq \alpha \oplus \alpha$ *where $\oplus$ denotes the Cantor sum.*
(3) *The Krull–Gabriel dimension* $\mathrm{KG}(R)$ *is not equal to* 1.

Puninski also shows that the isolation condition holds for such rings. There are classes of serial rings where the Cantor–Bendixson rank has been calculated and so, by Remark 2.13, the Krull–Gabriel dimension can be given exactly in terms of the Krull dimension of R. A ring R is called **Krull–Schmidt** if every finitely presented R-module is a finite direct sum of indecomposable modules with local endomorphism ring.

COROLLARY 3.2 ([**36**, Proposition 5.8, Corollary 5.9], [**35**], [**38**]). *Let R be either:*

- *a Krull–Schmidt uniserial ring;*
- *a Krull–Schmidt serial ring with finite Krull dimension; or*
- *a commutative valuation domain.*

If the Krull dimension of R is α, then the Krull–Gabriel dimension of R is $\alpha \oplus \alpha$.

3.2. Dedekind domains. Let R be an integral domain, then R is a **Dedekind domain** if every non-zero proper ideal can be written as a product of prime ideals. Examples include $\mathbb{Z}$ or $\mathrm{k}[T]$ for a field k.

THEOREM 3.3. *Let R be a Dedekind domain that is not a field. Then the Cantor–Bendixson rank of* $\mathrm{Zg}(R\text{-Mod})$ *is equal to the Krull–Gabriel dimension of R is equal to* 2.

PROOF. It follows from Ziegler's description of $\mathrm{Zg}(R\text{-Mod})$ [**45**] that the isolation condition is satisfied so, by Remark 2.13, the Krull–Gabriel dimension of R is equal to the Cantor–Bendixson rank of $\mathrm{Zg}(R\text{-Mod})$. Thus, by [**45**], both dimensions are equal to 2. □

REMARK 3.4. There is an explicit description of the Krull–Gabriel filtration of $\mathrm{F_{fp}}(R)$ in [**18**, Chapter 8] for R a Dedekind domain that is not a field. Following the notation in [**18**], for any R-module M, let $\overline{M}$ denote the unique functor that vanishes on all finite length modules and satisfies $\overline{M}(R) = 0$.

Let $\mathfrak{m}$ denote a maximal ideal in R. Then the functors $(R/\mathfrak{m}^n, -)/\mathrm{rad}(R/\mathfrak{m}^n, -)$ are shown to be simple in $\mathrm{F_{fp}}(R)$ and the images of the functors $\overline{(R/\mathfrak{m}, -)}$ are simple in $\mathrm{F_{fp}}(R)/\mathrm{F_{fp}}(R)_0$. One should note that the images of functors $\overline{R/\mathfrak{m}}$ are actually simple in $\mathrm{F_{fp}}(R)/\mathrm{F_{fp}}(R)_0$. Moreover the image of the functor $\overline{R}$ is simple in $\mathrm{F_{fp}}(R)/\mathrm{F_{fp}}(R)_1$. To see why this is the case, consider, for example, that there is a non-zero, non-surjective morphism

$$(- \otimes_R R/\mathfrak{m}^n) \to \overline{R/\mathfrak{m}}$$

and that $- \otimes_R R/\mathfrak{m}^n$ is a finitely presented functor for each $n \geq 1$. Moreover, the following is a chain of proper embeddings.

$$0 \to - \otimes_R R/\mathfrak{m} \to - \otimes_R R/\mathfrak{m}^2 \to - \otimes_R R/\mathfrak{m}^3 \to \dots.$$

3.3. Finite-dimensional algebras. For a finite-dimensional algebra R, the only known values of $\mathrm{KG}(R)$ are finite or undefined. We give examples of both cases and state some open questions and conjectures concerning the connections between Krull–Gabriel dimension and other notions of complexity in a module category.

The first examples we will consider are those with finite Krull–Gabriel dimension.

THEOREM 3.5 ([**2**, Corollary 3.14]). *Let R be a finite-dimensional algebra. Then* $\mathrm{KG}(R) = 0$ *if and only if R is of finite representation type.*

THEOREM 3.6 ([**20**, Corollary 11.4],[**17**, Theorem 3.6]). *Let R be an Artin algebra. Then* $\mathrm{KG}(R) \neq 1$.

The first example of Krull–Gabriel dimension being explicitly calculated in the functor category $\mathrm{F}_{\mathrm{fp}}(R)$ was carried out by Geigle in [**12**] for R a tame hereditary algebra of infinite representation type. He described both the simple functors at each level of the filtration and the factor categories $\mathcal{A}_{\alpha+1}/\mathcal{A}_{\alpha}$.

THEOREM 3.7 ([**12**]). *Let R be a tame hereditary algebra of infinite representation type. Then* $\mathrm{KG}(R) = 2$.

Since the Krull–Gabriel dimension of R is defined, we have that R satisfies the isolation condition. Thus, by Remark 2.13 the above result for tame hereditary algebras is implicit in the descriptions of the Ziegler spectrum given in [**30**] and [**39**].

The Krull–Gabriel dimension of many other algebras closely related to tame hereditary has been shown to equal 2. For example, [**13**] covers the tame tilted algebras, algebras which are stably equivalent to tame hereditary algebras and domestic one relation algebras. Other examples of $\mathrm{KG}(R) = 2$ can be found in [**9**], [**44**], [**26**], [**42**].

It has been known for some time that, for every $n \in \mathbb{N}$, there exists a domestic string algebra R of Krull–Gabriel dimension $n \geq 2$.

For example, the algebra with the quiver

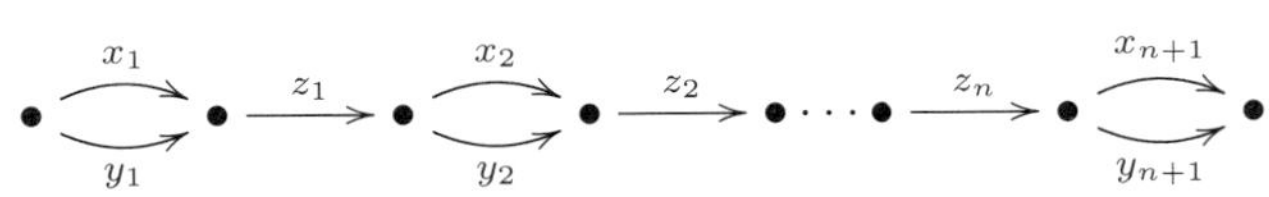

and relations $z_i x_i = y_{i+1} z_i = 0$ has Krull–Gabriel dimension $n + 2$.

This was shown in [**6**] by describing the Ziegler spectrum of these algebras and then proving explicitly that the isolation condition holds.

This result was proved simultaneously in [**40**] by making use of connections to the transfinite radical of mod-R.

DEFINITION 3.8. Let $\mathcal{I}$ be an ideal in mod-R. Then the finite powers of $\mathcal{I}$ are defined in the usual way:

$$\mathcal{I}^n := \left\{ \sum_{i=1}^{m} f_{in} \cdots f_{i1} \mid f_{ij} \in \mathcal{I} \text{ and } m < \omega \right\}.$$

Then for any limit ordinal λ let $\mathcal{I}^\lambda = \bigcap_{\beta<\lambda} \mathcal{I}^\beta$ and for any ordinal $\alpha := \lambda + n$ where λ is a limit ordinal and $n < \omega$ let $\mathcal{I}^\alpha := (\mathcal{I}^\lambda)^{n+1}$. We refer to the ideals $\mathcal{I}^\alpha$ for ordinals α as the **transfinite powers of** $\mathcal{I}$. Also let $\mathcal{I}^\infty := \bigcap_\alpha \mathcal{I}^\alpha$.

For an Artin algebra R, we will denote the Jacobson radical of mod-R by rad_R. That is, rad_R is the ideal of mod-R generated by the non-isomorphisms between indecomposable modules.

PROPOSITION 3.9. *Let R be an Artin algebra and let $\mathbf{X}_\alpha$ denote the closed subset of* $\mathrm{Zg}(R\text{-Mod})$ *containing points with Cantor–Bendixson rank greater than or equal to α.*

If R has Krull–Gabriel dimension β, then $\mathrm{rad}^{\omega\beta+n} = 0$ for some $n \in \mathbb{N}$. Moreover, for each successor ordinal α, every element of $\mathrm{rad}_R^{\omega\alpha}$ factors through a finite direct sum of modules in $\mathbf{X}_\alpha$.

PROOF. The first statement is [**23**, Corollary 8.14]. Since the Krull–Gabriel dimension is defined, the isolation condition holds and the Cantor–Bendixson rank is therefore defined. The latter result then follows immediately from [**32**] since, in any topological space where the Cantor–Bendixson rank is defined, the isolated points in the space are dense (see, for example, [**33**, Lem. 5.3.57]). □

OPEN QUESTION 3.10 ([**33**, Question 9.1.13]). If $\mathrm{rad}_R^{\omega\alpha+n} = 0$ for some ordinal α, is the Krull–Gabriel dimension of R less than or equal to α?

CONJECTURE 3.11 ([**41**]). *Let $n \geq 2$ and let R be a finite-dimensional algebra. Then the Krull–Gabriel dimension of R is equal to n if and only if $\mathrm{rad}_R^{\omega(n-1)} \neq 0$ and $\mathrm{rad}_R^{\omega n} = 0$.*

THEOREM 3.12. *Let R be a domestic string algebra over an algebraically closed field. Then the following statements hold.*

(1) *The Krull–Gabriel dimension of R is finite.*
(2) *Conjecture* 3.11 *holds.*

PROOF. The Krull–Gabriel dimension was shown to equal the length of the longest path in a finite graph known as the bridge quiver in [**25**]. In [**40**] Schröer has previously shown that this is also the value of n such that $\mathrm{rad}_R^{\omega(n-1)} \neq 0$ and $\mathrm{rad}_R^{\omega n} = 0$. □

There are also large classes of finite-dimensional algebras with undefined Krull–Gabriel dimension. The dividing line between algebras with defined and undefined Krull–Gabriel dimension is still unclear but it is conjectured to be connected with the representation type of the algebra.

THEOREM 3.13 ([**23**, Proposition 8.15]). *Let R be a finite-dimensional algebra. If R is of wild representation type, then $\mathrm{KG}(R) = \infty$.*

One might guess, given the above, that the Krull–Gabriel dimension of an algebra being undefined is equivalent to being of wild representation type. However there are examples of tame algebras with undefined Krull–Gabriel dimension. Examples include non-domestic string algebras [**40**] and non-domestic canonical algebras [**13**]. A different conjecture is therefore natural.

CONJECTURE 3.14 ([**34**, Conjecture 1.5]). *Let R be a finite-dimensional algebra. Then the following are equivalent.*

(1) *R is of tame domestic representation type.*
(2) *The Krull–Gabriel dimension of R is defined.*
(3) *The Krull–Gabriel dimension of R is finite.*

3.4. Compactly generated triangulated categories. In Example 1.14 we saw that when considering a compactly generated triangulated category $\mathcal{C}$ with compact generators $\mathcal{C}^{\mathrm{c}}$ we should analyse the category $(\text{mod-}\mathcal{C}^{\mathrm{c}})^{\mathrm{op}}$.

A Krull–Gabriel filtration of this category yields the same connections with the morphisms in $\mathcal{C}^{\mathrm{c}}$ (as in Section 2.2) and the Cantor–Bendixson rank of $\mathrm{Zg}(\mathcal{C})$ (as in Remark 2.11).

The first two results in the following theorem follow from [**5**]; the latter is explicitly contained in [**1**]. The third result follows from [**11**, Theorem 8.1] combined with the corresponding results for $\mathrm{KG}(R)$.

THEOREM 3.15. *Let* $\mathcal{A} = (\text{mod-}\mathcal{C}^{\mathrm{c}})^{\mathrm{op}}$.

(1) *Let R be a derived-discrete algebra and let* $\mathcal{C} = \mathrm{D}(R\text{-Mod})$ *with compact generators* $\mathcal{C}^{\mathrm{c}} \cong \mathrm{K}^{\mathrm{b}}(R\text{-proj})$. *Then*

$$\mathrm{KG}(\mathcal{A}) = \begin{cases} 1 & \text{if } gl.dim(R) = \infty \\ 2 & \text{if } gl.dim(R) < \infty \end{cases}$$

(2) *Let R be a derived-discrete algebra and let* $\mathcal{C} = \mathrm{K}(R\text{-Proj})$ *with compact generators* $\mathcal{C}^{\mathrm{c}} \cong \mathrm{D}^{\mathrm{b}}(R\text{-mod})$. *Then* $\mathrm{KG}(\mathcal{A}) = 2$.

(3) *Let R be a finite-dimensional hereditary algebra over an algebraically closed field and let* $\mathcal{C} = \mathrm{D}(R\text{-Mod})$ *with compact generators* $\mathcal{C}^{\mathrm{c}} \cong \mathrm{K}^{\mathrm{b}}(R\text{-proj})$. *Then*

$$\mathrm{KG}(\mathcal{A}) = \begin{cases} 0 & \text{if } R \text{ is of finite representation type} \\ 2 & \text{if } R \text{ is of tame representation type} \\ \infty & \text{if } R \text{ is of wild representation type} \end{cases}$$

References

[1] Kristin Krogh Arnesen, Rosanna Laking, David Pauksztello, and Mike Prest, *The Ziegler spectrum for derived-discrete algebras*, Adv. Math. **319** (2017), 653–698. MR3695887

[2] Maurice Auslander, *A functorial approach to representation theory*, Representations of algebras (Puebla, 1980), Lecture Notes in Math., vol. 944, Springer, Berlin-New York, 1982, pp. 105–179. MR672116

[3] Dagmar Baer, *Homological properties of wild hereditary Artin algebras*, Representation theory, I (Ottawa, Ont., 1984), Lecture Notes in Math., vol. 1177, Springer, Berlin, 1986, pp. 1–12, DOI 10.1007/BFb0075255. MR842456

[4] Apostolos Beligiannis, *Relative homological algebra and purity in triangulated categories*, J. Algebra **227** (2000), no. 1, 268–361, DOI 10.1006/jabr.1999.8237. MR1754234

[5] Grzegorz Bobiński and Henning Krause, *The Krull-Gabriel dimension of discrete derived categories*, Bull. Sci. Math. **139** (2015), no. 3, 269–282, DOI 10.1016/j.bulsci.2014.09.001. MR3336684

[6] Kevin Burke and Mike Prest, *The Ziegler and Zariski spectra of some domestic string algebras*, Algebr. Represent. Theory **5** (2002), no. 3, 211–234, DOI 10.1023/A:1016581927685. MR1921759

[7] William Crawley-Boevey, *Locally finitely presented additive categories*, Comm. Algebra **22** (1994), no. 5, 1641–1674, DOI 10.1080/00927879408824927. MR1264733

[8] William Crawley-Boevey, *Infinite-dimensional modules in the representation theory of finite-dimensional algebras*, Algebras and modules, I (Trondheim, 1996), CMS Conf. Proc., vol. 23, Amer. Math. Soc., Providence, RI, 1998, pp. 29–54. MR1648602

[9] Rolf Farnsteiner and Andrzej Skowroński, *Classification of restricted Lie algebras with tame principal block*, J. Reine Angew. Math. **546** (2002), 1–45, DOI 10.1515/crll.2002.043. MR1900992

[10] Pierre Gabriel, *Des catégories abéliennes* (French), Bull. Soc. Math. France **90** (1962), 323–448. MR0232821

[11] Grigory Garkusha and Mike Prest, *Triangulated categories and the Ziegler spectrum*, Algebr. Represent. Theory **8** (2005), no. 4, 499–523, DOI 10.1007/s10468-005-8147-2. MR2199207

[12] Werner Geigle, *The Krull-Gabriel dimension of the representation theory of a tame hereditary Artin algebra and applications to the structure of exact sequences*, Manuscripta Math. **54** (1985), no. 1-2, 83–106, DOI 10.1007/BF01171701. MR808682

[13] Werner Geigle, *Krull dimension and Artin algebras*, Representation theory, I (Ottawa, Ont., 1984), Lecture Notes in Math., vol. 1177, Springer, Berlin, 1986, pp. 135–155, DOI 10.1007/BFb0075263. MR842464

[14] Lorna Gregory, *Decidability of theories of modules over canonical algebras of tubular type*, available at arXiv:1603.03284.

[15] Laurent Gruson and Christian U. Jensen, *Modules algébriquement compacts et foncteurs* $\varprojlim^{(i)}$, C. R. Acad. Sci. Paris Sér. A-B **276** (1973), A1651–A1653. MR0320112

[16] Richard Harland and Mike Prest, *Modules with irrational slope over tubular algebras*, Proc. Lond. Math. Soc. (3) **110** (2015), no. 3, 695–720, DOI 10.1112/plms/pdu068. MR3342102

[17] Ivo Herzog, *The endomorphism ring of a localized coherent functor*, J. Algebra **191** (1997), no. 1, 416–426, DOI 10.1006/jabr.1997.6920. MR1444506

[18] Christian U. Jensen and Helmut Lenzing, *Model-theoretic algebra with particular emphasis on fields, rings, modules*, Algebra, Logic and Applications, vol. 2, Gordon and Breach Science Publishers, New York, 1989. MR1057608

[19] Henning Krause, *The spectrum of a locally coherent category*, J. Pure Appl. Algebra **114** (1997), no. 3, 259–271, DOI 10.1016/S0022-4049(95)00172-7. MR1426488

[20] Henning Krause, *Generic modules over Artin algebras*, Proc. London Math. Soc. (3) **76** (1998), no. 2, 276–306, DOI 10.1112/S0024611598000094. MR1490239

[21] Henning Krause, *Coherent functors in stable homotopy theory*, Fund. Math. **173** (2002), no. 1, 33–56, DOI 10.4064/fm173-1-3. MR1899046

[22] Henning Krause, *Smashing subcategories and the telescope conjecture—an algebraic approach*, Invent. Math. **139** (2000), no. 1, 99–133, DOI 10.1007/s002229900022. MR1728877

[23] Henning Krause, *The spectrum of a module category*, Mem. Amer. Math. Soc. **149** (2001), no. 707, x+125, DOI 10.1090/memo/0707. MR1803703

[24] Rosanna Laking, *String algebras in representation theory*, University of Manchester, 2016.

[25] Rosanna Laking, Mike Prest, and Gena Puninski, *Krull–Gabriel dimension of domestic string algebras*, Trans. Amer. Math. Soc., to appear.

[26] Piotr Malicki, *Krull dimension of tame generalized multicoil algebras*, Algebr. Represent. Theory **18** (2015), no. 3, 881–894, DOI 10.1007/s10468-015-9520-4. MR3357952

[27] J. C. McConnell and J. C. Robson, *Noncommutative Noetherian rings*, Revised edition, Graduate Studies in Mathematics, vol. 30, American Mathematical Society, Providence, RI, 2001. With the cooperation of L. W. Small. MR1811901

[28] Amnon Neeman, *Triangulated categories*, Annals of Mathematics Studies, vol. 148, Princeton University Press, Princeton, NJ, 2001. MR1812507

[29] N. Popescu, *Abelian categories with applications to rings and modules*, Academic Press, London-New York, 1973. London Mathematical Society Monographs, No. 3. MR0340375

[30] Mike Prest, *Ziegler spectra of tame hereditary algebras*, J. Algebra **207** (1998), no. 1, 146–164, DOI 10.1006/jabr.1998.7472. MR1643078

[31] Mike Prest, *Morphisms between finitely presented modules and infinite-dimensional representations*, Algebras and modules, II (Geiranger, 1996), CMS Conf. Proc., vol. 24, Amer. Math. Soc., Providence, RI, 1998, pp. 447–455. MR1648645

[32] Mike Prest, *Ideals in mod-R and the ω-radical*, J. London Math. Soc. (2) **71** (2005), no. 2, 321–334, DOI 10.1112/S0024610704006064. MR2122431

[33] Mike Prest, *Purity, spectra and localisation*, Encyclopedia of Mathematics and its Applications, vol. 121, Cambridge University Press, Cambridge, 2009. MR2530988

[34] Mike Prest and Jan Schröer, *Serial functors, Jacobson radical and representation type*, J. Pure Appl. Algebra **170** (2002), no. 2-3, 295–307, DOI 10.1016/S0022-4049(01)00133-5. MR1904849

[35] Gennadi Puninski, *Cantor-Bendixson rank of the Ziegler spectrum over a commutative valuation domain*, J. Symbolic Logic **64** (1999), no. 4, 1512–1518, DOI 10.2307/2586792. MR1780066

[36] Gena Puninski, *The Krull-Gabriel dimension of a serial ring*, Comm. Algebra **31** (2003), no. 12, 5977–5993, DOI 10.1081/AGB-120024862. MR2014911

[37] Gena Puninski and Mike Prest, *Ringel's conjecture for domestic string algebras*, Math. Z. **282** (2016), no. 1-2, 61–77, DOI 10.1007/s00209-015-1532-6. MR3448374

[38] Geert Reynders, *Ziegler spectra of serial rings with Krull dimension*, Comm. Algebra **27** (1999), no. 6, 2583–2611, DOI 10.1080/00927879908826583. MR1687254

[39] Claus Michael Ringel, *The Ziegler spectrum of a tame hereditary algebra*, Colloq. Math. **76** (1998), no. 1, 105–115. MR1611289

[40] Jan Schröer, *On the Krull-Gabriel dimension of an algebra*, Math. Z. **233** (2000), no. 2, 287–303, DOI 10.1007/PL00004799. MR1743438

[41] Jan Schröer, *The Krull-Gabriel dimension of an algebra—open problems and conjectures*, Infinite length modules (Bielefeld, 1998), Trends Math., Birkhäuser, Basel, 2000, pp. 419–424. MR1789229

[42] Andrzej Skowroński, *The Krull-Gabriel dimension of cycle-finite Artin algebras*, Algebr. Represent. Theory **19** (2016), no. 1, 215–233, DOI 10.1007/s10468-015-9571-6. MR3465899

[43] Bo Stenström, *Rings and modules of quotients*, Lecture Notes in Mathematics, Vol. 237, Springer-Verlag, Berlin-New York, 1971. MR0325663

[44] Michał Wenderlich, *Krull dimension of strongly simply connected algebras*, Bull. Polish Acad. Sci. Math. **44** (1996), no. 4, 473–480. MR1420962

[45] Martin Ziegler, *Model theory of modules*, Ann. Pure Appl. Logic **26** (1984), no. 2, 149–213, DOI 10.1016/0168-0072(84)90014-9. MR739577

Mathematisches Institut, Universität Bonn, Endenicher Allee 60 53115 Bonn, Germany

Email address: laking@math.uni-bonn.de

Contemporary Mathematics
Volume **705**, 2018
http://dx.doi.org/10.1090/conm/705/14203

On the K-theory of weighted projective curves

Helmut Lenzing

ABSTRACT. We present a largely self contained account on the K-theory of a weighted smooth projective curve over an algebraically closed field. In particular, we discuss the weighted version of divisor theory, Euler form, and Riemann-Roch theorem. This includes a treatment of the orbifold Euler characteristic.

CONTENTS

Introduction

Throughout we work over an algebraically closed base field k. This article deals with the K-theory of weighted smooth projective curves (weighted projective curves, for short). In this article, the term K-theory of a weighted projective curve refers to the study of the K_0-group, equipped with the Euler form, and of associated invariants. The K_1-group will be briefly mentioned. By definition, a weighted projective curve $\mathbb{X}$ is a pair (X, w), consisting of a smooth projective curve X and a weight function w which is an integral-valued positive function on X such that

2010 *Mathematics Subject Classification.* Primary: 18F30, 30F10, 14H55.

$w(x) > 1$ holds for only finitely many (closed) points $x_1, \ldots, x_t$ of X. We call the integers $p_i = w(x_i)$, $i = 1, \ldots, t$, the *weights* of $\mathbb{X}$, and write $\mathbb{X} = X\begin{pmatrix} p_1, p_2, \ldots, p_t \\ x_1, x_2, \ldots, x_t \end{pmatrix}$ or simply $\mathbb{X} = X\langle p_1, p_2, \ldots, p_t \rangle$ in contexts where the position of the weights does not matter.

Weighted projective curves, in particular weighted projective lines [**8**], where X is the projective line $\mathbf{P}^1$, occur in many different incarnations: as compact Riemann surfaces, equipped with a weight function [**15**], as finite group quotients of smooth projective curves (good characteristic assumed), as smooth projective curves equipped with a parabolic structure [**30**], [**5**], as holomorphic orbifolds, as one dimensional Deligne-Mumford curves (stacks) [**2**], as root stacks [**4**], [**31**], [**1**] and, for the base field of complex numbers, as finite (orbifold) group quotient of a compact Riemann surface, see [**15**].

Instead of dealing with any of the particular incarnations, our focus lies on the category of coherent sheaves associated with a weighted projective curve, which forms a bridge between the different incarnations. By work of Reiten and van den Bergh [**27**], see also [**23**], [**33**], [**19**] the category of coherent sheaves on a weighted projective curve $\mathbb{X}$ can be characterized to be a Hom-finite (over k) hereditary noetherian k-category $\mathcal{H}$ with Serre duality in the form $\mathrm{Ext}^1(A, B) = D(\mathrm{Hom}(B, \tau A))$, where τ is a self-equivalence of $\mathcal{H}$ and where the function field $K = k(\mathcal{H})$ of $\mathcal{H}$ has infinite k-dimension. Here, D denotes the usual duality for finite dimensional k-vector spaces. The category of coherent sheaves on (X, w) can, for instance, be constructed by means of a sheaf of hereditary orders over the underlying smooth projective curve X [**27**], see also [**12**], [**14**]. For the present article we apply instead the p-cycle construction of [**18**] to the category of coherent sheaves $\mathrm{coh}\, X$ which makes the K-theoretic implications immediately transparent.

We finally note that for the case of weighted projective lines there exists already detailed accounts on their K-theory, see [**17**], [**13**] which however do not cover divisor theory. For the study of weighted curves for non-algebraically closed base fields we recommend [**14**].

1. Coherent sheaves on a smooth projective curve

Let X be a smooth projective curve over k. By $\mathcal{H} = \mathrm{coh}\, X$ we denote its category of coherent sheaves and by $\mathcal{H}_0 = \mathrm{coh}_0 X$ its full subcategory of sheaves of finite length. Let $\mathcal{O}_X$ be a line bundle of X that we call structure sheaf of X. The k-dimension of $\mathrm{Ext}^1(\mathcal{O}_X, \mathcal{O}_X)$ is called the *genus* of X. An even more important invariant of X is the function field $K = k(X)$ of X given by the equivalence $\mathcal{H}/\mathcal{H}_0 = \mathrm{mod}(K)$, where $\mathrm{mod}(K)$ denotes the category of finite dimensional K-vector spaces and where $\widetilde{\mathcal{H}} = \mathcal{H}/\mathcal{H}_0$ is the Serre quotient [**7**] of $\mathcal{H}$ modulo its Serre subcategory $\mathcal{H}_0$. Each object in this quotient category has the form $\mathcal{O}_X^n$ for some integer n; moreover, the endomorphism ring of $\mathcal{O}_X$ in $\widetilde{\mathcal{H}}$ is a commutative field K, which is a function field K of one variable over k, and called the *function field* of X.

We next collect some fundamental facts about the category $\mathrm{coh}\, X$. By definition, a coherent sheaf E is called a *vector bundle* if E does not have any simple subobject. By $\mathrm{vect} X$ we denote the full subcategory of $\mathrm{coh}\, X$ formed by vector bundles. A *line bundle* is a vector bundle of rank one, and necessarily indecomposable.

THEOREM 1.1. *The category $\mathcal{H} = \operatorname{coh} X$ has the following properties.*

(1) *$\mathcal{H}$ is a noetherian abelian Hom-finite category having Serre duality in the form $\operatorname{Ext}^1(A,B) = D(\operatorname{Hom}(B,\tau A))$ for a self-equivalence τ of $\mathcal{H}$.*
(2) *$\mathcal{H}$ is also hereditary and has almost-split sequences with τ serving as the Auslander-Reiten translation.*
(3) *Each coherent sheaf is the direct sum $E \oplus U$ of a vector bundle E and a coherent sheaf U of finite length.*
(4) *The Auslander-Reiten translation τ, when restricted to the full subcategory $\operatorname{coh}_0 X$ of coherent sheaves of finite length, is isomorphic to the identity functor.*
(5) *Each indecomposable sheaf U of finite length ℓ is uniserial, that is, has a unique composition series with simple factors. Accordingly U is uniquely determined by its simple socle S and its length. Each simple sheaf S is concentrated in a uniquely determined point $x \in X$, yielding a bijection between points x and (isomorphism classes of) simple sheaves. Notation: $S = S_x$.*
(6) *Each vector bundle of rank r has a finite filtration $E = E_r \supset E_{r-1} \supset \ldots \supset E_1 \supset E_0 = 0$ with line bundle factors E_i/E_{i-1}.* □

For the proof we refer to [**10**] or, taking the view of hereditary categories, to [**19**]. Also the arguments of [**8**] apply. Here, we just remark that the rank of a coherent sheaf A may be defined as the length of A in the Serre quotient $\operatorname{coh} X/\operatorname{coh}_0 X$.

1.1. The Euler form. As for any Hom- and Ext-finite hereditary abelian category, the *Euler form* for $\operatorname{coh} X$ is the bilinear form on the Grothendieck group $\mathrm{K}_0(X) = \mathrm{K}_0(\operatorname{coh} X)$, given on classes of objects by the expression

$$\langle [A],[B]\rangle = \dim_k \operatorname{Hom}(A,B) - \dim_k \operatorname{Ext}^1(A,B), \tag{1.1}$$

where often we simply write $\langle A,B\rangle$. Due to Serre duality we further have

$$\langle b,a\rangle = -\langle a,\tau b\rangle \text{ for all members } a,b \text{ from } \mathrm{K}_0(X). \tag{1.2}$$

In particular, we have $\langle a,-\rangle = 0$ if and only if $\langle -,a\rangle = 0$.

LEMMA 1.2. *The kernel $N = \{a \in \mathrm{K}_0(X) \mid \langle a,-\rangle = 0\}$ of the Euler form (or of $\mathrm{K}_0(X)$) is generated by the differences $[S_x]-[S_y]$ ($x,y \in X$) of the classes of all pairs of simple sheaves.*

PROOF. We use the following properties

(1) The class of a vector bundle is a sum of classes of line bundles.
(2) Simple sheaves, concentrated in different points, are Hom-Ext-orthogonal.
(3) For each line bundle L the difference $[L]-[\mathcal{O}_X]$ belongs to the subgroup of $\mathrm{K}_0(X)$ generated by the classes of simple sheaves.

The proof is then straightforward. □

Clearly, the Euler form descends to the *reduced* (or *numerical*) Grothendieck group $\overline{\mathrm{K}}_0(X) = \mathrm{K}_0(X)/N$ and there yields a non-degenerate bilinear form. The class of a coherent sheaf A in $\overline{\mathrm{K}}_0(X)$ we denote by double brackets $[\![A]\!]$.

PROPOSITION 1.3. *The reduced Grothendieck group of $\operatorname{coh} X$ is the free abelian group on the class $[\![\mathcal{O}_X]\!]$ of the structure sheaf and the class $[\![S_x]\!]$ of any simple* *sheaf. Moreover, in this basis the Euler form is given by the following* Cartan matrix

$$\begin{pmatrix} \langle \mathcal{O}_X, \mathcal{O}_X \rangle & \langle \mathcal{O}_X, S_x \rangle \\ \langle S_x, \mathcal{O}_X \rangle & \langle S_x, S_x \rangle \end{pmatrix} = \begin{pmatrix} 1-g & 1 \\ -1 & 0 \end{pmatrix}, \tag{1.3}$$

where g is the genus of X.

If C denotes the Cartan matrix, the matrix $\Phi = -C^{-1}C^{tr}$ is called *Coxeter matrix*.

COROLLARY 1.4. *The Auslander-Reiten translation τ induces on the reduced Grothendieck group the Coxeter transformation, also denoted τ, given by the Coxeter matrix*

$$\begin{pmatrix} 1 & 0 \\ 2(g-1) & 1 \end{pmatrix} \tag{1.4}$$

whose characteristic polynomial, the Coxeter polynomial *of X, equals $(x-1)^2$. In particular, $[\![\tau\mathcal{O}_X]\!] = [\![\mathcal{O}_X]\!] + 2(g-1)[\![S_x]\!]$ where x is any point of X.* □

We note in this context that $2(1-g)$ is the *Euler characteristic* of X.

By Proposition 1.3 we obtain two "obvious" linear forms on the Grothendieck group, *rank* and *degree* given by the expressions

$$\mathrm{rk} = \langle -, S_x \rangle \text{ and } \deg = \langle \mathcal{O}_X, - \rangle - \langle \mathcal{O}_X, \mathcal{O}_X \rangle \mathrm{rk}, \tag{1.5}$$

respectively. We note first that the above K-theoretic definition of rank agrees for coherent sheaves with the sheaf theoretic one, given by the length of X in the Serre quotient $\mathrm{coh}\, X/\mathrm{coh}_0 X$: Clearly, the sheaf theoretic definition induces a linear form on $\overline{\mathrm{K}}_0(X)$ which agrees with the K-theoretic one on the basis $[\![\mathcal{O}_X]\!]$, $[\![S_x]\!]$ of $\overline{\mathrm{K}}_0(X)$.

Rank and degree have largely complementary properties:

PROPOSITION 1.5.

(1) *For each coherent sheaf A we have $\mathrm{rk}\, A \geq 0$; moreover, $\mathrm{rk}\, A = 0$ if and only if A has finite length.*
(2) *We have $\deg \mathcal{O}_X = 0$. Moreover, for each coherent sheaf A of finite length $\deg A$ equals the length of A. Hence for members of $\mathrm{coh}_0 X$ we have $\deg A \geq 0$ and $\deg A = 0$ if and only if $A = 0$.*

PROOF. The first assertion follows from the fact that the rank of a coherent sheaf X equals its length in the Serre quotient $\mathcal{H}/\mathcal{H}_0$. For the second assertion we use that $\mathrm{Hom}(\mathcal{O}_X, S_x) = k$ and, due to Serre duality, $\mathrm{Ext}^1(\mathcal{O}_X, S_x) = 0$ implying $\langle \mathcal{O}_X, S_x \rangle = 1$ for each simple sheaf S_x. □

PROPOSITION 1.6 (Riemann-Roch). *For any two coherent sheaves A, B on X we have*

$$\langle A, B \rangle = (1-g) \cdot \mathrm{rk}\, A \cdot \mathrm{rk}\, B + \begin{vmatrix} \mathrm{rk}\, A & \mathrm{rk}\, B \\ \deg A & \deg B \end{vmatrix}. \tag{1.6}$$

PROOF. It suffices to check the formula on the basis $[\![\mathcal{O}_X]\!]$, $[\![S_x]\!]$ of the reduced Grothendieck group $\overline{\mathrm{K}}_0(X)$, where its validity is obvious. □

We recall that in a hereditary abelian k-category $\mathcal{H}$ an object E is called *exceptional* if $\mathrm{End}(E) = k$ and $\mathrm{Ext}^1(E, E) = 0$.

COROLLARY 1.7. *The category* $\operatorname{coh} X$ *on a smooth projective curve* X *has an exceptional object only for genus zero, in which case exactly the line bundles are exceptional.*

PROOF. Assume E is exceptional in $\operatorname{coh} X$, then by the Riemann-Roch formula we obtain $1 = \langle E, E\rangle = (1-g)(\operatorname{rk} E)^2$. This implies $g = 0$ and $\operatorname{rk} E = 1$. Conversely, for genus zero, the structure sheaf, hence each line bundle, is exceptional. □

PROPOSITION 1.8. *A line bundle* L *is determined up to isomorphism by its class* $[L]$ *in* $\mathrm{K}_0(X)$.

PROOF. The proof is classical, and contained in the paper [**3**] that marks the beginning of algebraic K-theory. Let $\operatorname{Pic}(X)$ be the group of isomorphism classes of line bundles with respect to the tensor product of coherent sheaves on X. Following [**3**], mapping the class $[E]$ of a vector bundle E of rank r to the line bundle $\bigwedge^r E$, induces the *determinant homomorphism* $\det : \mathrm{K}_0(\operatorname{coh} X) \to \operatorname{Pic}(X)$, sending the class $[L]$ of a line bundle L to the isomorphism class of L, thus implying the claim. □

By Serre duality we have $\operatorname{Ext}^1(S_x, \mathcal{O}_X) = k$ for each x in X. There results a unique non-split exact sequence $0 \to \mathcal{O}_X \xrightarrow{\alpha_x} \mathcal{O}_X(x) \to S_x \to 0$, where $\mathcal{O}_X(x)$ is a line bundle.

THEOREM 1.9. *For a smooth projective curve we have the following alternative:*

(1) *For genus zero, all simple sheaves have the same class in* $\mathrm{K}_0(X)$. *Further, the space* $\operatorname{Hom}(\mathcal{O}_X, \mathcal{O}_X(x))$ *has dimension two for each* $x \in X$.
(2) *For genus greater zero, non-isomorphic simple sheaves have different classes in* $\mathrm{K}_0(X)$. *Further, for each* x *in* X, *the space* $\operatorname{Hom}(\mathcal{O}_X, \mathcal{O}_X(x))$ *has dimension one.*

PROOF. It suffices to deal with the second assertion. Assume that the simple sheaves S_x and S_y are non-isomorphic but have the same class in $\mathrm{K}_0(X)$. It follows that the line bundles $\mathcal{O}_X(x)$ and $\mathcal{O}_X(y)$ have the same class in $\mathrm{K}_0(X)$ and then are isomorphic by Proposition 1.8. We thus assume $\mathcal{O}_X(x) = \mathcal{O}_X(y)$. Since $u = \alpha_x$ and $v = \alpha_y$ have non-isomorphic cokernels S_x and S_y, the morphisms u and v are linearly independent over k. Hence the space $\operatorname{Hom}(\mathcal{O}_X, \mathcal{O}_X(x))$ has k-dimension at least two. From the exactness of the sequence $0 \to \operatorname{Hom}(\mathcal{O}_X, \mathcal{O}_X) \to \operatorname{Hom}(\mathcal{O}_X, \mathcal{O}_X(x)) \to \operatorname{Hom}(\mathcal{O}_X, S_x)$ we infer that $\operatorname{Hom}(\mathcal{O}_X, \mathcal{O}_X(x))$ has exactly dimension two. Therefore

$$2 - g = \langle \mathcal{O}_X, \mathcal{O}_X\rangle + \langle \mathcal{O}_X, S_x\rangle = \langle \mathcal{O}_X, \mathcal{O}_X(x)\rangle,$$

implying that $\operatorname{Ext}^1(\mathcal{O}_X, \mathcal{O}_X(x))$ has dimension g, the dimension of $\operatorname{Ext}^1(\mathcal{O}_X, \mathcal{O}_X)$. Invoking Serre duality it follows that $\operatorname{Hom}(\mathcal{O}_X, \tau\mathcal{O}_X)$ and $\operatorname{Hom}(\mathcal{O}_X(x), \tau\mathcal{O}_X)$ have the same dimension g.

For each non-zero morphism u from $\operatorname{Hom}(\mathcal{O}_X, \mathcal{O}_X(x))$ we obtain a non-split exact sequence

$$0 \to \mathcal{O}_X \xrightarrow{u} \mathcal{O}_X(x) \to S_u \to 0, \tag{1.7}$$

where S_u has finite length and degree one and hence is a simple sheaf. Using that $\operatorname{Ext}^1(S_u, \mathcal{O}_X)$ has dimension one, it follows that two non-zero members u and v of $\operatorname{Hom}(\mathcal{O}_X, \mathcal{O}_X(x))$ have isomorphic cokernels if and only if $ku = kv$.

Since $\mathrm{Hom}(\mathcal{O}_X, \mathcal{O}_X(x))$ has dimension two we obtain infinitely many pairwise non-isomorphic simple sheaves of the form S_u, $0 \neq u \in \mathrm{Hom}(\mathcal{O}_X, \mathcal{O}_X(x))$.

Assume now that $g \geq 1$ and hence $\mathrm{Hom}(\mathcal{O}_X, \tau\mathcal{O}_X)$ is non-zero. Fixing an embedding $j : \mathcal{O}_X \hookrightarrow \tau\mathcal{O}_X$ and applying $\mathrm{Hom}(\mathcal{O}_X, -)$ to the sequence (1.7) we see that u induces an epimorphism $\mathrm{Ext}^1(\mathcal{O}_X, \mathcal{O}_X) \to \mathrm{Ext}^1(\mathcal{O}_X, \mathcal{O}_X(x))$ which then is an isomorphism by equality of dimensions. By Serre duality this yields a bijection $\mathrm{Hom}(\mathcal{O}_X(x), \tau\mathcal{O}_X) \to \mathrm{Hom}(\mathcal{O}_X, \tau\mathcal{O}_X), h \mapsto h \circ u$. In particular, the inclusion $j : \mathcal{O}_X \to \tau\mathcal{O}_X$ factors into a product $j = h \circ u$ of two monomorphisms, implying that each simple module of the form S_u, $0 \neq u \in \mathrm{Hom}(\mathcal{O}_X, \mathcal{O}_X(x))$, embeds into the cokernel C of j. Since C has finite length and there are infinitely many non-isomorphic simples of the form S_u, this is impossible, thus contradicting our assumption $g > 0$. □

1.2. Shift action associated to a point. For $x \in X$ let S_x be the simple sheaf concentrated in x. For each vector bundle E on X we consider the S_x-universal extension

$$0 \to E \xrightarrow{x_E} E(x) \to \mathrm{Ext}^1(S_x, E) \otimes_k S_x \to 0 \tag{1.8}$$

which is uniquely defined up to isomorphism. This defines a self-equivalence σ_x : $\mathrm{vect}X \to \mathrm{vect}X$ which extends to a self-equivalence of $\mathrm{coh}\,X$, also denoted σ_x which, restricted to $\mathrm{coh}_0 X$, is (isomorphic to) the identity on $\mathrm{coh}_0 X$. For details we refer to [**19**, Section 10.3], see also [**21**, Section 3].

PROPOSITION 1.10. *The* divisor group $\mathrm{Div}(X)$, *defined as the free abelian group on* X, *acts as group of automorphisms on the category* $\mathrm{coh}\,X$ *by* $(\sum_{x\in X} d_x x)(E) = \left(\prod_{x\in X} \sigma_x^{d_x}\right)(E) =: E(D)$ *if* $D = \sum_{x\in X} d_x x$. *Moreover the following holds*

(1) *For each line bundle* L *and coherent sheaf of finite length* U *there is — up to isomorphism — a unique line bundle* $L(U)$ *such that* $[L(U)] = [L] + [U]$ *holds in* $\mathrm{K}_0(X)$.
(2) *The action of the divisor group* $\mathrm{Div}(X)$ *on the category* $\mathcal{L}(X)$ *of line bundles is transitive, up to isomorphism.*

PROOF. Due to the Hom-Ext-orthogonality for simple sheaves S_x, S_y with $x \neq y$, the equivalences σ_x and σ_y always commute. This immediately implies the first assertion.

Invoking a composition series for the finite length sheaf U, the class $[U]$ is a (finite) linear combination $\sum_{x\in X} d_x [S_x]$ of classes of simple sheaves. Hence $L(U) := (\prod_{x\in X} \sigma_x^{d_x})(L)$ is a line bundle with class $[L] + [U]$. By Proposition 1.8 $L(U)$ is uniquely determined by this property.

For the last assertion we consider line bundles L_0, L and a point x. For large n the Euler product $\langle L_0, L(nx)\rangle$ is > 0, yielding a non-zero morphism $L_0 \to L(nx)$ and then a short exact sequence $0 \to L_0 \to L(nx) \to V \to 0$, where the cokernel-term V has rank zero, and thus finite length. This implies $L(nx) = L_0(V)$ showing that L and L_0 are in the same $\mathrm{Div}(X)$-orbit, proving the last claim. □

COROLLARY 1.11. *Let* $\mathrm{K}_0'(X)$ *denote the subgroup of* $\mathrm{K}_0(X)$ *generated by the classes* $[S_x]$ *of all simple sheaves* S_x, $x \in X$. *Then* $\mathrm{K}_0(X) = \mathbb{Z}[\mathcal{O}_X] \oplus \mathrm{K}_0'(X)$.

PROOF. The subgroups $\mathbb{Z}[\mathcal{O}_X]$ and $\mathrm{K}_0'(X)$ have zero intersection. To show that their sum H equals $\mathrm{K}_0(X)$, it suffices by Theorem 1.1 to show that the class of each line bundle belongs to H. This is an immediate consequence of Proposition 1.10. □

1.3. The divisor sequence. Let X be a smooth projective curve. The next result is useful to treat the divisor theory of X.

PROPOSITION 1.12. *Let $f : L \to \bar{L}$ be a nonzero morphism between line bundles. Then the cokernel of f is a finite direct sum*

$$\operatorname{cok}(f) = \bigoplus_{x \in X} U_x,$$

where for each $x \in X$ the object U_x is either zero or else is an indecomposable finite length sheaf concentrated in x.

Further, as a member of $\operatorname{Hom}(L, \bar{L})$, *$f$ is determined by the class $[\operatorname{cok}(f)]_0$ in* $\mathrm{K}_0(\mathcal{H}_0)$ *up to a member of the multiplicative group k^* of k.*

PROOF. Assume that U_y is decomposable for some $y \in X$. Then U_y maps onto $S_y \oplus S_y$, yielding an epimorphism $\bar{L} \twoheadrightarrow S_y \oplus S_y$. Applying $\operatorname{Hom}(-, S_y)$ then yields a monomorphism $k \oplus k = \operatorname{Hom}(S_y^2, S_y) \to \operatorname{Hom}(\bar{L}, S_y) = k$, contradiction. This proves the first assertion.

Invoking this first result, the class $[\operatorname{cok}(f)]_0$ determines $U = \operatorname{cok}(f)$ up to isomorphism. We prove the second claim by induction on the length ℓ of U. For $\ell = 0$ nothing is to prove. If $\ell > 0$ we choose a simple subobject, say S_x, of U and analyze the pull-back diagram

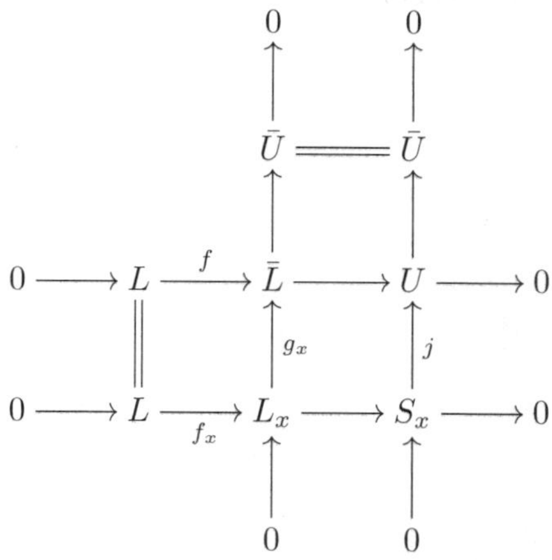

of η along inclusion $j : S_x \hookrightarrow U$. Since $\operatorname{Ext}^1(S_x, L) = k$, the line bundle L_x is unique up to isomorphism and f_x, as a member of $\operatorname{Hom}(L, L_x)$, is unique up to a scalar from k^*. By induction hypothesis, the same holds for g_x. The claim follows. □

The *function field* $k(X)$ of X can be viewed as the endomorphism ring of the structure sheaf $\mathcal{O}_X$ (or of any line bundle L) in the Serre quotient $\operatorname{coh} X / \operatorname{coh}_0 X$, where the latter category then can be identified with the category $\operatorname{mod}(k(X))$ of finite dimensional $k(X)$-vector spaces. By Quillen's localization sequence, see [**26**] the 'exact sequence' of abelian categories

$$0 \to \operatorname{coh}_0 X \to \operatorname{coh} X \to \operatorname{mod}(k(X)) \to 0$$

gives rise to a long exact sequence of higher K-groups, which in a truncated form leads to the two isomorphic exact sequences below

(1.9)

$$\begin{array}{ccccccccccccc} 0 \longrightarrow k^* & \longrightarrow & \mathrm{K}_1(\mathrm{mod}(k(X))) & \longrightarrow & \mathrm{K}_0(\mathrm{coh}_0 X) & \longrightarrow & \mathrm{K}_0(X) & \longrightarrow & \mathrm{K}_0(\mathrm{mod}(k(X))) & \longrightarrow & 0 \\ \| & & \| & & \| & & \| & & \| & & \\ 0 \longrightarrow k^* & \longrightarrow & k(X)^* & \xrightarrow{\mathrm{div}} & \mathrm{Div}(X) & \xrightarrow{\mathrm{nat}} & \mathrm{K}_0(X) & \xrightarrow{\mathrm{rk}} & \mathbb{Z} & \longrightarrow & 0 \end{array}$$

where specifically the lower sequence is called the *divisor sequence for* X. We note that the non-zero members of $k(X) = \mathrm{End}_{\widetilde{\mathcal{H}}}(\mathcal{O}_{\mathbb{X}})$ can be written as left fractions $s^{-1}u$, where L is a line bundle and $s : L \to \mathcal{O}_{\mathbb{X}}$, $u : L \to \mathcal{O}_{\mathbb{X}}$ are non-zero morphisms in $\mathrm{coh}\,\mathbb{X}$. We define the *divisor of* f as $\mathrm{div} f = [\mathrm{cok}(u)]_0 - [\mathrm{cok}(s)]_0 \in \mathrm{K}_0(\mathrm{coh}_0\mathbb{X})$.

THEOREM 1.13. *The divisor sequence for* X *is exact.*

PROOF. We put $\mathcal{H} = \mathrm{coh}\,X$, $\mathcal{H}_0 = \mathrm{coh}_0 X$, $\widetilde{\mathcal{H}} = \mathcal{H}/\mathcal{H}_0$ and $L_0 = \mathcal{O}_X$.

The natural homomorphism $\mathrm{End}_{\mathcal{H}}(L_0) \to \mathrm{End}_{\widetilde{\mathcal{H}}}(L_0)$, $f \mapsto \tilde{f}$, yields an interpretation of the embedding $k \hookrightarrow k(X)$ that we are going to use later on. In particular the sequence (1.9) is exact at k^*.

The sequence is exact at $k(X)^*$: By construction $\mathrm{div} f = 0$ holds for each $f \in k(X)$ represented by a non-zero member, that is, an isomorphism in $\mathrm{End}_{\mathcal{H}}(L_0)$. Assume conversely that f is represented by $f' : L' \to L_0$ with $0 \neq L' \subset L_0$ and that we have

$$0 = \mathrm{div} f = [\mathrm{cok}(f')]_0 - [\mathrm{cok}(j)]_0,$$

where $j : L' \hookrightarrow L_0$ is the inclusion. By Proposition 1.12 we conclude that there exists $\alpha \in k^*$ yielding a commutative diagram

$$\begin{array}{ccc} L_0 & \xrightarrow{\alpha} & L_0 \\ {\scriptstyle j}\uparrow & \nearrow{\scriptstyle f'} & \\ L' & & \end{array}$$

and $f = \alpha$ follows.

It is clear that composition $\mathrm{nat} \circ \mathrm{div}$ equals zero. Assume now that a is in $\mathrm{K}_0(\mathcal{H}_0)$, then $a = [V]_0 - [U]_0$ for suitable objects U and V from $\mathcal{H}_0$. We then get short exact sequences $0 \to L' \xrightarrow{s} L_0 \to U \to 0$ and $0 \to L'' \xrightarrow{a} L_0 \to V \to 0$ and such that $[L'] = [L'']$ in $\mathrm{K}_0(\mathcal{H})$. Since line bundles are determined by their class in $\mathrm{K}_0(X)$ we may assume that $L' = L''$. Putting $f = as^{-1}$ we obtain $f \in k(X)$ with $\mathrm{div} f = [V]_0 - [U]_0$, hence the exactness of the divisor sequence at $\mathrm{K}_0(\mathcal{H}_0)$.

Since the rank vanishes on objects of $\mathcal{H}_0$ composition $\mathrm{rk} \circ \mathrm{nat}$ is zero. Conversely assume that $u = [X] - [Y]$ has rank zero such that X and Y have the same rank r. By Corollary 1.11 $[X] = r[L_0] + x$ and $[Y] = r[L_0] + y$ where x and y hence also $u = x - y$ belong to the image of nat. This finishes the proof of the theorem. □

In particular, the image of the natural map $\mathrm{K}_0(\mathrm{coh}_0 X) \to \mathrm{K}_0(\mathrm{coh}\,X)$, $[U]_0 \mapsto [U]$, equals the cokernel of the divisor map $\mathrm{div} : k(X)^* \to \mathrm{K}_0(\mathrm{coh}_0 X) = \mathrm{Div}(X)$, the *divisor class group* $\mathcal{C}\ell(X)$ of X.

COROLLARY 1.14. *The map* $\mathcal{C}\ell(X) \oplus \mathbb{Z}[\mathcal{O}_X] \to \mathrm{K}_0(\mathrm{coh}\,X)$, $([D], n[\mathcal{O}_X] \mapsto [D] + n[\mathcal{O}_X]$, *yields an isomorphism* $\mathrm{K}_0(\mathrm{coh}\,X) = \mathcal{C}\ell(X) \oplus \mathbb{Z}[\mathcal{O}_X]$. □

REMARK 1.15. Let $\mathrm{Pic}(X)$ denote the *Picard group* of X, that is, the group of isomorphism classes of line bundles on X with respect to the tensor product. The action $\mathrm{Div}(X) \times \mathrm{Pic}(X) \to \mathrm{Pic}(X)$, $D \mapsto \mathcal{O}_X(D)$, from Proposition 1.10 induces an isomorphism between the class group $\mathcal{C}\ell(X)$ and the Picard group. The surjective degree function $\mathrm{Pic}(X) \to \mathbb{Z}$ further induces a splitting $\mathrm{Pic}(X) = \mathrm{Pic}_0(X) \oplus \mathbb{Z}$. It is known that $\mathrm{Pic}_0(X)$ is isomorphic to the Jacobian variety of X. Assume $g \geq 1$. For $k = \mathbb{C}$, the field of complex numbers, it is classical that the Jacobian variety is isomorphic to $(\mathbb{C}/(\mathbb{Z} \times \mathbb{Z}))^g$ [**10**], but also for an arbitrary algebraically closed field the Jacobian variety is a (nontrivial) divisible abelian group, see [**25**], which makes it impossible that $\mathrm{K}_0(X)$ is either free or finitely generated.

2. Coherent sheaves on a weighted projective curve

2.1. The category of p-cycles. Let $\mathcal{H}$ be a category of coherent sheaves on a smooth projective curve. Fixing a point x of X and an integer $p \geq 2$ we are going to form the category $\overline{\mathcal{H}}$ of p-cycles in x which will be viewed as the category of coherent sheaves on the weighted curve $\mathbb{X}$, having X as underlying smooth projective curve and x as its only weighted point (of weight p). Intuitively speaking the effect of the following construction is to form a p-th root of the natural transformation $x_E : E \to E(x)$ corresponding to the point x, relating to the construction of the Riemann surface of the p-th root function. This approach to weighted projective curves is particularly suitable for a K-theoretic analysis. Below we reproduce definition and main properties of p-cycles from [**18**].

DEFINITION 2.1. *A p-cycle E concentrated in x is a diagram*

$$\cdots \to E_n \xrightarrow{x_n} E_{n+1} \xrightarrow{x_{n+1}} E_{n+2} \xrightarrow{x_{n+2}} \cdots \longrightarrow E_{n+p} \xrightarrow{x_{n+p}} \cdots$$

which is p-periodic in the sense that $E_{n+p} = E_n(x)$, $x_{n+p} = x_n(x)$ and moreover $x_{n+p-1} \circ \cdots \circ x_n = x_{E_n}$ holds for each integer n.

A morphism $u : E \to F$ of p-cycles concentrated in x is a sequence of morphisms $u_n : E_n \to F_n$ which is p-periodic, i.e. satisfies $u_{n+p} = u_n$ for each n, and such that each diagram

$$\begin{array}{ccc} E_n & \xrightarrow{x_n} & E_{n+1} \\ u_n \downarrow & & \downarrow u_{n+1} \\ F_n & \xrightarrow{x_n} & F_{n+1} \end{array}$$

commutes. We denote p-cycles in the form $E_0 \xrightarrow{x_0} E_1 \xrightarrow{x_1} \cdots \to E_{p-1} \xrightarrow{x_{p-1}} E_0(x)$ and the category of all p-cycles concentrated in x by $\overline{\mathcal{H}} = \mathcal{H}\begin{pmatrix} p \\ x \end{pmatrix}$.

Obviously $\overline{\mathcal{H}}$ is an abelian category, where exactness and formation of kernels and cokernels has a pointwise interpretation. Moreover we have a full exact embedding

$$j : \mathcal{H} \hookrightarrow \overline{\mathcal{H}}, \qquad E \mapsto \bar{E} = E = \cdots = E \xrightarrow{x_E} E(x).$$

We therefore identify $\mathcal{H}$ with the resulting exact subcategory of $\overline{\mathcal{H}}$. We note that inclusion $j : \mathcal{H} \to \overline{\mathcal{H}}$ has a left adjoint ℓ and a right adjoint r which are both exact functors and are given by

$$\ell\left(E_0 \overset{x_0}{\to} E_1 \overset{x_1}{\to} \cdots \to E_{p-1} \overset{x_{p-1}}{\to} E_0(x)\right) = E_{p-1} \tag{2.1}$$

and

$$r\left(E_0 \overset{x_0}{\to} E_1 \overset{x_1}{\to} \cdots \to E_{p-1} \overset{x_{p-1}}{\to} E_0(x)\right) = E_0. \tag{2.2}$$

LEMMA 2.2. *The category $\overline{\mathcal{H}}$ is connected, abelian and noetherian. The simple objects (sheaves) of $\overline{\mathcal{H}}$ occur in two types:*

(1) *the simple sheaves S_y of $\mathcal{H}$ which are concentrated in a point $y \neq x$.*
(2) *the p simple sheaves*

$$\begin{array}{rc} S_1: & 0 \to 0 \to \cdots \to 0 \to S_x \to 0 \\ S_2: & 0 \to 0 \to \cdots \to S_x \to 0 \to 0 \\ \cdots & \cdots \\ S_{p-1}: & 0 \to S_x \to \cdots \to 0 \to 0 \to 0 \\ S_p: & S \to 0 \to \cdots \to 0 \to 0 \to S_x(x), \end{array}$$

said to be concentrated in x.

Each S_i is exceptional and $\mathrm{Ext}^1(S_i, S_{i+1}) = k$ for each $i \in \mathbb{Z}_p$.

If $\mathcal{S}$ is the system of simples $S_1, \ldots, S_{p-1}$ which are concentrated in x and different from S_p, then the extension closure $\langle \mathcal{S} \rangle$ of $\mathcal{S}$ is localizing in $\overline{\mathcal{H}}$. Moreover

a. *the quotient category $\overline{\mathcal{H}}/\langle \mathcal{S} \rangle \cong \mathcal{H}$ is equivalent to $\mathcal{H}$, the equivalence is induced by $r: \overline{\mathcal{H}} \to \mathcal{H}$.*
b. *the right perpendicular category $\mathcal{S}^\perp$ formed in $\overline{\mathcal{H}}$ is equivalent to $\mathcal{H}$.*

PROOF. Abelianness and noetherianness are obvious by pointwise consideration. It is straightforward from the construction and from the properties of $\mathcal{H}$ that the category $\mathcal{S}^\perp$ right perpendicular to $\mathcal{S}$ consists of exactly those p-cycles $E_0 \stackrel{x_0}{\to} E_1 \stackrel{x_1}{\to} \cdots \to E_{p-1} \stackrel{x_{p-1}}{\to} E_0(x)$ such that $x_0, \ldots, x_{p-2}$ are isomorphisms, hence — up to isomorphism — agree with the objects from $\mathcal{H}$. The remaining properties are straightforward to check. □

Note that the formation of categories of p-cycles can be done simultaneously at different points. Let therefore $x_1, x_2, \ldots, x_t$ be pairwise distinct points of X and $p_1, p_2, \ldots, p_t$ a corresponding sequence of weights, the iterated p-cycle construction leads to a category $\overline{\mathcal{H}} = \mathcal{H}\begin{pmatrix} p_1, p_2, \ldots, p_t \\ x_1, x_2, \ldots, x_t \end{pmatrix}$ that we view as category of coherent sheaves $\mathrm{coh}\,\mathbb{X}$ for the weighted projective curve given by the above weight data. Comparison with the categories of [**27**] yields the following:

THEOREM 2.3. *The category $\overline{\mathcal{H}} = \mathrm{coh}\,\mathbb{X}$ is again a hereditary noetherian k-category with Serre duality in the form $\mathrm{Ext}^1(A, B) = D(\mathrm{Hom}(B, \tau A))$. Moreover, the following properties hold:*

(1) *For each x in X there is a unique simple object (sheaf) S_x concentrated in x with the property $\mathrm{Hom}(\mathcal{O}_X, S_x) = k$.*
(2) *For $i = 1, \ldots, t$ we put $S_i = S_{x_i}$. Then the simple sheaves concentrated in x_i are exceptional and form a τ-orbit $(\tau^j S_i)_{j \in \mathbb{Z}_{p_i}}$ of period p_i. The remaining simples satisfy $\tau S_x = S_x$.*
(3) *The exceptional simple sheaves, different from $S_1, S_2, \ldots, S_t$ can be arranged to form an exceptional sequence $\mathcal{E}$ such that the right perpendicular category $\mathcal{E}^\perp$, formed in $\overline{\mathcal{H}}$, equals $\mathcal{H} = \mathrm{coh}\, X$.* □

COROLLARY 2.4. *The category $\mathcal{H}$ is a full, exact subcategory of $\overline{\mathcal{H}}$, and inclusion induces an equivalence of Serre quotients $\mathcal{H}/\mathcal{H}_0 = \overline{\mathcal{H}}/\overline{\mathcal{H}}_0$. Accordingly, $\mathcal{H}$ and $\overline{\mathcal{H}}$ share the same function field K and inclusion from $\mathcal{H}$ to $\overline{\mathcal{H}}$ is rank-preserving.* □

In particular, the structure sheaf $\mathcal{O}_X$ of $\mathcal{H}$ is a line bundle in $\overline{\mathcal{H}}$ which we use as *structure sheaf* for $\overline{\mathcal{H}}$: Notation: $\mathcal{O}_{\mathbb{X}}$. We further put $\operatorname{coh}\mathbb{X} = \overline{\mathcal{H}}$ and $\mathrm{K}_0(\mathbb{X}) = \mathrm{K}_0(\operatorname{coh}\mathbb{X})$.

By [**9**] and making use of Proposition 2.7 we get the following immediate consequence of the proposition.

COROLLARY 2.5. *The Grothendieck group* $\mathrm{K}_0(\mathbb{X})$ *decomposes into a direct sum*

$$\mathrm{K}_0(\mathbb{X}) = \mathrm{K}_0(X) \oplus \left(\bigoplus_{E\in\mathcal{E}} \mathbb{Z}[E] \right).$$

In particular, $\mathrm{K}_0(\mathbb{X}) = \mathrm{K}_0(X) \oplus \mathbb{Z}^n$ *with* $n = \sum_{i=1}^t (p_i - 1)$. *Moreover,* $\langle [E], [H] \rangle = 0$ *whenever* E *belongs to* $\mathcal{E}$ *and* H *belongs to* $\operatorname{coh} X$. □

2.2. The reduced (or numerical) Grothendieck group. Let $\mathbb{X} = (X, w)$ be a weighted projective curve. When speaking of the Grothendieck group $\mathrm{K}_0(\mathbb{X}) = \mathrm{K}_0(\operatorname{coh}\mathbb{X})$ we always assume that the Euler form is part of its structure. Factoring out the kernel of the Euler form, we obtain the reduced (or numerical) Grothendieck group $\overline{\mathrm{K}}_0(\mathbb{X})$, equipped with a non-degenerate Euler form.

LEMMA 2.6. *The kernel of the Euler form for* $\mathrm{K}_0(\mathbb{X})$ *is generated by the elements* $u_x - u_y$, $x, y \in X$, *where* $u_x = \sum_{j\in\mathbb{Z}_{w(x)}} [\tau^j S_x]$ *for the unique simple sheaf* S_x *concentrated in* x *with* $\operatorname{Hom}(\mathcal{O}_X, S_x) = k$. □

To analyze properties of the Grothendieck group $\mathrm{K}_0(\mathbb{X})$, or the *reduced Grothendieck group* $\overline{\mathrm{K}}_0(\mathbb{X}) = \mathrm{K}_0(\mathbb{X})/N$, the following proposition, going back to Hübner [**11**], compare [**19**], is quite useful. An exceptional sequence $E_1, E_2, \dots, E_n$ in $\operatorname{coh}\mathbb{X}$ consists of exceptional sheaves such that additionally $\operatorname{Ext}^j(E_h, E_i) = 0$ holds for $h > i$ and for $j = 0, 1$.

PROPOSITION 2.7. *Each exceptional sheaf* E *of* $\operatorname{coh}\mathbb{X}$ *is determined by its class* $[\![E]\!]$ *in the reduced Grothendieck group* $\overline{\mathrm{K}}_0(\mathbb{X})$. *In particular,* E *is determined by its class in* $\mathrm{K}_0(\mathbb{X})$. *Moreover, an exceptional sequence in* $\operatorname{coh} X$ *gives rise to a linearly independent sequence* $[\![E_1]\!], [\![E_2]\!], \dots, [\![E_n]\!]$ *in* $\overline{\mathrm{K}}_0(\mathbb{X})$ *(resp.* $[E_1], [E_2], \dots, [E_n]$ *in* $\mathrm{K}_0(\mathbb{X})$*), generating a direct summand in* $\overline{\mathrm{K}}_0(\mathbb{X})$ *(resp.* $\mathrm{K}_0(\mathbb{X})$*).*

PROOF. Let E and F be exceptional sheaves with $[\![E]\!] = [\![F]\!]$. Then $\langle [\![E]\!], [\![F]\!] \rangle = \langle [\![E]\!], [\![E]\!] \rangle = 1$, and hence there exists a non-zero morphism $f : E \to F$ whose kernel and image we denote by E' and F', respectively. We claim that f is an isomorphism and assume, for contradiction, that E' is non-zero. The assumption implies that $\operatorname{Hom}(F', E) = 0$: Otherwise there exists a nonzero morphism $g : F' \to E$ and the composition $E \twoheadrightarrow F' \xrightarrow{g} E$ yields a non-trivial endomorphism, hence an automorphism, of E. It then follows that the morphism $E \twoheadrightarrow F'$, induced by f, is an isomorphism, contradicting $E' \neq 0$. We have shown that the assumption $E' \neq 0$ implies $\operatorname{Hom}(F', E) = 0$. Further, since $\operatorname{coh}\mathbb{X}$ is hereditary the embedding $F' \hookrightarrow F$ induces an epimorphism $0 = \operatorname{Ext}^1(F, F) \twoheadrightarrow \operatorname{Ext}^1(F', F)$ implying $\operatorname{Ext}^1(F', F) = 0$. We thus obtain $\langle [\![F']\!], [\![E]\!] \rangle = -\dim \operatorname{Ext}^1(F', E) \leq 0$ and $\langle [\![F']\!], [\![F]\!] \rangle = \dim \operatorname{Hom}(F', F) \geq 0$. Since the classes of E and F in $\overline{\mathrm{K}}_0(\mathbb{X})$ agree, we get $0 = \langle [\![F']\!], [\![E]\!] \rangle = \langle [\![F']\!], [\![F]\!] \rangle = \dim \operatorname{Hom}(F', F)$, hence $\operatorname{Hom}(F', F) = 0$, contradicting $f \neq 0$. We have thus shown that there exists a monomorphism $f : E \to F$. Similarly, there exists a monomorphism $g : F \to E$ yielding non-zero endomorphisms, hence automorphisms, $f \circ g$ and $g \circ f$ of F resp. E. Therefore f and g are isomorphisms and the first claim follows. The last assertion uses that $\langle E_i, E_i\rangle = 1$ for each i and $\langle E_j, E_i\rangle = 0$ for $j > i$. □

2.3. Attaching tubes. An abelian group V, equipped with a (usually non-symmetric) bilinear form $\langle -,-\rangle$ and a $\mathbb{Z}$-linear automorphism τ is called a *bilinear group* if the identity $\langle y, x\rangle = -\langle x, \tau y\rangle$ holds for all members x, y of V. A bilinear group V is called a *bilinear lattice* if V is a free abelian group of finite rank and additionally the bilinear form is non-degenerate. For weighted projective lines, the process of attaching tubes from [**17**], that is described below, has been the key technique to understand the structure of the Grothendieck group.

A p-tube is a pair (T, s) where T is a bilinear group equipped with an automorphism τ of finite period p, $s \in T$ satisfies $\langle s, s\rangle = 1$ and, moreover, the τ-orbit $(\tau^j s)_{s\in\mathbb{Z}_p}$ is a $\mathbb{Z}$-basis of T. We further request

$$\langle \tau^i s, \tau^j s\rangle = \begin{cases} 1 & \text{if } j = i \text{ in } \mathbb{Z}_p \\ -1 & \text{if } j = i+1 \text{ in } \mathbb{Z}_p \\ 0 & \text{else,} \end{cases} \tag{2.3}$$

We note that $s_0 = \sum_{j\in\mathbb{Z}_p} s_j$ lies in the kernel of T, that is $\langle s_0, -\rangle$ so T is not a bilinear lattice. Moreover, $\langle s_0, s_0\rangle = 0$.

Now let V denote the bilinear group $\mathrm{K}_0(X)$ for a smooth projective curve X and let S be a simple sheaf on X. We are going to define a bilinear group $\bar{V}$ by attaching a p-tube T_p at $[S]$. First we define a bilinear form on $V \oplus T$ by putting

(1) $\langle v_1, v_2\rangle = \langle v_1, v_2\rangle_V$ for members v_1, v_2 from V and $\langle u_1, u_2\rangle = \langle u_1, u_2\rangle_T$ for members u_1, u_2 from T,

(2)
$$-\langle \tau^{j-1} s, v\rangle = \langle v, \tau^j s\rangle = \begin{cases} \langle v, [S]\rangle_V, & \text{if } j = 1 \text{ in } \mathbb{Z}_p, v \in V \\ 0, & \text{otherwise.} \end{cases}$$

As is easily checked, the subgroup U generated by $\sum_{j\in\mathbb{Z}_p} \tau^j s - [S]$ belongs to the kernel of the bilinear form on $V\oplus T$, thus yielding a bilinear group $\bar{V} = V \oplus T/U$. By definition, $\bar{V}$ is the *bilinear group obtained by attaching a p-tube* in $[S]$. Similarly, starting with the bilinear lattice $\overline{\mathrm{K}}_0(X)$ and the class $[\![S]\!]$ of a simple sheaf on X we obtain by attaching a p-tube to $[\![S]\!]$ a bilinear lattice $\bar{V}$.

For a given sequence of weights $p_1, p_2, \ldots, p_t$ and pairwise non-isomorphic simple sheaves $S_1, S_2, \ldots, S_t$ the above construction extends to yield an extension of $\mathrm{K}_0(X)$, respectively $\overline{\mathrm{K}}_0(X)$, by simultaneously attaching a p_i-tube in $[S_i]$, respectively $[\![S_i]\!]$, for each $i = 1, 2, \ldots, t$.

THEOREM 2.8. *Let $\mathbb{X}$ be a weighted projective curve, with weights $p_1, p_2, \ldots, p_t$ put in the points $x_1, x_2, \ldots, x_t$ of the smooth projective curve X underlying $\mathbb{X}$. Then Grothendieck group $\mathrm{K}_0(\mathbb{X})$ of $\operatorname{coh}\mathbb{X}$ arises from $\mathrm{K}_0(X)$ by inserting a p_i-tube T_i in the class $[S_i]$ of the simple sheaf S_i concentrated in x_i, for each $i = 1, 2, \ldots, t$.*

PROOF. For each $i = 1, 2, \ldots t$ let S_i denote the unique simple sheaf on $\mathbb{X}$ concentrated in x_i with the property $\operatorname{Hom}(\mathcal{O}_{\mathbb{X}}, S_i) = k$. Further let $\mathcal{E}$ denote the system of all remaining exceptional simple sheaves ordered into an exceptional sequence. We thus obtain a semi-orthogonal decomposition $\langle \operatorname{coh} X, \mathcal{E}\rangle$, yielding a decomposition of $\mathrm{K}_0(\mathbb{X})$ into $\mathrm{K}_0(X)$ and the subgroup of the classes spanned by the linearly independent system of classes E with E from $\mathcal{E}$. From this the proof of the claim is straightforward. $\square$

Corollary 2.9. *The reduced Grothendieck group* $\overline{\mathrm{K}}_0(\mathbb{X})$ *is the bilinear lattice arising from* $\overline{\mathrm{K}}_0(X)$ *by inserting* p_i*-tubes in the classes* $[\![S_i]\!]$, $i = 1, \dots, t$. *Moreover, the Coxeter transformation, induced by* τ *on* $\overline{\mathrm{K}}_0(\mathbb{X})$ *yields as characteristic polynomial the Coxeter polynomial* $(x-1)^2 \prod_{i=1}^{t} v_{p_i}$, *where* $v_n = (x^n - 1)/(x-1)$.

Proof. For the last assertion we use that, according to [**17**], each attachment of a p-tube in $[\![S_x]\!]$ for a non-weighted point x yields an additional factor v_p for the Coxeter polynomial. $\square$

We note that the weight sequence of $\mathbb{X}$ can be recovered from its Coxeter polynomial. However, the genus of the underlying curve X does not influence the Coxeter polynomial. With the above notations, we put $a = [\![\mathcal{O}_{\mathbb{X}}]\!]$, $s_i = [\![S_i]\!]$, $s_0 = [\![S_x]\!]$ for any ordinary simple sheaf from $\operatorname{coh} X$. Then the elements a, $(\tau^j s_i)_{j \in \mathbb{Z}_{p_i}}$, and s_0 form a generating system for $\overline{\mathrm{K}}_0(\mathbb{X})$.

2.4. Orbifold Euler characteristic and weighted Riemann-Roch. We recall that for a smooth projective curve X the *Euler characteristic* (or *Euler number*) χ_X is given by the expression $\chi_X = 2\langle \mathcal{O}_X, \mathcal{O}_X \rangle = 2(1 - g_X)$, where g_X is the genus of X, defined as the k-dimension of $\operatorname{Ext}^1(\mathcal{O}_X, \mathcal{O}_X)$. Assume now that $\mathbb{X} = X\langle p_1, p_2, \dots, p_t \rangle$ is a weighted projective curve with structure sheaf $\mathcal{O}_{\mathbb{X}}$. Then $\operatorname{Ext}^1(\mathcal{O}_{\mathbb{X}}, \mathcal{O}_{\mathbb{X}})$ has dimension g_X, so the *homological genus* does not reflect the weights. To define an Euler characteristic $\chi_{\mathbb{X}}$ of $\mathbb{X}$ that keeps notice of the weights, we use the *weighted Euler form*, defined as an average of $\bar{p} = lcm(p_1, p_2, \dots, p_t)$ twisted Euler forms:

$$\langle\!\langle x, y \rangle\!\rangle = \frac{1}{\bar{p}} \left(\sum_{j=0}^{\bar{p}-1} \langle \tau^j x, y \rangle \right). \tag{2.4}$$

Here, $\tau = \tau_{\mathbb{X}}$ denotes the Auslander-Reiten translation for $\operatorname{coh} \mathbb{X}$. We then define the *orbifold Euler characteristic* of $\mathbb{X}$ as $\chi_{\mathbb{X}} = \frac{2}{\bar{p}} \langle\!\langle \mathcal{O}_{\mathbb{X}}, \mathcal{O}_{\mathbb{X}} \rangle\!\rangle$. Clearly, for non-weighted smooth projective curves the orbifold Euler characteristic agrees with the ordinary Euler characteristic.

Theorem 2.10. *Let* $\mathbb{X} = X\langle p_1, p_2, \dots, p_t \rangle$ *be a weighted projective curve with underlying smooth projective curve* X. *Then the orbifold Euler characteristic of* $\mathbb{X}$ *is given by the expression*

$$\chi_{\mathbb{X}} = \frac{2}{\bar{p}} \langle\!\langle \mathcal{O}_{\mathbb{X}}, \mathcal{O}_{\mathbb{X}} \rangle\!\rangle = \chi_X - \sum_{i=1}^{t} \left(1 - \frac{1}{p_i} \right). \tag{2.5}$$

Proof. We work in the reduced Grothendieck group $\overline{\mathrm{K}}_0(\mathbb{X})$. We put $a = [\![\mathcal{O}_{\mathbb{X}}]\!]$, $s_0 = [\![S_x]\!]$ where S_x is the simple sheaf concentrated in a non-weighted point, $s_i = [\![S_i]\!]$, where S_i is the simple sheaf concentrated in x_i with $\operatorname{Hom}(\mathcal{O}_{\mathbb{X}}, S_i) = k$ $(i = 1, 2, \dots, t)$. Then we have

$$\tau a - a = -\sum_{i=1}^{t} s_i + (t - \chi_X) s_0, \tag{2.6}$$

where $\chi_X = 2\langle a, a \rangle = 2(1-g)$ is the Euler characteristic of the underlying smooth projective curve X. The validity of formula (2.6) is easily checked by forming for both sides the Euler product with each member of the generating system $a, s_0, \tau^j s_i$,

$i = 1, 2, \ldots, t$ and using non-degeneracy of the Euler form on $\overline{\mathrm{K}}_0(\mathbb{X})$. Then, applying $\sum_{j=0}^{\bar{p}-1} \tau^j$ to (2.6) implies

$$\tau^{\bar{p}} a - a = \bar{p}\, \delta\, s_0, \text{ where } \delta = \sum_{j=1}^{t} (1 - \frac{1}{p_i}) - \chi_X. \tag{2.7}$$

Finally, we obtain

$$2\langle\!\langle a, a\rangle\!\rangle + \bar{p}\delta = \langle\!\langle a, a\rangle\!\rangle + \langle\!\langle a, \tau^{\bar{p}} a\rangle\!\rangle = \frac{1}{\bar{p}} \left(\sum_{j=0}^{\bar{p}-1} \langle \tau^j a, a\rangle - \sum_{j=0}^{\bar{p}-1} \langle \tau^{(\bar{p}-1)-j} a, a\rangle \right) = 0, \tag{2.8}$$

proving the claim. □

Also for a weighted projective curve $\mathbb{X}$ we have two "obvious" linear forms *rank* and *degree* on the Grothendieck group $\mathrm{K}_0(\mathbb{X}) = \mathrm{K}_0(\mathrm{coh}\,\mathbb{X})$ given on coherent sheaves by the expressions

$$\mathrm{rk}\, A = \langle\!\langle A, S_x\rangle\!\rangle \text{ and } \deg A = \langle\!\langle \mathcal{O}_{\mathbb{X}}, A\rangle\!\rangle - \langle\!\langle \mathcal{O}_{\mathbb{X}}, \mathcal{O}_{\mathbb{X}}\rangle\!\rangle \mathrm{rk}\, A. \tag{2.9}$$

Again this K-theoretic rank agrees with the sheaf-theoretic rank, defined for a coherent sheaf A as its length in the Serre quotient $\mathrm{coh}\,\mathbb{X}/\mathrm{coh}_0\mathbb{X}$.

THEOREM 2.11 (Riemann-Roch). *Assume b, c are members of $\mathrm{K}_0(\mathbb{X})$, where $\mathbb{X} = X\langle p_1, p_2, \ldots, p_t\rangle$ then*

$$\langle\!\langle b, c\rangle\!\rangle = \langle\!\langle \mathcal{O}_X, \mathcal{O}_X\rangle\!\rangle \cdot \mathrm{rk}\, b \cdot \mathrm{rk}\, c + \frac{1}{\bar{p}} \cdot \begin{vmatrix} \mathrm{rk}\, b & \mathrm{rk}\, c \\ \deg b & \deg c \end{vmatrix}. \tag{2.10}$$

PROOF. Passing to the reduced Grothendieck group $\overline{\mathrm{K}}_0(\mathbb{X})$ one checks the equality of both sides of (2.10) on the generating system $a, s_0, \tau^j s_i$, $i = 1, 2, \ldots, t$, $j \in \mathbb{Z}_{p_i}$. □

REMARK 2.12. The factor $1/\bar{p}$ in front of the determinant in the Riemann-Roch formula is due to the fact that we have normalized the degree in order to be always an integer. One may argue in favour of a different normalization of the degree such that simple sheaves concentrated in ordinary points always get degree one and simple sheaves concentrated in a point of weight p get degree $1/p$. With this normalization the factor $1/\bar{p}$ disappears.

2.5. Impact of the Euler characteristic. In this section we collect a number of facts showing the importance of the Euler characteristic. The following trichotomy is known for weighted projective lines and has a proof along similar lines for weighted projective curves.

THEOREM 2.13. *Let $\mathbb{X} = X\langle p_1, p_2, \ldots, p_t\rangle$ and let $\chi_{\mathbb{X}} = \chi_X - \sum_{i=1}^{t}(1 - 1/p_i)$ be its orbifold Euler characteristic. We have the following three cases:*

(1) *If $\chi_{\mathbb{X}} > 0$, then $X = \mathbf{P}^1$ and the weight sequence is one of $\langle p, q\rangle$ $(p, q \geq 1)$, $\langle 2, 2, n\rangle$ $(n \geq 2)$, $\langle 2, 3, 3\rangle$, $\langle 2, 3, 4\rangle$ or $\langle 2, 3, 5\rangle$. In these cases $\mathrm{coh}\,\mathbb{X}$ has tame domestic representation type.*

(2) *If $\chi_{\mathbb{X}} = 0$, then $\mathbb{X}$ is either $\mathbf{P}^1\langle 2, 3, 6\rangle$, $\mathbf{P}^1\langle 2, 4, 4\rangle$, $\mathbf{P}^1\langle 3, 3, 3\rangle$, $\mathbf{P}^1\langle 2, 2, 2, 2\rangle$ or else $\mathbb{X}$ is a non-weighted smooth elliptic curve. In these cases $\mathrm{coh}\,\mathbb{X}$ has tame tubular representation type.*

(3) *In all remaining cases $\mathbb{X}$ has negative Euler characteristic $\chi_{\mathbb{X}} < 0$ and wild representation type.* □

For the base field of complex numbers more specific assertions can be made, because then the trichotomy agrees with the classical division of plane geometry into euclidean (elliptic), spherical (parabolic) and hyperbolic geometry, provided we exclude weighted projective lines $\mathbf{P}^1(p,q)$ with weights $p \neq q$, $p,q \geq 1$. For the arising concepts *orbifold fundamental group*, *universal orbifold cover* and for the proof we refer to [**15**].

THEOREM 2.14. *Let $\mathbb{X}$ be a weighted projective curve over $\mathbb{C}$. Then the fundamental group $\pi_1^{orb}(\mathbb{X})$ acts on the universal orbifold cover $\widetilde{\mathbb{X}}$ as group of deck transformations (the members of* $\mathrm{Aut}(\mathbb{X})$ *commuting with $\pi : \widetilde{\mathbb{X}} \to \mathbb{X}$). This action represents $\mathbb{X}$ as orbifold quotient*

$$\mathbb{X} = \widetilde{\mathbb{X}}/\pi_1^{orb}(\mathbb{X}).$$

Assuming, moreover, that $\mathbb{X}$ is not isomorphic to any $\mathbf{P}^1\langle p_1,p_2\rangle$ with $p_1 \neq p_2$, we have the following trisection:

(1) spherical*: If $\chi_{\mathbb{X}} > 0$ then $\mathbb{X}$ admits a Kähler metric of constant curvature $+1$. Further $\widetilde{\mathbb{X}} = \mathbf{P}^1$, $\pi_1^{orb}(\mathbb{X})$ is a finite polyhedral group and $\mathbb{X}$ is one of $\mathbf{P}^1\langle n,n\rangle$, $\mathbf{P}^1\langle 2,2,n\rangle$, or $\mathbf{P}^1\langle 2,3,p\rangle$ with $p = 3,4,5$.*
(2) parabolic*: If $\chi_{\mathbb{X}} = 0$ then $\mathbb{X}$ admits a Kähler metric of constant curvature 0. Further $\widetilde{\mathbb{X}} = \mathbb{C}$, and $\mathbb{X}$ is either a smooth elliptic curve or else a weighted projective line of tubular type $\langle 2,3,6\rangle$, $\langle 2,4,4\rangle$, $\langle 3,3,3\rangle$ or $\langle 2,2,2,2\rangle$.*
(3) hyperbolic*: If $\chi_{\mathbb{X}} < 0$ then $\mathbb{X}$ admits a Kähler metric of constant curvature -1, $\widetilde{\mathbb{X}} = \mathbb{H}$, and $\mathbb{X}$ has hyperbolic type.* □

We only remark that the proof of the theorem is based on the general Riemann mapping theorem, that is, the uniformization theorem of Poincaré and Koebe [**6**] stating that a simply connected Riemann surface is holomorphically isomorphic to either $\mathbf{P}^1$, $\mathbb{C}$ or $\mathbb{H}$, where $\mathbb{H}$ denotes the open upper complex half-plane. It also uses that $\mathbf{P}^1$, $\mathbb{C}$ or $\mathbb{H}$ have Kähler metrics with constant curvature $+1$, 0 and -1, respectively, that then descend to $\mathbb{X}$. For further details we refer to [**15**].

Still assuming the base field of complex numbers, the Euler characteristic develops its full power with the Riemann-Hurwitz theorem implying, in particular, that the trisection — discussed above — is preserved under finite orbifold group quotients. If G is a finite group of automorphisms of $\mathbb{X} = (X,w)$, then there are only finitely many orbits Gx in X/G having a non-trivial stabilizer G_x (which is necessarily cyclic). Putting $\bar{w}(Gx) = w(x) \cdot |G_x|$ defines $\mathbb{X}/G = (X/G, \bar{w})$ as a weighted projective curve, called the *orbifold quotient*, or just quotient, of $\mathbb{X}$ by G. By definition the automorphism group $\mathrm{Aut}(\mathbb{X})$ of a weighted projective curve $\mathbb{X} = (X,w)$ is the subgroup of all automorphisms of X that commute with w.

THEOREM 2.15 (Riemann-Hurwitz). *Let $\mathbb{X}$ be a weighted projective curve and G a finite subgroup of the automorphism group* $\mathrm{Aut}(\mathbb{X})$ *of $\mathbb{X}$. Then $\mathbb{X}/G$ is again a weighted projective curve having Euler characteristic $\chi_{\mathbb{X}/G} = \chi_{\mathbb{X}}/|G|$.*

PROOF. For a proof we may refer to [**32**, Proposition 13.3.4]. It is perhaps more instructive to invoke the orbifold version of Gauss-Bonnet's theorem [**28**]: If $\mathbb{X}$ is a weighted projective line with one or two weights, then the assertion is directly checked. Otherwise, invoking Theorem 2.14, we may assume that $\mathbb{X}$ has constant curvature $\kappa \in \{+1, 0, -1\}$, a property that being local is preserved under passage to the quotient $\mathbb{Y} = \mathbb{X}/G$. Now, the orbifold version of Gauss-Bonnet's theorem states that the Euler characteristic of $\mathbb{X}$ (resp. $\mathbb{Y}$) is given by the expression $\frac{1}{2\pi}\kappa\cdot|\mathbb{X}|$ (resp. $\frac{1}{2\pi}\kappa\cdot|\mathbb{Y}|$), where $|\mathbb{X}|$ (resp. $|\mathbb{Y}|$) denotes the area of $\mathbb{X}$ (resp. $|\mathbb{Y}|$) with respect to the Kählerian metric induced by the holomorphic structure. Clearly, we now obtain $|\mathbb{Y}| = \frac{|\mathbb{X}|}{|G|}$, implying the claim. □

2.6. Shift action, weighted divisor group and weighted Picard group. An early instance of weighted divisor theory is [**16**, Section 7] which — in retrospect — deals with the divisor theory of weighted projective lines of tame domestic representation type, that is, of those having positive Euler characteristic. We are going to extend this treatment to weighted projective curves, in general.

Let (X, w) be a weighted projective curve. By definition, the *weighted divisor group*, or simply divisor group, $\mathrm{Div}(\mathbb{X})$ of $\mathbb{X}$ is defined as the free abelian group on X. We first show that $\mathrm{Div}(\mathbb{X})$ acts on the category $\mathrm{coh}\,\mathbb{X}$ as a group of self-equivalences. The approach is similar to Section 1.2 but, due to the weights, a bit more sophisticated: For each point of x let S_x be the unique simple sheaf on $\mathbb{X}$ concentrated in x which satisfies $\mathrm{Hom}(\mathcal{O}_{\mathbb{X}}, S_x) = k$. If p_x is the weight of x, then the simple sheaves, concentrated in x, form a τ-orbit $(\tau^j S_x)_{j\in\mathbb{Z}_{p_x}}$ of cardinality p_x. For each vector bundle E we form the universal extension

$$(2.11)\qquad 0 \to E \xrightarrow{x_E} E(x) \to E_x \to 0, \text{ where } E_x = \bigoplus_{j\in\mathbb{Z}_{p_x}} \mathrm{Ext}^1(\tau^j S_x, E)\otimes\tau^j S_x.$$

By construction E_x is a semi-simple sheaf, that is, a finite direct sum of simple sheaves. As in Section 1.2 the assignment $E \mapsto E(x)$ leads to a self-equivalence $\sigma_x : \mathrm{vect}\mathbb{X} \to \mathrm{vect}\mathbb{X}$ that extends to a self-equivalence of $\mathrm{coh}\,\mathbb{X}$, compare [**19**], which we denote by the same symbol σ_x, and which is called *point-shift by x*. Notation: $\sigma_x E = E(x)$ for E in $\mathrm{coh}\,\mathbb{X}$.

Remark 2.16. When extended to the bounded derived category the point shift σ_x is a special instance of what is termed a *tubular mutation* in [**24**] or otherwise a *spherical twist* or *Seidel-Thomas twist* [**29**].

Specializing to a line bundle L, the point-shift construction leads to an exact sequence $0 \to L \to L(x) \to L_x \to 0$, where L_x is the unique simple sheaf concentrated in x which satisfies $\mathrm{Ext}^1(L_x, L) = k$. We recall that all the other simple sheaves $\tau^j L_x$, $0 \neq j \in \mathbb{Z}_{p_x}$, that are concentrated in x, satisfy $\mathrm{Ext}^1(\tau^j L_x, L) = 0$.

In this section, the next section and the appendix we are going to encounter four conceptually different incarnations of the *Picard group* $\mathrm{Pic}(\mathbb{X})$ of $\mathbb{X}$. The first, and classical one, builds the Picard group as the set of all isomorphism classes of line bundles with the abelian group structure given by the tensor product of $\mathrm{coh}\,\mathbb{X}$ discussed in the appendix (Section A). The other three incarnations are byproducts of divisor theory and just rely on the structure of $\mathrm{coh}\,\mathbb{X}$ as a noetherian, hereditary abelian category with Serre duality.

For the second incarnation of the Picard group let $\mathrm{Pic}(\mathbb{X})$ be the pointed set of isomorphism classes of line bundles with base element $\mathcal{O}_{\mathbb{X}}$.

Let U be a finite length sheaf that is a homomorphic image of the structure sheaf $\mathcal{O}_{\mathbb{X}}$. Then $U = \bigoplus_{x\in X} U_x$, where U_x is either zero or else an indecomposable sheaf concentrated in x with top S_x. We associate to U the divisor $|U| = \sum_{x\in\mathbb{X}} \ell(U_x)x$ in $\mathrm{Div}(\mathbb{X})$, where ℓ denotes the length. The exact inclusion $\mathrm{coh}\,X \hookrightarrow \mathrm{coh}\,\mathbb{X}$ induces an equivalence of the Serre quotients $\mathrm{coh}\,X/\mathrm{coh}_0 X$ and $\mathrm{coh}\,\mathbb{X}/\mathrm{coh}_0\mathbb{X}$. Hence the *function field* $K = k(\mathbb{X})$ of $\mathbb{X}$, defined the endomorphism ring of $\mathcal{O}_{\mathbb{X}}$ in the Serre quotient $\operatorname{coh}\mathbb{X}/\operatorname{coh}_0\mathbb{X}$, is isomorphic to the function field $k(X)$ of X. For a non-zero member $f = s^{-1}u$ of $k(\mathbb{X})$ we define its divisor $\operatorname{div}_{\mathbb{X}}(f)$ as $|\operatorname{cok}(u)| - |\operatorname{cok}(s)|$. By definition, the *divisor class group* $\mathcal{C}\ell(\mathbb{X})$ of $\mathbb{X}$ is the cokernel of the divisor map $\operatorname{div}_{\mathbb{X}} : k(\mathbb{X})^* \to \operatorname{Div}(\mathbb{X})$.

PROPOSITION 2.17. *The divisor group* $\operatorname{Div}(\mathbb{X})$ *of* $\mathbb{X}$ *acts on the category* $\operatorname{coh}\mathbb{X}$ *of coherent sheaves on* $\mathbb{X}$, *preserving its full subcategories* $\operatorname{coh}_0\mathbb{X}$ *of finite length,* $\operatorname{vect}\mathbb{X}$ *of vector bundles and* $\mathcal{L}(\mathbb{X})$ *of line bundles on* $\mathbb{X}$.

The action of the group $\operatorname{Div}(\mathbb{X})$ *on the set* $\operatorname{Pic}(\mathbb{X})$ *of isomorphism classes of line bundles is transitive and the assignment* $D = \sum_{x\in X} d_x x \mapsto \mathcal{O}_{\mathbb{X}}(D) = \prod_{x\in X} \sigma_x^{d_x}\mathcal{O}_{\mathbb{X}}$ *induces a bijection* $\mathcal{C}\ell(\mathbb{X}) \to \operatorname{Pic}(\mathbb{X})$.

PROOF. For the first statement, we refer to [**19**, Section 10.3] and [**21**, Section 3]. We are next going to show that any two line bundles L, $\bar{L}$ belong to the same $\operatorname{Div}(\mathbb{X})$-orbit. Let x_0 be an ordinary point, then $\langle L, \bar{L}(nx_0)\rangle > 0$ for large n. Accordingly, we obtain an exact sequence $0 \to L \to \bar{L}(nx_0) \to U \to 0$ with U having finite length ℓ. If $\ell = 0$, we are done; otherwise we choose a simple subobject S of U which yields a line bundle L' with $L \subset L' \subseteq \bar{L}$ together with an exact sequence $0 \to L \to L' \to S \to 0$. If S is concentrated in x, then $L' = L(x)$ and $\bar{L}/L(x)$ has length $\ell - 1$. By induction $L(x)$ and $\bar{L}$ belongs to the same $\operatorname{Div}(\mathbb{X})$-orbit showing that the action of $\operatorname{Div}(\mathbb{X})$ on $\operatorname{Pic}(\mathbb{X})$ is transitive.

It remains to show that the stabilizer group of divisors D with $\mathcal{O}_{\mathbb{X}}(D) = \mathcal{O}_{\mathbb{X}}$ agrees with the image of the divisor map $\operatorname{div}_{\mathbb{X}}$: Let D a divisor with $\mathcal{O}_{\mathbb{X}}(D) = \mathcal{O}_{\mathbb{X}}$. We write D as $D_+ - D_-$, where $D_+ = \sum_{x\in X} n_x x$ (resp. $D_- = \sum_{x\in X} m_x x$) with all $n_x, m_x \geq 0$. The assumption $\mathcal{O}_{\mathbb{X}}(D) = \mathcal{O}_{\mathbb{X}}$ translates into $\mathcal{O}_{\mathbb{X}}(-D_+) = \mathcal{O}_{\mathbb{X}}(-D_-)$ yielding short exact sequences $0 \to \mathcal{O}_{\mathbb{X}}(-D_+) \xrightarrow{u} \mathcal{O}_{\mathbb{X}} \to U \to 0$ and $0 \to \mathcal{O}_{\mathbb{X}}(-D_-) \xrightarrow{s} \mathcal{O}_{\mathbb{X}} \to V \to 0$. This yields a member $f = s^{-1}u$ of $k(\mathbb{X})^*$ with divisor $\operatorname{div}_{\mathbb{X}}(f) = |U| - |V| = D_+ - D_- = D$. □

Using the bijection $\mathcal{C}\ell(\mathbb{X}) \to \operatorname{Pic}(\mathbb{X})$, $[D] \mapsto \mathcal{O}_{\mathbb{X}}(D)$, to equip the pointed set $\operatorname{Pic}(\mathbb{X})$ with the structure of an abelian group, we obtain the second incarnation for the Picard group. Moreover, this leads to the weighted divisor sequence of $\mathbb{X}$:

COROLLARY 2.18 (Weighted divisor sequence). *The sequence*

$$1 \to k^* \to k(\mathbb{X})^* \xrightarrow{\operatorname{div}_{\mathbb{X}}} \operatorname{Div}(\mathbb{X}) \xrightarrow{\varphi} \operatorname{Pic}(\mathbb{X}) \to 0 \tag{2.12}$$

is exact and identifies $\operatorname{Pic}(\mathbb{X})$ *with the divisor class group of* $\mathbb{X}$. □

We next introduce a third incarnation of the Picard group by defining the *Picard group* $\operatorname{Pic}(\mathbb{X})$ of $\mathbb{X}$ as the subgroup of $\operatorname{Aut}(\operatorname{coh}\mathbb{X})$, the group of isomorphism classes of self-equivalences of $\operatorname{coh}\mathbb{X}$, which is generated by all (isomorphism classes of) point-shifts σ_x, $x \in X$. We note that $\operatorname{Aut}(\mathbb{X})$, defined as the subgroup of automorphisms of X commuting with the weight function, can be identified with the subgroup of (isomorphism classes) of self-equivalences of $\operatorname{coh}\mathbb{X}$, that are fixing the structure sheaf $\mathcal{O}_{\mathbb{X}}$. This follows as in [**22**].

Since the shift actions σ_x and σ_y commute for each pair x, y of points from X, the Picard group is commutative. Its role is clarified by the following statement.

PROPOSITION 2.19. *The group* $\operatorname{Aut}(\operatorname{coh}\mathbb{X})$ *is generated by its subgroups* $\operatorname{Aut}(\mathbb{X})$ *and* $\operatorname{Pic}(\mathbb{X})$. *More precisely,* $\operatorname{Aut}(\operatorname{coh}\mathbb{X}) = \operatorname{Aut}(\mathbb{X}) \ltimes \operatorname{Pic}(\mathbb{X})$.

PROOF. Let $\mathcal{L}(\mathbb{X})$ be denote the full subcategory of $\operatorname{coh}\mathbb{X}$ formed by all line bundles. Since automorphisms of $\operatorname{coh}\mathbb{X}$ preserve the rank, the action of $G = \operatorname{Aut}(\operatorname{coh}\mathbb{X})$ restricts to an action of G on $\mathcal{L}(\mathbb{X})$. This action is transitive (up to isomorphism). Moreover, the stabilizer group of $\mathcal{O}_{\mathbb{X}}$ is just $\operatorname{Aut}(\mathbb{X})$. This proves that $\operatorname{Aut}(\mathcal{L}(\mathbb{X}))$ is generated by $\operatorname{Pic}(\mathbb{X})$ and $\operatorname{Aut}(\mathbb{X})$. Now, each self-equivalence of $\mathcal{L}(\mathbb{X})$ extends uniquely to a self-equivalence of $\operatorname{coh}\mathbb{X}$, implying the claim on the structure of $\operatorname{Aut}(\operatorname{coh}\mathbb{X})$. □

The Picard group is the dominant part of $\operatorname{Aut}(\operatorname{coh}\mathbb{X})$, since in most cases $\operatorname{Aut}(\mathbb{X})$ is finite. Indeed only for a non-weighted elliptic curve or for a weighted projective line $\mathbb{X}$ with at most two weighted points, the group $\operatorname{Aut}(\mathbb{X})$ is infinite: weighted projective lines with at least three weighted points have only finitely many automorphisms [**22**]. Also, for a point of a smooth elliptic curve X, there are only finitely many automorphisms of X that are fixing x, implying that elliptic curves that are non-trivially weighted have a finite automorphism group. Finally we use that by Harnack's theorem the automorphism group of a smooth projective curve X of negative Euler characteristic is always finite.

Next, we discuss the relationship between the divisor groups $\operatorname{Div}(X)$ and $\operatorname{Div}(\mathbb{X})$ respectively the relationship between their class groups $\mathcal{C}\ell(X)$ and $\mathcal{C}\ell(\mathbb{X})$, that is of the Picard groups $\operatorname{Pic}(X)$ and $\operatorname{Pic}(\mathbb{X})$.

Let H be an abelian group together with a sequence $(h_1, h_2, \ldots, h_t)$ of members of H and a corresponding sequence $(p_1, p_2, \ldots, p_t)$ of positive integers. Let further $(e_1, \ldots, e_t)$ be the standard basis of $\mathbb{Z}^t$ and U be the subgroup of $H \oplus \mathbb{Z}^t$ generated by the elements $p_i e_i - h_i$. We denote the factor group $\bar{H} = (H \oplus \mathbb{Z}^t)/U$ by $\bar{H} = H[\frac{h_1}{p_1}, \frac{h_2}{p_2}, \ldots, \frac{h_t}{p_t}]$, where $\frac{h_i}{p_i}$ denotes the coset Ue_i from $\bar{H}$, and call $\bar{H}$ the group obtained from H by *adjoining the universal fractions* $\frac{h_i}{p_i}$, $i = 1, 2, \ldots, t$.

THEOREM 2.20 (Universal fractions). *Let* $\mathbb{X} = (X, w)$ *be a weighted projective curve and* $x_1, x_2, \ldots, x_t$ *its weighted points with* $p_i = w(x_i)$. *Then* $\operatorname{Div}(\mathbb{X})$ *(respectively* $\mathcal{C}\ell(\mathbb{X})$*) arises from* $\operatorname{Div}(X)$ *(resp. from* $\mathcal{C}\ell(\mathbb{X})$*) by adjoining the universal fractions* $\frac{x_i}{p_i}$, $i = 1, \ldots, t$. *Accordingly, also* $\operatorname{Pic}(\mathbb{X})$ *arises from* $\operatorname{Pic}(X)$ *by adjoining universal fractions.*

PROOF. We interpret $\operatorname{coh}X$ as the full exact subcategory of $\operatorname{coh}\mathbb{X}$ obtained as the right perpendicular category to the system $\mathcal{E}$ of exceptional sheaves $(\tau^j S_{x_i})$, $j = 1, \ldots, p_i - 1$, $i = 1, \ldots, t$, compare [**9**]. We observe that each simple sheaf S_x of $\operatorname{coh}X$ under this full, exact embedding $j : \operatorname{coh}X \to \operatorname{coh}\mathbb{X}$ is sent to the unique indecomposable sheaf U_x of $\operatorname{coh}\mathbb{X}$ that is concentrated in x, has length $w(x)$ and satisfies $\operatorname{Hom}(\mathcal{O}_{\mathbb{X}}, U_x) = k$. Further, the structure sheaf $\mathcal{O}_X$ of $\operatorname{coh}X$, when viewed as a member of $\operatorname{coh}\mathbb{X}$ takes the role of the structure sheaf there (notation: $\mathcal{O}_{\mathbb{X}}$). Passing to divisor groups j induces an inclusion $\operatorname{Div}(X) \to \operatorname{Div}(\mathbb{X})$, $\sum_{x \in X} n_x x \mapsto \sum_{x \in X} w(x) n_x x$ for the divisor groups. As is easily checked, the following diagram with exact rows is commutative and has also exact columns

$$\begin{array}{ccccccccc} & & & & 0 & & 0 & & \\ & & & & \downarrow & & \downarrow & & \\ 0 & \longrightarrow & K^*/k^* & \xrightarrow{\mathrm{div}_X} & \mathrm{Div}(X) & \longrightarrow & \mathcal{C}\ell(X) & \longrightarrow & 0 \\ & & \| & & \downarrow & & \downarrow & & \\ 0 & \longrightarrow & K^*/k* & \xrightarrow{\mathrm{div}_{\mathbb{X}}} & \mathrm{Div}(\mathbb{X}) & \longrightarrow & \mathcal{C}\ell(\mathbb{X}) & \longrightarrow & 0 \\ & & & & \downarrow & & \downarrow & & \\ & & & & \prod_{i=1}^t \mathbb{Z}_{p_i} & = & \prod_{i=1}^t \mathbb{Z}_{p_i} & & \\ & & & & \downarrow & & \downarrow & & \\ & & & & 0 & & 0 & & \end{array}$$

which immediately implies the claim. □

EXAMPLE 2.21. If $\mathbb{X}$ is a weighted projective line $\mathbf{P}^1\langle p_1, p_2, \ldots, p_t\rangle$ then the Picard group of $\mathbb{X}$, written additively, is obtained by adjoining universal fractions as $\mathrm{Pic}(\mathbb{X}) = \mathbb{Z}[\frac{1}{p_1}, \frac{1}{p_2}, \ldots, \frac{1}{p_t}]$. This group is obviously isomorphic to the rank-one abelian group $\mathbb{L}(p_1, p_2, \ldots, p_t)$ on generators $\vec{x}_1, \vec{x}_2, \ldots, \vec{x}_t$ with relations $p_1\vec{x}_1 = p_2\vec{x}_2 = \ldots, p_t\vec{x}_t$ showing up in the theory of weighted projective lines [**8**]. These groups usually have torsion; thus a universal fraction $\frac{1}{p}$ should not be confused with the rational number given by the same expression.

2.7. The localization sequence. Following the approach in Section 1.3 the exact localization K-theory sequence, due to Quillen [**26**], transforms the 'exact sequence' $0 \to \mathrm{coh}_0\mathbb{X} \to \mathrm{coh}\,\mathbb{X} \to \mathrm{mod}(k(X)) \to 0$ of abelian categories to an exact sequence

$$1 \to k(\mathbb{X})^*/k^* \xrightarrow{\mathrm{div}} \mathrm{K}_0(\mathrm{coh}_0\mathbb{X}) \xrightarrow{\mathrm{nat}} \mathrm{K}_0(\mathrm{coh}\,\mathbb{X}) \xrightarrow{\mathrm{rk}} \mathbb{Z} \to 0, \tag{2.13}$$

where div sends a fraction $f = s^{-1}u$ to $[\mathrm{cok}(u)]_0 - [\mathrm{cok}(s)]_0$, and nat sends the class $[U]_0$ from $\mathrm{K}_0(\mathrm{coh}_0\mathbb{X})$ to the class $[U]$ in $\mathrm{K}_0(\mathrm{coh}\,\mathbb{X})$. We are now going to relate this sequence to the divisor theory for $\mathbb{X}$, yielding a fourth incarnation of the Picard group $\mathrm{Pic}(\mathbb{X})$. We recall that $\mathrm{K}_0'(\mathbb{X})$ is the kernel of the rank homomorphism $\mathrm{K}_0(\mathbb{X}) \to \mathbb{Z}$, and also the subgroup of $\mathrm{K}_0(\mathbb{X})$ generated by the classes of simple sheaves. Further let V (resp. V_0) be the subgroup of $\mathrm{K}_0'(\mathbb{X})$ (resp. $\mathrm{K}_0(\mathrm{coh}_0\mathbb{X})$) generated by all elements $[\tau S_x] - [S_x]$ (resp. $[\tau^j S_x]_0 - [S_x]_0$), $j = 1, \ldots, w(x)$, where x runs through all weighted points of $\mathbb{X}$ and S_x is a simple sheaf concentrated in x.

PROPOSITION 2.22. *The following diagram is commutative with exact rows and columns*

$$\begin{array}{ccccccccc} & & & & 0 & & 0 & & \\ & & & & \downarrow & & \downarrow & & \\ & & & & V_0 & = & V & & \\ & & & & \downarrow & & \downarrow & & \\ 1 & \longrightarrow & k(\mathbb{X})^*/k^* & \xrightarrow{\mathrm{div}} & \mathrm{K}_0(\mathrm{coh}_0\mathbb{X}) & \xrightarrow{\mathrm{nat}} & \mathrm{K}_0'(\mathbb{X}) & \longrightarrow & 0 \\ & & \| & & \downarrow & & \downarrow & & \\ 1 & \longrightarrow & k(\mathbb{X})^*/k^* & \xrightarrow{\mathrm{div}_{\mathbb{X}}} & \mathrm{Div}(\mathbb{X}) & \longrightarrow & \mathcal{C}\ell(\mathbb{X}) & \longrightarrow & 0 \\ & & & & \downarrow & & \downarrow & & \\ & & & & 0 & & 0 & & \end{array} \tag{2.14}$$

PROOF. The commutativity of the diagram is easily checked. Moreover, by Proposition 2.7, the induced map from V_0 to V is an isomorphism. $\square$

COROLLARY 2.23. *We have a direct decomposition* $\mathrm{K}_0(\mathbb{X}) = \mathrm{K}_0'(\mathbb{X}) \oplus \mathbb{Z}[\mathcal{O}_{\mathbb{X}}]$. *Moreover, there is an exact sequence* $0 \to U \to \mathrm{K}_0'(\mathbb{X}) \to \mathcal{C}\ell(\mathbb{X}) \to 0$, *representing* $\mathcal{C}\ell(\mathbb{X})$, *hence* $\mathrm{Pic}(\mathbb{X})$, *as a factor group* $\mathrm{K}_0(\mathrm{coh}\,\mathbb{X})/(U \oplus \mathbb{Z}[\mathcal{O}_{\mathbb{X}}])$ *of the Grothendieck group* $\mathrm{K}_0(\mathbb{X})$. $\square$

We thus obtain a fourth incarnation of the Picard group as a factor group of the Grothendieck group of coh $\mathbb{X}$. As the weighted projective lines show, it is—unlike in the non-weighted case— usually not possible to represent the Picard group as a subgroup of the Grothendieck group.

Appendix A. Multiplicative structure

Let $\mathcal{H}$ be the category of coherent sheaves on a weighted projective curve $\mathbb{X} = X\begin{pmatrix} p_1, p_2, \ldots, p_t \\ x_1, x_2, \ldots, x_t \end{pmatrix}$ obtained by simultaneous p-cycle construction (for distinct points) from the category coh X of coherent sheaves on the smooth projective curve X. We sketch in this appendix how to introduce a $\mathbb{Z}$-graded sheaf theory such that the attached category of coherent sheaves is equivalent to $\mathcal{H}$ and, moreover, equips $\mathcal{H}$ with a commutative tensor product satisfying $\mathcal{O}_{\mathbb{X}}(x) \otimes \mathcal{O}_{\mathbb{X}}(y) = \mathcal{O}_{\mathbb{X}}(x+y)$, functorially. In particular, this turns the isomorphism classes of line bundles on $\mathbb{X}$ into a commutative group, the *Picard group* $\mathrm{Pic}(\mathbb{X})$ of $\mathbb{X}$. Moreover, the graded sheaf theory allows to form the *determinant homomorphism* $\det : \mathrm{K}_0(\mathrm{coh}\,\mathbb{X}) \to \mathrm{Pic}(\mathbb{X})$, sending the class $[E]$ of a vector bundle E of rank r to its p-th exterior power $\bigwedge^p E$.

The procedure follows [**27**], which the reader should consult for details. Let x_0 be an ordinary point. We form the divisor $D = mx_0 + \sum_{i=1}^{t} a_i x_i$ with $a_i = \pm 1$. Functoriality of the shift functor $X \mapsto X(D)$ turns the direct sum

$$R = \bigoplus_{n \in \mathbb{Z}} \mathrm{Hom}(\mathcal{O}_{\mathbb{X}}, \mathcal{O}_{\mathbb{X}}(nD))$$

into a $\mathbb{Z}$-graded algebra with finite dimensional components $R_n = \mathrm{Hom}(\mathcal{O}_{\mathbb{X}}, \mathcal{O}_{\mathbb{X}}(nD))$. The components R_n can be naturally seen as subspaces of the function field $k(\mathbb{X}) = k(X)$, identifying R by means of the map $\sum_{n\in\mathbb{Z}} r_n \mapsto \sum_{n\in\mathbb{Z}} r_n Y^n$ with a subalgebra of the polynomial algebra $k(X)[Y]$ in one variable over $k(X)$. Taking noetherianness of $\mathcal{H}$ into account, this shows that R is a positively $\mathbb{Z}$-graded affine k-algebra, provided we choose m sufficiently large.

PROPOSITION A.1. *The following properties hold:*

(1) *For each A in $\mathcal{H}$, the direct sum* $F(A) = \bigoplus_{n \geq 0} \mathrm{Hom}(\mathcal{O}_{\mathbb{X}}(-nD), A)$ *is a finitely generated positively graded R-module.*
(2) *The attachment $A \mapsto F(A)$ induces an equivalence between $\mathcal{H}$ and the Serre quotient $\mathcal{C}$ of the category* $\mathrm{mod}^{\mathbb{Z}_{\geq 0}}(R)$ *of finitely generated positively $\mathbb{Z}$-graded modules modulo its Serre subcategory of graded modules of finite length.*
(3) *The graded tensor product $\otimes$ (resp. graded r-th exterior power $\bigwedge^r$) induce corresponding operations on $\mathcal{C}$ and therefore $\mathcal{H}$, having the usual properties.*
(4) *The set of* $\mathrm{Pic}(\mathbb{X})$ *of isomorphism classes of line bundles is a group with respect to the tensor product operation.*

PROOF. Concerning the first two statements, we refer to [27]. We just remark that the key point is that, by the choice of the divisor D, we obtain for any simple sheaf S an *infinite* set of positive integers n such that $\mathrm{Hom}(\mathcal{O}_{\mathbb{X}}(-nD), S)$ is non-zero.

Concerning the third item, the proofs from [8] can be copied. □

COROLLARY A.2. *Each line bundle L on $\mathbb{X}$ is determined, up to isomorphism, by its class $[L]$ in the Grothendieck group.*

PROOF. Due to the multiplicative structure on $\mathrm{coh}\,\mathbb{X}$, the classical proof of Proposition 1.8 turns over. □

Unlike exceptional objects, line bundles are not determined by their class in the reduced Grothendieck group $\overline{\mathrm{K}}_0(\mathbb{X})$.

We illustrate Proposition A.1 by a couple of examples.

EXAMPLE A.3. Assume first that $\mathbb{X}$ is a weighted projective line with three weighted points x_1, x_2, x_3 having weights (p_1, p_2, p_3). Further let x_0 be an ordinary point. We put $D = x_0 - \sum_{i=1}^{3} x_i$. For the next two assertions we refer to Theorem 2.6, respectively Proposition 5.5 from [20].

(1) If $\mathbb{X}$ has positive Euler characteristic, the ring
$$R = \bigoplus_{n\geq 0} \mathrm{Hom}(\mathcal{O}_{\mathbb{X}}, \mathcal{O}_{\mathbb{X}}(-nD))$$
is a $\mathbb{Z}$-graded simple singularity. For instance, the weight sequence $(2,3,3)$ yields the algebra $R = k[x,y,z]/(z^2 + y^3 + x^2 z)$ where x, y, z have degrees $3, 4, 6$, respectively.

(2) If $\mathbb{X}$ has negative Euler characteristic, the ring
$$R = \bigoplus_{n\geq 0} \mathrm{Hom}(\mathcal{O}_{\mathbb{X}}, \mathcal{O}_{\mathbb{X}}(nD))$$
is a hypersurface only for 14 weight triples, among them $(2,3,8)$ in which case $R = k[x,y,z]/(z^2 + x^5 + xy^3)$ and where x, y, z have degrees $6, 8, 15$, respectively.

EXAMPLE A.4. Let E be an elliptic curve and let e be a (closed) point of E.

(1) With $D = e$ we deduce from Proposition 1.6 that $R_n = \mathrm{Hom}(\mathcal{O}_X, \mathcal{O}_X(nD))$ has dimension 1 for $n = 0$ and dimension n for $n \geq 1$. If the base field is algebraically closed of characteristic different from 2 and 3, then for a proper choice of $x \in R_1$, $y \in R_2$ and $z \in R_3$ the algebra $R = \bigoplus_{n\geq 0} R_n$ has the form $R = k[x,y,z]/(x^6 + y^3 + z^2 + axyz)$, where x, y, z get degrees 1,2,3, respectively, and where the parameter a relates to the j-invariant of E by the formula
$$j(E) = \frac{a^{12}}{a^6 - 432}.$$
In particular, we have $a^6 \neq 432$.

(2) Next, we consider the weighted elliptic curve $\mathbb{E} = E\langle 2\rangle$, where the point e obtains weight 2. Again, we put $D = e$. Now, a straightforward calculation in the reduced Grothendieck group yields for the dimension of $S_n = \mathrm{Hom}(\mathcal{O}_{\mathbb{X}}, \mathcal{O}_{\mathbb{X}}(nD))$ the values 1,1,1,1 for $n = 0,1,2,3$ and for each $n \geq 4$ the value $[\frac{n}{2}]$, the largest integer contained in $\frac{n}{2}$. Keeping the assumption on the base field, then for a proper choice of $x \in S_1$, $y \in S_4$ and

$z \in S_6$, a calculation, almost identical to the previous one, yields that the algebra $S = \bigoplus_{n \geq 0} S_n$ gets the form $S = k[x, y, z]/(x^{12} + y^3 + z^2 + axyz)$ with x, y, z getting degrees $1, 4, 6$, respectively. Moreover, the relationship between the parameter a and the j-invariant of $\mathbb{E}$ is the same as in (1). We note that S is obtained from the $\mathbb{Z}$-graded algebra R from (1) by *root construction*, that is, by adjoining a square root of x to R and, correspondingly, adjoining the universal fraction $\frac{1}{2}$ to the grading group $\mathbb{Z}$.

Acknowledgements

I thank the referee for constructive and thoughtful comments. Thanks further go to Dirk Kussin for spotting a missing factor in the formula for the orbifold Euler characteristic and also to Mattia Talpo for providing a number of references to root stacks and parabolic sheaves.

References

[1] Tarig Abdelgadir and Kazushi Ueda, *Weighted projective lines as fine moduli spaces of quiver representations*, Comm. Algebra **43** (2015), no. 2, 636–649, DOI 10.1080/00927872.2013.842245. MR3274027

[2] Kai Behrend and Behrang Noohi, *Uniformization of Deligne-Mumford curves*, J. Reine Angew. Math. **599** (2006), 111–153, DOI 10.1515/CRELLE.2006.080. MR2279100

[3] Armand Borel and Jean-Pierre Serre, *Le théorème de Riemann-Roch* (French), Bull. Soc. Math. France **86** (1958), 97–136. MR0116022

[4] Niels Borne, *Fibrés paraboliques et champ des racines* (French), Int. Math. Res. Not. IMRN **16** (2007), Art. ID rnm049, 38, DOI 10.1093/imrn/rnm049. MR2353089

[5] Niels Borne and Angelo Vistoli, *Parabolic sheaves on logarithmic schemes*, Adv. Math. **231** (2012), no. 3-4, 1327–1363, DOI 10.1016/j.aim.2012.06.015. MR2964607

[6] Henri Paul de Saint-Gervais, *Uniformization of Riemann surfaces*, Heritage of European Mathematics, European Mathematical Society (EMS), Zürich, 2016. Revisiting a hundred-year-old theorem; Translated from the 2010 French original [MR2768303] by Robert G. Burns; The name of Henri Paul de Saint-Gervais covers a group composed of fifteen mathematicians: Aurélien Alvarez, Christophe Bavard, François Béguin, Nicolas Bergeron, Maxime Bourrigan, Bertrand Deroin, Sorin Dumitrescu, Charles Frances, Étienne Ghys, Antonin Guilloux, Frank Loray, Patrick Popescu-Pampu, Pierre Py, Bruno Sévennec and Jean-Claude Sikorav. MR3494804

[7] Pierre Gabriel, *Des catégories abéliennes* (French), Bull. Soc. Math. France **90** (1962), 323–448. MR0232821

[8] Werner Geigle and Helmut Lenzing, *A class of weighted projective curves arising in representation theory of finite-dimensional algebras*, Singularities, representation of algebras, and vector bundles (Lambrecht, 1985), Lecture Notes in Math., vol. 1273, Springer, Berlin, 1987, pp. 265–297, DOI 10.1007/BFb0078849. MR915180

[9] Werner Geigle and Helmut Lenzing, *Perpendicular categories with applications to representations and sheaves*, J. Algebra **144** (1991), no. 2, 273–343, DOI 10.1016/0021-8693(91)90107-J. MR1140607

[10] Robin Hartshorne, *Algebraic geometry*, Springer-Verlag, New York-Heidelberg, 1977. Graduate Texts in Mathematics, No. 52. MR0463157

[11] T. Hübner. Exzeptionelle Vektorbündel und Reflektionen an Kippgarben Über projektiven gewichteten Kurven. Dissertation, Paderborn, 1996.

[12] Osamu Iyama and Boris Lerner, *Tilting bundles on orders on* $\mathbb{P}^d$, Israel J. Math. **211** (2016), no. 1, 147–169, DOI 10.1007/s11856-015-1263-8. MR3474959

[13] Dirk Kussin, *On the K-theory of tubular algebras*, Colloq. Math. **86** (2000), no. 1, 137–152. MR1799893

[14] Dirk Kussin, *Weighted noncommutative regular projective curves*, J. Noncommut. Geom. **10** (2016), no. 4, 1465–1540, DOI 10.4171/JNCG/264. MR3597149

[15] Helmut Lenzing, *Weighted projective lines and Riemann surfaces*, Proceedings of the 49th Symposium on Ring Theory and Representation Theory, Symp. Ring Theory Represent. Theory Organ. Comm., Shimane, 2017, pp. 67–79. MR3644095

[16] Helmut Lenzing, *Curve singularities arising from the representation theory of tame hereditary algebras*, Representation theory, I (Ottawa, Ont., 1984), Lecture Notes in Math., vol. 1177, Springer, Berlin, 1986, pp. 199–231, DOI 10.1007/BFb0075266. MR842467

[17] Helmut Lenzing, *A K-theoretic study of canonical algebras*, Representation theory of algebras (Cocoyoc, 1994), CMS Conf. Proc., vol. 18, Amer. Math. Soc., Providence, RI, 1996, pp. 433–454. MR1388066

[18] Helmut Lenzing, *Representations of finite-dimensional algebras and singularity theory*, Trends in ring theory (Miskolc, 1996), CMS Conf. Proc., vol. 22, Amer. Math. Soc., Providence, RI, 1998, pp. 71–97. MR1491919

[19] Helmut Lenzing, *Hereditary categories*, Handbook of tilting theory, London Math. Soc. Lecture Note Ser., vol. 332, Cambridge Univ. Press, Cambridge, 2007, pp. 105–146, DOI 10.1017/CBO9780511735134.006. MR2384609

[20] H. Lenzing, *Rings of singularities*, Bull. Iranian Math. Soc. **37** (2011), no. 2, 235–271. MR2890586

[21] Helmut Lenzing and José Antonio de la Peña, *Concealed-canonical algebras and separating tubular families*, Proc. London Math. Soc. (3) **78** (1999), no. 3, 513–540, DOI 10.1112/S0024611599001872. MR1674837

[22] Helmut Lenzing and Hagen Meltzer, *The automorphism group of the derived category for a weighted projective line*, Comm. Algebra **28** (2000), no. 4, 1685–1700, DOI 10.1080/00927870008826922. MR1747349

[23] Helmut Lenzing and Idun Reiten, *Hereditary Noetherian categories of positive Euler characteristic*, Math. Z. **254** (2006), no. 1, 133–171, DOI 10.1007/s00209-006-0938-6. MR2232010

[24] Hagen Meltzer, *Tubular mutations*, Colloq. Math. **74** (1997), no. 2, 267–274. MR1477569

[25] David Mumford, *Abelian varieties*, Tata Institute of Fundamental Research Studies in Mathematics, vol. 5, Published for the Tata Institute of Fundamental Research, Bombay; by Hindustan Book Agency, New Delhi, 2008. With appendices by C. P. Ramanujam and Yuri Manin; Corrected reprint of the second (1974) edition. MR2514037

[26] Daniel Quillen, *Higher algebraic K-theory. I*, Algebraic K-theory, I: Higher K-theories (Proc. Conf., Battelle Memorial Inst., Seattle, Wash., 1972), Springer, Berlin, 1973, pp. 85–147. Lecture Notes in Math., Vol. 341. MR0338129

[27] I. Reiten and M. Van den Bergh, *Noetherian hereditary abelian categories satisfying Serre duality*, J. Amer. Math. Soc. **15** (2002), no. 2, 295–366, DOI 10.1090/S0894-0347-02-00387-9. MR1887637

[28] Ichirô Satake, *The Gauss-Bonnet theorem for V-manifolds*, J. Math. Soc. Japan **9** (1957), 464–492, DOI 10.2969/jmsj/00940464. MR0095520

[29] Paul Seidel and Richard Thomas, *Braid group actions on derived categories of coherent sheaves*, Duke Math. J. **108** (2001), no. 1, 37–108, DOI 10.1215/S0012-7094-01-10812-0. MR1831820

[30] C. S. Seshadri, *Fibrés vectoriels sur les courbes algébriques* (French), Astérisque, vol. 96, Société Mathématique de France, Paris, 1982. Notes written by J.-M. Drezet from a course at the École Normale Supérieure, June 1980. MR699278

[31] Mattia Talpo, *Moduli of parabolic sheaves on a polarized logarithmic scheme*, Trans. Amer. Math. Soc. **369** (2017), no. 5, 3483–3545, DOI 10.1090/tran/6747. MR3605978

[32] W. P. Thurston. The Geometry and Topology of Three-Manifolds. Electronic version 1.1 — March 2002, available at http://www.msri.org/publications/books/gt3m/PDF/13.pdf.

[33] Adam-Christiaan van Roosmalen, *Numerically finite hereditary categories with Serre duality*, Trans. Amer. Math. Soc. **368** (2016), no. 10, 7189–7238, DOI 10.1090/tran/6569. MR3471089

University of Paderborn, Institute for mathematics, D-33098 Paderborn, Germany
Email address: `helmut@math.uni-paderborn.de`

Contemporary Mathematics
Volume **705**, 2018
http://dx.doi.org/10.1090/conm/705/14202

Finite-dimensional algebras arising as blocks of finite group algebras

Markus Linckelmann

Abstract. This mostly expository paper, based on three introductory lectures on block theory given at the ICRA 2016, starts with a list of properties which set blocks of finite group algebras apart from finite-dimensional algebras in general. This is followed by a description of basic algebras of blocks of finite groups in low dimension, illustrating the expectation, that there should be very few Morita equivalence classes of blocks of finite group algebras. We describe various invariants of block algebras some of which cannot be associated with arbitrary finite-dimensional algebras, such as defect groups and fusion systems of blocks. We explore a range of techniques, questions, and conjectures, which attempt to shed light on why so few finite-dimensional algebras arise as blocks of finite groups.

Introduction

Block theory explores the structural connections between finite groups and their group algebras over fields and p-local rings. Few indecomposable finite-dimensional algebras over a field k arise as direct factors of a finite group algebra. If the characteristic of k does not divide the order of G, then the group algebra kG is semisimple. In that case, any indecomposable direct factor of kG is a finite-dimensional simple algebra, hence a matrix algebra over a finite-dimensional division k-algebra. We will therefore focus on the case where k is a field of prime characteristic p.

In Section 1, we describe a sample list of properties which set indecomposable factors of finite group algebras over k apart within the universe of finite-dimensional k-algebras. Section 2 contains a classification of all possible split basic algebras of blocks of finite groups up to dimension 12. Some of the arguments refer to concepts developed later, such as the notion of defect groups, introduced in Section 3. In Section 4 we describe some conjectures, such as Donovan's conjecture, which formalise the intuition that there should be few block algebras of finite groups. The three sections 5, 6, and 7 are devoted to introducing background material on Hochschild cohomology, and then relate Hochschild cohomology in degree 0 and 1 to the structure of a block algebra. Pursuing this theme, we consider Gerstenhaber's notion of *integrable derivations* in the sections 8 and 9. The concept of *separably equivalent algebras*, described in Section 10, formalises the relationship between a block algebra of a finite group and its defect group algebras. This concept is applied in Section 11 to calculating the finitistic and dominant dimensions of categories

2010 *Mathematics Subject Classification.* 20C20, 16E40.

of cohomological Mackey functors associated with a block. Section 12 considers connections between the algebra structure of a block and its fusion systems.

CONTENTS

1. Properties of blocks of finite group algebras

Let G be a finite group and k a field. Following terminology going back to Brauer, an indecomposable direct algebra factor B of kG is called a *block of kG*. The unit element of B is a primitive idempotent in $Z(kG)$, and the correpondence $B \mapsto 1_B$ is a bijection between the set of blocks of kG and the set of primitive idempotents in $Z(kG)$. Some authors define a block of kG to be a primitive idempotent b in $Z(kG)$, and the associated indecomposable algebra factor B of kG with b as unit element is the algebra $B = kGb$, called the *block algebra* associated with the block b. This notion of blocks extends to finite group algebras over a commutative Noetherian ring in the obvious way.

Let B a block algebra of kG. The following five properties illustrate different structural and numerical properties of block algebras of finite groups.

1.1. The block algebra B has a semisimple lift to characteristic zero. To state this more precisely, let $\mathcal{O}$ be a complete discrete valuation ring with residue field $k = \mathcal{O}/J(\mathcal{O})$ and with field of fractions K of characteristic zero. Then KG is semisimple, as mentioned before. Moreover, the blocks of $\mathcal{O}G$ and of kG correspond bijectively to each other via the canonical surjections $\mathcal{O}G \to kG$. Thus $\mathcal{O}G$ has a unique block $\hat{B}$ such that

$$k \otimes_{\mathcal{O}} \hat{B} = B$$

and then the K-algebra $K \otimes_{\mathcal{O}} \hat{B}$ is semisimple, since it is a (no longer necessarily indecomposable) direct factor of KG. Since it is easy to construct finite-dimensional algebras which do not have semisimple lifts to characteristic zero in this sense, this is a first property which narrows down the pool of possible algebras arising as block algebras of finite groups over k.

1.2. The Cartan matrix C of the block algebra B is positive definite, and its determinant is a power of p. Recall that C is the square matrix defined by

$C = (c_{ij})_{i,j \in I}$, where $\{U_i\}_{i \in I}$ is a set of representatives of the isomorphism classes of projective indecomposable B-modules for some indexing set I, and where c_{ij} is the number of composition factors in a composition series of U_j which are isomorphic to $S_i = U_i/\mathrm{rad}(U_i)$. If k is a splitting field for G, then $c_{ij} = \dim_k(\mathrm{Hom}_B(U_i, U_j))$ for all i, $j \in I$. One can chose for I a set of representatives of the conjugacy classes of primitive idempotents in B and then set $U_i = Bi$ for $i \in I$. The Cartan matrix is defined modulo fixing a total order on the idexing set I, hence up to a simultaneous permutation of its rows and columns. The Cartan matrix of B is indecomposable; that is, there is no choice of a total order on I such that C can be written as a block diagonal matrix in a nontrivial way (this statement is true for any indecomposable finite-dimensional k-algebra). If k is a splitting field for G, then the largest elementary divisor of the Cartan matrix of B is equal to the order of the *defect groups of* B, a fundamental block invariant due to Brauer, which we will review in Section 3.

1.3. The block algebra B is *symmetric*; that is, B is isomorphic, as a B-B-bimodule, to its k-dual $B^\vee = \mathrm{Hom}_k(B, k)$. This follows from the fact that kG itself is symmetric; an easy verification shows that the map sending a linear map $t : kG \to k$ to the element $\sum_{x \in G} t(x^{-1})x$ in kG is a bimodule isomorphism $(kG)^\vee \cong kG$. This argument remains true with k replaced by any commutative ring. In particular, the corresponding block $\hat{B}$ of $\mathcal{O}G$ is a symmetric $\mathcal{O}$-algebra. The symmetry of B implies that its Cartan matrix is a symmetric matrix, with diagonal entries at least 2 whenever B is not simple.

1.4. The block algebra *is defined over a finite subfield* $\mathbb{F}$ of k. That is, there is a finite dimensional $\mathbb{F}$-algebra B' such that $B \cong k \otimes_{\mathbb{F}} B'$. Equivalently, B has a k-basis X such that the multiplicative constants of B with respect to this basis are all contained in $\mathbb{F}$. This follows from the fact that $kG = k \otimes_{\mathbb{F}_p} \mathbb{F}_pG$ is defined over the prime field $\mathbb{F}_p$, hence some finite extension $\mathbb{F}$ of $\mathbb{F}_p$ is a splitting field of $\mathbb{F}_pG$, and thus any block idempotent of kG is contained in $\mathbb{F}G$ for some finite subfield $\mathbb{F}$ of k. The field $\mathbb{F}$ depends in general on B, so may be larger than $\mathbb{F}_p$. By a result of Cliff, Plesken, and Weiss in [**14**], if k is a splitting field for B, then the center $Z(B)$ of B is always defined over the prime field $\mathbb{F}_p$.

1.5. The Hochschild cohomology $HH^*(B) = \mathrm{Ext}^*_{B \otimes_k B^{\mathrm{op}}}(B, B)$ is finitely generated, and for any finitely generated B-module U, the algebra $\mathrm{Ext}^*_B(U, U)$ is finitely generated as a module over $HH^*(B)$ via the canonical algebra homomorphism

$$HH^*(B) \to \mathrm{Ext}^*_B(U, U)$$

induced by the functor $- \otimes_B U$. The analogous statements hold for the block algebra $\hat{B}$ over $\mathcal{O}$. This follows from the Evens-Venkov theorem on finite generation of finite group cohomology.

2. Basic algebras of dimension at most 12 of blocks

The purpose of this section is to illustrate that indeed very few algebras arise as block algebras. We describe a list of basic algebras containing all basic algebras of dimension at most 12 of block algebras of finite groups over an algebraically closed field. This list is finite in each dimension, even across all primes if the dimension is at least 2. The finiteness of this list in each dimension is predicted as a consequence of *Donovan's conjecture*, and is known to hold for principal blocks; see Section 4 for details. In order to describe this list, we freely use some block theoretic terminology - such as defect groups - which will be reviewed in subsequent sections. Background material on Brauer tree algebras in the context of block theory can be found in Alperin's book [**1**].

Let k be an algebraically closed field of prime characteristic p, let G a finite group, and B a block of kG. Let A be a basic algebra of B.

2.1. We have $\dim_k(A) = 1$ if and only if B is isomorphic to a matrix algebra over k. If p does not divide $|G|$, then every block of kG is a matrix algebra. If p divides $|G|$, then kG is not semisimple, but it is still possible for some blocks to be matrix algebras. The number $z(kG)$ of blocks of kG which are matrix algebras plays a crucial role in the formulation of *Alperin's weight conjecture*. The group theoretic version of Alperin's weight conjecture states that the number $\ell(kG)$ of isomorphism classes of simple kG-modules should be equal to $\sum_P z(kN_G(P)/P)$, where P runs over a set of representatives of the conjugacy classes of p-subgroups of G. Alperin's weight conjecture admits a refinement to blocks of finite group algebras. There are many reformulations of Alperin's weight conjecture, such as the alternating sum version of Knörr and Robinson [**42**].

2.2. If $\dim_k(A) = 2$, then $p = 2$ and $A \cong kC_2$. Indeed, up to isomorphism, the only indecomposable k-algebra of dimension 2 is the uniserial algebra $k[x]/(x^2)$. Its Cartan matrix is (2). Property 1.2 implies that if B is Morita equivalent to $k[x]/(x^2)$, then $p = 2$. In that case we have $k[x]/(x^2) \cong kC_2$.

2.3. If $\dim_k(A) = 3$, then $p = 3$ and $A \cong kC_3$. This follows from the fact that, up to isomorphism, the only indecomposable symmetric 3-dimensional k-algebra of dimension 3 is the uniserial algebra $k[x]/(x^3)$. Arguing as in the previous case, we get that $p = 3$, and hence $kC_3 \cong k[x]/(x^3)$.

2.4. If $\dim_k(A) = 4$, then $p = 2$ and A is isomorphic to either kC_4 or to $k(C_2 \times C_2)$. Since A is a symmetric indecomposable k-algebra of dimension 4, it follows (e. g. by considering 1.2) that A is local, hence has the Cartan matrix (4), implying $p = 2$. One then uses the structure theory of blocks with a defect group of order 4. Note that there are infinitely many other isomorphism classes of 4-dimensional indecomposable algebras for arbitrary p, such as the algebras $C(\lambda) = k\langle x, y\rangle/(x^2, y^2, xy - \lambda yx)$, where $\lambda \in k^\times$. For $\lambda \neq 1$ these algebras are selfinjective but not symmetric, so cannot arise as block algebras.

2.5. If $\dim_k(A) = 5$, then $p = 5$ and $A \cong kC_5$. As in the previous example, A is local, with Cartan matrix is (5), so is the basic algebra of a block with a cyclic defect group C_5 with a single isomorphism class of simple modules. Thus B is nilpotent, hence Morita equivalent to kC_5.

2.6. If $\dim_k(A) = 6$, then $p = 3$ and $A \cong kS_3$. To see this, observe first, that A cannot be local, because the Cartan matrix would have determinant 6, which is not a prime power. The only possible Cartan matrix of an indecomposable 6-dimensional symmetric k-algebra with at least two isomorphism classes of simple modules is the matrix

$$\begin{pmatrix} 2 & 1 \\ 1 & 2 \end{pmatrix}$$

Its determinant is 3, so B has the cyclic defect group C_3 and two isomorphism classes of simple modules. Thus A is a Brauer tree algebra of a tree with two edges and no exceptional muliplicity. Up to isomorphism, this is kS_3.

2.7. If $\dim_k(A) = 7$, then either $p = 7$ and $A \cong kC_7$, or $p = 5$ and A is a Brauer tree algebra of a tree with two edges, exceptional multiplicity 2 and exceptional vertex at one end of the tree. The second case arises as the principal block algebra of kA_5.

Indeed, if A is local, then the Cartan matrix of A is (7), hence $p = 7$ and $A \cong kC_7$, since B is a nilpotent block with defect group C_7. If A is not local, then there is exactly one possible Cartan matrix, namely

$$\begin{pmatrix} 3 & 1 \\ 1 & 2 \end{pmatrix}$$

The determinant of this matrix is 5, so B has a defect group isomorphic to C_5, and hence A is a Brauer tree algebra of a tree with two edges and exceptional multiplicity 2. The exceptional vertex cannot be in the middle, since that would yield an algebra of dimension 10.

2.8. If $\dim_k(A) = 8$, then either $p = 2$ and $A \cong kP$ for some group P of order 8, or $p = 7$ and A is a Brauer tree algebra of a tree with two edges, exceptional multiplicity 3 and exceptional vertex at one end of the tree.

Indeed, if A is local, then its Cartan matrix is (8), hence B has a defect group P of order 8. The structure theory of blocks with a defect group of order 8 implies that if A is local, then B is nilpotent, and hence isomorphic to a defect group algebra. If A is not local, then the possible Cartan matrices are

$$\begin{pmatrix} 3 & 1 \\ 1 & 3 \end{pmatrix}, \begin{pmatrix} 4 & 1 \\ 1 & 2 \end{pmatrix}$$

The determinants of these matrices are 8 and 7, respectively. The first matrix has elementary divisors 8 and 1, so would have to be the Cartan matrix of a block with a defect group of order 8, but no such block has two isomorphism classes of simple modules. The second matrix implies that $p = 7$ and that a defect group is isomorphic to C_7, so A is a Brauer tree algebra with two edges and exceptional multiplicity 3. Arguing as in the last case, the exceptional vertex cannot be in the middle, since that would yield an algebra of dimension 14.

There are, however, infinitely many isomorphism classes of local symmetric 8-dimensional algebras; see Example 2.14 below.

2.9. Suppose that $\dim_k(A) = 9$. Then $p = 3$ and the defect groups of B have order 9. If A is local, then A is isomorphic to one of the algebras kC_9, $k(C_3 \times C_3)$, or to the quantum complete intersection

$$k\langle x, y\rangle/(x^3, y^3, xy + yx)$$

The latter algebra can be viewed as a deformation of the group algebra

$$k(C_3 \times C_3) \cong k[x, y]/(x^3, y^3)$$

in which the two generators x and y anticommute. It arises as a nonprincipal block, described in the example 2.15 below. If A is not local, then the Cartan matrix of A is equal to

$$\begin{pmatrix} 5 & 1 \\ 1 & 2 \end{pmatrix}.$$

Either A is a Brauer tree algebra of a tree with two edges, exceptional multiplicity 4, and exceptional vertex at one end of the tree, or the defect groups of B are isomorphic to $C_3 \times C_3$, in which case A is uniquely determined by the endomorphism algebras of the two projective indecomposable A-modules, and these endomorphism algebras are isomorphic to $k[x,y]/(xy, x^2 - y^3)$ and $k[x]/(x^3)$. (This is the only algebra on this list for which we do not know whether it actually arises as a basic algebra of a block.)

To see this, consider first that case where A is local. Then B has defect groups isomorphic to either C_9 or $C_3 \times C_3$. If a defect group is C_9, then B is nilpotent, hence $A \cong kC_9$. If a defect group is $C_3 \times C_3$, then Kiyota's Table 1 in [**41**] implies that either B is nilpotent, or that B has inertial quotient $C_2 \times C_2$. In the nilpotent case, A is a group algebra, and in the second case, it follows from Kessar [**36**] that A is the quantum complete intersection as stated.

Suppose that A is not local. The only possible Cartan matrices for a basic indecomposable symmetric 9-dimensional k-algebra are

$$\begin{pmatrix} 4 & 1 \\ 1 & 3 \end{pmatrix}, \begin{pmatrix} 5 & 1 \\ 1 & 2 \end{pmatrix}, \begin{pmatrix} 3 & 2 \\ 2 & 2 \end{pmatrix}$$

Their determinants are 11, 9, 2, respectively. This rules out the last matrix, because every block with a defect group of order 2 is nilpotent, hence has a local basic algebra. The first matrix would lead to a block with a cyclic defect group C_{11} and exceptional multiplicity 5 - but this would force the Cartan matrix to have a diagonal entry 6, so this case is impossible. The second matrix has elementary divisors 9 and 1, so implies that B has a defect group of order 9. If a defect group is cyclic, we get the Brauer tree algebra as stated. If not, then B is a block with defect group $C_3 \times C_3$ and two isomorphism classes of simple modules. It follows from [**41**, Table 1] that $|\mathrm{Irr}(B)|$ is either 6 or 9. If it were 9, then A would be commutative, hence local, contradicting our current assumption. Thus $|\mathrm{Irr}(B)| = 6$, which yields the decomposition matrix as stated. Write $1_A = i + j$, with i, j primitive orthogonal idempotents. Then A is determined by iAi, jAj, and the bimodule structures of iAj and jAi. The latter two spaces are 1-dimensional, so they are uniquely determined as bimodules, and hence A is determined by iAi and jAj. By the Cartan matrix of A, the dimensions of these two local algebras are 5 and 3, respectively (up to exchanging i and j, if necessary). This forces $jAj \cong k[x]/(x^3)$. In order to determine the structure of iAi, note first that iAi is commutative by [**45**, B. Theorem]. We determine next the dimension of J/J^2, where $J = J(iAi)$. This dimension cannot be 1, because then iAi would be uniserial, so A would be the Brauer tree algebra considered before. This dimension cannot be 3, because this would imply that the Loewy length of A is 3 - but this is impossible by Okuyama's classification in [**59**] of such blocks. So this dimension is 2. Then J^2 has dimension 2, so is strictly bigger than $\mathrm{soc}(iAi)$, and hence $J^3 \neq \{0\}$. Since iAi is commutative, it is isomorphic to a quotient of $k[x,y]$. Identify x and y with their images in iAi. The annihilator of J^2 contains J^2 and has dimension 3, so some nonzero linear combination of x and y annihilates J^2. Thus we may assume $xJ^2 = \{0\}$. Since iAi is symmetric, this implies also $J^2x = \{0\}$. Thus $xyJ = \{0\}$, hence $Jxy = \{0\}$. Thus all words of length three in x and y are zero in A, except possibly y^3. Since J^3 is nonzero, this implies $y^3 \neq 0$, so $\mathrm{soc}(iAi) = ky^3$. Since $xJ^2 = \{0\}$, both x and y annihilate xy, hence $xy \in \mathrm{soc}(iAi)$, or equivalently $xy = \mu y^3$ for some

$\mu \in k$. We may modify x by any element in J^2 without affecting any of the previous properties of x. Replacing x by $x - \mu y^2$, we get now in addition that $xy = 0$. Since $xJ \neq \{0\}$, it follows that $x^2 \neq 0$. Thus x^2 is a nonzero scalar multiple of y^3, and hence, after rescaling one of x or y (using that k is algebraically closed) we may assume that $x^2 = y^3$. The structure of iAi is therefore as stated.

By [**41**, Table 1] again, this case can only arise if B has an inertial quotient E which is either of order 2 acting freely on $P \setminus \{1\}$, or isomorphic to D_8, with nontrivial Külshammer-Puig 2-class, but as mentioned above, we do not know whether this case actually does arise.

2.10. Suppose that $\dim_k(A) = 10$. Then $p = 5$ and $A = kD_{10}$, where $D_{10} = C_5 \rtimes C_2$ is dihedral of order 10, or $p = 11$ and A is a Brauer tree algebra of a tree with two edges, exceptional multiplicity 5, and exceptional vertex at one end of the tree.

To see this, observe first that A is not local, since otherwise its Cartan matrix (10) would not have p-power determinant in that case. Suppose that A has two isomorphism classes of simple modules. The only possible 2×2-Cartan matrices of a basic indecomposable symmetric 10-dimensional algebra with p-power determinant are

$$\begin{pmatrix} 3 & 2 \\ 2 & 3 \end{pmatrix}, \begin{pmatrix} 6 & 1 \\ 1 & 2 \end{pmatrix}, \begin{pmatrix} 4 & 2 \\ 2 & 2 \end{pmatrix}$$

The first of these has determinant 5, hence arises as a Brauer tree algebra with two edges, exceptional multiplicity 2, and exceptional vertex in the middle. This algebra is isomorphic to kD_{10}. The second matrix has determinant 11, and is the Cartan matrix of the Brauer tree algebra as stated in that case. The third has determinant 4, with elementary divisors 2, 2, so does not occur.

Suppose that A has three isomorphism classes of simple modules. Up to permutation of rows and columns, the only possible Cartan matrices are

$$\begin{pmatrix} 2 & 0 & 1 \\ 0 & 2 & 1 \\ 1 & 1 & 2 \end{pmatrix}, \begin{pmatrix} 2 & 1 & 0 \\ 1 & 2 & 1 \\ 0 & 1 & 2 \end{pmatrix}$$

The determinant of both matrices is 4, which rules them out, because, by [**17**, Theorem 4] or [**18**, Theorem 4], the Klein four blocks with three isomorphism classes of simple modules have basic algebras of dimension 12 or 18.

2.11. Suppose that $\dim_k(A) = 11$. If A is local, then $p = 11$ and $A \cong kC_{11}$. If A has two isomorphism classes of simple modules, then $p = 13$ and A is a Brauer tree algebra of a tree with two edges, exceptional multiplicity 6, and exceptional vertex at one end of the tree, or $p = 2$ and A is the basic algebra of kS_4. If A has three isomorphism classes of simple modules, then $p = 7$ and A is a Brauer tree algebra, with a line having three edges as tree, exceptional multiplicity 2, and exceptional vertex at the end of the line.

The local case follows as in previous cases. The possible 2×2-Cartan matrices are

$$\begin{pmatrix} 7 & 1 \\ 1 & 2 \end{pmatrix}, \begin{pmatrix} 6 & 1 \\ 1 & 3 \end{pmatrix}, \begin{pmatrix} 5 & 1 \\ 1 & 4 \end{pmatrix}, \begin{pmatrix} 5 & 2 \\ 2 & 2 \end{pmatrix}, \begin{pmatrix} 4 & 2 \\ 2 & 3 \end{pmatrix}$$

with determinants 13, 17, 19, 6, 8, respectively. The first can arise from a block, and yields the Brauer tree algebra as stated. The matrices with determinant 17,

19 cannot arise from blocks, since their multiplicities would have to be 8, 9, respectively, and the last matrix with determinant 8 is the Cartan matrix of S_4 at $p = 2$. The only possible 3×3-Cartan matrices are

$$\begin{pmatrix} 3 & 1 & 0 \\ 1 & 2 & 1 \\ 0 & 1 & 2 \end{pmatrix}, \begin{pmatrix} 3 & 1 & 1 \\ 1 & 2 & 0 \\ 1 & 0 & 2 \end{pmatrix}.$$

their determinants are 7 and 8, respectively. The first matrix implies that $p = 7$ and that the defect groups are isomorphic to C_7, which yields the Brauer tree algebra as stated. The second matrix cannot occur as a Cartan matrix of a block with defect groups of order 8. To see this, denote by S, T, U three pairwise nonisomorphic simple A-modules with projective covers P_S, P_T, P_U of composition lengths 5, 3, 3, respectively. The Cartan matrix implies that P_T is uniserial with composition series T, S, T, and that P_U is uniserial with composition series U, S, U. In particular, T and U have no nontrivial extensions by each other. Moreover, $T \oplus U$ is a summand of $J(P_S)/J^2(P_S)$. If $T \oplus U = J(P_S)/J^2(P_S)$, then the Cartan matrix implies that $J^2(P_S)$ is uniserial of length 2, with both composition factors S. This uniserial module would have to be a quotient of P_S, but then S would have to appear in $J(P_S)/J^2(P_S)$. It follows that $J^2(P_S) = \mathrm{soc}(P_S)$, and that $J(P_S)/\mathrm{soc}(P_S) = S \oplus T \oplus U$. But then A has Loewy length 3, which is not possible by Okuyama's results in [**59**].

2.12. If $\dim_k(A) = 12$, then $p = 2$, and $A \cong kA_4$.

Indeed, since 12 is not a prime, the algebra A cannot be local. The possible 2×2-Cartan matrices with positive determinant are

$$\begin{pmatrix} 8 & 1 \\ 1 & 2 \end{pmatrix}, \begin{pmatrix} 7 & 1 \\ 1 & 3 \end{pmatrix}, \begin{pmatrix} 6 & 1 \\ 1 & 4 \end{pmatrix}, \begin{pmatrix} 5 & 1 \\ 1 & 5 \end{pmatrix},$$
$$\begin{pmatrix} 6 & 2 \\ 2 & 2 \end{pmatrix}, \begin{pmatrix} 5 & 2 \\ 2 & 3 \end{pmatrix}, \begin{pmatrix} 4 & 2 \\ 2 & 4 \end{pmatrix}$$

with determinants 15, 20, 23, 24, 8, 11, 12, respectively. Four of these do not have prime power determinants. The matrices with determinant 23 and 11 are not the Cartan matrices of a Brauer tree algebra with two edges. The matrix with determinant 8 has elementary divisors 4 and 2, but cannot be the Cartan matrix of a block with a defect group of order 4. The only candidate for a 3×3-Cartan matrix is

$$\begin{pmatrix} 2 & 1 & 1 \\ 1 & 2 & 1 \\ 1 & 1 & 2 \end{pmatrix}.$$

The determinant of this matrix is 4, thus $p = 2$, and the only possibility of a block with a defect group of order 4 and this Cartan matrix is the algebra kA_4.

REMARK 2.13. The methods used up to this point are not sufficient to determine all possible 13-dimensional basic algebras of blocks of finite groups. There are candidate Cartan matrices with determinants 16 and 27, but we do not seem to have enough information about block algebras with defect groups of these orders. See Schaps [**70**] for a substantial data base of blocks.

EXAMPLE 2.14 (Rickard, 2015). For $\lambda \in k^\times$, define

$$A(\lambda) = k\langle x, y, z \mid x^2 = y^2 = z^2 = 0,\ xy = \lambda yx,\ yz = \lambda zy,\ zx = \lambda xz\rangle$$

The image in $A(\lambda)$ of the set $\{1, x, y, z, xy, xz, yz, xyz\}$ is a k-basis, and xyz spans the socle of $A(\lambda)$. Any linear form on $A(\lambda)$ which vanishes on all basis elements except on xyz is a symmetrising form, so $A(\lambda)$ is symmetric. Set $J = J(A(\lambda))$. If $\lambda \neq -1$, then, modulo J^2, the only elements in J which square to zero are the scalar multiples of x, y, z. Thus any two elements u, v in J which have linearly independent images in J/J^2 and which square to zero satisfy $uv = \lambda vu$ or $uv = \lambda^{-1}vu$. Therefore the isomorphism class of $A(\lambda)$ determines the pair $\{\lambda, \lambda^{-1}\}$. This implies that there are infinitely many isomorphism classes of 8-dimensional local symmetric algebras. One can show that $A(\lambda)$ is the trivial extension algebra of the 4-dimensional selfinjective algebra $C(\lambda) = k\langle x, y \mid x^2 = y^2 = 0,\ xy = \lambda yx\rangle$.

EXAMPLE 2.15. The 9-dimensional quantum complete intersection arises as the basic algebra of a block in the following situation. Suppose that $p = 3$. Set $P = C_3 \times C_3$ and $G = P \rtimes Q_8$, where $Z(Q_8)$ acts trivially on P, and where the two copies of C_2 in $C_2 \times C_2 \cong Q_8/Z(Q_8)$ act by inversion on the two copies in C_3 with respect to some decomposition $P = C_3 \times C_3$. Set $E = C_2 \times C_2$. The group algebra kG has two blocks. The principal block is isomorphic to $k(P \rtimes E)$, and this is a basic algebra. The unique nonprincipal block is isomorphic to the twisted group algebra $k_\alpha(P \rtimes E)$, where α is a 2-cocycle of E representing the central extension

$$1 \longrightarrow C_2 \longrightarrow Q_8 \longrightarrow E \longrightarrow 1$$

extended to $P \rtimes E$ via the surjection $P \rtimes E \to E$. The algebra $k_\alpha(P \rtimes E)$ is not basic; its basic algebras are isomorphic to the 9-dimensional quantum complete intersection

$$k\langle x, y\rangle/(x^3, y^3, xy + yx)$$

See [**5**] and [**29**] for general construction principles of blocks with a single isomorphism class of simple modules.

REMARK 2.16. It is not known in general, which local algebras can occur as basic algebras of blocks of finite groups. By a result of Okuyama and Tsushima in [**60**], the commutative local algebras that can occur in this way are exactly the group algebras of abelian p-groups.

REMARK 2.17. Most algebras in the above list have finite representation type. The algebras of tame represenation type in the above list are $k(C_2 \times C_2)$ in dimension 4, kQ_8 and kD_8 in dimension 8, the basic algebra of kS_4 in dimension 11, and kA_4 in dimension 12 (with $p = 2$). The algebras of wild representation type in the above list are $k(C_2 \times C_4)$ and $k(C_2 \times C_2 \times C_2)$ in dimension 8 (with $p = 2$), and the three algebras in dimension 9 with noncyclic defect (with $p = 3$; for the third of these three algebras it is not known whether it actually arises as a block).

3. Defect groups of block algebras

The defect groups of a block algebra B of kG, a concept introduced by Brauer in the 1940s, are an analogue of the Sylow p-subgroups of G. Defect groups are invariant under the passage from B to its lift $\hat{B}$ over the ring $\mathcal{O}$. In order to cover both cases simultaneously, we assume in this section that $\mathcal{O}$ is a complete local principal ideal domain with residue field k of prime characteristic p, allowing the case $\mathcal{O} = k$.

Let G be a finite group. As mentioned in the last section, if B is a block algebra of $\mathcal{O}G$, then $B = \mathcal{O}Gb$ for a unique primitive idempotent b in $Z(\mathcal{O}G)$, and then $b = 1_B$. In particular, B is also an indecomposable $\mathcal{O}G$-$\mathcal{O}G$-bimodule.

Since the ring $\mathcal{O}$ is local, it has 1 as its unique idempotent, and hence the augmentation homomorphism $\mathcal{O}G \to \mathcal{O}$ sends all but one primitive idempotent in $\mathbb{Z}(\mathcal{O}G)$ to zero. Denote by b_0 the unique primitive idempotent of $Z(\mathcal{O}G)$ which is not contained in the kernel of the augmentation homomorphism $\mathcal{O}G \to \mathcal{O}$. Then the corresponding block algebra $B_0 = \mathcal{O}Gb_0$ is called the *principal block algebra* of $\mathcal{O}G$, and b_0 is called the *principal block idempotent* of $\mathcal{O}G$. We write $B_0(\mathcal{O}G)$ for the principal block algebra of $\mathcal{O}G$ in case the group G needs to be specified.

If P is a finite p-group, then $\mathcal{O}P$ is local, hence indecomposable as an algebra and as an $\mathcal{O}P$-$\mathcal{O}P$-bimodule. In other words, $\mathcal{O}P$ is equal to its principal block algebra. If P is a p-subgroup of G, then $\mathcal{O}P$ is an indecomposable direct summand of $\mathcal{O}G$ as an $\mathcal{O}P$-$\mathcal{O}P$-bimodule, since we clearly have an $\mathcal{O}P$-$\mathcal{O}P$-bimodule decomposition $\mathcal{O}G = \mathcal{O}P \oplus \mathcal{O}[G \setminus P]$. Note that if $x \in C_G(P)$, then $\mathcal{O}Px$ is also a direct summand of $\mathcal{O}G$ isomorphic to $\mathcal{O}P$ as an $\mathcal{O}P$-$\mathcal{O}P$-bimodule, and hence the multiplicity of $\mathcal{O}P$ as a direct summand of $\mathcal{O}G$ in a decomposition as $\mathcal{O}P$-$\mathcal{O}P$-bimodule is equal to $|PC_G(P) : P|$. Since $\mathcal{O}G$ is direct sum of its blocks regarded as $\mathcal{O}G$-$\mathcal{O}G$-bimodules, it follows from the Krull-Schmidt theorem that $\mathcal{O}P$ is isomorphic to a direct summand, as an $\mathcal{O}P$-$\mathcal{O}P$-bimodule, of some block B of $\mathcal{O}G$.

DEFINITION 3.1 (Brauer, 1940s). Let G be a finite group and B a block algebra of $\mathcal{O}G$. A *defect group of B* is a maximal p-subgroup P of G such that $\mathcal{O}P$ is isomorphic to a direct summand of B as an $\mathcal{O}P$-$\mathcal{O}P$-bimodule.

There are many equivalent reformulations of this definition.

PROPOSITION 3.2. *Let G be a finite group and B a block of $\mathcal{O}G$. The following are equivalent for a subgroup P of G.*

(a) *P is a defect group of B.*
(b) *P is a minimal subgroup such that B is isomorphic to a direct summand of $B \otimes_{\mathcal{O}P} B$ as a B-B-bimodule.*
(c) *P is a p-subgroup such that $\mathcal{O}P$ is isomorphic to a direct summand of B as an $\mathcal{O}P$-$\mathcal{O}P$-bimodule, and B is isomorphic to a direct summand of $B \otimes_{\mathcal{O}P} B$ as a B-B-bimodule.*

This can be found in many standard sources; for a proof using only the $\mathcal{O}P$-$\mathcal{O}P$-bimodule structure of $\mathcal{O}G$ see [**51**, §3].

PROPOSITION 3.3 (Basic properties of defect groups). *Let G be a finite group.*

(i) *The defect groups of a block B of $\mathcal{O}G$ form a G-conjugacy class of p-subgroups of G.*
(ii) *The defect groups of the principal block B_0 of $\mathcal{O}G$ are the Sylow p-subgroups of G.*
(iii) *The defect groups of a block of $\mathcal{O}G$ contain all normal p-subgroups of G, or equivalently, the largest normal p-subgroup $O_p(G)$ of G is contained in all defect groups of all blocks of $\mathcal{O}G$.*
(iv) *The defect groups are invariant under the canonical passage from $\mathcal{O}G$ to kG; that is, a block B of $\mathcal{O}G$ and the corresponding block $k \otimes_{\mathcal{O}} B$ of kG have the same defect groups.*

A defect group of a block B of kG need not be equal to a Sylow p-subgroup of G, but there are some restrictions as to which p-subgroups of G can be defect groups of blocks. By a result of Green, a defect group P of B is always equal to the intersection of two Sylow p-subgroups of G, and $N_G(P)/P$ has no nontrivial normal p-subgroup. The defect groups of a block of symmetric group S_n are isomorphic to the Sylow p-subgroups of S_m for some positive integer $m \leq n$.

EXAMPLES 3.4. Let G be a finite group, B a block algebra of $\mathcal{O}G$ and P a defect group of B.

(1) We have $P = 1$ if and only if B is a separable $\mathcal{O}$-algebra; if we assume k to be algebraically closed, this is the case if and only if $B \cong M_n\mathcal{O})$ for some positive integer n.
(2) P is cyclic if and only if $k \otimes_{\mathcal{O}} B$ has finite representation type. In that case, assuming that k is a splitting field, the algebra $k \otimes_{\mathcal{O}} B$ is a Brauer tree algebra of a Brauer tree with at most $p-1$ edges and with exceptional multiplicity $\frac{|P|-1}{e}$, where e is the number of isomorphism classes of simple $k \otimes_{\mathcal{O}} B$-modules.

The following example illustrates a key aspect of the expected relationship between defect groups and block algebras.

EXAMPLE 3.5. Suppose that $p = 2$ and that k is algebraically closed. Let G be a finite group and B a block algebra of $\mathcal{O}G$.

(a) If B has a Klein four defect group, then B is Morita equivalent to either $\mathcal{O}P$, or $\mathcal{O}A_4$, or $B_0(\mathcal{O}A_5)$.
(b) If B has a quaternion defect group of order 8, then B is Morita equivalent to one of $\mathcal{O}P$, or $\mathcal{O}\tilde{A}_4$, or $B_0(\mathcal{O}\tilde{A}_5)$.

The quaternion group of order 8 is a central extension of a Klein four group by a cyclic group of order 2. What this example illustrates is that the block algebras with a quaternion defect group are obtained from extensions of the block algebra of Klein four defect blocks. In other words: what happens 'locally' (at the level of defect groups) also happens 'globally' (at the level of the block algebras themselves). This is a fundamental principle in block theory, with group theoretic applications including the Brauer-Suzuki theorem, which implies that a finite group with a quaternion group as a Sylow 2-subgroup cannot be simple. Most conjectures in block theory revolve around a 'local to global' principle, asking in what way the algebra structure of B is related to invariants of its defect groups. For the relationship of numerical invariants of blocks to their defect groups, as well as a long list of further references, see Sambale [**69**].

4. Some finiteness conjectures

Finiteness conjectures in block theory state that block invariants should be determined up to finitely many possibilities by their defect groups. By a theorem of Brauer and Feit [**8**], the number of ordinary irreducible characters of a block is bounded in terms of the defect groups - see 6.1 below for details. In particular, the defect groups of a block algebra bound the number of isomorphism classes of modular simple modules, or equivalently, the size of the Cartan matrix. Let k be an algebraically closed field of prime characteristic p. The existence of a bound for the number of irreducible characters of a block in terms of its defect groups would be a consequence of the following conjecture.

CONJECTURE 4.1 (Donovan's conjecture, 1970s). *For a fixed finite p-group P, there are only finitely many Morita equivalence classes of block algebras of finite group algebras over k with defect groups isomorphic to P.*

Donovan's conjecture is, for instance, known if P is a cyclic p-group (this follows from work of Dade [**15**], Janusz [**32**] and Kupisch [**48**]), if P a Klein four or generalised dihedral 2-group (Erdmann [**17**], [**18**], [**19**]), or if P is an elementary abelian 2-group (Eaton-Kessar-Külshammer-Sambale [**16**]; unlike in the previous cases, this requires the classification of finite simple groups). Although any given block algebra of a finite group algebra kG can be realised over some finite subfield of k, it is essential to require k to be algebraically closed in Donovan's conjecture, since blocks of finite group algebras over a finite field with a fixed defect group may have arbitrarily large finite splitting fields.

Since the Cartan matrix of a block is determined by its Morita equivalence class, Donovan's conjecture would have the following consequence.

CONJECTURE 4.2 (Weak Donovan conjecture). *For a fixed finite p-group P, the set of entries of Cartan matrices of block algebras of finite group algebras over k with defect groups isomorphic to P is bounded above.*

The dimension of a basic algebra is the sum of its Cartan invariants, and the Cartan invariants of a block algebra bound the order of a defect group (as the largest elementary divisor). Thus Donovan's conjecture implies the following conjecture.

CONJECTURE 4.3. *For any positive integer n there are finitely many isomorphism classes of basic algebras of dimension n which are Morita equivalent to a block algebra of a finite group algebra over k.*

This conjecture holds for principal blocks; see the Remark 4.6 below.

Let G be a finite group. If τ is an automorphism of the field k then τ extends uniquely to a ring automorphism of kG, abusively still denoted by τ, fixing the elements of G in kG. If B is a block algebra of kG then so is its image $\tau(B)$ under this ring automorphism. The blocks B and $\tau(B)$ have the same defect groups. Although B and $\tau(B)$ are isomorphic as rings, they need not be Morita equivalent. We denote by $m(G, B)$ the number of pairwise Morita inequivalent blocks of kG of the form $\tau^t(B)$, where τ is the Frobenius automorphism $\lambda \mapsto \lambda^p$ of k and where t runs over the set of nonnegative integers. Note that the principal block B_0 is fixed under τ, and hence $m(G, B_0) = 1$. Certain quantum complete intersections provided the first examples in [**6**] of ring isomorphic but not Morita equivalent block algebras over k. In those examples we have $m(G, B) = 2$. There are currently no known examples with $m(G, B) \geq 3$.

CONJECTURE 4.4 (Kessar's rationality conjecture [**34**]). *For any finite p-group P there is a positive integer $m(P)$, depending only on the isomorphism class of P, such that for any finite group G and any block B of kG with defect groups isomorphic to P, we have $m(G, B) \leq m(P)$.*

The rationality conjecture covers the gap between Donovan's conjecture and the Weak Donovan conjecture.

THEOREM 4.5 ([**34**, Theorem 1.4]). *The Weak Donovan Conjecture* 4.2 *and Kessar's rationality conjecture* 4.4 *are together equivalent to Donovan's conjecture* 4.1.

REMARK 4.6. The key step in the proof of this theorem in [**34**] consists of showing that a field of definition of a basic algebra of a block B of kG is bounded in terms of $m(G, B)$ and the Cartan matrix of the block. Thus Conjecture 4.4 implies conjecture 4.3. Principal blocks are defined over $\mathbb{F}_p$, so basic algebras of principal blocks are defined over a finite field bounded in terms of their Cartan matrices, and hence the conjectures 4.4 and 4.3 hold for principal blocks.

REMARK 4.7. The Weak Donovan conjecture 4.2 and Conjecture 4.3 are also together equivalent to Donovan's conjecture 4.1. To see this, one observes that 4.3 and 4.2 imply together the rationality conjecture 4.4, and then one applies Theorem 4.5.

REMARK 4.8. The isomorphism class of a split finite-dimensional k-algebra is determined by the isomorphism class of its basic algebras as well as the dimensions of its simple modules. Thus Conjecture 4.3 implies that for a fixed positive integer n there are only finitely many isomorphism classes of n-dimensional k-algebras which are Morita equivalent to block algebras of finite groups.

Donovan stated Conjecture 4.1 for blocks of finite group algebras over the field k, but the conjecture makes, of course, sense for blocks over $\mathcal{O}$. However, the answers to any of the following questions are not known in general.

QUESTION 4.9. *For B a block of $\mathcal{O}G$, does the Morita equivalence class of $k \otimes_{\mathcal{O}} B$ determine the Morita equivalence class of B, and, if not uniquely, at least up to finitely many possibilities?*

The cases that have been calculated point to a positive answer. Donovan's conjecture asks to what extent a defect group of a block determines its algebra structure. Equally open is the reverse question:

QUESTION 4.10. *To what extent does the Morita equivalence class of a block algebra B determine the isomorphism class of its defect groups?*

It is known that the Morita equivalence class of a block B of kG determines the order $|P|$ of a defect group P of B (as the largest elementary divisor of the Cartan matrix of B), the rank of P (as the Krull dimension of $HH^*(B)$), and the exponent of P (by a result of Külshammer [**44**, Satz J]). In particular, the Morita equivalence class of B determines whether B has elementary abelian defect groups. The strongest version of the previous question remains, however, open.

QUESTION 4.11. *Does the Morita equivalence class of a block algebra B determine the isomorphism class of its defect groups?*

Since a finite p-group algebra $\mathcal{O}P$ is a block with P as a defect group, this question contains the *modular isomorphism problem* as a sub-problem: given two finite p-groups P, Q such that $kP \cong kQ$ as k-algebras, does this imply a group isomorphism $P \cong Q$? This seems to be a difficult question. See the article by Hertweck and Soriano [**27**] for an overview.

The proximity of the previous question to the modular isomorphism problem suggests that the choice of the base ring may be of fundamental importance in this context. Over k the modular isomorphism problem seems out of reach, but it is known to be true over the ring $\mathcal{O}$ if the characteristic of $\mathcal{O}$ is zero, by a result of Roggenkamp and Scott in [**67**]. In order to stay away from the modular isomorphism problem, it seems therefore good advice to reformulate the last question for blocks of $\mathcal{O}G$ with $\mathcal{O}$ a complete discrete valuation ring having k as a residue field and a field of fractions K of characteristic zero. This is also relevant for the next far weaker - and still open - question.

QUESTION 4.12. *Does the Morita equivalence class of a block algebra B determine whether its defect groups are abelian?*

If we phrase this question for blocks over the ring $\mathcal{O}$, then Brauer's Height Zero conjecture would imply that this is true, since this conjecture characterises blocks with abelian defect groups as blocks all of whose irreducible characters have height zero. Using the classification of finite simple groups, Kessar and Malle proved in [**40**] one implication: if the defect groups of B are abelian, then its irreducible characters have height zero. The converse implication is still open, and if true, would imply a positive answer to the previous question, because heights of characters are invariant under Morita equivalences.

5. Hochschild cohomology background

We briefly depart from our overall assumption that k is a field, and allow k to be an arbitrary commutative ring. Let A be a k-algebra. For convenience, we assume that A is projective as a k-module. The Hochschild cohomology of A is the graded algebra

$$HH^*(A) = \mathrm{Ext}^*_{A\otimes_k A^{\mathrm{op}}}(A, A)$$

Without the assumption on A being projective as a k-module, we would need to define $HH^*(A)$ as a relative Ext-algebra. We sketch some of the fundamental properties of Hochschild cohomology, and refer to the sources mentioned below for precise statements.

THEOREM 5.1 (Gerstenhaber [**23**, Corollary 1]). *Let k be a commutative ring and A be a k-algebra which is projective as a k-module. The algebra $HH^*(A)$ is graded-commutative; that is, of any two nonnegative integers m, n and any homogeneous elements $\zeta \in HH^m(A)$, $\eta \in HH^n(A)$ we have $\eta\zeta = (-1)^{mn}\zeta\eta$.*

Thus if k is a field of characteristic 2, then $HH^*(A)$ is commutative, and if k is a field of odd prime characteristic p, then the even part $HH^{\mathrm{even}}(A)$ of $HH^*(A)$ is commutative, while all homogeneous elements of odd degree square to zero. There is more structure on $HH^*(A)$.

THEOREM 5.2 (Gerstenhaber [**23**, Theorem 4]). *Let k be a commutative ring and A be a k-algebra which is projective as a k-module. Then $HH^*(A)$ is a graded Lie algebra of degree -1. More precisely, there is a Lie bracket $[-,-]$ on $HH^*(A)$ such that for $\zeta \in HH^m(A)$, $\eta \in HH^n(A)$ we have $[\zeta, \eta] \in HH^{m+n-1}(A)$.*

See [**23**] for more precise statements, notably for a more precise description of the relations between the associative product and the Lie bracket in $HH^*(A)$ in [**23**, §8]. Note that in particular $HH^1(A)$ is a Lie algebra with this bracket. We will describe the Lie algebra structure on $HH^1(A)$ in greater detail below. If k is a field of prime characteristic p, then there is a p-restricted Lie algebra structure:

THEOREM 5.3 (Zimmermann [**78**, 4.3]). *Suppose that k is a field of prime characteristic p, and let A be a finite-dimensional k-algebra. If $p = 2$, then $HH^{>0}(A)$ is a 2-restricted Lie algebra. If p is odd, then $HH^{\text{odd}}(A)$ is a p-restricted Lie algebra.*

The p-power map of the restricted Lie algebra structure in the above theorem is described in [**78**] using a description of Hochschild cohomology in terms of coderivations from Stasheff [**71**].

In general, Hochschild cohomology is not functorial, in the sense that an algebra homomorphism $A \to B$ need not induce a homomorphism $HH^*(A) \to HH^*(B)$. But upon restricting attention to the 2-category of symmetric k-algebras with 1-morphisms bounded complexes of bimodules which are finitely generated projective as left and right modules, Hochschild cohomology becomes a functor from this 2-category to the category of graded k-modules. The key fact why symmetric algebras are well suited in this context is the following well-known adjunction property.

Let A, B symmetric k-algebras, and let M be an A-B-bimodule which is finitely generated projective as a left A-module and as a right B-module. Set $M^\vee = \mathrm{Hom}_k(M, k)$. Then the functors

$$M \otimes_B - : \mathrm{mod}(B) \to \mathrm{mod}(A)$$
$$M^\vee \otimes_A - : \mathrm{mod}(A) \to \mathrm{mod}(B)$$

are left and right adjoint to each other. Following [**49**], any choice of adjunction isomorphisms

$$\mathrm{Hom}_A(M \otimes_B -, -) \cong \mathrm{Hom}_B(-, M^\vee \otimes_A -)$$
$$\mathrm{Hom}_B(M^\vee \otimes_A -, -) \cong \mathrm{Hom}_A(-, M \otimes_B -)$$

gives rise to a graded k-module homomorphism

$$t_M : HH^*(B) \to HH^*(A)$$

sending $\eta : B \to B[n]$ in $HH^n(B)$ to the element $\mathrm{tr}_M(\eta)$ in $HH^n(A)$ given by the composition

$$A \longrightarrow M \otimes_B M^\vee \xrightarrow{\mathrm{Id} \otimes \eta \otimes \mathrm{Id}} M \otimes_B M^\vee[n] \longrightarrow A[n]$$

where the first and last map are the relevant adjunction unit and counit, and where the map in the middle uses the obvious identifications $M \otimes_B M^\vee = M \otimes_B B \otimes_B M^\vee$ and $M \otimes_B B[n] \otimes_B M^\vee = M \otimes_B M^\vee[n]$.

The map tr_M is graded k-linear, but need not be compatibel with the graded-commutative algebra structure or the Lie algebra structure.

EXAMPLE 5.4 ([**49**, 4.6, 4.7]). Let G be a finite group and H a subgroup of G. There is a canonical embedding $H^*(G; k) \to HH^*(kG)$ obtained as follows. Consider $G \cong \Delta G = \{(x, x) \mid x \in G\} \leq G \times G$. The canonical isomorphism $\mathrm{Ind}_{\Delta G}^{G \times G}(k) \cong kG$ implies that the functor $\mathrm{Ind}_{\Delta G}^{G \times G}$ induces a graded algebra homomorphism $H^*(\Delta G; k) \to HH^*(kG)$. Together with the identification $H^*(G; k) \cong H^*(\Delta G; k)$ this yields a graded algebra homomorphism

$$\delta_G : H^*(G; k) \to HH^*(kG) \ .$$

One verifies that this is split injective, with retraction given by the canonical map $HH^*(kG) \to H^*(G;k)$ induced by the functor $- \otimes_{kG} k$. The usual transfer and restriction maps tr_H^G and res_H^G in group cohomology make the following diagrams commutative:

$$\begin{array}{ccccc} H^*(H;k) & \xrightarrow{\mathrm{tr}_H^G} & H^*(G;k) & \xrightarrow{\mathrm{res}_H^G} & H^*(H,k) \\ \downarrow{\scriptstyle \delta_H} & & \downarrow{\scriptstyle \delta_G} & & \downarrow{\scriptstyle \delta_H} \\ HH^*(kH) & \xrightarrow[\mathrm{tr}_M]{} & HH^*(kG) & \xrightarrow[\mathrm{tr}_{M^\vee}]{} & HH^*(kH) \end{array}$$

For C another symmetric k-algebra and N a B-C-bimodule which is finitely generated projective as a left and right module we have

$$\mathrm{tr}_{M\otimes_B N} = \mathrm{tr}_M \circ \mathrm{tr}_N$$

which shows that transfer maps are functorial on the 1-morphisms in this 2-category. The definition of tr_M extends verbatim with M replaced by a bounded complex consisting of bimodules which are finitely generated projective as left and right modules (one needs to verify that the above standard adjunction extends to bounded complexes, which is a formality).

6. $HH^0(B)$

We assume in this section that $\mathcal{O}$ is a complete discrete valuation ring with algebraically closed residue field k of prime characteristic p and field of fractions K of characteristic zero.

Let G be a finite group and let B be a block algebra of $\mathcal{O}G$. Assume that K contains a primitive $|G|$-th root of unity; this hypothesis implies that K is a splitting field for KG. The degree zero Hochschild cohomology of B is

$$HH^0(B) = \mathrm{End}_{B\otimes_\mathcal{O} B^{\mathrm{op}}}(B) \cong Z(B)$$

This is so far true for an arbitrary $\mathcal{O}$-algebra. It is a special property of finite group algebras, that the canonical map $Z(\mathcal{O}G) \to Z(kG)$ is surjective, and hence the map $Z(B) \to Z(k \otimes_\mathcal{O} B)$ is surjective as well. Thus we have

$$\dim_K(K \otimes_\mathcal{O} B) = \mathrm{rk}_\mathcal{O}(Z(B)) = \dim_k(Z(k \otimes_\mathcal{O} B)) \ .$$

Since K is a splitting field of characteristic zero, the algebra KG is a direct product of finitely many matrix algebras which are in bijection with the ordinary irreducible characters $\mathrm{Irr}(G)$. That is, we have

$$KG = \prod_{\chi\in\mathrm{Irr}(G)} KGe(\chi)$$

where $e(\chi)$ is a primitive idempotent in $Z(KG)$ and where $KGe(\chi)$ is a matrix algebra over K, isomorphic to $\mathrm{End}_K(X)$, where X is a simple KG-module with character χ. Since B is an algebra factor of $\mathcal{O}G$, it follows that $K \otimes_\mathcal{O} B$ is an algebra factor of KG. Thus there is a unique subset $\mathrm{Irr}(B)$ of $\mathrm{Irr}(G)$ such that

$$K \otimes_\mathcal{O} B = \prod_{\chi\in\mathrm{Irr}(B)} KGe(\chi)$$

Since each factor $KGe(\chi)$ is a matrix algebra, it follows that $Z(KGe(\chi))$ has dimension 1. Thus

$$\mathrm{rk}_{\mathcal{O}}(Z(B)) = \dim_k(Z(k \otimes_{\mathcal{O}} B)) = |\mathrm{Irr}(B)|$$

In other words, the degree zero Hochschild cohomology of B determines the number of ordinary irreducible characters of B. A fundamental result fo Brauer and Feit shows that this number is bounded in terms of the defect groups of B.

THEOREM 6.1 ([**8**]). *Let d be the nonnegative integer such that the defect groups of B have order p^d. Then*

$$|\mathrm{Irr}(B)| \leq \frac{1}{4}p^{2d} + 1 \ .$$

Morita equivalent algebras have isomorphic centers, so Donovan's conjecture would imply that the center of a block over k is determined up to finitely many possibilities by the defect groups. It was noted by Kessar that combining the Brauer-Feit theorem with the earlier mentioned result by Cliff-Plesken-Weiss shows that this consequence of Donovan's conjecture is indeed true:

THEOREM 6.2 ([**35**, Corollary 1.3]). *For a fixed finite p-group P, there are only finitely many isomorphism classes of commutative k-algebras which are isomorphic to $Z(B)$ for some block algebra B of a finite group algebra kG with defect groups isomorphic to P.*

If P is cyclic or a Klein four group, one can calculate all possible algebras $Z(B)$, with B a block of kG with defect group isomorphic to P. If $P \cong C_2 \times C_2 \times C_2$, one can still classify $Z(B)$, but this requires the classification of finite simple groups; see [**37**]. Recently, Landrock and Sambale calculated $Z(B)$ in some cases where P is elementary abelian of order 16.

While the Brauer-Feit theorem shows that $|\mathrm{Irr}(B)|$ is bounded in terms of the order $|P|$ of a defect group P of the block B, the reverse is open:

QUESTION 6.3 (Brauer's problem 21). *Is it possible to bound $|P|$ in terms of $|\mathrm{Irr}(B)|$?*

Külshammer [**46**] showed that the answer is positive if G is p-solvable. Using this and Zelmanov's solution of the restricted Burnside problem, Külshammer and Robinson [**47**] showed that in general this would be a consequence of the Alperin-McKay conjecture. Brauer's problem 21 is known if $|\mathrm{Irr}(B)|$ is equal to 1 or 2, but not in general for $|\mathrm{Irr}(B)| = 3$; see [**69**, Ch. 15].

7. $HH^1(B)$

Let A be an algebra over a commutative ring k. A *derivation on A* is a k-linear map $f : A \to A$ satisfying the *Leibniz rule*

$$f(ab) = f(a)b + af(b)$$

for all a, $b \in A$. For instance, the differential operator $\frac{d}{dx}$ is a derivation on the polynomial algebra $k[x]$. The set $\mathrm{Der}(A)$ of derivations on A is a k-submodule of $\mathrm{End}_k(A)$.

If $c \in A$, then the map $[c, -] : A \to A$ defined by $[c, a] = ca - ac$ for all $a \in A$ is a derivation. Any derivation of this form is called an *inner derivation.* The set $\mathrm{IDer}(A)$ of inner derivation on A is a k-submodule of $\mathrm{Der}(A)$.

The algebra $\mathrm{End}_k(A)$ is a Lie algebra, with the bracket $[-,-]$ given by $[f,g] = f \circ g - g \circ f$ for all linear endomorphisms f, g of A. The k-submodule $\mathrm{Der}(A)$ is a Lie subalgebra of $\mathrm{End}_k(A)$; indeed, one verifies that if f, g are derivations on A, then so is $[f,g]$. However, $f \circ g$ need not be a derivation even if both f, g are. In other words, $\mathrm{Der}(A)$ need not be a subalgebra of the associative algebra $\mathrm{End}_k(A)$. The k-module $\mathrm{IDer}(A)$ is a Lie ideal in $\mathrm{Der}(A)$; indeed, one verifies that $[f,[c,-]] = [f(c),-]$ for any $f \in \mathrm{Der}(A)$ and any $c \in A$. Thus $\mathrm{Der}(A)/\mathrm{IDer}(A)$ is a Lie algebra. We have a canonical isomorphism

$$HH^1(A) \cong \mathrm{Der}(A)/\mathrm{IDer}(A)$$

and the Lie algebra structure on $HH^1(A)$ induced by this isomorphism is the Gerstenhaber bracket restricted to $HH^1(A)$.

The k-modules $\mathrm{Der}(A)$ and $\mathrm{IDer}(A)$ are $Z(A)$-submodules of $\mathrm{End}_k(A)$ via $(z \cdot f)(a) = zf(a)$, for any $f \in \mathrm{End}_k(A)$, $z \in Z(A)$, $a \in A$. Thus $HH^1(A)$ inherits a $Z(A)$-module structure. This is the same as the product in $HH^*(A)$ in degree 0 and 1, via the identifications $HH^0(A) = Z(A)$ and $HH^1(A) = \mathrm{Der}(A)/\mathrm{IDer}(A)$.

If k is a field of prime characteristic p, then for any derivation f on A, then map $f^{[p]} = f \circ f \circ \cdots \circ f$ (p times) is again a derivation. The correspondence $f \mapsto f^{[p]}$ induces a map on $HH^1(A)$, through which $HH^1(A)$ becomes a *p-restricted Lie algebra.*

Let k be a field of prime characteristic p. In the context of block algebras of finite groups, calculating examples suggests that the restricted Lie algebra $HH^1(B)$ determines many fundamental invariants of B, where B is a block of a finite group algebra kG, for some finite group G. We start with a basic question regarding the relationship between the structure of the restricted Lie algebra $HH^1(B)$ and the associative algebra B.

QUESTION 7.1. *Which simple Lie algebras arise as $HH^1(B)$, where B is a block of a finite group algebra for some finite group G?*

Finite-dimensional simple Lie algebras over an algebraically closed field of characteristic zero are classified in terms of Dynkin diagrams. In positive characteristic, the situation is far more complicated (see [**72**], [**73**]). The first simple Lie algebra in prime characteristic which does not have a finite-dimensional analogue in characteristic zero is the *Witt algebra*

$$W_1 = HH^1(kC_p) = \mathrm{Der}(kC_p)$$

where C_p is a cyclic group of order p, and where the second equality uses the fact that kC_p is commutative, so has no nonzero inner derivations. The algebra W_1, discovered by Witt in the 1930s, has dimension p. Through the identification $kC_p = k[x]/(x^p)$, the algebra W_1 has as a basis the set of derivations $\{f_i \mid 0 \leq i \leq p-1\}$ given by $f_i(x) = x^i$; note that any derivation of $k[x]/(x^p)$ is determined by its value on the generator x, here identified to its image in $k[x]/(x^p)$. A short calculation shows that

$$[f_i, f_j] = (j-i)f_{i+j-1}$$

whenever $0 \leq i+j-1 \leq p-1$, and $[f_i, f_j] = 0$ otherwise. If $p \geq 3$, then W_1 is a simple Lie algebra. Jacobson generalised this to elementary abelian p-groups.

THEOREM 7.2 (Jacobson [**31**, Theorem 1]). *Let P be a finite elementary abelian p-group of order at least 3. Then $HH^1(kP)$ is a simple Lie algebra.*

For blocks with a single isomorphism class, no other modular simple Lie algebras can arise. For the remainder of this section, we assume that k is an algebraically closed field of prime characteristic p.

THEOREM 7.3 (Linckelmann-Rubio y Degrassi [**56**, Theorem 1.1]). *Let G be a finite group and let B be a block algebra of kG having a unique isomorphism class of simple modules. Then $HH^1(B)$ is a simple Lie algebra if and only if B is nilpotent with an elementary abelian defect group P of order at least* 3. *In that case, we have a Lie algebra isomorphism $HH^1(B) \cong HH^1(kP)$.*

We do not know whether the hypothesis on B to have a single isomorphism class of simple modules is necessary in Theorem 7.3. There is a far more basic question which is open at present:

QUESTION 7.4. *Is it true that for any block B of a finite group algebra kG with a nontrivial defect group we have $HH^1(B) \neq \{0\}$?*

The hypothesis on B having a nontrivial defect group is necessary. Indeed, if B has the trivial group as a defect group, then B is a matrix algebra over k, and hence $HH^1(B)$ is zero in that case.

EXAMPLE 7.5. If B is a block of a symmetric group algebra kS_n with a nontrivial abelian defect group P, then $HH^1(B) \neq \{0\}$. Indeed, as a consequence of work of Chuang-Rouquier [**13**], proving Broué's abelian defect conjecture for blocks of symmetric groups, we have $HH^1(B) \cong HH^1(k(P \rtimes E))$, where E is the inertial quotient of B. An easy calculation using the standard centraliser decomposition of Hochschild cohomology shows that this is nonzero.

REMARK 7.6. If $HH^1(B)$ is nonzero for a block B of a finite group kG, then in particular $HH^1(kG)$ is nonzero. A result of Fleischmann, Janiszczak, and Lempken [**22**] on centralisers in finite groups, using the classification of finite simple groups, implies that $HH^1(kG)$ is indeed nonzero for any finite group G of order divisible by p. Gerstenhaber and D. J. Green [**25**] showed that the nonvanishing of $HH^1(kG)$ would also follow from the Donald-Flanigan conjecture. This conjecture predicts that kG admits a semisimple 1-parameter deformation, although some experts seem to believe that this conjecture is not true for all finite groups (with the quaternion group as a likely counterexample).

REMARK 7.7. If $\mathcal{O}$ is a complete discrete valuation ring of characteristic zero, then it is easy to see that $HH^1(\mathcal{O}G)$ is zero for all finite groups G. Thus $HH^1(B)$ is zero for all blocks of finite group algebras over $\mathcal{O}$. This is the reason why the results in this section are formulated over k.

If the quiver of B has a loop, or equivalently, if some simple B-module has a nonsplit selfextension, then $HH^1(B)$ is nonzero. We have the following more precise results for symmetric algebras.

THEOREM 7.8 ([**7**, Theorem 3.1]). *Let A be a symmetric k-algebra and let E be a maximal semisimple subalgebra. Let $f\colon A \to A$ be an E-E-bimodule homomorphism satisfying $E + J(A)^2 \subseteq \ker(f)$ and $\mathrm{Im}(f) \subseteq \mathrm{soc}(A)$. Then f is a derivation on A in $\mathrm{soc}_{Z(A)}(\mathrm{Der}(A))$, and if $f \neq 0$, then f is an outer derivation of A. In particular, we have*

$$\sum_S \dim_k(\mathrm{Ext}^1_A(S,S)) \leq \dim_k(\mathrm{soc}_{Z(A)}(HH^1(A)))$$

where in the sum S runs over a set of representatives of the isomorphism classes of simple A-modules.

COROLLARY 7.9 ([**7**, Corollary 3.2]). *Let A be a local symmetric k-algebra. Let $f\colon A \to A$ be a k-linear map satisfying $1 + J(A)^2 \subseteq \ker(f)$ and $\mathrm{Im}(f) \subseteq \mathrm{soc}(A)$. Then f is a derivation on A in $\mathrm{soc}_{Z(A)}(\mathrm{Der}(A))$, and if $f \neq 0$, then f is an outer derivation of A. In particular, we have*

$$\dim_k(J(A)/J(A)^2) \leq \dim_k(\mathrm{soc}_{Z(A)}(HH^1(A))) \ .$$

The above results have been used in [**7**] to calculate the Lie algebra structure of $HH^1(B)$ for rank 2 quantum complete intersections.

8. Integrable derivations

Let A be a finite-dimensional algebra over a field k. Let Φ be an automorphism of the $k[[t]]$-algebra $A[[t]] \cong k[[t]] \otimes_k A$. By $k[[t]]$-linearity, any such automorphism is determined by its effect on the copy of A in degree zero of $A[[t]]$. Suppose that Φ induces the identity on A via the canonical surjection $A[[t]] \to A$ sending t to 0. This is equivalent to requiring that for all $a \in A$ we have

$$\Phi(a) = a + d_1(a)t + d_2(a)t^2 + \cdots$$

where $d_i : A \to A$ is a k-linear map for $i \geq 1$. An easy verification, using that $\Phi(ab) = \Phi(a)\Phi(b)$ for all a, $b \in A$, shows that the linear map $d_1 : A \to A$ obtained in this way is in fact a derivation on A. Any derivation arising in this way is called integrable; more precisely:

DEFINITION 8.1 (Gerstenhaber [**24**]). A derivation d on A is called *integrable* if there exist a $k[[t]]$-algebra automorphism Φ of $A[[t]]$ and linear maps $d_i : A \to A$, $i \geq 2$, satisfying $\Phi(a) = a + d(a) + d_2(a) + \ldots$ for all $a \in A$.

The integrable derivations on A form a k-subspace of $\mathrm{Der}(A)$. The image in $HH^1(A)$ of this space is denoted by $HH_{\mathrm{int}}(A)$.

QUESTION 8.2. *When is $HH_{\mathrm{int}}(A)$ is a sovable Lie subalgebra of $HH^1(A)$?*

The reason for asking this is that the examples from block theory, such as in [**7**], where it is possible to calculate these spaces suggest that $HH_{\mathrm{int}}(B)$ is indeed a solvable Lie subalgebra of $HH^1(B)$, for B a block of a finite group algebra over an algebraically closed field of prime characteristic. This is an open problem in general.

REMARK 8.3. If $k = \mathbb{C}$, then every derivation on A is integrable. Given a derivation d on A, one constructs Φ by setting

$$\Phi = \sum_{n \geq 0} \frac{d^n}{n!} t^n$$

where $d^n = d \circ d \circ \cdots \circ d$, the n-fold composition of d with itself, with the convention $d^0 = \mathrm{Id}_A$. Moreover, $HH^1(A)$ is the tangent space of the outer automorphism group $\mathrm{Out}(A)$ as an algebraic gorup. This is not true in general for algebras over fields of positive characteristic.

For the purpose of applications in block theory, we extend Gerstenhaber's definition in the obvious way to arbitrary complete discrete valuation rings instead of $k[[t]]$.

Let $\mathcal{O}$ be a complete discrete valuation ring, let $\pi \in \mathcal{O}$ such that $J(\mathcal{O}) = \pi\mathcal{O}$, and suppose that $k = \mathcal{O}/J(\mathcal{O})$ is algebraically closed of prime characteristic p. Let A be an $\mathcal{O}$-algebra which is free of finite rank as an $\mathcal{O}$-module.

Note the change of notation: the algebras A and $\bar{A} = A/\pi A$ correspond to $A[[t]]$ and A, respectively, at the beginning of this section. Moreover, A need not be determined by $\bar{A}$ in general, unlike in the case $\mathcal{O} = k[[t]]$.

DEFINITION 8.4. Let n be a positive integer. We denote by $\mathrm{Aut}_n(A)$ the subgroup of $\mathrm{Aut}(A)$ consisting of all $\mathcal{O}$-algebra automorphisms of A which induce the identity on $A/\pi^n A$. We denote by $\mathrm{Out}_n(A)$ the image of $\mathrm{Aut}_n(A)$ in the outer automorphism group $\mathrm{Out}(A) = \mathrm{Aut}(A)/\mathrm{Inn}(A)$.

In what follows, we will assume that the canonical map $Z(A) \to Z(A/\pi^n A)$ is surjective. If $\mathcal{O} = k[[t]]$ and $A = \bar{A}[[t]]$, this hypothesis holds trivially. The following easy observation is the reason for this hypothesis.

LEMMA 8.5 ([**53**, 3.1]). *Let A be an $\mathcal{O}$-algebra which is free of finite rank as an $\mathcal{O}$-module, and let n be a positive integer. Suppose that the canonical map $Z(A) \to Z(A/\pi^n A)$ is surjective. The group $\mathrm{Out}_n(A)$ is equal to the kernel of the canonical map $\mathrm{Out}(A) \to \mathrm{Out}(A/\pi^n A)$.*

In order to extend the notion of integrable derivations to this situation, we will need the following elementary facts.

PROPOSITION 8.6 ([**53**, 3.2]). *Let A be an $\mathcal{O}$-algebra which is free of finite rank as an $\mathcal{O}$-module. Let n be a positive integer. Suppose that the canonical map $Z(A) \to Z(A/\pi^n A)$ is surjective. Let $\alpha \in \mathrm{Aut}_n(A)$, and let $\mu : A \to A$ be the unique linear map satisfying $\alpha(a) = a + \pi^n\mu(a)$ for all $a \in A$.*

(i) *The map $\bar{\mu} : A/\pi^n A \to A/\pi^n A$ induced by μ is a derivation. The class of $\bar{\mu}$ in $HH^1(A/\pi^n A)$ depends only on the class of α in $\mathrm{Out}(A)$.*
(ii) *If α is an inner automorphism of A, then α is induced by conjugation with an element of the form $c = 1 + \pi^n d$ for some $d \in A$, and we have $\bar{\mu} = [\bar{d}, -]$; in particular, $\bar{\mu}$ is an inner derivation in that case.*
(iii) *The map sending α to the class of $\bar{\mu}$ induces a group homomorphism $\mathrm{Out}_n(A) \to HH^1(A/\pi^n A)$, with kernel equal to $\mathrm{Out}_{2n}(A)$. In particular, $HH^1_A(A/\pi^n A)$ is a subgroup of $HH^1(A/\pi^n A)$, and we have a short exact sequence of groups*

$$1 \longrightarrow \mathrm{Out}_{2n}(A) \longrightarrow \mathrm{Out}_n(A) \longrightarrow HH^1_A(A/\pi^n A) \longrightarrow 1$$

Gerstenhaber's definition of integrable derivation takes the following form.

DEFINITION 8.7 ([**53**, 3.3]). Let A be an $\mathcal{O}$-algebra which is free of finite rank as an $\mathcal{O}$-module, and let n be a positive integer. A derivation δ on $A/\pi^n A$ is called *A-integrable* if there is an algebra automorphism α of A and an $\mathcal{O}$-linear endomorphism μ of A such that $\alpha(a) = a + \pi^n\mu(a)$ for all $a \in A$ and such that δ is equal to the map induced by μ on $A/\pi^n A$. A class $\eta \in HH^1(A/\pi^n A)$ is called *A-integrable* if it can be represented by an A-integrable derivation. We denote by $HH^1_A(A/\pi^n A)$ the set of A-integrable classes in $HH^1(A/\pi^n A)$.

The set $HH^1_{\mathcal{A}}(A/\pi^n A)$ depends on the algebra A as well as the ring $\mathcal{O}$. If $\mathcal{A} = \bar{A}[[t]]$, where $\bar{A} = A/\pi A$, then $HH^1_{\mathcal{A}}(\bar{A}) = HH^1_{\mathrm{int}}(\bar{A})$. The canonical map

$$A/\pi^n A \to \bar{A}$$

induces a map $HH^1(A/\pi^n A) \to HH^1(\bar{A})$. We denote by $HH^1_n(\bar{A})$ the image in $HH^1(\bar{A})$ of $HH^1_{\mathcal{A}}(A/\pi^n A)$ under this map.

9. Invariance properties of integrable derivations

Stable equivalences induced by tensoring with suitable bimodules abound in block theory. The problem with stable equivalences is that we lack control over which invariants they preserve. A conjecture of Auslander and Reiten from the 1970s predicts that stably equivalent finite-dimensional algebras over a field should have the same number of isomorphism classes of nonprojective simple modules. This conjecture, if true, would go a long way towards determining block invariants in many cases that are currently beyond reach. As it turns out, the subgroups of integrable derivations have good invariance properties under stable equivalences.

Let $\mathcal{O}$ be a complete discrete valuation ring, let $\pi \in \mathcal{O}$ such that $J(\mathcal{O}) = \pi\mathcal{O}$, and suppose that $k = \mathcal{O}/J(\mathcal{O})$ is algebraically closed of prime characteristic p. Hochschild cohomology is invariant under Morita and derived equivalences. The following result shows that Morita equivalences preserve the subspaces of intergrable derivations in the case where $\mathcal{O} = k[[t]]$.

THEOREM 9.1 (Farkas-Geiss-Marcos [**21**]). *Let A be a finite-dimensional k-algebra. The subspace $HH^1_{\mathrm{int}}(A)$ is invariant under Morita equivalences.*

We do not know in general whether $HH^1_{\mathrm{int}}(A)$ is invariant under derived equivalences in general, but for symmetric algebras, we will show that this is the case, as a consequence of a stronger result, showing that $HH^1_{\mathrm{int}}(A)$ is invariant under stable equivalences *of Morita type*, and we will do this is in the slightly more general setting of arbitrary complete discrete valuation rings. We briefly review the notion of stable equivalences of Morita type.

DEFINITION 9.2 (Broué [**11**, §5 A]). Let A and B be $\mathcal{O}$-algebras, free of finite rank over $\mathcal{O}$. An A-B-bimodule M and a B-A-bimodule N are said to induce a *stable equivalence of Morita type*, if M, N are finitely generated projective as left and as right modules, and if we have isomorphisms

$$M \otimes_B N \cong A \oplus Y$$

$$N \otimes_A M \cong B \oplus Z$$

as A-A-bimodules and B-B-bimodules, respectively, for some projective A-A-bimodule Y and some projective B-B-bimodule Z.

In the situation of this definition, the functors $M \otimes_B -$ and $N \otimes_A -$ induce equivalences between the relatively $\mathcal{O}$-stable categories of A-modules and of B-modules, and if A and B are symmetric, then we may choose $N = M^\vee$. See [**11**, §5] for more details.

EXAMPLES 9.3.

(a) Suppose that $p = 2$. Set $A = \mathcal{O}A_5$ and $B = \mathcal{O}A_4$. Consider B as a subalgebra of A, and set $M = A_B$ and $N = {}_BA$. Then M and N induce a stable equivalence of Morita type between A and B.

(b) Let G be a finite group with a cyclic Sylow p-subgroup P. Let H be the normaliser in G of the unique subgroup of order p of P. Set $A = \mathcal{O}G$ and $B = \mathcal{O}H$. Then $M = A_B$ and $N = {}_BA$ induce a stable equivalence of Morita type between A and B. In this case, it is well-known that kG and kH have the same number of isomorphism classes of nonprojective simple modules, thanks to the fact that kG and kH have finite representation type. This example extends to blocks.

(c) Let G be a finite group and B a block of $\mathcal{O}G$ with a nontrivial cyclic defect group P. Let H be the normaliser in G of the unique subgroup of order p of P. By a result of Brauer, there is a unique block C of $\mathcal{O}H$ having P as a defect group such that C is isomorphic to a direct summand of B as an $\mathcal{O}H$-$\mathcal{O}H$-bimodule. Set $b = 1_B$ and $c = 1_C$. Then the B-C-bimodule $M = b\mathcal{O}Gc$ and its dual, which is isomorphic to $c\mathcal{O}Gb$, induce a stable equivalence of Morita type between B and C. One can show that there is a Morita equivalence between $\mathcal{O}H$ and $\mathcal{O}(P \rtimes E)$, where E is the common inertial quotient of B and of C. This example is again a special case of a more general situation.

(d) (Puig [**64**]) Let G be a finite group and B a block of $\mathcal{O}G$ with an abelian defect group and inertial quotient E acting freely on $P \setminus \{1\}$. Then there is a stable equivalence of Morita type between B and $\mathcal{O}(P \rtimes E)$.

The following result shows that stable equivalences of Morita type preserve integrable derivations, as well as the automorphism groups which induce these, in a compatible way.

THEOREM 9.4 ([**53**, 5.1]). *Let A, B be $\mathcal{O}$-algebras which are free of finite rank as $\mathcal{O}$-modules, such that the k-algebras $k \otimes_{\mathcal{O}} A$ and $k \otimes_{\mathcal{O}} B$ are indecomposable non-simple selfinjective with separable semisimple quotients. Let n be a positive integer. Suppose that the canonical maps $Z(A) \to Z(A/\pi^n A)$ and $Z(B) \to Z(B/\pi^n B)$ are surjective. Let M be an A-B-bimodule and N a B-A-bimodule inducing a stable equivalence of Morita type between A and B. For any $\alpha \in \mathrm{Aut}_n(A)$ there is $\beta \in \mathrm{Aut}_n(B)$ such that ${}_{\alpha^{-1}}M \cong M_\beta$ as A-B-bimodules, and the correspondence $\alpha \mapsto \beta$ induces a group isomorphism $\mathrm{Out}_n(A) \cong \mathrm{Out}_n(B)$ making the following diagram of groups commutative:*

$$\begin{array}{ccc} \mathrm{Out}_n(A) & \xrightarrow{\cong} & \mathrm{Out}_n(B) \\ \downarrow & & \downarrow \\ HH^1_A(A/\pi^n A) & \xrightarrow{\cong} & HH^1_B(B/\pi^n B) \end{array}$$

where the vertical maps are the group homomorphisms from Proposition 8.6 (iii) and where the lower horizontal isomorphism is induced by the functor $N \otimes_A - \otimes_A M$.

REMARK 9.5. By a result of Rickard [**66**, 2.2], a derived equivalence between $\mathcal{O}$-selfinjective algebras induces a stable equivalence of Morita type, and hence the above result implies that under the hypotheses in 9.4, the groups $HH^1_A(A/\pi^n A)$ are invariant under derived equivalences.

By [**43**, Theorem 10.7], a stable equivalence of Morita type between finite-dimensional symmetric algebras preserves the Lie algebra structure on Hochschild cohomology. The following result shows that stable equivalences of Morita type between finite-dimensional symmetric algebras over a field k of prime characteristic preserve furthermore the p-power map on integrable derivations. Given a finite-dimensional k-algebra A and a positive integer, we use the notation introduced at the end of the last section; that is, we write $HH^1_n(A)$ for the image of $HH^1_{A[[t]]}(A[[t]]/t^nA[[t]])$ in $HH^1(A)$ under the canonical map

$$HH^1(A[[t]]/t^nA[[t]]) \to HH^1(A)$$

THEOREM 9.6 (Rubio y Degrassi [**68**, 1.1]). *Let A, B be finite-dimensional selfinjective algebras over a field k of prime characteristic p. Suppose that the semisimple quotients of A and B are separable. Let M be an A-B-bimodule and N a B-A-bimodule inducing a stable equivalence of Morita type between A and B. Let n be a positive integer. The p-power map on $HH^1(A)$ sends $HH^1_n(A)$ to $HH^1_{np}(A)$, and we have a commutative diagram of k-linear maps*

$$\begin{array}{ccc} HH^1_n(A) & \xrightarrow{\cong} & HH^1_n(B) \\ {\scriptstyle [p]}\downarrow & & \downarrow{\scriptstyle [p]} \\ HH^1_{np}(A) & \xrightarrow{\cong} & HH^1_{np}(B) \end{array}$$

where the horizontal isomorphisms are induced by the functor $N \otimes_A - \otimes_A M$, and where the vertical maps are the p-power maps.

The proofs of the theorems 9.4 and 9.6 require an interpretation of integrable derivations in terms of certain Bockstein homomorphisms.

10. Separably equivalent algebras

DEFINITION 10.1 ([**33**, 6.1], [**50**, 3.1]). Two algebras A, B over a commutative ring k are called *separably equivalent* if there exist an A-B-bimodule M and a B-A-bimodule N, both finitely generated projective as left and right modules, such that A is isomorphic to a direct summand of $M \otimes_B N$ as an A-A-bimodule, and B is isomorphic to a direct summand of $N \otimes_A M$ as a B-B-bimodule.

EXAMPLE 10.2. Let $\mathcal{O}$ be a complete local principal ideal domain with residue field k of prime characteristic p. Let G be a finite group, B is a block algebra of $\mathcal{O}G$, and P a defect group of B. Then B and $\mathcal{O}P$ are separably equivalent. Indeed, B is isomorphic to a direct summand of $B \otimes_{\mathcal{O}P} B$, and $\mathcal{O}P$ is isomorphic to a direct summand of $B \cong B \otimes_B B$, hence the bimodules $B_{\mathcal{O}P}$ and ${}_{\mathcal{O}P}B$ realise a separable equivalence between B and $\mathcal{O}P$.

REMARKS 10.3.

(a) In the situation of Definition 10.1, M and N are progenerators as one-sided modules. Indeed, N is finitely generated as a left B-module, hence isomorphic to a direct summand of a free module B^n of finite rank n. Thus $M \otimes_B N$ is isomorphic, as a left A-module, to a direct summand of $M \otimes_B B^n \cong M^n$. Since A is isomorphic to a direct summand of $M \otimes_B N$, it follows that A is isomorphic to a direct summand of M^n as a left A-module. This shows that M is a progenerator as a left A-module, and the remaining cases follow similarly.

(b) If A and B are Morita equivalent algebras, or if there is a stable equivalence of Morita type between A and B, then A and B are separably equivalent.

(c) If A and B are separably equivalent symmetric algebras over a commutative ring k, then one can choose the bimodules M and N realising a separable equivalence such that $N \cong M^\vee = \mathrm{Hom}_k(M, k)$. In that case, the four adjunction units/counits $A \to M \otimes_B M^\vee$, $M^\vee \otimes_A M \to B$, $B \to M^\vee \otimes_A M$, and $M \otimes_B M^\vee \to A$ are split bimodule homomorphisms.

(d) The notion of separable equivalence is transitive. In particular, if two block algebras B, B' of finite group algebras $\mathcal{O}G$, $\mathcal{O}G'$ with defect groups P, P', respectively, are Morita equivalent or derived equivalent or stably equivalent of Morita type, then their defect group algebras $\mathcal{O}P$, $\mathcal{O}P'$ are separably equivalent.

The main motivation in [**50**] for considering the notion of separable equivalence is that it preserves certain cohomological dimensions. For the remainder of this section we denote by k an algebraically closed field. Following terminology introduced in [**20**], a finite-dimensional k-algebra A is said to satisfy the property (Fg), if $HH^*(A)$ is Noetherian, and if $\mathrm{Ext}^*_A(U, U)$ is Noetherian as a module over $HH^*(A)$ via the algebra homomorphism $HH^*(A) \to \mathrm{Ext}^*_A(U, U)$ induced by the functor $- \otimes_A U$, for any finitely generated A-module U. As mentioned in property 1.5, block algebras of finite group algebras over k satisfy the property (Fg), as a consequence of the Evens-Venkov theorem.

THEOREM 10.4. *Let A, B be separably equivalent symmetric k-algebras. Then* $\mathrm{Ext}^*_{A \otimes_k A^{\mathrm{op}}}(A, U)$ *is Noetherian as an $HH^*(A)$-module for any finitely generated A-A-bimodule U if and only if* $\mathrm{Ext}_{B \otimes_k B^{\mathrm{op}}}(B, V)$ *is Noetherian as an $HH^*(B)$-module for any finitely generated B-B-bimodule V. In that case, the Krull dimensions of $HH^*(A)$ and of $HH^*(B)$ are equal.*

Thus the Krull dimension of $HH^*(B)$ of a block B of a finite group algebra over k with defect group P is equal to the Krull dimension of $HH^*(kP)$, which is well-known to be equal to the rank of P. Another application of Theorem 10.4 is the following result.

THEOREM 10.5 ([**50**, Theorem 1.1]). *Suppose that* $\mathrm{char}(k) = 0$. *Let $\mathcal{H}$ be an Iwahori-Hecke algebra of a finite Coxeter group (W, S) with non-zero parameter q in k. Suppose that all irreducible components of W are of type A, B, D, and suppose in addition that if W involves a component of type B or D then the order of q in $k^\times$ is not even. Then, for any finitely generated $\mathcal{H}$-$\mathcal{H}$-bimodule M, the $HH^*(\mathcal{H})$-module* $\mathrm{Ext}^*_{\mathcal{H} \otimes_k \mathcal{H}^0}(\mathcal{H}, M)$ *is Noetherian. In particular, $HH^*(\mathcal{H})$ is finitely generated as a k-algebra.*

It is possible to add some precisions regarding Krull dimensions; see [**50**, Theorem 1.2]. Condition (Fg) remains open in general for Iwahori-Hecke algebras of exceptional types or Iwahori-Hecke algebras over fields of positive characteristic.

Morita equivalent algebras have isomorphic Hochschild cohomology algebras. Thus Donovan's conjecture, if true, would imply that for a fixed finite p-group P, there are only finitely many isomorphism classes of graded-commutative algebras of the form $HH^*(B)$, where B runs over the blocks of finite group algebras over k with defect groups isomorphic to P. This, in turn, would imply the following result.

THEOREM 10.6 ([**38**, Theorem 2]). *Suppose that* $\mathrm{char}(k) = p > 0$. *Let* P *be a finite* p*-group. There are only finitely many power series in* $\mathbb{Z}[[t]]$ *which are equal to Hilbert series* $\sum_{n\geq 0} \dim_k(HH^n(B))t^n$ *of Hochschild cohomology algebra* $HH^*(B)$ *of a block* B *of a finite group algebra over* k *with defect groups isomorphic to* P.

For the proof of Theorem 10.6, one first bounds $\dim_k(HH^*(B))$ in terms of P, arguing by induction over n. For $n = 0$ this is the Brauer-Feit theorem 6.1. Then one bounds the degrees of generators and relations in $HH^*(B)$ in terms of P. Here the key ingredient is the Castelnuovo-Mumford regularity, and more precisely, Symonds' proof in [**74**] of Benson's regularity conjecture [**3**], [**4**], stating that $\mathrm{reg}(H^*(G;k)) = 0$. This implies that $\mathrm{reg}(HH^*(B)) \leq 0$, and therefore yields bounds on generators and relations by a theorem of Serre.

REMARK 10.7. Whether there are actually only finitely many isomorphism classes of Hochschild cohomology algebras of blocks with a given defect group remains at this point an open problem; as mentioned before, this would be a consequence of Donovan's conjecture. It would suffice to bound the size of the field generated by the coefficients in the relations between a suitably chosen set of generators of $HH^*(B)$. Evidence in that direction includes the aforementioned work of Cliff, Plesken and Weiss [**14**], showing that $HH^0(B) \cong Z(B)$ has always a k-basis with multiplicative constants in the prime field $\mathbb{F}_p$. Since principal blocks are defined over $\mathbb{F}_p$, we can conclude that there are only finitely many isomorphism classes of Hochschild cohomology algebras of principal blocks with a given defect group. Theorem 4.5 implies that Donovan's conjecture is in fact equivalent to bounding Cartan invariants and fields of definition of a block in terms of its defect groups.

REMARK 10.8. It is also true that fixing the Hilbert series

$$\sum_{n\geq 0} \dim_k(HH^n(B))t^n$$

of the Hochschild cohomology of a block B determines the defect groups P up to finitely many isomorphism classes. See [**38**, Theorem 3]. This is a weaker version of Brauer's problem 21 mentioned above, stipulating that P should be determined up to finitely many possibilities by the dimension of $HH^0(B)$ alone.

QUESTION 10.9. *Suppose that* A *and* B *are separably equivalent symmetric* k*-algebras. Does this imply that* $HH^*(A)$ *and* $HH^*(B)$ *have the same regularity?*

If true, this would simplify the proof, in [**39**], of the fact that $\mathrm{reg}(HH^*(B)) = 0$ for any block B of a finite group algebra over k.

REMARK 10.10. Two separably equivalent finite-dimensional k-algebras have the same representation type (cf. [**50**, 3.5], [**61**]). The definition of separable equivalence extends to categories, and two separably equivalent triangulated categories have the same dimension (cf. [**50**, 3.6]). In particular, if A and B are separably equivalent finite-dimensional k-algebras, then their bounded derived categories have the same dimension, and if A, B are in addition selfinjective, then their stable categories have the same dimension, as triangulated categories. See [**61**] for more properties of separable equivalences.

11. Finitistic and dominant dimensions

Let $\mathcal{A}$ be an abelian category with enough projective objects. The *projective dimension* $\mathrm{pdim}(U)$ of an object U in $\mathcal{A}$ is the smallest nonnegative integer n for which there exists a projective resolution which is zero in all degrees larger than n, provided that U has a bounded projective resolution, and $\text{pdim}(U) = \infty$ otherwise. The *finitistic dimension of* $\mathcal{A}$ is the supremum of the set of integers $\text{pdim}(U)$, with U running over all objects in $\mathcal{A}$ having a finite projective dimension (and with the convention that the finitistic dimension is ∞ if this supremum is unbounded). The left (resp. right) finitistic dimension of a finite-dimensional algebra A over a field is the finitistic dimension of the category $\text{mod}(A)$ (resp. $\text{mod}(A^{\text{op}})$) of finitely generated left (resp. right) A-modules. The - to date open - finitistic dimension conjecture states that the left or right finitistic dimension of A is finite, for any finite-dimensional algebra A. See [**30**] for an overview and further references on this conjecture. Finitistic dimensions are preserved under separable equivalences. Let k be a field.

PROPOSITION 11.1. *Let A and B be finite-dimensional k-algebras. If A and B are separably equivalent, then the left (resp. right) finitistic dimensions of A and B are equal.*

PROOF. Suppose that the left finitistic dimension of A is finite and equal to n. It suffices to show that the left finitistic dimension of B is at most n. Let M be an A-B-bimodule and N a B-A-bimodule inducing a separable equivalence between A and B; that is, M and N are finitely generated projective as one-sided modules, A is isomorphic to a direct summand of the A-A-bimodule $M \otimes_B N$, and B is isomorphic to a direct summand to the B-B-bimodule $N \otimes_A M$. Let V be a finitely generated B-module such that $\text{pdim}(V)$ is finite. The functor $M \otimes_B -$ sends a projective resolution of V to a projective resolution of $M \otimes_B V$. Thus $\text{pdim}(M \otimes_B V)$ is finite, hence less or equal to n by the assumptions on A. Applying the functor $N \otimes_B -$ to a projective resolution of $M \otimes_B V$ shows that $\text{pdim}(N \otimes_A M \otimes_B V) \leq n$. By the assumptions, V is isomorphic to a direct summand of $N \otimes_A M \otimes_B V$, and hence also $\text{pdim}(V) \leq n$. The result follows. □

Let k be a field of prime characteristic p. Tambara calculated the finitistic dimension of cohomological Mackey functors on a finite group G. To state this, we need the *p-sectional rank of* G; this is the largest integer n such that G has a subquotient isomorphic to an elementary abelian p-group of rank n. Clearly $s(G) = s(P)$, where P is a Sylow p-subgroup of G.

THEOREM 11.2 (Tambara [**76**]). *Let G be a finite group of order divisible by p. The finitistic dimension of the category of cohomological Mackey functors on G with coefficients in* $\text{mod}(k)$ *is equal to* $s(G) + 1$.

Tambara's proof exploits implicitly the fact that kG is separably equivalent to kP, where P is a Sylow p-subgroup of G, combined with a result of Yoshida, which describes cohomological Mackey functors as modules over a certain endomorphism algebra. The following result (which holds regardless of the characteristic of k) extracts the underlying formalism.

THEOREM 11.3 ([**52**, 3.1]). *Let A, B be symmetric k-algebras. Let M be an A-B-bimodule such that M and $M^\vee$ induce a separable equivalence between A and B. Let U be a finitely generated A-module and V a finitely generated B-module. Suppose that $M^\vee \otimes_A U \in$* $\text{add}(V)$ *and that $M \otimes_B V \in$* $\text{add}(U)$*. Then the right finitistic dimensions of* $\text{End}_A(U)$ *and of* $\text{End}_B(V)$ *are equal.*

This theorem is used in [**52**] to extend Tambara's theorem to the category coMack(B) of cohomological Mackey functors on a finite group G associated with a block B of kG.

THEOREM 11.4 ([**52**, Theorem 1.6]). *Let G be a finite group and B a block of kG having a nontrivial defect group P. The finitistic dimension of* coMack(G,b) *is equal to* $1 + s(P)$.

REMARK 11.5. The word 'nontrivial' is missing in the statement of [**52**, Theorem 1.6], but this hypothesis is necessary.

The sectional rank of a defect group appears also in a conjecture by Malle and Robinson, stated in 2015, and one may speculate if there is a connection.

CONJECTURE 11.6 (Malle and Robinson [**57**]). *Let G be a finite group, B a block of kG, and P a defect group of B. The number $\ell(B)$ of isomorphism classes of simple B-modules is at most $p^{s(P)}$.*

Following Tachikawa [**75**], the *dominant dimension of a finite-dimensional k-algebra A*, which we will denote by ddim(A), is the largest nonnegative integer d such that there exists an injective resolution

$$0 \longrightarrow A \longrightarrow I^0 \longrightarrow I^1 \longrightarrow \cdots$$

of A as a right A-module with the property that I^n is projective for $0 \leq n \leq d-1$, provided there is such an integer. If for any injective resolution I of A the term I^0 is not projective, then $\mathrm{ddim}(A) = 0$, and if there exists an injective resolution I of A such that I^n is projective for all $n \geq 0$, then we adopt the convention $\mathrm{ddim}(A) = \infty$. By a result of Müller [**58**, Theorem 4] the dominant dimension is equal to the obvious left module analogue.

THEOREM 11.7 ([**55**, Theorem 1.2]). *Let A, B be symmetric k-algebras. Let M be an A-B-bimodule inducing a separable equivalence between A and B. Let U be a finitely generated A-module such that A is isomorphic to a direct summand of U, and let V a finitely generated B-module such that B is isomorphic to a direct summand of V. Suppose that $M^\vee \otimes_A U \in \mathrm{add}(V)$ and that $M \otimes_B V \in \mathrm{add}(U)$. Then the dominant dimensions of $\mathrm{End}_A(U)$ and of $\mathrm{End}_B(V)$ are equal.*

This is used in [**55**] to determine the dominant dimension of cohomological Mackey functor categories of blocks.

THEOREM 11.8 ([**55**, Theorem 1.1]). *Let G be a finite group, k a field of prime characteristic p and B a block of kG with a nontrivial defect group P. The dominant dimension of* coMack(B) *is equal to* 2.

12. Fusion and algebra structure of blocks

Let G be a finite group, p a prime, and P a Sylow p-subgroup. The *fusion system of G on P* is the category $\mathcal{F} = \mathcal{F}_P(G)$, defined as follows.

The objects of $\mathcal{F}$ are the subgroups of P. For any two subgroups Q, R, the set $\mathrm{Hom}_\mathcal{F}(Q,R)$ of morphisms in $\mathcal{F}$ from Q to R consists of all group homomorphisms induced by conjugation in G; more precisely, $\mathrm{Hom}_\mathcal{F}(Q,R)$ is the set of group homomorphisms $\varphi : Q \to R$ such that there exists an element $x \in G$ satisfying $\varphi(u) = xux^{-1}$ for all $u \in Q$. In particular, x satisfies ${}^xQ \leq R$. Endomorphisms of

Q in $\mathcal{F}$ are induced by conjugation with elements in $N_G(Q)$, and hence we have a canonical isomorphism

$$\mathrm{Aut}_{\mathcal{F}}(Q) \cong N_G(Q)/C_G(Q) \ .$$

The fusion system $\mathcal{F}_P(G)$ of G on P captures - in addition to the structure of P - the way in which P is embedded in G in a category theoretic language.

If H is a subgroup of G containing P, then $\mathcal{F}_P(H)$ is a subcategory of $\mathcal{F}_G(P)$, with both categories having the same objects. If $\mathcal{F}_P(H) = \mathcal{F}_P(G)$, then H is said to *control fusion of G in P*. Classical results on control of fusion include the following.

THEOREM 12.1 (Burnside). *Suppose that the finite group G has an abelian Sylow p-subgroup P. Then $\mathcal{F}_P(G) = \mathcal{F}_P(N_P(G))$.*

In other words, if G has an abelian Sylow p-subgroup P, then any homomorphism between subgroups of P induced by conjugation in G is in fact induced by in element in $N_G(P)$, and hence any such homomorphism between subgroups extends to an automorphism of P.

THEOREM 12.2 (Frobenius). *Let G be a finite group and P a Sylow p-subgroup. We have $\mathcal{F}_P(G) = \mathcal{F}_P(P)$ if and only if $G = O_{p'}(G) \rtimes P$.*

The *focal subgroup of G in P* is the group $\mathfrak{foc}(G) = P \cap [G, G]$. D. G. Higman's focal subgroup theorem states that the focal subgroup of G in P depends only on $\mathcal{F}$. More precisely:

THEOREM 12.3 ([**28**]). *We have $\mathfrak{foc}(G) = \langle \varphi(u)u^{-1} \mid Q \leq P, u \in Q, \varphi \in \mathrm{Aut}_{\mathcal{F}}(Q)\rangle$.*

Work of Alperin and Broué [**2**], significantly extending ideas of Brauer, paved the way for associating similar fusion systems with any block of a finite group, replacing Sylow p-subgroups by defect groups.

Let $\mathcal{O}$ be a complete discrete valuation ring of characteristic zero with a residue field k of prime characteristic p.

Let G be a finite group. Given a block B of kG or of $\mathcal{O}G$ with a defect group P, one can define a *fusion system* $\mathcal{F}_P(B)$ of the block B on its defect group P as a category with similar properties to that of a fusion system of a group. There is a slight abuse of notation here, since $\mathcal{F}_P(B)$ is defined only up to conjugation by $N_G(P)$. Without going into technical detail, the objects of $\mathcal{F}_P(B)$ are the subgroups of P, and the morphisms are some (but not necessarily all) group homomorphisms between subgroups of P which are induced by conjugation in G. In other words, $\mathcal{F}_P(B)$ is a not necessarily full subcategory of $\mathcal{F}_S(G)$, where S is a Sylow p-subgroup of G containing P. If B is the principal block of kG, then P is a Sylow p-subgroup of G, and we have $\mathcal{F}_P(B) = \mathcal{F}_P(G)$. As in the case of defect groups, the fusion system of a block B of $\mathcal{O}G$ on P is equal to that of the corresponding block of kG.

The formal similarities between fusion systems of finite groups on Sylow p-subgroups and of blocks on defect groups led Puig to the definition of an abstract *saturated fusion system on a finite p-group P* (in the terminology of [**10**]), extending the category theoretic approach in [**62**].

REMARK 12.4. While fusion systems of a block B on a defect group P of B can be defined without any hypothesis on k, one of the axioms of a saturated fusion system (namely the *Sylow axiom*) holds only if k is large enough in the sense that block idempotents of $kC_G(P)$ associated with B remain blocks upon extending the field k by any finite extension. This holds, for instance, if k is a splitting field for $C_G(P)$.

In view of Frobenius' theorem 12.2, Broué and Puig [**12**] introduced *nilpotent blocks* as the blocks B satisfying $\mathcal{F}_P(B) = \mathcal{F}_P(P)$, for some (hence any) defect group P of B. Puig proved in [**63**, 1.6 Main Theorem] that nilpotent blocks are Morita equivalent to their defect group algebras, and he proved in [**65**, Theorem 8.2] that the converse holds for blocks over $\mathcal{O}$.

For an arbitrary saturated fusion system $\mathcal{F}$ on a finite p-group P we define its *focal subgroup* as the subgroup $\mathfrak{foc}(\mathcal{F})$ of P generated by all elements of the form $\varphi(u)u^{-1}$, where Q runs over the subgroups of P, $u \in Q$, and $\varphi \in \mathrm{Aut}_{\mathcal{F}}(Q)$. Clearly $\mathfrak{foc}(\mathcal{F})$ is a normal subgroup of P, invariant under $\mathrm{Aut}_{\mathcal{F}}(P)$, and $P/\mathfrak{foc}(\mathcal{F})$ is abelian. Alperin's fusion theorem implies that $\mathfrak{foc}(\mathcal{F})$ is generated by the set of elements of the form $\varphi(u)u^{-1}$, where $u \in P$ and $\varphi \in \mathrm{Hom}_{\mathcal{F}}(\langle u \rangle, P)$. By Higman's focal subgroup theorem, if $\mathcal{F} = \mathcal{F}_P(G)$ for some finite group G and Sylow p-subgroup P, then $\mathfrak{foc}(G) = \mathfrak{foc}(\mathcal{F})$.

In the following result we assume that $\mathcal{O}$ is a finite extension of the ring of p-adic integers. We use the notation $\mathrm{Out}_1(B)$ and $\mathrm{Out}_m(B)$ introduced in 8.4; that is, $\mathrm{Out}_m(B)$ is the image in $\mathrm{Out}(B)$ of the group of $\mathcal{O}$-algebra automorphisms of B which induce the identity on $B/\pi^m B$, where π is an element in $\mathcal{O}$ satisfying $\pi\mathcal{O} = J(\mathcal{O})$. The hypothesis in the next statement that $\mathcal{O}^\times$ contains a primitive $|G|$-th root of unity implies in particular that k is a splitting field for the subgroups of G, and hence the fusion systems of blocks of kG are saturated.

THEOREM 12.5 ([**54**, Theorem 1.1]). *Let G be a finite group, B a block of $\mathcal{O}G$ with a nontrivial defect group P, and let $\mathcal{F} = \mathcal{F}_P(B)$ be a fusion system of B on P. Suppose that $\mathcal{O}^\times$ contains a primitive $|G|$-th root of unity. There is a canonical injective group homomorphism*

$$\Phi : \mathrm{Hom}(P/\mathfrak{foc}(\mathcal{F}), \mathcal{O}^\times) \to \mathrm{Out}_1(B)$$

Moreover, Φ restricts to an isomorphism

$$\mathrm{Hom}(P/\mathfrak{foc}(\mathcal{F}), \mu) \cong \mathrm{Out}_m(B)$$

where μ is the subgroup of order p of $\mathcal{O}^\times$ and where m is the p-adic valuation of $1 - \zeta_p$, for some primitive p-th root of unity ζ_p in $\mathcal{O}^\times$.

The statement in [**54**] is slightly more precise; the class of Φ in $\mathrm{Out}_1(B)$ is characterised by the property that the action of Φ on $\mathrm{Irr}(B)$ is the action $\mathrm{Hom}(P/\mathfrak{foc}(\mathcal{F}), \mathcal{O}^\times)$ via the Broué-Puig $*$-construction. The proof requires results of Hertweck-Kimmerle [**26**] and Weiss [**77**].

REMARK 12.6. The group $P/\mathfrak{foc}(\mathcal{F})$ has a topological interpretation: by [**9**, Theorem 2.5] this group is the abelianisation of the fundamental group of the p-completed nerve of a centric linking system of $\mathcal{F}$.

The point of Theorem 12.5 is that the left side in the isomorphism depends on the fusion system of B and the right side on the $\mathcal{O}$-algebra structure of B. This leads to the following consequence (where we implicitly assume the ring $\mathcal{O}$ to contain sufficiently high order roots of unity):

COROLLARY 12.7 ([**54**, Corollary 1.5]). *The rank of $P/\mathfrak{foc}(\mathcal{F})$ is invariant under stable equivalences of Morita type between block algebras over $\mathcal{O}$.*

In view of the questions surrounding the nonvanishing of Hochschild cohomology, we mention the following sufficient criterion.

PROPOSITION 12.8. *Let G be a finite group, B a block of kG with a defect group P and fusion system $\mathcal{F}$. Suppose that k is a splitting field for $kC_G(P)$. If $\mathfrak{foc}(\mathcal{F}) < P$, then $HH^1(B) \neq \{0\}$.*

PROOF. The hypothesis on k ensures that $\mathcal{F}$ is a saturated fusion system. Suppose that $\mathfrak{foc}(P)$ is a proper subgroup of P. Thus there exists a nonzero group homomorphism $\zeta : P \to k$ with kernel containing $\mathfrak{foc}(\mathcal{F})$. Then ζ is $\mathcal{F}$-stable; that is, we have $\zeta(u) = \zeta(\varphi(u))$ for any $u \in P$ and any $\varphi : \langle u \rangle \to P$ in $\mathcal{F}$. In other words, ζ is a nonzero class in the block cohomology $H^1(B)$, defined as the set of $\mathcal{F}$-stable elements in $H^1(P;k)$ (cf. [**49**, Definition 5.1]). By [**49**, Theorem 5.6] (which also requires the hypothesis on k being large enough) there is an injective algebra homomorphism $H^*(B) \to HH^*(B)$, whence the result. □

References

[1] J. L. Alperin, *Local representation theory*, Cambridge Studies in Advanced Mathematics, vol. 11, Cambridge University Press, Cambridge, 1986. Modular representations as an introduction to the local representation theory of finite groups. MR860771

[2] J. Alperin and M. Broué, *Local methods in block theory*, Ann. of Math. (2) **110** (1979), no. 1, 143–157, DOI 10.2307/1971248. MR541333

[3] D. Benson, *Dickson invariants, regularity and computation in group cohomology*, Illinois J. Math. **48** (2004), no. 1, 171–197. MR2048221

[4] D. J. Benson, *On the regularity conjecture for the cohomology of finite groups*, Proc. Edinb. Math. Soc. (2) **51** (2008), no. 2, 273–284, DOI 10.1017/S0013091505001203. MR2465907

[5] D. J. Benson and E. L. Green, *Non-principal blocks with one simple module*, Q. J. Math. **55** (2004), no. 1, 1–11, DOI 10.1093/qjmath/55.1.1. MR2043003

[6] D. Benson and R. Kessar, *Blocks inequivalent to their Frobenius twists*, J. Algebra **315** (2007), no. 2, 588–599, DOI 10.1016/j.jalgebra.2007.03.044. MR2351880

[7] D. Benson, R. Kessar, and M. Linckelmann, *On blocks of defect two and one simple module, and Lie algebra structure of HH^1*, J. Pure Appl. Algebra **221** (2017), no. 12, 2953–2973, DOI 10.1016/j.jpaa.2017.02.010. MR3666734

[8] R. Bauer and W. Feit, *On the number of irreducible characters of finite groups in a given block*, Proc. Nat. Acad. Sci. U.S.A. **45** (1959), 361–365. MR0106246

[9] C. Broto, N. Castellana, J. Grodal, R. Levi, and B. Oliver, *Extensions of p-local finite groups*, Trans. Amer. Math. Soc. **359** (2007), no. 8, 3791–3858, DOI 10.1090/S0002-9947-07-04225-0. MR2302515

[10] C. Broto, R. Levi, and B. Oliver, *The homotopy theory of fusion systems*, J. Amer. Math. Soc. **16** (2003), no. 4, 779–856, DOI 10.1090/S0894-0347-03-00434-X. MR1992826

[11] M. Broué, *Equivalences of blocks of group algebras*, Finite-dimensional algebras and related topics (Ottawa, ON, 1992), NATO Adv. Sci. Inst. Ser. C Math. Phys. Sci., vol. 424, Kluwer Acad. Publ., Dordrecht, 1994, pp. 1–26, DOI 10.1007/978-94-017-1556-0_1. MR1308978

[12] M. Broué and L. Puig, *A Frobenius theorem for blocks*, Invent. Math. **56** (1980), no. 2, 117–128, DOI 10.1007/BF01392547. MR558864

[13] J. Chuang and R. Rouquier, *Derived equivalences for symmetric groups and $\mathfrak{sl}_2$-categorification*, Ann. of Math. (2) **167** (2008), no. 1, 245–298, DOI 10.4007/annals.2008.167.245. MR2373155

[14] G. H. Cliff, W. Plesken, and A. Weiss, *Order-theoretic properties of the center of a block*, The Arcata Conference on Representations of Finite Groups (Arcata, Calif., 1986), Proc. Sympos. Pure Math., vol. 47, Amer. Math. Soc., Providence, RI, 1987, pp. 413–420. MR933378

[15] E. C. Dade, *Blocks with cyclic defect groups*, Ann. of Math. (2) **84** (1966), 20–48, DOI 10.2307/1970529. MR0200355

[16] C. W. Eaton, R. Kessar, B. Külshammer, and B. Sambale, *2-blocks with abelian defect groups*, Adv. Math. **254** (2014), 706–735, DOI 10.1016/j.aim.2013.12.024. MR3161112

[17] K. Erdmann, *Blocks whose defect groups are Klein four groups*, J. Algebra **59** (1979), no. 2, 452–465, DOI 10.1016/0021-8693(79)90140-6. MR543263

[18] K. Erdmann, *Blocks whose defect groups are Klein four groups: a correction*, J. Algebra **76** (1982), no. 2, 505–518, DOI 10.1016/0021-8693(82)90228-9. MR661869

[19] K. Erdmann, *Blocks of tame representation type and related algebras*, Lecture Notes in Mathematics, vol. 1428, Springer-Verlag, Berlin, 1990. MR1064107

[20] K. Erdmann, M. Holloway, R. Taillefer, N. Snashall, and Ø. Solberg, *Support varieties for selfinjective algebras*, K-Theory **33** (2004), no. 1, 67–87, DOI 10.1007/s10977-004-0838-7. MR2199789

[21] D. R. Farkas, C. Geiss, and E. N. Marcos, *Smooth automorphism group schemes*, Representations of algebras (São Paulo, 1999), Lecture Notes in Pure and Appl. Math., vol. 224, Dekker, New York, 2002, pp. 71–89. MR1884808

[22] P. Fleischmann, I. Janiszczak, and W. Lempken, *Finite groups have local non-Schur centralizers*, Manuscripta Math. **80** (1993), no. 2, 213–224, DOI 10.1007/BF03026547. MR1233482

[23] M. Gerstenhaber, *The cohomology structure of an associative ring*, Ann. of Math. (2) **78** (1963), 267–288, DOI 10.2307/1970343. MR0161898

[24] M. Gerstenhaber, *On the deformation of rings and algebras*, Ann. of Math. (2) **79** (1964), 59–103, DOI 10.2307/1970484. MR0171807

[25] M. Gerstenhaber and D. J. Green, *A group-theoretic consequence of the Donald-Flanigan conjecture*, J. Algebra **166** (1994), no. 2, 356–363, DOI 10.1006/jabr.1994.1156. MR1279262

[26] M. Hertweck and W. Kimmerle, *On principal blocks of p-constrained groups*, Proc. London Math. Soc. (3) **84** (2002), no. 1, 179–193, DOI 10.1112/S002461150101317X. MR1863399

[27] M. Hertweck and M. Soriano, *On the modular isomorphism problem: groups of order* 2^6, Groups, rings and algebras, Contemp. Math., vol. 420, Amer. Math. Soc., Providence, RI, 2006, pp. 177–213, DOI 10.1090/conm/420/07976. MR2279240

[28] D. G. Higman, *Focal series in finite groups*, Canadian J. Math. **5** (1953), 477–497. MR0058597

[29] M. Holloway and R. Kessar, *Quantum complete rings and blocks with one simple module*, Q. J. Math. **56** (2005), no. 2, 209–221, DOI 10.1093/qmath/hah036. MR2143498

[30] B. Zimmermann Huisgen, *The finitistic dimension conjectures—a tale of* 3.5 *decades*, Abelian groups and modules (Padova, 1994), Math. Appl., vol. 343, Kluwer Acad. Publ., Dordrecht, 1995, pp. 501–517. MR1378224

[31] N. Jacobson, *Classes of restricted Lie algebras of characteristic p. II*, Duke Math. J. **10** (1943), 107–121. MR0007749

[32] G. J. Janusz, *Indecomposable modules for finite groups*, Ann. of Math. (2) **89** (1969), 209–241, DOI 10.2307/1970666. MR0244307

[33] L. Kadison, *On split, separable subalgebras with counitality condition*, Hokkaido Math. J. **24** (1995), no. 3, 527–549. MR1357029

[34] R. Kessar, *A remark on Donovan's conjecture*, Arch. Math. (Basel) **82** (2004), no. 5, 391–394, DOI 10.1007/s00013-004-4880-8. MR2061445

[35] R. Kessar, *On isotypies between Galois conjugate blocks*, Buildings, finite geometries and groups, Springer Proc. Math., vol. 10, Springer, New York, 2012, pp. 153–162, DOI 10.1007/978-1-4614-0709-6_7. MR3074865

[36] R. Kessar, *On blocks stably equivalent to a quantum complete intersection of dimension 9 in characteristic 3 and a case of the Abelian defect group conjecture*, J. Lond. Math. Soc. (2) **85** (2012), no. 2, 491–510, DOI 10.1112/jlms/jdr047. MR2901075

[37] R. Kessar, S. Koshitani, and M. Linckelmann, *Conjectures of Alperin and Broué for 2-blocks with elementary abelian defect groups of order 8*, J. Reine Angew. Math. **671** (2012), 85–130. MR2983198

[38] R. Kessar and M. Linckelmann, *On the Hilbert series of Hochschild cohomology of block algebras*, J. Algebra **371** (2012), 457–461, DOI 10.1016/j.jalgebra.2012.07.020. MR2975405

[39] R. Kessar and M. Linckelmann, *On the Castelnuovo-Mumford regularity of the cohomology of fusion systems and of the Hochschild cohomology of block algebras*, Groups St Andrews 2013, London Math. Soc. Lecture Note Ser., vol. 422, Cambridge Univ. Press, Cambridge, 2015, pp. 324–330. MR3495664

[40] R. Kessar and G. Malle, *Quasi-isolated blocks and Brauer's height zero conjecture*, Ann. of Math. (2) **178** (2013), no. 1, 321–384, DOI 10.4007/annals.2013.178.1.6. MR3043583

[41] M. Kiyota, *On 3-blocks with an elementary abelian defect group of order* 9, J. Fac. Sci. Univ. Tokyo Sect. IA Math. **31** (1984), no. 1, 33–58. MR743518

[42] R. Knörr and G. R. Robinson, *Some remarks on a conjecture of Alperin*, J. London Math. Soc. (2) **39** (1989), no. 1, 48–60, DOI 10.1112/jlms/s2-39.1.48. MR989918

[43] S. Koenig, Y. Liu, and G. Zhou, *Transfer maps in Hochschild (co)homology and applications to stable and derived invariants and to the Auslander–Reiten conjecture*, Trans. Amer. Math. Soc. **364** (2012), no. 1, 195–232, DOI 10.1090/S0002-9947-2011-05358-4. MR2833582

[44] B. Külshammer, *Bemerkungen über die Gruppenalgebra als symmetrische Algebra. II* (German), J. Algebra **75** (1982), no. 1, 59–69, DOI 10.1016/0021-8693(82)90063-1. MR650409

[45] B. Külshammer, *Symmetric local algebras and small blocks of finite groups*, J. Algebra **88** (1984), no. 1, 190–195, DOI 10.1016/0021-8693(84)90097-8. MR741939

[46] B. Külshammer, *Landau's theorem for p-blocks of p-solvable groups*, J. Reine Angew. Math. **404** (1990), 189–191, DOI 10.1515/crll.1990.404.189. MR1037437

[47] B. Külshammer and G. R. Robinson, *Alperin-McKay implies Brauer's problem* 21, J. Algebra **180** (1996), no. 1, 208–210, DOI 10.1006/jabr.1996.0062. MR1375576

[48] H. Kupisch, *Unzerlegbare Moduln endlicher Gruppen mit zyklischer p-Sylow-Gruppe.* (German), Math. Z. **108** (1969), 77–104, DOI 10.1007/BF01114463. MR0241549

[49] M. Linckelmann, *Transfer in Hochschild cohomology of blocks of finite groups*, Algebr. Represent. Theory **2** (1999), no. 2, 107–135, DOI 10.1023/A:1009979222100. MR1702272

[50] M. Linckelmann, *Finite generation of Hochschild cohomology of Hecke algebras of finite classical type in characteristic zero*, Bull. Lond. Math. Soc. **43** (2011), no. 5, 871–885, DOI 10.1112/blms/bdr024. MR2854558

[51] M. Linckelmann, *Cohomology of block algebras of finite groups*, Representations of algebras and related topics, EMS Ser. Congr. Rep., Eur. Math. Soc., Zürich, 2011, pp. 189–250. MR2931899

[52] M. Linckelmann, *On equivalences for cohomological Mackey functors*, Represent. Theory **20** (2016), 162–171, DOI 10.1090/ert/482. MR3503952

[53] M. Linckelmann, *Integrable derivations and stable equivalences of Morita type.* Preprint (2015) Proc. Edinburgh Math. Soc., to appear.

[54] M. Linckelmann, *On automorphisms and focal subgroups of blocks*, preprint (2015).

[55] M. Linckelmann, *The dominant dimension of cohomological Mackey functors.* Preprint (2016).

[56] M. Linckelmann and LL. Rubio y Degrassi, *Block algebras with HH^1 a simple Lie algebra.* Preprint (2016).

[57] G. Malle and G. R. Robinson, *On the number of simple modules in a block of a finite group*, J. Algebra **475** (2017), 423–438, DOI 10.1016/j.jalgebra.2016.10.022. MR3612478

[58] B. J. Müller, *The classification of algebras by dominant dimension*, Canad. J. Math. **20** (1968), 398–409, DOI 10.4153/CJM-1968-037-9. MR0224656

[59] T. Okuyama, *On blocks of finite groups with radical cube zero*, Osaka J. Math. **23** (1986), no. 2, 461–465. MR856899

[60] T. Okuyama and Y. Tsushima, *Local properties of p-block algebras of finite groups*, Osaka J. Math. **20** (1983), no. 1, 33–41. MR695615

[61] S. F. Peacock, *Separable equivalence, complexity and representation type*, J. Algebra **490** (2017), 219–240, DOI 10.1016/j.jalgebra.2017.06.039. MR3690332

[62] L. Puig, *Structure locale dans les groupes finis* (French), Bull. Soc. Math. France Suppl. Mém. **47** (1976), 132. MR0450410

[63] L. Puig, *Nilpotent blocks and their source algebras*, Invent. Math. **93** (1988), no. 1, 77–116, DOI 10.1007/BF01393688. MR943924

[64] L. Puig, *Une correspondance de modules pour les blocs à groupes de défaut abéliens* (French, with English summary), Geom. Dedicata **37** (1991), no. 1, 9–43, DOI 10.1007/BF00150403. MR1094226

[65] L. Puig, *On the local structure of Morita and Rickard equivalences between Brauer blocks*, Progress in Mathematics, vol. 178, Birkhäuser Verlag, Basel, 1999. MR1707300

[66] J. Rickard, *Derived categories and stable equivalence*, J. Pure Appl. Algebra **61** (1989), no. 3, 303–317, DOI 10.1016/0022-4049(89)90081-9. MR1027750

[67] K. Roggenkamp and L. Scott, *Isomorphisms of p-adic group rings*, Ann. of Math. (2) **126** (1987), no. 3, 593–647, DOI 10.2307/1971362. MR916720

[68] L. Rubio y Degrassi, *Invariance of the restricted p-power map on integrable derivations under stable equivalences*, J. Algebra **469** (2017), 288–301, DOI 10.1016/j.jalgebra.2016.08.002. MR3563015

[69] B. Sambale, *Blocks of finite groups and their invariants*, Lecture Notes in Mathematics, vol. 2127, Springer, Cham, 2014. MR3289382

[70] M. Schaps, http://u.math.biu.ac.il/ mschaps/math.html

[71] J. Stasheff, *The intrinsic bracket on the deformation complex of an associative algebra*, J. Pure Appl. Algebra **89** (1993), no. 1-2, 231–235, DOI 10.1016/0022-4049(93)90096-C. MR1239562

[72] H. Strade, *Simple Lie algebras over fields of positive characteristic. I*, De Gruyter Expositions in Mathematics, vol. 38, Walter de Gruyter & Co., Berlin, 2004. Structure theory. MR2059133

[73] H. Strade, *Simple Lie algebras over fields of positive characteristic. II*, De Gruyter Expositions in Mathematics, vol. 42, Walter de Gruyter & Co., Berlin, 2009. Classifying the absolute toral rank two case. MR2573283

[74] P. Symonds, *On the Castelnuovo-Mumford regularity of the cohomology ring of a group*, J. Amer. Math. Soc. **23** (2010), no. 4, 1159–1173, DOI 10.1090/S0894-0347-10-00666-1. MR2669710

[75] H. Tachikawa, *On dominant dimensions of* QF-*3 algebras*, Trans. Amer. Math. Soc. **112** (1964), 249–266, DOI 10.2307/1994293. MR0161888

[76] D. Tambara, *Homological properties of the endomorphism rings of certain permutation modules*, Osaka J. Math. **26** (1989), no. 4, 807–828. MR1040426

[77] A. Weiss, *Rigidity of p-adic p-torsion*, Ann. of Math. (2) **127** (1988), no. 2, 317–332, DOI 10.2307/2007056. MR932300

[78] A. Zimmermann, *Fine Hochschild invariants of derived categories for symmetric algebras*, J. Algebra **308** (2007), no. 1, 350–367, DOI 10.1016/j.jalgebra.2006.08.021. MR2290926

City, University of London, London EC1V 0HB, United Kingdom

Contemporary Mathematics
Volume **705**, 2018
http://dx.doi.org/10.1090/conm/705/14194

Kronecker modules generated by modules of length 2

Claus Michael Ringel

ABSTRACT. Let Λ be a ring and $\mathcal{N}$ a class of Λ-modules. A Λ-module is said to be generated by $\mathcal{N}$ provided that it is a factor module of a direct sum of modules in $\mathcal{N}$. The semi-simple Λ-modules are just the Λ-modules which are generated by the Λ-modules of length 1. It seems that the modules which are generated by the modules of length 2 (we call them bristled modules) have not attracted the interest they deserve.

In this paper we deal with the basic case of the Kronecker modules, these are the (finite-dimensional) representations of an n-Kronecker quiver, where n is a natural number. We show that for $n \geq 3$, there is an abundance of bristled Kronecker modules.

1. Introduction

1.1. Let Λ be a ring and $\mathcal{N}$ a class of Λ-modules. A Λ-module is said to be generated by $\mathcal{N}$ provided that it is a factor module of a direct sum of modules in $\mathcal{N}$. A Λ-module M is said to be $\mathcal{N}$*-saturated* provided that $\mathrm{Ext}^1(N, M) = 0$ for all modules $N \in \mathcal{N}$. Of course, *the class of modules which are both generated by $\mathcal{N}$ and $\mathcal{N}$-saturated is closed under extensions. In case Λ is hereditary, the class of modules which are both generated by $\mathcal{N}$ and $\mathcal{N}$-saturated is also closed under factor modules.* We consider the case where $\mathcal{N}$ is a class of modules of length 2. The modules which are generated by modules of length 2 will be called *bristled* modules; a *bristle* is by definition an indecomposable module of length 2, and we denote by $\mathcal{B}$ the set of isomorphism classes of bristles (we should stress that a bristled module is not necessarily the sum of its bristle submodules — it is the sum of submodules which are either bristles or simple).

Given any artin algebra Λ, we denote by $\mathrm{mod}\,\Lambda$ the category of all (finitely generated) Λ-modules and by τ the Auslander-Reiten translation in $\mathrm{mod}\,\Lambda$.

1.2. We focus the attention to the n-Kronecker algebras and their modules, the n-Kronecker modules (for convenience, we sometimes will refer to these algebras and their modules just as the Kronecker algebras and the Kronecker modules). Despite of the importance of the Kronecker algebras, not much is known about

---

2010 *Mathematics Subject Classification.* Primary 16G20; Secondary 16G60, 16D90, 16G70, 15A22.

Key words and phrases. Kronecker quivers, Kronecker algebras, Kronecker modules, modules of length 2, bristles, bristled modules, the n-regular tree, preprojective and preinjective modules, Auslander-Reiten components, modules over artin algebras, tame and wild.

Kronecker modules. Our approach to look at the bristled Kronecker modules may be seen as one of the first attempts to get insight into the structure of Kronecker modules in general. On the other hand, the Appendix B will outline in which way the study of bristled Kronecker modules is an essential step for understanding bristled Λ-modules for any finite-dimensional k-algebra, where k is an algebraically closed field.

Let k be a field. The n-Kronecker quiver $K(n)$ has two vertices, denoted by 1 and 2, and n arrows $\alpha_i \colon 1 \to 2$, with $1 \leq i \leq n$.

$$1 \quad \begin{matrix} \xrightarrow{\ \alpha_1\ } \\ \vdots \\ \xrightarrow[\ \alpha_n\ ]{} \end{matrix} \quad 2$$

The n-Kronecker algebra $kK(n)$ is the path algebra of $K(n)$, the n-Kronecker modules are the representations of $K(n)$ (over k). Kronecker modules are written in the form $M = (M_1, M_2; \alpha_1, \ldots, \alpha_n)$, where M_1, M_2 are two vector spaces and $\alpha_i \colon M_1 \to M_2$ are linear transformations, for $1 \leq i \leq n$.

Note that $kK(n)$ is a hereditary artin algebra. Given any hereditary artin algebra Λ, an indecomposable Λ-module is called *preinjective* provided that it is of the form $\tau^t I$ for some injective Λ-module I and some $t \geq 0$.

An n-Kronecker module which is a bristle is of the form $B(\lambda) = (k, k; \lambda_1, \ldots, \lambda_n)$, where $(\lambda_1, \ldots, \lambda_n)$ is a non-zero element of k^n; the bristles $B(\lambda)$ and $B(\lambda')$ are isomorphic if and only if $(\lambda'_1, \ldots, \lambda'_n)$ is a scalar multiple of $(\lambda_1, \ldots, \lambda_n)$ (thus the set $\mathcal{B}$ of isomorphism classes of bristles may be identified with the projective space $\mathbb{P}^{n-1}$).

1.3. The essential part of the paper concerns the class of bristled Kronecker modules which are $\mathcal{B}$-saturated. As we have mentioned, this class of modules is closed under factor modules and extensions. As we will see, there is an abundance of such modules (but note that for $n \geq 2$, the bristles themselves do not belong to this class of modules, since they are not $\mathcal{B}$-saturated). Actually, it seems to be appropriate to select a suitable finite set $\mathcal{B}'$ of (isomorphism classes of) bristles and to look at the (slightly smaller) class of modules which are generated by $\mathcal{B}'$ and which are $\mathcal{B}$-saturated (also this class is closed under factor modules and extensions). Here is the main theorem.

Main Theorem. *Let $n \geq 3$. There is a set $\mathcal{B}_0$ of $n+2$ bristles in* $\operatorname{mod} kK(n)$ *with the following properties:*

(a) *Any preinjective module is generated by $\mathcal{B}_0$ and is $\mathcal{B}$-saturated.*

(b) *If M is any module, then there is a number $t(M)$ such that all the modules $\tau^t M$ with $t \geq t(M)$ are generated by $\mathcal{B}_0$ and are $\mathcal{B}$-saturated.*

For the choice of the set $\mathcal{B}_0$, see section 4.2. As we will see in the Addendum 4.23, the number $n+2$ is best possible: *There is no set of $n+1$ bristles which generates all the preinjective modules.*

As a consequence of Main Theorem, we see that for $n \geq 3$ there exists a regular module which generates all the preinjective modules, for example the direct sum of the modules in $\mathcal{B}_0$ (and there are also indecomposable regular modules with this property, see 4.21). In 1994, Kerner [**3**] had asked whether such a module does exist. In 2014, Kerner gave a non-constructive proof for the existence of such a module [**4**]; here is now an explicit example. Note that for $n = 2$, any regular module generates only finitely many indecomposable preinjective modules, see 4.22 and Appendix C.

Using duality, we see that also the corresponding dual statements hold:

(a)* *Any preprojective module M is cogenerated by $\mathcal{B}_0$ and $\mathcal{B}$-cosaturated.*

(b)* *If M is any module, then there is a number $t^*(M)$ such that all the modules $\tau^{-t}M$ with $t \geq t^*(M)$ are cogenerated by $\mathcal{B}_0$ and are $\mathcal{B}$-cosaturated.*

Of course, a module M is said to be *cogenerated* by a class $\mathcal{N}$ of modules provided that it is a submodule of a direct sum of modules in $\mathcal{N}$ (note that the modules considered here are modules of finite length; otherwise we would have to take direct products instead of direct sums). And M is said to be $\mathcal{N}$*-cosaturated* provided $\operatorname{Ext}^1(M, N) = 0$ for all modules N in $\mathcal{N}$.

1.4. Outline of the paper. Section 2 contains some preliminaries: we fix the notation and recall some general properties of the categories $\operatorname{mod} kK(n)$. Of course, the structure of the categories $\operatorname{mod} kK(1)$ and $\operatorname{mod} kK(2)$ is well-known (whereas only little is known about the categories $\operatorname{mod} kK(n)$ with $n \geq 3$).

The proof of theorem 1.3 will be presented in section 4 and some related results can be found in section 5. In particular, we will show in section 5 that an indecomposable $\mathcal{B}$-saturated module which is not simple is faithful. Section 4 (and most parts of section 5) deal with n-Kronecker modules, where $n \geq 3$. The cases $n = 1$ and $n = 2$ are discussed in section 3.

There are three appendices. The first is a short introduction to the paper [**6**]. The set of bristle submodules of a Kronecker module M without simple injective submodules is in a natural way a projective variety. In [**6**] we show that any projective variety occurs in this way. Appendix B deals with arbitrary artin algebras Λ and their bristled modules. We are going to point out in which way the study of bristled Λ-modules can be reduced to the study of bristled Kronecker modules. As we have mentioned already, the existence of a regular module generating all preinjective modules holds true for $K(n)$ with $n \geq 3$, but not for $K(2)$. It does not hold for any tame hereditary artin algebras, see the Appendix C.

1.5. Acknowledgment. A weaker version of the Main Theorem (as well as the result of [**6**]) were first presented at the conference in honor of Jerzy Weyman's 60th birthday, April 2015, at the University of Connecticut. The weaker version used a set of $2n-1$ bristles instead of the set $\mathcal{B}_0$ with the optimal cardinality $n+2$; it was written up at a stay at the IPM in Isfahan, October 2016 and the author is grateful to the Iranian hospitality. The present version was finished in April 2017. The author is indebted to various mathematicians for questions and comments.

2. Preliminaries

The aim of this section is to recall some general properties of the categories $\operatorname{mod} kK(n)$ of the n-Kronecker modules, see for example [**5**].

Given any acyclic quiver Q and a vertex x of Q, we write $S(x)$, $P(x)$ and $I(x)$ for the corresponding indecomposable representation of Q which is simple, projective, or injective, respectively. Thus, $S(x)_x = k$ and $S(x)_y = 0$ for any vertex $y \neq x$. The representation $P(x)$ is the projective cover, the representation $I(x)$ the injective envelope of $S(x)$.

2.1. There are two simple n-Kronecker modules, namely the injective module $S(1) = I(1) = (k, 0; 0, \ldots, 0)$ and the projective module $S(2) = P(2) = (0, k; 0, \ldots, 0)$.

If $M = (M_1, M_2; \alpha_1, \ldots, \alpha_n)$ is an n-Kronecker module, we denote by $\mathbf{dim}\, M = (\dim M_1, \dim M_2)$ (or also by $\binom{\dim M_1}{\dim M_2}$) its *dimension vector*, it is an element of the Grothendieck group $K_0(kK(n)) = \mathbb{Z}^2$.

On $K_0(kK(n))$, there is defined the bilinear form $\langle -, - \rangle = \langle -, - \rangle_n$ via

$$\langle \begin{pmatrix} a \\ b \end{pmatrix}, \begin{pmatrix} a' \\ b' \end{pmatrix} \rangle = aa' + bb' - nab',$$

and one knows that

$$\langle \mathbf{dim}\, M, \mathbf{dim}\, M' \rangle = \dim \operatorname{Hom}(M, M') - \dim \operatorname{Ext}^1(M, M'),$$

for any pair M, M' of n-Kronecker modules.

On $K_0(kK(n))$, the Coxeter transformation $\Phi\colon K_0(kK(n)) \to K_0(kK(n))$ is defined via $\Phi(a, b) = (n^2a - nb - a, na - b)$. It mimics the Auslander-Reiten translation as follows: Let M be an indecomposable n-Kronecker module. If M is non-projective, then

$$\mathbf{dim}\, \tau M = \Phi(\mathbf{dim}\, M).$$

If M is projective, then, of course, $\tau M = 0$ (and we have $\Phi(\mathbf{dim}\, P(i)) = -\,\mathbf{dim}\, I(i)$ for $i = 1, 2$).

For $n \geq 1$, a module is bristled if and only if it is generated by the bristles and $S(2)$; and an indecomposable bristled module is either isomorphic to $S(2)$ or else generated by the set $\mathcal{B}$ of bristles.

2.2. The Auslander-Reiten quiver of $\operatorname{mod} kK(1)$**.** There are just three isomorphism classes of indecomposable 1-Kronecker modules, namely $S(2) = P(2)$, $B = P(1) = I(2)$ and $I(1) = S(1)$, and the Auslander-Reiten quiver of $\operatorname{mod} kK(1)$ is of the form

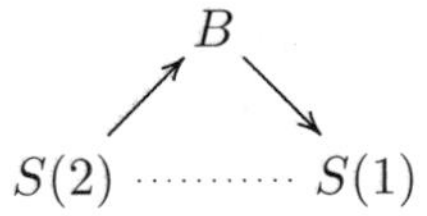

(we write B instead of $P(1) = I(2)$ in order to stress that this module is a bristle; of course, it is the only bristle).

2.3. The Auslander-Reiten quiver of $\operatorname{mod} kK(n)$ **with** $n \geq 2$**.** We recall that the indecomposable representations of $K(n)$ fall into three classes: there are the preprojective ones, the regular ones and the preinjective ones.

There are countably many (isomorphism classes of) indecomposable preinjective modules, we denote these modules by I_t with $t \in \mathbb{N}_0$, as follows: $I_0 = S(1) = I(1)$, $I_1 = I(2)$, and $I_t = \tau I_{t-2}$ for $t \geq 2$, thus they are ordered in such a way that $\operatorname{Hom}(I_t, I_{t'}) = 0$ if and only if $t < t'$.

Similarly, there are the indecomposable preprojective modules $P_0 = S(2) = P(2), P_1 = P(1), P_2, P_3, \ldots$, but this notation will not be needed in an essential way in the paper.

The Auslander-Reiten quiver looks as follows:

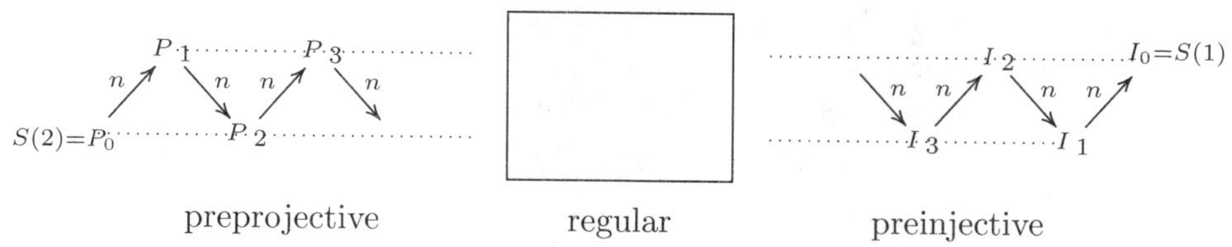

where we have endowed the arrows by the number n in order to indicate that the corresponding bimodule of irreducible maps is n-dimensional. Note that there are no non-zero maps from a regular module to a preprojective module, or from a preinjective module to a preprojective or a regular module. All non-zero maps in the preprojective component and in the preinjective component go from left to right. The components in the regular part are of type $\mathbb{A}_\infty$. For $n = 2$, they are regular tubes; for $n \geq 3$ they are of the form $\mathbb{Z}A_\infty$.

It is important to be aware that the bristles are regular. Thus, any indecomposable bristled module different from $S(2)$ is generated by bristles and therefore is regular or preinjective.

3. The bristled modules for $n = 1$ and $n = 2$.

3.1. The case $n = 1$. As we have mentioned, there is (up to isomorphism) a unique bristle, namely $B = P(1) = I(2)$; this bristle is projective and injective. All the 1-Kronecker modules are $\mathcal{B}$-saturated bristled modules.

3.2. The case $n = 2$. Recall that we have indexed the (isomorphism classes of the) bristles by the elements of the projective space $\mathbb{P}^{n-1} = \mathbb{P}^1$. For $c \in k$, let $B_c = B(1 : c)$, and let $B_\infty = B(0 : 1)$.

Proposition. *Let $n = 2$. Let J be a subset of $k \cup \{\infty\}$. Then I_t is generated by the bristles B_c with $c \in J$ if and only if $|J| \geq t + 1$.*

Proof. Note that the dimension vector of I_t is $(t+1, t)$. First, assume that I_t is generated by the bristles B_c with $c \in J$. Since $\dim \operatorname{Hom}(B_c, M) = 1$ for all bristles B_c, it follows that I_t is a factor module of the module $\bigoplus_{c \in J} B_c$ with dimension vector $(|J|, |J|)$, therefore $|J| \geq t + 1$.

On the other hand, let $I_t = (M_1, M_2; \alpha_1, \alpha_2)$. We can assume that M_1 has the basis $e_0, e_1, \ldots, e_t$, that M_2 has the basis $e'_1, \ldots, e'_t$, that $\alpha_1(e_i) = e'_{i+1}$ for $0 \leq i < t$ and $\alpha_1(e_t) = 0$ and that $\alpha_2(e_0) = 0$ and $\alpha_2(e_i) = e'_i$ for $1 \leq i \leq n$, thus M may be visualized as follows:

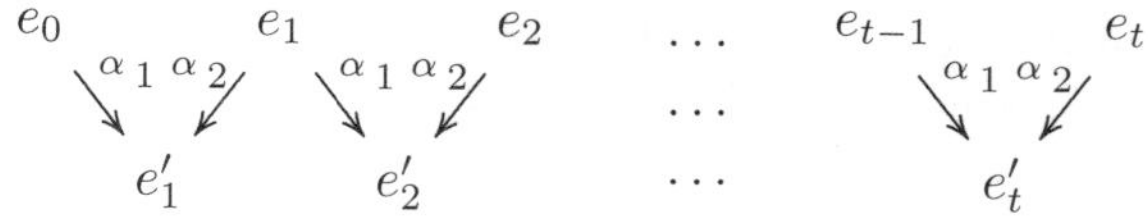

For any $c \in k$, the element $m_c = \sum_{i=0}^t c^i e_i$ in M_1 is sent under α_1 to $\alpha_1(m_c) = \sum_{i=1}^t c^{i-1} e'_i$ and under α_2 to $\alpha_2(m_c) = \sum_{i=1}^t c^i e'_i = c \cdot \alpha_1(m_c)$. This shows that m_c generates a bristle of the form B_c. The Vandermonde determinant shows that for $t + 1$ pairwise different elements c in k, the corresponding elements m_c in M_1 are linearly independent, thus form a basis. But this implies that M is a bristled module.

There is one additional submodule of M which is a bristle, namely the submodule generated by $m_\infty = e_t$; it is isomorphic to B_∞. Given t pairwise different scalars $c_1, \dots, c_t$ in k, then also the elements $m_{c_1}, \dots, m_{c_t}, m_\infty$ are linearly independent: the determinant of the $(t+1) \times (t+1)$ matrix whose rows are the vectors $m_{c_1}, \dots, m_{c_t}, m_\infty$ is the product of the $t \times t$ Vandermonde determinant for the scalars $c_1, \dots, c_t$ and the determinant of the 1×1-matrix with entry 1, thus non-zero. □

3.3. Corollary. *Let $n = 2$ and $q = |k|$. An indecomposable module M different from $S(2)$ is bristled if and only if M is one of the bristles B_c with $c \in k \cup \{\infty\}$ or $M = I_t$ is preinjective with $0 \le t \le q$.*

The preinjective bristled modules are $\mathcal{B}$-saturated, the module $S(2)$ and the bristles are not $\mathcal{B}$-saturated.

Proof. Let M be an indecomposable bristled module. First, assume that M is regular. If the regular socle U of M is a bristle B, then $\dim \operatorname{Hom}(B, M) = 1$ and $\operatorname{Hom}(B', M) = 0$ for any bristle not isomorphic to B, thus $M = U = B$. Or else $\operatorname{Hom}(B, M) = 0$ for any bristle B, thus $M = 0$ (since M is generated by bristles), a contradiction. This shows that the only indecomposable regular bristled modules are the bristles.

It remains to decide whether the bristled modules are $\mathcal{B}$-saturated or not. For $M = S(2)$, we have $\operatorname{Ext}^1(B, M) \neq 0$ for any bristle B. For $M = B_c$, we have $\operatorname{Ext}^1(B_c, M) \neq 0$. This shows that the bristled modules $S(2)$ and B_c are not $\mathcal{B}$-saturated. On the other hand, we have $\operatorname{Ext}^1(X, Y) = 0$ for X regular and Y preinjective. This implies that the preinjective modules are $\mathcal{B}$-saturated modules. It follows that the only $\mathcal{B}$-saturated bristled modules are the modules $M = I_t$ with $0 \le t \le q$. □

3.4. Remark. There is the following consequence: *For $n = 2$, the set of indecomposable bristled modules is characterized by their dimension vectors.*

Let us mention already here that for $n \ge 3$, the set of indecomposable bristled modules (even the set of faithful ones) is not characterized by their dimension vectors, as already the dimension vector $(3, 2)$ shows:

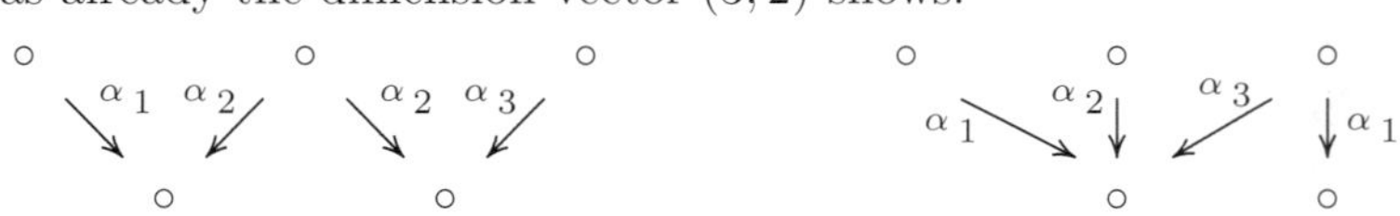

bristled | not bristled

4. Proof of the main theorem.

In this section, we will assume that $n \ge 3$. We will identify the index set $\{1, 2, \dots, n\}$ of the arrow set of $K(n)$ with $\mathbb{Z}/n\mathbb{Z}$, thus sometimes we will write α_{n+1} instead of α_1.

4.1. We will use the following notation for some bristles (this convention deviates from the previous notation, but there should be no confusion). If $1 \le r \le n$, let $B(r)$ be defined by $B(r)_1 = B(r)_2 = k$ with α_i being the identity map in case $i = r$, and the zero map otherwise. If $r \neq s$ are numbers between 1 and n, let $B(r, s)$ be defined by $B(r, s)_1 = B(r, s)_2 = k$ with α_i being the identity map in case $i \in \{r, s\}$, and the zero map otherwise.

4.2. Let $\mathcal{B}_0$ be the set of (the isomorphism classes of) the bristles $B(n-1), B(n)$ and the bristles $B(r, r+1)$ with $1 \le r \le n$ (recall that $B(n, n+1)$ is just $B(n, 1)$). Note that the cardinality of $\mathcal{B}_0$ is $n+2$. Let $\mathcal{B}'_0$ be obtained from $\mathcal{B}_0$ by deleting $B(n-1, n)$, thus the cardinality of $\mathcal{B}'_0$ is $n+1$.

4.3. For the proof of the assertion (b) of Theorem 1.3, we may assume that M is indecomposable. In case M is preprojective, nothing has to be shown since $\tau^t M = 0$ for t large. The case that M is preinjective is considered in assertion (a). Finally, there is the case that M is regular. Since the class of modules which are generated by $\mathcal{B}_0$ and $\mathcal{B}$-saturated is closed under extensions, it is sufficient to consider just the elementary modules as introduced by Crawley-Boevey (see [**3**] and [**4**], or section 4.14 below). The proof of (a) dealing with the preinjective modules and of the statement (b) for elementary modules proceeds in several steps and will be completed in 4.20.

4.4. *The preinjective module I_2 is generated by $\mathcal{B}'_0$, thus by $\mathcal{B}_0$.*

For the proof of 4.4 (and also for the proof of 4.6) we are going to use the universal cover $\widetilde{K}(n)$ of the n-Kronecker quiver $K(n)$ (see for example [**1**] and [**2**]) in order to exhibit sufficiently many bristles as submodules of the modules I_2 and $\tau B(1)$. Note that the quiver $\widetilde{K}(n)$ is the n-regular tree with bipartite orientation. The representations of $\widetilde{K}(n)$ will be called $\widetilde{K}(n)$-modules (we may consider them as modules over the corresponding path algebra; we should note that the path algebra of such a quiver with infinitely many vertices does not have a unit element, but at least sufficiently many primitive idempotents).

The quiver $\widetilde{K}(n)$ is a "labeled" quiver: The projection $\pi\colon \widetilde{K}(n) \to K(n)$ sends the arrow α of $\widetilde{K}(n)$ to an arrow $\pi(\alpha)$ of $K(n)$, this is one of $\alpha_1, \dots, \alpha_n$, and we use $\pi(\alpha)$ (or its index i with $1 \le i \le n$) as a label for α. By abuse of notation, we sometimes denote α by its label $\pi(\alpha)$. We also denote by π the corresponding push-down functor $\operatorname{mod} \widetilde{K}(n) \to \operatorname{mod} K(n)$.

Proof of 4.4. Let $M = I_2$ and let X be a representation of $\widetilde{K}(n)$ with $\pi(X) = M$, thus there is a vertex z of $\widetilde{K}(n)$ such that the support Q of X consists of the vertices in $\widetilde{K}(n)$ with distance at most 2 from z and $\dim X_z = n-1$, whereas $\dim X_a = 1$ for the remaining vertices a in Q.

Here is, in the case $n = 4$, the quiver Q as well as the dimension vector of X

Q

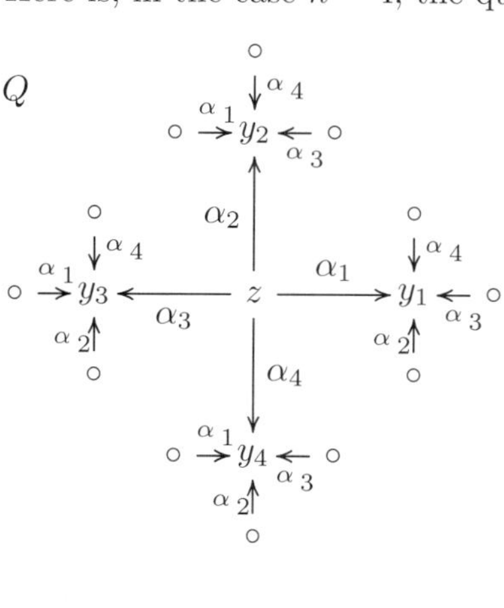

X

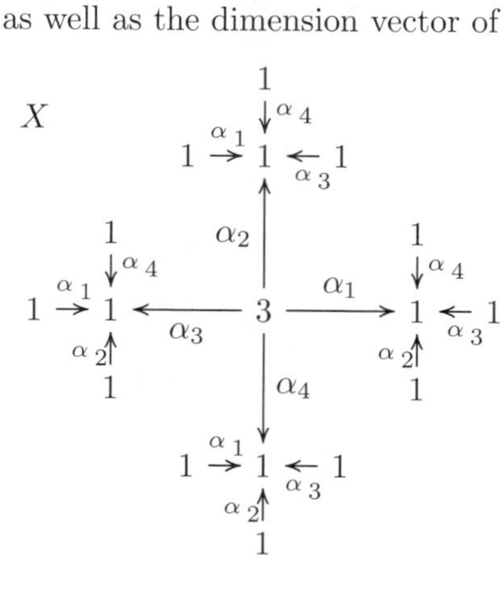

Always, we denote by y_j the neighbor of z such that the arrow $z \to y_j$ has label j. Recall that a *leaf* of a tree is a vertex with at most one neighbor. The leaves of Q are just the vertices which have distance 2 from the center z; in the picture on the left, they are indicated by small circles $\circ$. If x is a leaf of Q and the arrow starting in x has label α_i and ends in y_j, we write $x = x(j, i)$ and we say that x is *of type* i

In order to find bristle submodules in M, we will exhibit some submodules of X; all these submodules are indecomposable and their support is a path in Q.

For $1 \leq j \leq n$, let $Q(j)$ be the full subquiver of Q with vertices y_j and $x(j, i)$ where $1 \leq i \leq n$ and $i \neq j$. Note that $Q(j)$ is a quiver with one sink and $n-1$ sources (thus it is a so-called subspace quiver).

For $1 \leq j \leq n$, let $Y(j) = X|Q(j)$. Since the subquiver $Q(j)$ of Q is closed under successors, we see that $Y(j)$ is a submodule of X. Since the subquivers $Q(j)$ are pairwise disjoint, the submodule $Y(j)$ form a direct sum $Y = \bigoplus_j Y(j)$ and the factor module X/Y is nothing else than X_z, considered as a direct sum of copies of the simple representation $S(z)$.

Let $N(j) = \pi(Y(j))$ and $N = \bigoplus_j N(j)$; these are submodules of M. The factor module M/N is a direct sum of copies of $S(1)$ and may be identifies with M_z. We are going to look on the one hand at N, and will show that any $N(j)$ is generated by bristles which are denoted by $D(x)$ and $E(x)$, where x is a leaf in $Q(j)$, see (I) below. On the other hand, we have to find bristle submodules $M(i, j)$ of M which are not contained in N, see (II) below.

For any leaf x, let $P(x)$ be the indecomposable projective representation of Q corresponding to the vertex x. Note that $P(x)$ is a bristle, and, of course, also $D(x) = \pi(P(x))$ is a bristle. If x is of type i, the bristle $D(x)$ is isomorphic to $B(i)$.

(I) The bristle submodules $D(x)$ and $E(x)$ of M. We are going to show the following:

For any j, the submodule $N(j)$ is generated by bristle submodules which belong to $\mathcal{B}_0'$.

Proof. Fix some j and write $[i] = x(j, i)$. In order to reduce the number of brackets, we will write $P[i]$ instead of $P(x(j, i))$ and so on.

Let $G(j)$ be the set of leaves in $Q(j)$ of types $n-1$ and n. If $x \in G(j)$, then $D(x) = \pi(P(x))$ belongs to $\mathcal{B}_0'$. If $j \in \{n-1, n\}$, then $G(j)$ is a single leave, otherwise $G(j)$ consists of two leaves.

We consider in $Q(j)$ subquivers with three vertices, namely those of the following kind:

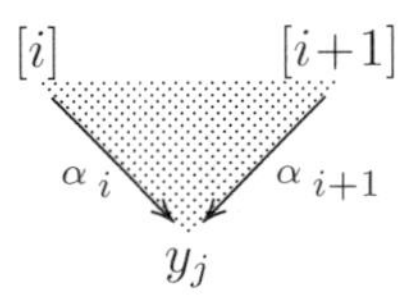

such that $i \neq n$–1 (note that we also have that $i \neq j$ and $i+1 \neq j$, thus $i \neq j-1$) and we denote by $V[i] = V(x(j, i))$ the indecomposable kQ-module with this support $\{y_j, [i], [i+1]\}$. Let $H(j)$ be the set of all $1 \leq i \leq n$ with $i \notin \{j-1, j, n-1\}$. Note that $i \in H(j)$ if and only if $V[i]$ is defined. We claim that

$$Y(j) = \sum_{i \in G(j)} P[i] + \sum_{i \in H(j)} V[i]. \tag{1}$$

Any $V[i]$ has two bristle submodules, namely $P[i]$ and $P[i+1]$. The Kronecker module $\pi(V[i])$ has the bristle submodules $D[i]=\pi(P[i])$ and $D[i+1]=\pi(P[i+1])$, but actually many more bristle submodules: any non-zero element in the top of $\pi(V[i])$ generates a bristle. We are interested in the submodule $E[i]\subset\pi(V[i])$ which is isomorphic to $B(i,i+1)$, thus belongs to $\mathcal{B}'_0$ (since $i\neq n-1$). We claim that

$$N(j)=\pi(Y(j))=\sum_{i\in G(j)}D[i]+\sum_{i\in H(j)}E[i]. \tag{2}$$

For the proof of (1) and (2) we distinguish the cases whether j is one of the numbers $n-1,n$ or not.

Consider first the case $j\notin\{n-1,n\}$. Then $G(j)=\{n-1,n\}$ and $H(j)=\{j+1,j+2,\dots,n-2,n,1,\dots,j-2\}$. The support of the modules $V[i]$ with $i\in H(j)$ are the various shaded triangles in the following picture:

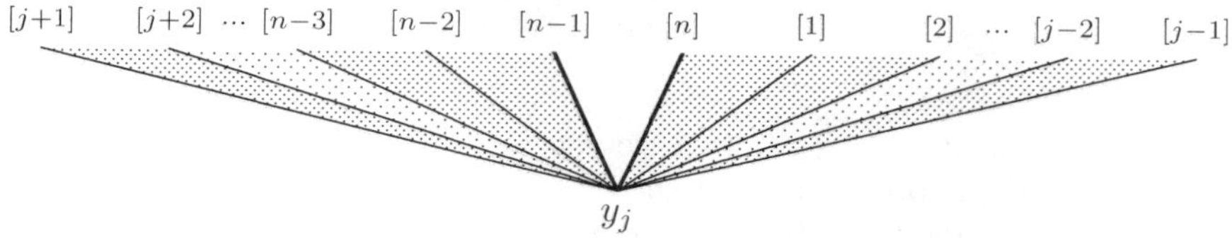

Let us show (1) and (2) in this case. Consider first (1). Of course, the submodules mentioned on the right side of (1) have support in $Q(j)$, thus they are submodules of $Y(j)$. What we have to show is that for any leaf $x=[i]$ of $Q(j)$, the module $P(x)$ is included in the right side. We only have to deal with the vertices $[i]$, where $1\le i\le n-2$ (and, of course, $i\neq j$). First, let $1\le i\le j-1$. We know that $P[i]$ is contained in $V[i-1]$ and $i-1$ is either equal to n or $1\le i-1\le j-2$, thus always $i-1$ belongs to $H(j)$. Second, let $j+1\le i\le n-2$. In this case, we use that $P[i]$ is contained in $V[i]$ and that i itself belongs to $H(j)$. This completes the proof of (1).

The equality (1) implies that

$$\pi(Y(j))=\sum_{i\in G(j)}\pi(P[i])+\sum_{i\in H(j)}\pi(V[i]),$$

and $\pi(P[i])=D[i]$. It remains to be seen that for any $i\in H(j)$, the submodule $\pi(V[i])$ is contained in the right hand side R of (2). For $i=n$, we have $\pi(V[n])=D[n]+E[n]$ and n belongs both to $G(j)$ and to $H(j)$. Next, let $1\le i\le j-2$. We use induction on i in order to see that $\pi(V[i])$ is contained in R. Now

$$\pi(V[i])=\pi(P[i])+E[i]\subset\pi(V[i-1])+E[i].$$

For $i=1$, we know already that $\pi(V[i-1])=\pi(V[n])$ belongs to R; for $i>1$, we know by induction that $\pi V[i-1]\subset R$. Thus, we see that let all $1\le i\le j-2$, the module $\pi(V[i])$ is contained in R.

Second, we have to show that $\pi(V[i])$ is contained in R, for $j+1\le i\le n-2$. This time, we use induction going downwards, starting with $i=n-2$, and the fact that $\pi(V[i])=E[i]+\pi(P[i+1])$. For the start $i=n-2$ of the induction, we have $i+1=n-1$ and this number belongs to $G(j)$, thus $\pi(P[i+1])=D[i+1]=D[n-1]$ belongs to R. For $i<n-2$, we have $\pi(P[i+1])\subset\pi(V[i+1]$ and $\pi(V[i+1]\subseteq R$, by induction. This completes the proof of (2) in the case that $j\notin\{n-1,n\}$.

Now, we deal with the two cases where $j \in \{n-1, n\}$. In both cases, the proof will be similar to the previous proof, thus we only describe the setting in detail.

If $j = n-1$, then $G(j) = \{n\}$ and $H(j) = \{n, 1, \ldots, n-3\}$. The support of the modules $V[i]$ with $i \in H(j)$ are the various shaded triangles in the following picture:

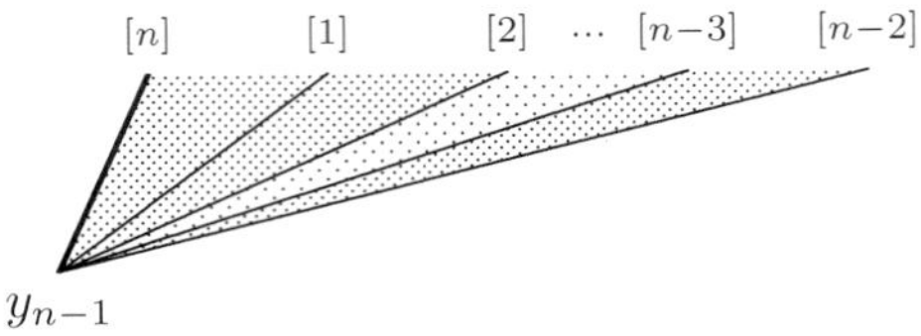

For $j = n$, we have $G(j) = \{n-1\}$ and $H(j) = \{1, \ldots, n-2\}$. The support of the modules $V[i]$ with $i \in H(j)$ are the various shaded triangles in the following picture:

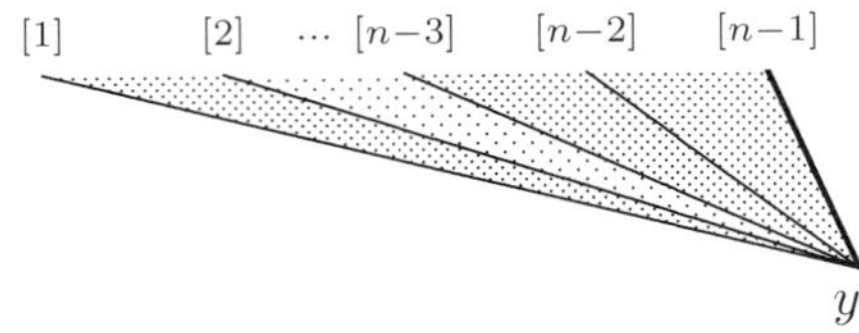

As we have mentioned, in both case $j = n-1$ and $j = n$, the proof of the equalities (1) and (2) proceeds as in the case when $j \notin \{n-1, n\}$, but using just one induction and not two. In the case $j = n-1$, we have to use the upgoing induction, in the case $j = n$ the downgoing induction.

(II) The bristle submodules $M(i,j)$ of M. For any pair $i \neq j$ with $1 \leq i, j \leq n$, there is a (unique) path $p(i,j) = \alpha_i^{-1}\alpha_j\alpha_i^{-1}\alpha_j$ in Q between $x(j,i)$ and $x(i,j)$:

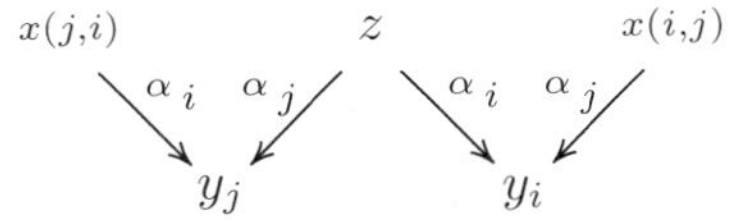

Let $W(i,j)$ be the indecomposable kQ-module whose support is the path $p(i,j)$.

Observe that $W(i,j)$ is a submodule of X; it is defined by

$$W(i,j)_z = \bigcap_{s \notin \{i,j\}} \mathrm{Ker}(\alpha_s)$$

(this is a one-dimensional subspace of X_z) and $W(i,j)_a = X_a$ for the remaining vertices a on the path $p(i,j)$.

Let us look at $\pi(W(i,j))$. We claim that the n-Kronecker module $\pi(W(i,j))$ contains a bristle $M(i,j)$ which is isomorphic to $B(i,j)$. The top of $\pi(W(i,j))$ is given by

$$W(i,j)_{x(j,i)} \oplus W(i,j)_z \oplus W(i,j)_{x(i,j)} = k \oplus k \oplus k.$$

The element $(1,1,1) \in k^3$ is mapped under both α_i and α_j to $(1,1) \in W(i,j)_{y_i} \oplus W(i,j)_{y_j}$. This shows that the $kK(n)$-submodule $M(i,j)$ of $\pi(W(i,j))$ generated by $(1,1,1) \in W(i,j)_{x(j,i)} \oplus W(i,j)_z \oplus W(i,j)_{x(i,j)}$ is isomorphic to $B(i,j)$. Altogether, we see that

$$\pi(W(i,j)) = \pi(P(x(j,i))) + M(i,j) + \pi(P(x(i,j))).$$

Since both $\pi(P(x(j,i)))$ and $\pi(P(x(i,j)))$ are contained in N, it follows that

$$\text{(3)} \qquad \pi(W(i,j)) \subset N + M(i,j).$$

The pairs (i, j) which we will consider are those of the form $(i, i+1)$. Let I be a subset of $\{1, 2, \ldots, n\}$ of cardinality $n-1$, thus obtained from $\{1, 2, \ldots, n\}$ by deleting one of its elements. We will use the submodules $W(i, i+1)$ with $i \in I$.

For example, for $n = 4$, and $I = \{1, 2, 3\}$, here are the dimension vectors of the submodules $W(i, i+1)$ of X with $i \in I$ (as well as the corresponding paths ${\alpha_i}^{-1}\alpha_{i+1}{\alpha_i}^{-1}\alpha_{i+1}$):

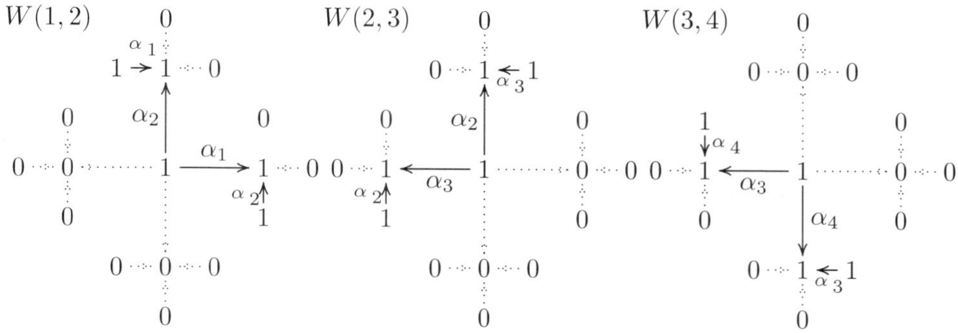

${\alpha_1}^{-1}\alpha_2{\alpha_1}^{-1}\alpha_2$ ${\alpha_2}^{-1}\alpha_3{\alpha_2}^{-1}\alpha_3$ ${\alpha_3}^{-1}\alpha_4{\alpha_3}^{-1}\alpha_4$

Here is the essential assertion: *If $I \subset \{1, 2, \ldots, n\}$ has cardinality $n-1$, then*

$$M = N + \sum_{i \in I} M(i, i+1). \tag{4}$$

PROOF OF (4). We can assume that $I = \{1, 2, \ldots, n-1\}$. First, we show that there is a vector space decomposition

$$X_z = \bigoplus_{i \in I} W(i, i+1)_z. \tag{5}$$

For the proof of (5), we recall that the arrow of type α_i starting in z ends in the vertex y_i. The Auslander-Reiten sequence in $\operatorname{mod} kQ$ ending in $S(z)$ is of the form

$$0 \longrightarrow X \xrightarrow{\ \iota\ } \bigoplus_{1 \le i \le n} I(y_i) \xrightarrow{[\pi(1), \ldots, \pi(n)]} S(z) \longrightarrow 0$$

where the support of $I(y_i)$ consists of y_i and its neighbors, where $\pi(i)\colon I(y_i) \to S(z)$ is the canonical projection. We can assume that ι is an inclusion map. At the vertex z, there is the corresponding exact sequence of vector spaces

$$0 \longrightarrow X_z \xrightarrow{\ \iota_z\ } \bigoplus_{1 \le i \le n} I(y_i)_z \xrightarrow{[\pi(1)_z, \ldots, \pi(n)_z]} S(z)_z \longrightarrow 0$$

with $I(y_i)_z = k$, $S(z) = k$ and $\pi(i)_z$ the identity map. Thus X_z is the kernel of the map

$$[\pi(1)_z, \ldots, \pi(n)_z] = [1, \ldots, 1]\colon k^n \to k$$

and therefore generated by the elements $e(i) - e(i+1)$, with $1 \le i < n$; here $e(i)$ is the canonical generator of $k = I(y_i)_z$. Of course, $e(i) - e(i+1)$ belongs to the kernel of $\alpha_s\colon X_z \to X_{a_s}$ for $s \notin \{i, j\}$, thus to $W(i, i+1)_z$. This shows that

$$X_z = \bigoplus_{1 \le i < n} \langle e(i) - e(i+1) \rangle = \bigoplus_{1 \le i < n} W(i, i+1)_z.$$

This completes the proof of (5).

As a consequence of (5), we have

$$X = Y + \sum_{i \in I} W(i, i+1).$$

We apply π and obtain

$$M = \pi(X) = N + \sum_{i \in I} \pi(W(i, i+1)).$$

Now we use (3) in order to see that

$$M = N + \sum_{i \in I} \pi(W(i, i+1)) \subseteq N + \sum_{i \in I} M(i, i+1) \subseteq M,$$

thus we obtain (4). □

In order to complete the proof of 4.4, we choose $I = \{1, 2, \dots, n-2, n\}$. Then all the submodules $M(i, i+1)$ with $i \in I$ belong to $\mathcal{B}'_0$. We know already by (I) that N is generated by $\mathcal{B}'_0$. Thus the equality (4) shows that M is generated by $\mathcal{B}'_0$. □

4.5. Remark: Book-keeping. For the sake of book-keeping, let us mention for any bristle type the number of bristles used for generating $M = I_2$. The number of leaves of Q of types $n-1$ and n is $n-1$. Thus, as bristles of the form $D(x)$ we have used $n-1$ bristles isomorphic to $B(n-1)$ as well as $n-1$ bristles isomorphic to $B(n)$. Let $I = \{1, 2, \dots, n-2, n\}$. For any $i \in I$, there are $n-2$ modules of the form $V[i]$, namely with support inside $Q(j)$, where $j \notin \{i, i+1\}$, thus we have used $n-2$ bristles of the form $E[i]$ and these bristles are isomorphic to $B(i, i+1)$. Finally, for $i \in I$, we needed the module $M(i, i+1)$. Now, $M(i, i+1)$ again is isomorphic to $B(i, i+1)$. Altogether, we see that for any bristle $B \in \mathcal{B}'_0$, the number of bristles isomorphic to B used in order to generate I_2 is $n-1$.

Therefore, the total number of bristles used for generating I_2 is $(n+1)(n-1) = n^2 - 1$. Since $\mathbf{dim}\, I_2 = (n^2-1, n)$, the module $\operatorname{top} I_2$ is of length $n^2 - 1$, thus a minimal number of bristle submodules which generate I_2 consists of precisely $n^2 - 1$ bristles. We also should mention that for any bristle B,

$$\dim \operatorname{Hom}(B, I_2) = \langle \begin{pmatrix} 1 \\ 1 \end{pmatrix}, \begin{pmatrix} n^2 - 1 \\ n \end{pmatrix} \rangle = n^2 - 1 + n - n^2 = n - 1$$

(since $\operatorname{Ext}^1(B, I_2) = 0$). Looking at any minimal set of bristle submodules which generate I_2, this set can contain at most $n-1$ bristles isomorphic to B.

4.6. *The module $\tau B(1)$ is generated by $\mathcal{B}_0$ and it has a proper submodule isomorphic to $B(1)$.*

PROOF. The proof of the first assertion is similar to the considerations in 4.4 dealing with I_2. Actually, one may consider $\tau B(1)$ as a submodule of I_2, generated by some of the bristles constructed above.

Here, we consider a representation X' of $\widetilde{K}(n)$ with $\pi(X') = \tau B(1)$. We denote by Q' its support: it is obtained from Q defined above as follows: Deleting the arrow $z \to y_1$, we obtain two connected components and we delete the component which contains y_1 (this is just $Q(1)$ as considered in 4.4). Thus Q' consists of the vertices a of $\widetilde{K}(n)$ which have distance at most 2 from z, and such that the path from z to a does not start with the arrow with label α_1. We have $\dim X'_z = n-2$, and $\dim X'_a = 1$ for the remaining vertices a in Q'.

The module X' can be described as a submodule of X as follows: X'_z is the kernel of the map $X_z \to X_{y_1}$, and $X'_a = X_a$ for the remaining vertices a in Q'.

As in the proof of 4.4, we first consider the subquivers $Q(j)$ and the restriction of X' to $Q(j)$. Of course, now we deal only with $j \neq 1$. Actually, the restrictions of X' to $Q(j)$ is just $Y(j)$ as considered in 4.4. We recall that all the modules $Y(j)$ are generated by $\mathcal{B}_0$.

Second, we need bristles which are not contained in $\pi(Y')$, where $Y' = \bigoplus_{j=2}^n Y(j)$. We start with the submodules $W(i,i+1) \subset X'$, where $2 \le i \le n-1$ and take the corresponding submodules $M(i,i+1)$ of $\pi(W(i,i+1))$. As we know, $M(i,i+1)$ is isomorphic to $B(i,i+1) \in \mathcal{B}_0$.

Since $\tau B(1) = \pi(Y') + \sum_{i=2}^{n-1} M(i,i+1)$, we see that $\tau B(1)$ is generated by $\mathcal{B}_0$. This yields the first assertion.

For the proof of the second assertion, just take any submodule $\pi(P(x(j,1))$ with $j \neq 1$. Of course, one may also refer to the Auslander-Reiten formula which yields non-zero homomorphisms (and this embeddings) $B(1) \to \tau B(1)$, since obviously $\mathrm{Ext}^1(B(1), B(1)) \neq 0$. □

Here are, for $n = 4$, on the left the dimension vector of X', in the middle and on the right those of $W(2,4)$ and $W(3,4)$, respectively.

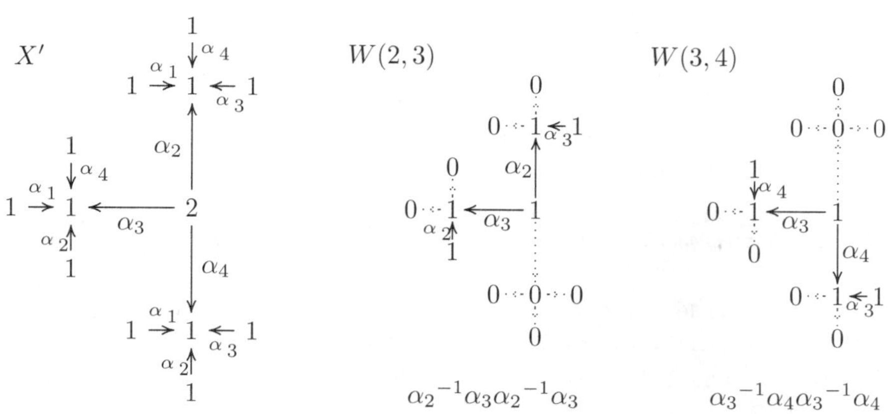

4.7. Remark: Again the book-keeping. The number of leaves of Q' of type $n-1$ is $n-2$, as is the number of leaves of type n. Thus, as bristles of the form $D(x)$ we have used $n-2$ bristles isomorphic to $B(n-1)$ as well as $n-2$ bristles isomorphic to $B(n)$. The number of bristles $E(x)$ which we have used are as follows: We have used $n-2$ bristles $E(x)$ isomorphic to $B(1,2)$, we have used $n-3$ bristles $E(x)$ isomorphic to $B(i,i+1)$, for any i with $2 \le i \le n-2$, and we have used $n-2$ bristles $E(x)$ isomorphic to $B(n,1)$. In addition, there are the bristles $M(i,i+1)$ isomorphic to $B(i,i+1)$. Altogether we have used $n-2$ bristles of the form $B(n-1), B(n)$ and $B(i,i+1)$ with $1 \le i \le n-2$ and with $i = n$ and a single bristle of the form $B(n-1,n)$.

In 4.6 we have presented a set of cardinality $n+2$ in order to generate $\tau B(1)$. As we will see in section 4.24, there are bristle sets of cardinality $n+1$ which generate $\tau B(1)$, but they are not helpful for the remaining proof of 1.3.

4.8. Corollary to 4.6. *If B is a bristle, then τB is generated by bristles, but is not a bristle, and $\tau^{-1}B$ is cogenerated by bristles and is not a bristle.*

PROOF. Using an automorphism of the Kronecker algebra, we may shift B to $B(1)$, apply 4.6 and shift back. This shows that τB is generated by bristles. Using duality, it follows that τ^{-1} is cogenerated by bristles. □

4.9. *If B, B' are non-isomorphic bristles, then* $\operatorname{Ext}^1(B, \tau B') = 0$.

PROOF. This follows from the isomorphism $\operatorname{Ext}^1(B, \tau B') \simeq D\operatorname{Hom}(B', B)$, where D denotes k-duality, since non-isomorphic bristles are Hom-orthogonal. □

4.10. *Let M be an indecomposable module and U a submodule of M isomorphic to $B(1)$. If $\tau(M/U)$ is generated by $\mathcal{B}_0$, then τM is generated by $\mathcal{B}_0$.*

PROOF. We apply τ to the exact sequence

$$0 \to U \to M \to M/U \to 0$$

and obtain the exact sequence

$$0 \to \tau U \to \tau M \to \tau(M/U) \to 0.$$

Consider any homomorphism $f\colon B \to \tau(M/U)$ with $B \in \mathcal{B}_0$. Since B is not isomorphic to $B(1)$, 4.9 shows that $\operatorname{Ext}^1(B, \tau B(1)) = 0$, thus f can be lifted to τM. And 4.6 asserts that τU is generated by $\mathcal{B}_0$. Since by assumption also $\tau(M/U)$ is generated by $\mathcal{B}_0$, it follows that τM is generated by $\mathcal{B}_0$. □

4.11. *All preinjective modules are generated by $\mathcal{B}_0$ and are $\mathcal{B}$-saturated.*

PROOF. First, let us show that I_t with $t \geq 0$ is generated by $\mathcal{B}_0$. We use induction on t. The assertion is trivial for $t = 0$. The module I_1 is obviously generated by $B(2), \dots, B(n)$ and $B(1,2)$, thus by $\mathcal{B}_0$. Next, 4.4 asserts that I_2 is generated by $\mathcal{B}_0$. Thus, we can assume that $t \geq 3$.

Clearly, $\operatorname{Hom}(B(1), I_s) \neq 0$ for any $s \geq 0$, thus $\operatorname{Hom}(B(1), I_{t-2}) \neq 0$. Take a nonzero homomorphism $f\colon B(1) \to I_{t-2}$. Since $t-2 \geq 1$, the map f has to be a monomorphism and its cokernel C will be a direct sum of preinjective modules of the form I_s with $s < t-2$. We apply τ to the exact sequence

$$0 \to B(1) \to I_{t-2} \to C \to 0$$

and obtain the exact sequence

$$0 \to \tau B(1) \to I_t \to \tau C \to 0.$$

Now τC is a direct sum of modules of the form I_{s+2} with $s+2 < t$. By induction, C is generated by $\mathcal{B}_0$, thus 4.10 shows that I_t is generated by $\mathcal{B}_0$.

Of course, all preinjective modules I_t satisfy $\operatorname{Ext}^1(X, I_t) = 0$, for X regular. Since bristles are regular, we see that I_t is $\mathcal{B}$-saturated. □

4.12. *Let M be an indecomposable module which is generated by bristles, but not a bristle. Then* $\mathrm{Hom}(M, B) = 0$ *for any bristle B, thus τM is $\mathcal{B}$-saturated.*

PROOF. Since M is a generated by bristles, there are bristles $B_1, \dots, B_t$ and a surjective map $f \colon \bigoplus B_i \to M$. Assume there exists a non-zero map $g \colon M \to B$. Then $gf_i \neq 0$ for some i. But a non-zero map between bristles is an isomorphism, thus g is a split monomorphism. Since M is indecomposable, g has to be an isomorphism, thus M is a bristle. This is the first assertion. The vector space isomorphism $\mathrm{Ext}^1(B, \tau M) \simeq D\underline{\mathrm{Hom}}(M, B)$ shows that $\mathrm{Ext}^1(B, \tau M) = 0$ for any bristle B, thus τM is $\mathcal{B}$-saturated. □

We also will need the dual assertion of 4.12:

4.13. *Let M be an indecomposable module which is cogenerated by bristles, but not a bristle. Then* $\mathrm{Hom}(B, M) = 0$ *for any bristle B.*

4.14. Let us consider now elementary modules as defined by Crawley-Bovey and studied by Lukas and Kerner (see for example [**3**]): a regular module M is said to be *elementary* provided that for any non-zero regular submodule U of M, the factor module M/U is preinjective.

Let M be an elementary module with $\mathrm{Hom}(B(1), M) \neq 0$. *Then $\tau^t M$ is generated by $\mathcal{B}_0$ for all $t \geq 1$.*

PROOF. Let M be an elementary module with $\mathrm{Hom}(B(1), M) \neq 0$. A non-zero homomorphism $f \colon B(1) \to M$ has to be injective, since M has no submodule isomorphic to $S(1)$. Since M is elementary, the cokernel C of f has to be preinjective. Thus, we deal with an exact sequence of the form

$$0 \to B(1) \to M \to C \to 0.$$

If we apply τ, we obtain the exact sequence

$$0 \to \tau B(1) \to \tau M \to \tau C \to 0.$$

Now τC is preinjective, thus according to 4.11, generated by $\mathcal{B}_0$. According to 4.10, also τM is generated by B_0.

Now, the module τM is again elementary, and it has $\tau B(1)$ as a submodule. According to 4.6, $\tau B(1)$ has a submodule isomorphic to $B(1)$. It follows that $\mathrm{Hom}(B(1), \tau M) \neq 0$. This shows, that τM satisfies again the assumption, thus also $\tau^2 M$ is generated by $\mathcal{B}_0$. Altogether we see that we can use induction. □

Since bristles are elementary modules, we see:

4.15. Corollary. *If B is a bristle, then the module $\tau^t B$ is generated by $\mathcal{B}_0$, for all $t \geq 1$.*

4.16. Corollary. *If B is a bristle, then the module $\tau^{-t} B$ is cogenerated by bristles, for all $t \geq 1$.*

Corollary 4.16 follows from 4.15 by using duality.

4.17. *Any τ-orbit contains at most one bristle. If B, B' are bristles, then we have* $\mathrm{Hom}(\tau^t B, B') = 0$ *for* $t \geq 1$.

PROOF. Assume B and $\tau^t B$ are bristles, for some $t \geq 1$. This implies that $\binom{1}{1}$ is an eigenvector of Φ^t with eigenvalue 1, where Φ is the Coxeter transformation for $K(n)$. It is well-known (and easy to see) that 1 is not an eigenvalue of Φ^t for any $t \geq 1$. This yields the first assertion.

For the proof of the second assertion, let $t \geq 1$. We know that $\tau^t B$ is generated by bristles, thus there are maps $f_i \colon B_i \to \tau^t B$ with B_i a bristle, such that $(f_i)_i \colon \bigoplus B_i \to \tau^t B$ is surjective. Assume there exists a non-zero map $g \colon \tau^t B \to B'$. Then $gf_i \neq 0$ for some i. A non-zero map between bristles is an isomorphism, thus gf_i is an isomorphism and therefore $g \colon \tau^t B \to B'$ is split epi. Since $\tau^t B$ is indecomposable, we see that g is an isomorphism, thus $\tau^t B$ is a bristle. But the τ-orbit of τ cannot contain a second bristle. □

4.18. *Let M be an elementary module with* $\mathrm{Hom}(B(1), M) \neq 0$. *Then $\tau^t M$ is $\mathcal{B}$-saturated for all $t \geq 2$.*

PROOF. Let M be an elementary module with $\mathrm{Hom}(B(1), M) \neq 0$. Let $t \geq 2$. By 4.14 we know that $\tau^{t-1} M$ is generated by bristles. If we know that $\tau^{t-1} M$ is not a bristle, then we can apply 4.12 in order to see that $\tau^t M$ is $\mathcal{B}$-saturated.

Thus assume that $B = \tau^{t-1} M$ is a bristle. Then $M = \tau^{-t+1} B$ is cogenerated by bristles, according to Corollary 4.16. Since $\mathrm{Hom}(B(1), M) = 0$, 4.13 shows that M has to be a bristle. But according to the first assertion 4.17, we know that any τ-orbit contains at most one bristle. □

4.19. Corollary. *If B is a bristle, then $\tau^t B$ is $\mathcal{B}$-saturated for all $t \geq 2$.*

Note that the assumption $t \geq 2$ is essential, since τB is not $\mathcal{B}$-saturated: we obviously have $\mathrm{Ext}^1(B, \tau B) \neq 0$.

4.20. Proof of Theorem 1.3. Part (a) has been established in 4.11. Thus it remains to deal with part (b). As we have mentioned already, it is sufficient to look at a module M which is regular.

First, let us consider the special case of an elementary module M. It has been shown by D. Baer that for any pair M', M'' of regular modules there is some number $s(M', M'')$ such that $\mathrm{Hom}(M', \tau^s M'') \neq 0$ for all $s \geq s(M', M'')$ (see for example [**3**], 10.7). Thus $\mathrm{Hom}(B(1), \tau^s M) \neq 0$ for all $s \geq s(B(1), M)$. According to 4.14 and 4.18, we know that $\tau^s M$ is generated by $\mathcal{B}_0$ and is $\mathcal{B}$-saturated for all $s \geq s(B(1), M) + 2$.

Now any regular module M has a filtration by elementary modules, say

$$0 = M_0 \subset M_1 \subset \cdots \subset M_m = M$$

with all M_i/M_{i-1} for $1 \leq i \leq m$ being elementary. Let

$$t(M) = \max_i s(B(1), M_i/M_{i-1}) + 2.$$

Let $t \geq t(M)$. The module $\tau^t M$ has the filtration

$$0 = \tau^t M_0 \subset \tau^t M_1 \subset \cdots \subset \tau^t M_m = \tau^t M$$

and the factors are $\tau^t M_i / \tau^t M_{i-1} = \tau^t (M_i/M_{i-1})$. These are elementary modules which are generated by $\mathcal{B}_0$ and are $\mathcal{B}$-saturated. Since the class of modules which are generated by $\mathcal{B}_0$ and $\mathcal{B}$-saturated is closed under extensions, it follows that also $\tau^t M$ is generated by $\mathcal{B}_0$ and $\mathcal{B}$-saturated. This completes the proof. □

4.21. Corollary. *There is a regular module X which generates all the preinjective modules, for example $X = \bigoplus_{B\in\mathcal{B}_0} B$. There are also indecomposable regular modules which generate all preinjective modules.*

This provides an explicit answer to a question raised by Kerner in 1994 (Problem 10.9 in [**3**]): to exhibit a regular module X which generates all the preinjective modules. An existence proof for such a module was given by Kerner in [**4**].

PROOF. The first assertion is a direct consequence of 1.3.

For the second assertion, let X be a suitable extensions $0 \to B(1) \to \overline{X} \to X \to 0$ with $\overline{X}$ indecomposable (such an extension exists, since $\mathcal{B}_0 \cup \{B(1)\}$ is a set of pairwise non-isomorphic bristles, and pairwise non-isomorphic bristles are pairwise orthogonal, and $\operatorname{Ext}^1(B, B') \neq 0$ for any pair B, B' of bristles). Such a module $\overline{X}$ has dimension vector $(n+2, n+2)$. □

4.22. Remark. Let us stress that the existence of a (finitely generated!) regular module which generates all preinjective modules is a special feature for the cases $n \geq 3$: It is a finiteness result in the wild cases which does not hold in the tame case $n = 2$, see Appendix C.

4.23. Addendum to the Main Theorem. *Let $\mathcal{B}'$ be a set of $n+1$ bristles and let $t \geq 3$. The preinjective module I_t cannot be generated by $\mathcal{B}'$.* Thus, the number $n+2$ occurring in the Main Theorem is best possible.

PROOF. First, let $t = 3$. We use the Euler form $\langle -, - \rangle$ on the Grothendieck group $K_0(\operatorname{mod}\Lambda)$. Note that I_3 has the dimension vector $\mathbf{dim}\, I_3 = (n^3 - 2n, n^2 - 1)$. We have

$$\left\langle \begin{pmatrix} 1 \\ 1 \end{pmatrix}, \begin{pmatrix} n^3 - 2n \\ n^2 - 1 \end{pmatrix} \right\rangle = (n^3 - 2n) + (n^2 - 1) - n(n^2 - 1) = n^2 - n - 1.$$

Since for any bristle B, we have $\operatorname{Ext}^1(B, I_3) = 0$, it follows that

$$\dim \operatorname{Hom}(B, I_3) = n^2 - n - 1.$$

Let X be the direct sum of $n+1$ pairwise non-isomorphic bristles. It follows that there is a right X-approximation of the form $f\colon X^t \to I_3$ with $t \leq n^2 - n - 1$. Now

$$\dim \operatorname{top} X^t = (n+1)t \leq (n+1)(n^2 - n - 1) = n^3 - 2n - 1.$$

On the other hand,

$$\dim \operatorname{top} I_3 = n^3 - 2n,$$

thus f cannot be surjective. This shows that X does not generate I_3.

Now consider any $t \geq 3$. Assume that I_t is generated by $\mathcal{B}'$. Since I_t generates I_3, it follows that $\mathcal{B}'$ generates I_3. But we have shown already that this is not possible. □

4.24. Let us look at optimal ways to generate $\tau B(1)$, using bristle submodules. *Let $\mathcal{B}'$ be a set of $n+1$ bristles. If $\tau B(1)$ is generated by $\mathcal{B}'$, then $B(1)$ has to belong to $\mathcal{B}'$. The set $\mathcal{B}'_1$ given by $B(1)$ as well as the bristles $B(i, i+1)$ for $1 \leq i \leq n$ generates $\tau B(1)$.*

Proof. We have $\mathbf{dim}\, \tau B(1) = (n^2 - n - 1, n - 1)$. If B is a bristle, then

$$\langle \begin{pmatrix} 1 \\ 1 \end{pmatrix}, \begin{pmatrix} n^2 - n - 1 \\ n - 1 \end{pmatrix} \rangle = (n^2 - n - 1) + (n - 1) - n(n - 1) = n - 2.$$

Since bristles are bricks and pairwise orthogonal, the Auslander-Reiten formula shows that $\dim \operatorname{Ext}^1(B(1), B(1)) = 1$ and $\dim \operatorname{Ext}^1(B, B(1)) = 0$, provided B is not isomorphic to $B(1)$. It follows that $\dim \operatorname{Hom}(B(1), \tau B(1)) = n - 1$, but $\dim \operatorname{Hom}(B, B(1)) = n - 2$ in case B is not isomorphic to $B(1)$.

Let $\mathcal{B}'$ be a set of bristles of cardinality $n + 1$ and assume that $\mathcal{B}'$ does not contain $B(1)$. Since $\dim \operatorname{Hom}(B, \tau B(1)) = n-2$ for any $B \in \mathcal{B}'$, there is a right $\mathcal{B}'$-approximation of the form $f \colon X^t \to \tau B(1)$ with $t \leq n - 2$. We have $\dim \operatorname{top} X^t = (n + 1)t \leq (n + 1)(n - 2) = n^2 - n - 2$, and $\dim \operatorname{top} \tau B(1) = n^2 - n - 1$. As a consequence, f cannot be surjective. This shows that $\mathcal{B}'$ cannot generate $\tau(B(1)$.

The proof of the second assertion is similar to the proofs presented in 4.4 and 4.6. □

As we have seen in the proof, the existence of sufficiently many bristles in $\tau B(1)$ which are isomorphic to $B(1)$ is related to the non-vanishing of $\operatorname{Ext}^1(B(1), \tau B(1))$. Of course, the only non-trivial extension is given by the middle term $\mu(B(1))$ of the Auslander-Reiten sequence ending in $B(1)$. We refer to the end of section 5.1 for a closer look at $\mu(B(1))$.

4.25. The use of the universal cover $\widetilde{K}(n)$ of $K(n)$. We have used the universal cover $\widetilde{K}(n)$ of $K(n)$ in order to exhibit bristle submodules of some n-Kronecker modules M.

(1) Any leaf x in the support of the covering module yields the bristle module $P(x)$ for $\widetilde{K}(n)$, provided that x is a source. But there may be other bristle submodules of M of the form $\pi(U)$, where U is a $\widetilde{K}(n)$-bristle. Let us look at the covering module M for I_3 and, on the left, its support quiver:

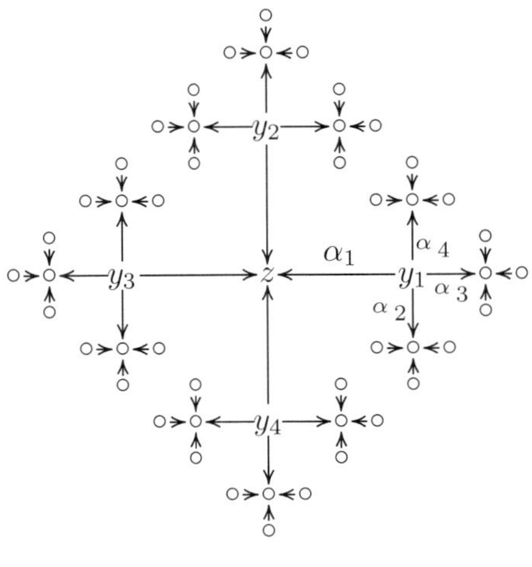

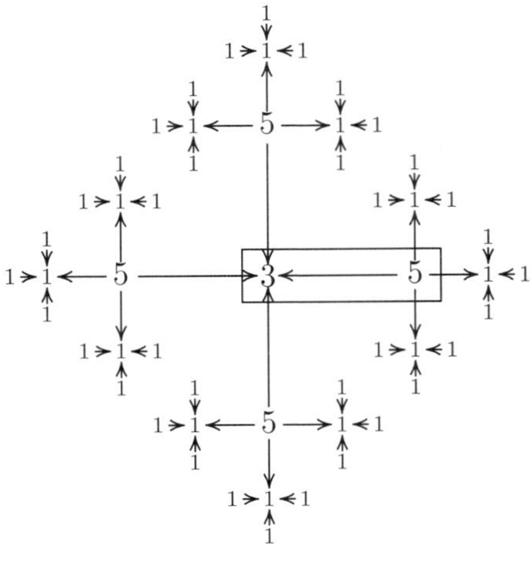

Consider in the subspace M_{y_1} the intersection A of the three maps with labels $\alpha_1, \alpha_2, \alpha_3$. This subspace A is mapped under α_1 injectively to M_z, thus it generates a direct sum of two bristles of type $B(1)$ (see also (B.4) in the appendix).

Since $\dim \operatorname{Hom}(B(1), I_3) = 11$, we see that we obtain sufficiently many bristles of type $B(1)$ by looking at the 9 leaves of type 1 and two of the bristles generated by elements of A.

(2) The universal cover X of I_2 displays all the bristles of type $B(i)$ of I_2 (with $1 \le i \le n$) in a nice way: as coming from bristles of $\widetilde{I}_2$ given by leaves in Q. Namely, the number of leaves x in Q which are starting points of arrows with label α_i is $n-1$, and, as we have mentioned already in 4.5, we have $\dim \operatorname{Hom}(B(i), I_2) = n-1$.

(3) The universal cover X of I_2 allows also to trace the bristles of type

$$B(1,2,\dots,n) = (k,k;1,1,\dots,1).$$

For $1 \le j \le n$, let $Y(j)$ be the sum of the submodules $P(x)$ of X, where x is a leaf and a neighbor of y_j. Then $\pi(Y(j))$ has dimension vector $\binom{n-1}{1}$, and $X/\sum Y(j)$ is a direct sum of $n-1$ copies of $S(z)$, and canonically isomorphic to X_z. On the one hand, we have $\dim \operatorname{Hom}(B(1,2,\dots,n), I_2) = n-1$, on the other hand $\operatorname{Hom}(B(1,2,\dots,n), Y(j)) = 0$ for all j. This shows that X has a submodule U isomorphic to $B(1,2,\dots,n)^{n-1}$ such that the composition of the canonical maps $U \to X \to X/\sum Y(j)$ is surjective.

5. Further considerations.

5.1. Proposition. *Let M be an indecomposable bristled module, not a bristle and not simple projective. If τM is a bristled module, then the middle term $\mu(M)$ of the Auslander-Reiten sequence ending in M is a bristled module.*

If $n \ge 2$, then the middle term $\mu(B)$ of the Auslander-Reiten sequence ending in a bristle B is not a bristled module.

PROOF. Let M be an indecomposable bristled module, not a bristle and not simple projective. Then M is not projective, thus we can consider the middle term $\mu(M)$ of the Auslander-Reiten sequence ending in M. Since the canonical map $\mu(M) \to M$ is right almost split, all the maps $B \to M$ with B a bristle can be lifted to $\mu(M)$. Together with the maps $B \to \tau(M)$, where B is a bristle, we obtain enough maps from bristles to $\mu(M)$ in order to see that $\mu(M)$ is generated by bristles.

Now consider $\mu(B)$, where B is a bristle and let $\epsilon\colon \mu(B) \to B$ be the canonical map. Assume that $\mu(B)$ is bristled. Then $\mu(B)$ is a factor module of a direct sum of bristles, thus there is a map $f\colon B' \to \mu(B)$ with B' a bristle such that $\epsilon f \neq 0$. But any non-zero map between bristles is an isomorphism, thus ϵf is an isomorphism and therefore ϵ is a split epimorphism. This is impossible, thus $\mu(B)$ cannot be bristled. □

Let us have a closer look at $\mu(B)$, say for $B = B(1)$. It can be realized as a submodule of I_2, as follows. As in section 4.4, we consider the universal cover $\widetilde{K}(n)$ of $K(n)$ with push-down functor π and denote by X a representation of $\widetilde{K}(n)$ with $\pi(X) = I_2$. Let X'' be the submodule of X generated by the submodules $Y(j)$ with $2 \le j \le n$ and the subspace X_z. Thus we have

$$X' \subset X'' \subset X,$$

with X' as defined in 4.6 (and $\pi(X') = \tau B(1)$). In order to see that $\pi(X'') = \mu(B(1))$, we only have to observe that X'' is indecomposable and that X''/X' is an indecomposable length 2 module whose support is an arrow with label α_1, thus $\pi(X''/X')$ is isomorphic to $B(1)$.

For example in case $n = 4$, the dimension vector of X'' looks as follows:

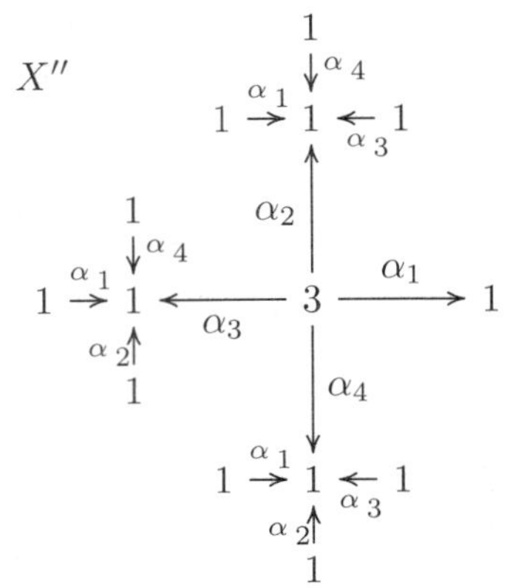

5.2. Let $n \geq 3$. The Auslander-Reiten component containing a bristle B has the following shape

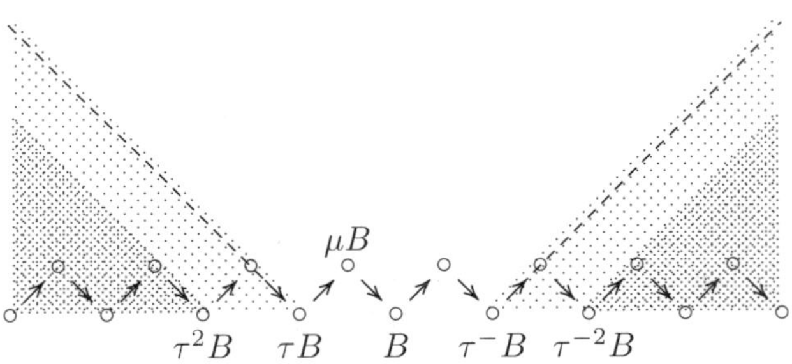

On the left hand side, we see the shaded cocone of all modules with a path in the Auslander-Reiten quiver ending in τB, all these modules are bristled modules; the darker shading on the left marks the region of the modules which are in addition $\mathcal{B}$-saturated (the assertions follow from 4.15 and 4.19 and the closure under extensions).

On the right hand side, we see the shaded cone of all modules with a path in the Auslander-Reiten quiver starting at $\tau^- B$. By duality, all these modules are cogenerated by bristles. Here, the darker shading indicates that the corresponding modules M are in addition $\mathcal{B}$-cosaturated.

The remaining modules M in the component (those which are neither predecessors of τB nor successors of $\tau^- B$) all have bristle submodules as well as bristle factor modules (since the northeast arrows correspond to monomorphisms, the southeast arrows to epimorphisms); thus, with the exception of B itself, are neither bristled nor cobristled.

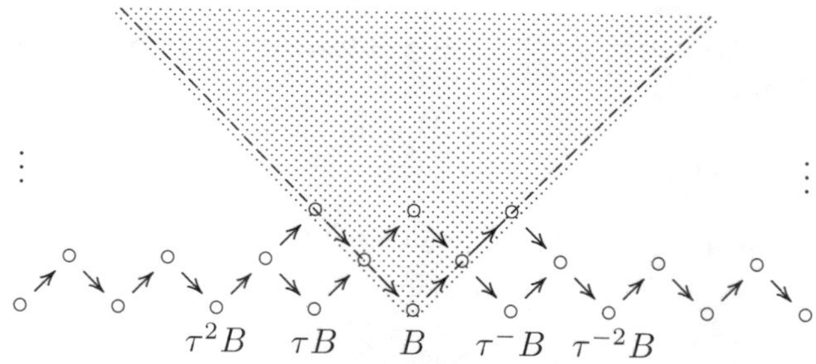

5.3. *Let $n \geq 2$. If M is a module with a sectional path ending in a bristle B, then M is not $\mathcal{B}$-saturated.* (In particular, if μB is the middle term of the Auslander-Reiten sequence ending in B, then μB is not $\mathcal{B}$-saturated.)

PROOF. The epimorphism $M \to B$ yields a surjection $\operatorname{Ext}^1(B, M) \to \operatorname{Ext}^1(B, B)$. Calculating $\langle \mathbf{dim}\, B, \mathbf{dim}\, B\rangle$, we see that $\dim \operatorname{Ext}^1(B, B) = n - 1$. $\square$

5.4. Actually, it is easy to see that for any bristle B, we have $\dim \operatorname{Ext}^1(B, \mu B) = n - 1$. Namely, the Auslander-Reiten sequence ending in B yields an exact sequence

$$\operatorname{Hom}(B, \mu B) \to \operatorname{Hom}(B, B) \to \operatorname{Ext}^1(\tau B, B) \to \operatorname{Ext}^1(\mu B, B) \to \operatorname{Ext}^1(B, B) \to 0.$$

The first map cannot be surjective, since otherwise the canonical projection $\mu B \to B$ would split. Since $\operatorname{End}(B) = k$, we see that the first map is the zero map, thus the second map is injective. Again using that $\operatorname{End}(B) = k$, we have $\dim \operatorname{Ext}^1(\tau B, B) = 1$. Thus, the second map is an isomorphism and therefore the map $\operatorname{Ext}^1(\mu B, B) \to \operatorname{Ext}^1(B, B)$ is bijective. As we have mentioned in the proof above, one may use the bilinear form $\langle -, -\rangle$ in order to see that $\dim \operatorname{Ext}^1(B, B) = n - 1$. Thus, also $\dim \operatorname{Ext}^1(\mu B, B) = n - 1$.

The following picture shows for $n = 4$ and $B = B(1)$ an indecomposable representation X''' of the universal cover $\widetilde{K}(4)$ with submodule X'' such that $\pi(X'') = \mu(B(1))$ and $\pi(X'''/X'')$ is isomorphic to $B(1)^3$. Thus, the universal cover exhibits nicely a basis of $\operatorname{Ext}^1(B(1), \mu(B(1)))$.

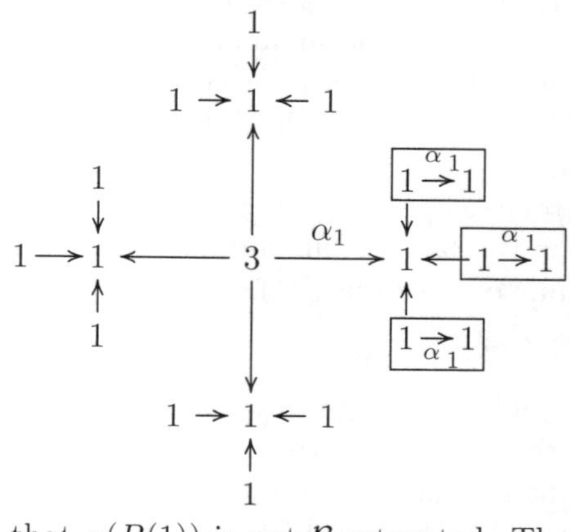

We have just seen that $\mu(B(1))$ is not $\mathcal{B}$-saturated. The following proposition shows that there are many modules, even many bristled modules, which are not $\mathcal{B}$-saturated.

5.5. Proposition. *An indecomposable $\mathcal{B}$-saturated module is simple or faithful.*

PROOF. Proof. For $n = 1$, all indecomposable modules are simple or faithful. Thus, we may assume that $n \geq 2$. Let M be a non-faithful indecomposable n-Kronecker module. Let as assume that M is annihilated by α_n. We also assume that M is different from $S(1)$. The following lemma shows that M is not $\mathcal{B}$-saturated. □

5.6. Lemma. *If M is a $K(n)$-module which is annihilated by α_n, then*

$$\dim \operatorname{Ext}^1(B(1), M) \geq \dim M_2.$$

PROOF. We may consider both $B(1)$ and M also as $(n-1)$-Kronecker modules. We calculate $\langle \mathbf{dim}\, B(1), \mathbf{dim}\, M, \rangle_t$ for $t = n$ and for $t = n - 1$. We have

$$\begin{aligned} \langle \mathbf{dim}\, B(1), \mathbf{dim}\, M\rangle_n &= \dim M_1 - (n-1)\dim M_2, \\ \langle \mathbf{dim}\, B(1), \mathbf{dim}\, M\rangle_{n-1} &= \dim M_1 - (n-2)\dim M_2, \end{aligned}$$

thus the difference is

$$D = \langle \mathbf{dim}\, B(1), \mathbf{dim}\, M\rangle_{n-1} - \langle \mathbf{dim}\, B(1), \mathbf{dim}\, M\rangle_n = \dim M_2.$$

Since there is no difference whether we calculate Hom for $K(n)$ or for $K(n-1)$), the difference D is also

$$D = -\dim \operatorname{Ext}^1_{K(n-1)}(B(1), M)) + \dim \operatorname{Ext}^1_{K(n)}(B(1), M).$$

Altogether, we see that

$$\dim \operatorname{Ext}^1_{K(n)}(B(1), M)) = \dim \operatorname{Ext}^1_{K(n-1)}(B(1), M)) + \dim M_2 \geq \dim M_2. \quad \square$$

Appendix A. The bristle variety $\beta(M)$ of a Kronecker module M.

We assume here that k is algebraically closed.

Let $M = (M_1, M_2, \alpha_1, \ldots, \alpha_n)$ be a Kronecker module. First, we deal with the case where M has no direct summand of the form $S(1)$, thus the intersection of the kernels of the maps α_i is zero. Let $\mathbb{P}M_1$ be the projective space corresponding to M_1, this is the set of all 1-dimensional subspaces of M_1, endowed with the Zariski topology. If m is a non-zero element of M_1, we write $\langle m\rangle$ for the k-subspace generated by m (this is an element of $\mathbb{P}M_1$). Let $\beta(M)$ be the subset of $\mathbb{P}M_1$ of all $\langle u\rangle$ such that u generates a bristle submodule in M_1 (or, equivalently, such that $\alpha_1(u), \ldots, \alpha_n(u)$ generate a one-dimensional subspace of M_2). It is easy to see that $\beta(M)$ is closed in the Zariski topology, thus *$\beta(M)$ is a projective variety,* we call it the *bristle variety* of M.

For a general n-Kronecker module M, the *bristle variety* of M is by definition $\beta(M/M')$, where M' is the largest submodule of M generated by $S(1)$ (so that M/M' has no submodule of the form $S(1)$).

(A.1) Theorem. *Any projective variety occurs as a bristle variety.*

The proof of Theorem (A.1) is given in [**6**]. Note that this theorem is a variant of a sequence of results by Zimmermann-Huisgen, Hille, Reineke and Van den Bergh which concern the realization of projective varieties say as quiver Grassmannians. As a consequence of Theorem (A.1), the following is shown in [**7**]: *given any connected wild acyclic quiver Q with at least 3 vertices, then any projective variety occurs as a quiver Grassmannian for a suitable representation of Q.*

A Kronecker module $M = (M_1, M_2, \alpha_1, \ldots, \alpha_n)$ is nothing else than a matrix pencil (as soon as we choose bases of the vector spaces M_1, M_2), and generators of bristle submodules can be considered as (generalized) eigenvectors. Indeed, if $n = 2$ and α_1 is the identity map of a vector space $V = M_1 = M_2$, then a submodule of the 2-Kronecker module $(V, V; 1, \phi)$ is a bristle of type $B_c = B(1 : c)$ if and only if any generator of U in M_1 is an eigenvector of ϕ with eigenvalue c.

Appendix B. Bristled modules for arbitrary artin algebras.

Let Λ be an artin algebra. Here, we collect some properties of the bristled Λ-modules and show in which way the study of bristled Λ-modules can be reduced to the study of bristled Kronecker modules. Given a Λ-module M, let $|M|$ be its length.

(B.1) Proposition. *If M is an indecomposable bristled Λ-module M then $|\operatorname{top} M| \geq |\operatorname{soc} M|$, with equality only in case M is simple or a bristle.*

PROOF. A bristled module has Loewy-length at most 2. It follows that any indecomposable bristled Λ-module M is either simple or it satisfies $\operatorname{soc} M = \operatorname{rad} M$. Of course, if M is simple, then $|\operatorname{top} M| = 1 = |\operatorname{soc} M|$.

Thus, let us assume that M is a bristled module with $\operatorname{soc} M = \operatorname{rad} M$. If $\operatorname{top} M$ has length t, there is a surjective map $f\colon \bigoplus_{i=1}^{t} B_i \to M$ with bristles B_i. The kernel of f has to be contained in the socle of $\bigoplus B_i$, let s be its length. Then $|\operatorname{soc} M| = t - s \leq t$, this is the first assertion. If $|\operatorname{soc} M| = t$, then $s = 0$, thus f is an isomorphism. If M is indecomposable, then we must have that $t = 1$, thus M is a bristle. □

Recall that a semi-simple Λ-module is said to be *homogeneous* provided that all its simple submodules are isomorphic.

(B.2) Proposition. *The socle of an indecomposable bristled module M is homogeneous.*

PROOF. Let M be an indecomposable bristled module. Note that M has Loewy length at most 2, since M is bristled. If M has Loewy length 1, then M is simple, therefore nothing has to be shown. Thus, we assume that M has Loewy length 2. This means that $\operatorname{rad} M = \operatorname{soc} M$.

Let $S_1, \ldots, S_n$ be the simple Λ-modules and write $\operatorname{soc} M = \bigoplus_{i=1}^{n} H_i$, where H_i is the sum of all submodules of M isomorphic to S_i.

By assumption, M is generated by bristles, thus M is the sum of images of some maps $B \to M$, where B is a bristle. Since $\operatorname{rad} M = \operatorname{soc} M$, it follows that M is generated by the images of some maps $B \to M$ which are in addition injective, thus M is the sum of submodules U_j which are isomorphic to bristles, say $M = \sum_{j \in J} U_j$. We may assume that J is minimal, or, equivalently, that $M/\operatorname{soc} M = \bigoplus (U_j + \operatorname{soc} M)/\operatorname{soc} M$.

Let M_i be the sum of the submodules U_j such that $\operatorname{soc} U_j$ is isomorphic to S_i. Now $\operatorname{soc} U_j$ is a submodule of $\operatorname{soc} M$ and isomorphic to S_i, thus $\operatorname{soc} U_j \subseteq H_i$. It follows easily that $M = \bigoplus_{i=1}^{n} M_i$. Since we assume that M is indecomposable, we have $M = M_i$ for some i. Of course, the socle of M_i is just H_i, thus homogeneous. □

(B.3) We say that a module M is the *socle amalgamation of the modules* $M_1, \dots, M_t$ provided the modules $M_1, \dots, M_t$ are submodules of M containing the socle $\operatorname{soc} M$ of M and $M/\operatorname{soc} M$ is the direct sum $\bigoplus_i M_i/\operatorname{soc} M$.

It follows from (B.2) that *any indecomposable bristled module M is the socle amalgamation of bristled modules $M_1, \dots, M_t$, such that* $\operatorname{soc} M$ *as well all the modules $M_i/\operatorname{soc} M$ are homogeneous.*

Namely, $\overline{M} = M/\operatorname{soc} M$ is semi-simple, thus we may write $\overline{M}$ as a direct sum $\bigoplus_i \overline{M}_i$ of pairwise Hom-orthogonal homogeneous modules $\overline{M}_i$. Let M_i be the preimage of $\overline{M}_i$ under the canonical projection $M \to \overline{M}$.

Thus, the first aim for understanding bristle modules seems to be to investigate bristled modules M such that both $\operatorname{soc} M$ and $M/\operatorname{soc} M$ are homogeneous.

In case Λ is a finite-dimensional k-algebra, where k is an algebraically closed field, it is sufficient to look just at the Kronecker algebras. Namely, we may assume that M is faithful and that Λ is basic. Under these assumptions, Λ is a basic radical-square-zero algebra with at most two simple modules. Since k is algebraically closed, only two cases are possible. If Λ has two simple modules, then Λ is already a Kronecker algebra. If Λ has only one simple module, then Λ is of the form $k[x_1, \dots, x_n]/(x_1, \dots, x_n)^2$, and therefore stably equivalent to the Kronecker algebra $kK(n)$. Under such a stable equivalence, the indecomposable bristled Λ-modules correspond bijectively to the indecomposable bristled $kK(n)$-modules.

(B.4) Let us assume now that Λ is the path algebra of a finite acyclic quiver Q without multiple arrows. In this case, the support of a bristle B is an arrow. Attaching to any arrow α the thin indecomposable module $B(\alpha)$ with support α, we obtain a bijection between the arrows and the (isomorphism classes of the) bristles. There is an easy numerical criterion to detect (at least some) bristle submodules of a kQ-module M: If $\alpha\colon y \to z$ is an arrow and $z_1 = z, z_2, \dots, z_t$ are the terminal vertices of the arrows starting in y, then $\dim \operatorname{Hom}(B(\alpha), M) \ge \dim M_y - \sum_{i\ge 2} \dim M_{z_i}$. (For the proof, one just has to look at $\langle \mathbf{dim}\, B(\alpha), \mathbf{dim}\, M\rangle$.) Of course, if B has no submodule isomorphic to $S(y)$, then $\dim \operatorname{Hom}(B(\alpha), M) = b$ implies that M has a submodule isomorphic to $B(\alpha)^b$.

In order to see that in general not all bristles are detected in this way, consider the following $\widetilde{\mathbb{D}}_5$-quiver shown below on the left. There exists an indecomposable representation M with dimension vector presented below on the right, which has a submodule $B(\alpha)$. whereas $\dim M_y - \sum_{i\ge 2} \dim M_{z_i} = 0$.

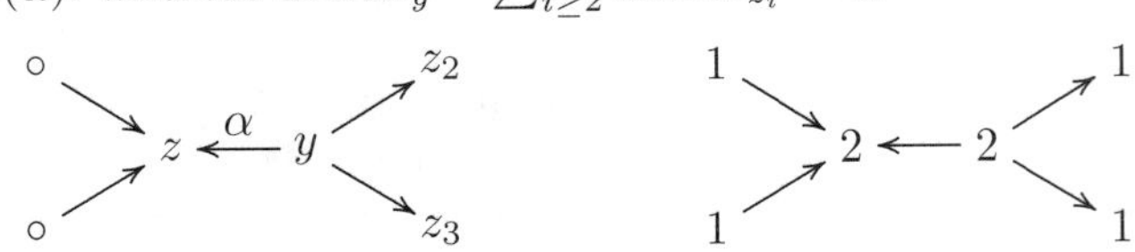

(B.5) Using the universal covering $\widetilde{K}(n)$ of $K(n)$, we have shown in the paper that all the preinjective $K(n)$-modules are bristled. But we should note that the non-injective preinjective $\widetilde{K}(n)$-modules themselves are not bristled.

Proposition. *Let Y be an indecomposable preinjective $\widetilde{K}(n)$-module. If Y is injective, then Y is bristled. If Y is not injective, then Y is not bristled.*

PROOF. It is obvious that the indecomposable injective $\widetilde{K}(n)$-modules are bristled. Thus, we only have to consider the remaining indecomposable preinjective $\widetilde{K}(n)$-modules.

First, we consider the $\widetilde{K}(n)$-module $X = X(z) = \widetilde{\tau}S(z)$, where z is a source of $\widetilde{K}(n)$ (and $\widetilde{\tau}$ the Auslander-Reiten translation for $\widetilde{K}(n)$); note that X has been discussed already in 4.4. As before, we denote by $\alpha_i\colon z \to y_i$ the arrows starting at z. The thin indecomposable representation of $\widetilde{K}(n)$ with support $\alpha_i\colon z \to y_i$ will be denoted by B_i. Note that the modules B_i are the only bristles with top $S(z)$.

We claim that $\mathrm{Hom}(B_i, X) = 0$. Denote by Q' the full subquiver of $\widetilde{K}(n)$ with vertices $z, y_1, \dots, y_n$. The restriction X' of X to Q' is indecomposable (and not injective). Since B_i and all its factor modules are injective, when considered as representations of Q', we see that $\mathrm{Hom}(B_i, X') = 0$, thus also $\mathrm{Hom}(B_i, X) = 0$.

If $\mathcal{B}(X)$ is the maximal bristled submodule of X, then the top of $\mathcal{B}(X)$ has no composition factor of the form $S(z)$. As a consequence, $\mathcal{B}(X)$ is a proper submodule of X. Thus X is not bristled.

If Y is an indecomposable preinjective representation of $\widetilde{K}(n)$ which is not injective, then Y has a factor module of the form $X(z)$ for some z. Since $X(z)$ is not bristled, also Y cannot be bristled. □

Appendix C. A non-finiteness assertion for tame algebras.

(C.1) Proposition. *Let Λ be a tame hereditary artin algebra. There is no regular module which generates infinitely many indecomposable modules. In particular, there is no regular or preinjective module which generates all preinjective modules.*

PROOF. Obviously, a preinjective module generates only finitely many indecomposable modules (and these are preinjective again).

If R is simple regular and $t \in \mathbb{N}$, then the only regular modules generated by $R[t]$ are modules in the tube of R of regular length at most t, thus only finitely many isomorphism classes. This shows that a regular module generates only finitely many indecomposable regular modules.

Finally, let R be indecomposable regular and M indecomposable preinjective. Then $\dim \mathrm{Hom}(R, M) \le rs$ where r is the regular length of R and s is the maximal length of a simple regular module in the tube of R (use τ in order to reduce to the case that M is indecomposable injective; under this shift, r and s do not change). It follows that M is a factor module of R^{rs}, thus of bounded length. □

(C.2) In the case of the 2-Kronecker quiver, one may strengthen the assertion as follows:

Proposition. *Let $\Lambda = kK(2)$ and M in* $\mathrm{mod}\,\Lambda$. *Then the following assertions are equivalent:*

(i) *M generates infinitely many indecomposable modules.*
(ii) *M generates all indecomposable modules with the exception of at most finitely many preprojective modules.*
(iii) *M has an indecomposable preprojective direct summand.*

If the preprojective module P_t is a direct summand of M and t is minimal, then the only indecomposable modules which are not generated by M are the modules $P_0, \dots, P_{t-1}$.

The proof is left to the reader. □

References

[1] Philipp Fahr and Claus Michael Ringel, *A partition formula for Fibonacci numbers*, J. Integer Seq. **11** (2008), no. 1, Article 08.1.4, 9. MR2377570

[2] Philipp Fahr and Claus Michael Ringel, *Categorification of the Fibonacci numbers using representations of quivers*, J. Integer Seq. **15** (2012), no. 2, Article 12.2.1, 12. MR2872450

[3] Otto Kerner, *Representations of wild quivers*, Representation theory of algebras and related topics (Mexico City, 1994), CMS Conf. Proc., vol. 19, Amer. Math. Soc., Providence, RI, 1996, pp. 65–107. MR1388560

[4] Otto Kerner, *More representations of wild quivers*, Expository lectures on representation theory, Contemp. Math., vol. 607, Amer. Math. Soc., Providence, RI, 2014, pp. 35–55, DOI 10.1090/conm/607/12087. MR3204865

[5] Claus Michael Ringel, *Tame algebras and integral quadratic forms*, Lecture Notes in Mathematics, vol. 1099, Springer-Verlag, Berlin, 1984. MR774589

[6] Ringel, C. M., *The eigenvector variety of a matrix pencil.* Lin. Algebra and Appl. 531 (2017) 447-458, DOI 10.1016/j.laa.2017.05.004

[7] Ringel, C. M., *Quiver Grassmannians for Wild Acyclic Quivers.* Proceedings Amer. Math. Soc. March 2017. DOI 10.1090/proc/13882

Fakultät für Mathematik, Universität Bielefeld, D-33501 Bielefeld, Germany and Department of Mathematics, Shanghai Jiao Tong University, Shanghai 200240, PR China

Email address: `ringel@math.uni-bielefeld.de`

Contemporary Mathematics
Volume **705**, 2018
http://dx.doi.org/10.1090/conm/705/14200

Noetherian properties in representation theory

Steven V Sam

ABSTRACT. The goal of this expository article, based on a lecture I gave at the 2016 ICRA, is to explain some recent applications of "categorical symmetries" in topology and algebraic geometry with an eye toward twisted commutative algebras as a unifying framework. The general idea is to find an action of a category on the object of interest, prove some niceness property, like finite generation, and then deduce consequences from the general properties of the category. The key in these cases is to prove that the representation theory of this category is locally noetherian, and we will discuss an outline for such proofs.

1. Introduction

In the past few years, a number of stabilization phenomena, which uses representation theory in an essential way, have been observed in many different areas, including topology, algebraic geometry, group theory, and combinatorics. One striking feature is that these phenomena have been discovered independently by several groups of researchers working in these different areas, and while the language used by each is different, there are many underlying commonalities. Notably, one finds what we will call a "categorical symmetry" of the objects in question, and uses finite generation of this symmetry to deduce the desired stabilization of uniformity results.

The goal of this article is to present two such examples (one from topology and one from algebraic geometry) and to explain a possible unifying framework (twisted commutative algebras, aka tca's). While tca's are rather complicated objects at first glance, we hope that this presentation will motivate their study. We end with a brief discussion on future work that needs to be done.

This is by no means meant to be a survey of the subject. For that, one might consult [**Dr1**] or [**Fa**]. Instead, we aim to get the reader as quickly as possible to the philosophy behind "representation stability" and to get a feeling for how some of the results can be proven in some concrete examples.

Acknowledgements. This paper is based on a talk given at the 2016 International Conference on Representations of Algebras (ICRA) at Syracuse University,

2010 *Mathematics Subject Classification.* 14M99, 16P40, 18A25.

SS was partially supported by NSF grant DMS-1500069.

and I thank the organizers for the opportunity to present. I also thank Eric Ramos and Andrew Snowden for reading a draft of this article.

2. Two examples

2.1. Cohomology of configuration spaces. Let X be a "nice" (connected, orientable, and with the homotopy type of a finite CW complex) manifold of dimension ≥ 2. For $n \geq 0$, define the nth **configuration space** by

$$\mathrm{Conf}_n(X) = \{(x_1, \dots, x_n) \in X^n \mid x_i \neq x_j \text{ for } i \neq j\}.$$

Let $\mathbf{k}$ be a field. The following result can be found in [**Ch**] (when $\mathrm{char}(\mathbf{k}) = 0$) and [**CEFN**] (in general).

THEOREM 2.1 (Church, Church–Ellenberg–Farb–Nagpal). *With the notation above, the function*

$$n \mapsto \dim_{\mathbf{k}} \mathrm{H}^i(\mathrm{Conf}_n(X); \mathbf{k})$$

is given by a polynomial function in n for $n \gg 0$.

The symmetric group S_n acts freely on $\mathrm{Conf}_n(X)$, and one can also study the cohomology of the quotient, though the behavior depends on the characteristic. The following result can be found in [**Ch**] (when $\mathrm{char}(\mathbf{k}) = 0$) and [**Na**] (when $\mathrm{char}(\mathbf{k}) > 0$).

THEOREM 2.2 (Church, Nagpal). *Keep the notation above.*

- *When* $\mathrm{char}(\mathbf{k}) = 0$, $\dim_{\mathbf{k}} \mathrm{H}^i(\mathrm{Conf}_n(X)/S_n; \mathbf{k})$ *is constant for $n \gg 0$.*
- *When* $\mathrm{char}(\mathbf{k}) = p > 0$, $\dim_{\mathbf{k}} \mathrm{H}^i(\mathrm{Conf}_n(X)/S_n; \mathbf{k})$ *is periodic with period a power of p for $n \gg 0$.*

REMARK 2.3. These results can be extended to more general classes of spaces, see [**Pe**] and [**To**]. □

2.2. Syzygies of secants of Veronese embeddings. Let $\mathbf{k}$ be a field of characteristic 0. Let $B = \bigoplus_{n \geq 0} B_n$ be a $\mathbf{Z}$-graded $\mathbf{k}$-algebras with $B_0 = \mathbf{k}$, which is generated in degree 1, and satisfies $\dim_{\mathbf{k}} B_1 < \infty$. Define the dth Veronese subring to be

$$B^{(d)} = \bigoplus_{n \geq 0} B_{nd}.$$

This is also a $\mathbf{Z}$-graded $\mathbf{k}$-algebra which is generated in degree 1 and $\dim_{\mathbf{k}} B_1^{(d)} = \dim_{\mathbf{k}} B_d < \infty$. In particular, we have a surjection $\mathrm{Sym}(B_d) \to B^{(d)}$ where Sym denotes the symmetric algebra. Denote the kernel by $I_B^{(d)}$. The general philosophy is that $B^{(d)}$ becomes better behaved with respect to various homological invariants as $d \to \infty$ (see [**Ba**] and [**ERT**] for examples related to Koszulity).

The tensor product $\mathrm{Sym}(V) \otimes \mathrm{Sym}(V)$ is also an algebra via the multiplication

$$(p_1 \otimes p_2)(q_1 \otimes q_2) = p_1 q_1 \otimes p_2 q_2$$

on simple tensors (and extended linearly). We define a map

$$\Delta\colon \mathrm{Sym}(V) \to \mathrm{Sym}(V) \otimes \mathrm{Sym}(V)$$

on the degree 1 piece V by $v \mapsto v \otimes 1 + 1 \otimes v$ and extend it to higher degrees by requiring that Δ is a ring homomorphism. Given ideals $I, J \subset \mathrm{Sym}(V)$, their **join** is the ideal $I \star J$, which is the kernel of the composition

$$\mathrm{Sym}(V) \xrightarrow{\Delta} \mathrm{Sym}(V) \otimes \mathrm{Sym}(V) \to \mathrm{Sym}(V)/I \otimes \mathrm{Sym}(V)/J.$$

Finally, define the **secant ideals** of I by $\mathrm{Sec}^1 I = I$ and $\mathrm{Sec}^r I = \mathrm{Sec}^{r-1} I \star I$. The following is the main result of [**Sa1**].

THEOREM 2.4 (Sam). *Fix B as above, and an integer $r \geq 1$. There exists a constant $C_B(r)$ such that for all d, the ideal $\mathrm{Sec}^r I_B^{(d)}$ is generated in degrees $\leq C_B(r)$.*

For a geometric interpretation of this statement, consider the cases when $I, J \subseteq \mathrm{Sym}(V)$ are the homogeneous prime ideals defining projective varieties $V(I), V(J)$ in projective space $\mathbf{P}(V^*)$. Given a point v in the vector space V^*, let $[v] \in \mathbf{P}(V^*)$ denote the line it spans. Then

$$V(I \star J) = \overline{\{[x+y] \mid [x] \in V(I), [y] \in V(J)\}}.$$

To be more precise, we are taking the set of all lines spanned by elements of the form $x + y$ where the line spanned by x is in $V(I)$ and the line spanned by y is in $V(J)$, and the overline denotes Zariski closure. Also, if B is the homogeneous coordinate ring of $X \subset \mathbf{P}(V^*)$, then $V(I_B^{(d)}) \subset \mathbf{P}((\mathrm{Sym}^d V)^*)$ is the image of X under the dth Veronese map (the dth power map).

Phrased this way, Theorem 2.4 says that given an embedded projective variety $X \subseteq \mathbf{P}^n$, there is a bound $C_X(r)$ so that the full ideal of the rth secant variety of any Veronese re-embedding of X is generated in degree $\leq C_X(r)$.

REMARK 2.5. The above leads to a relaxation of the problem: instead of finding degree bounds for the generators of the full ideal of the rth secant variety, look for degree bounds for generators of any ideal whose radical is the full ideal of the rth secant variety (i.e., just find equations that define the same set as the rth secant variety). In fact, analogous theorems for this set-theoretic relaxation exist for the rth secant variety of arbitrary Segre products of projective spaces [**DK**] and for the rth secant of arbitrary Grassmannians [**DE**]. □

REMARK 2.6. The framework behind the theorem above can also be used to give bounds for higher-order syzygies of secant varieties, see [**Sa2**]. An earlier result for Segre varieties can be found in [**Sn**]. □

3. Twisted commutative algebras

Both results mentioned above are formal consequences of some algebraic structure being finitely generated. A possible framework for both structures is given by twisted commutative algebras. See [**SS1**] for an introduction.

3.1. The definition. Let $A = \bigoplus_{n \geq 0} A_n$ be a $\mathbf{Z}_{\geq 0}$-graded, associative, unital A_0-algebra such that each A_n has a linear action of the symmetric group S_n. Embed $S_n \times S_m$ as a subgroup of S_{n+m} by having S_n act on $1, \dots, n$ and S_m act on $n+1, \dots, n+m$ in the natural way. Then A is a **twisted commutative algebra (tca)** if, for all n and m,

- the multiplication map
$$A_n \otimes A_m \to A_{n+m}$$
is $S_n \times S_m$-equivariant with respect to the embedding just specified, and
- $\tau(xy) = yx$ where $x \in A_n$, $y \in A_m$, and $\tau \in S_{n+m}$ is the permutation that swaps $\{1, \dots, n\}$ and $\{n+1, \dots, n+m\}$ in order, i.e., $\tau(i) = m + i$ if $1 \leq i \leq n$ and $\tau(n+j) = j$ for $1 \leq j \leq m$.

A basic example is when E is a free $\mathbf{k}$-module, and $A_n = E^{\otimes n}$ (by convention, $A_0 = \mathbf{k}$). The action of S_n is by permuting tensor factors, and multiplication is concatenation of tensors. This is the free tca generated in degree 1 by E, and we will denote it by $\mathrm{Sym}(E\langle 1\rangle)$.

We can also define modules over tca's. Let A be a tca. An A-module M is a graded abelian group $M = \bigoplus_{n\geq 0} M_n$ such that each M_n is a linear S_n-representation, M is a graded A-module in the usual sense, and such that the multiplication map

$$A_n \otimes M_m \to M_{n+m}$$

is $S_n \times S_m$-equivariant. A module is **finitely generated** if it can be generated by finitely many elements $m_1, \dots, m_n$ under the action of A and the symmetric groups. We are interested in the category of A-modules, and specifically, in formal properties of finitely generated A-modules.

3.2. Categorical model. For the tca $A = \mathrm{Sym}(E\langle 1\rangle)$, we can give an alternate model for the category of A-modules. For simplicity, suppose that $\dim_{\mathbf{k}} E = d < \infty$ and choose a basis $E \cong \mathbf{k}^d$.

Define a category $\mathbf{FI}_d$ whose objects are finite sets S and whose morphisms $S \to T$ consist of pairs (f, g) where $f\colon S \to T$ is an injective function and $g\colon T \setminus f(S) \to \{1, \dots, d\}$ is an arbitrary function ("d-coloring"). Given another morphism $(f', g')\colon T \to U$, the composition $(f'', g'')\colon S \to U$ is obtained by taking $f'' = f' \circ f$ and $g''(u) = g'(u)$ for $u \in U \setminus f''(S)$. When $d = 1$, the function g is superfluous, and $\mathbf{FI}_1 = \mathbf{FI}$ is simply the category of finite sets and injective functions. Note that $\mathbf{FI}_d$ is equivalent to the full subcategory of finite sets of the form $[n] = \{1, \dots, n\}$ (with $[0] = \varnothing$).

Given a commutative ring $\mathbf{k}$, an **$\mathbf{FI}_d$-module** is a functor $M\colon \mathbf{FI}_d \to \mathrm{Mod}_{\mathbf{k}}$. Concretely, we have a $\mathbf{k}$-module $M(S)$ for every finite set S, and $\mathbf{k}$-linear maps $M(f, g)\colon M(S) \to M(T)$ for every $\mathbf{FI}_d$-morphism $(f, g)\colon S \to T$. Given the remark above, the data of M is completely encoded by the sequence of $\mathbf{k}$-modules $M_n = M([n])$ together with transition maps $M_n \to M_m$ given by the $\mathbf{FI}_d$-morphisms between $[n]$ and $[m]$. When $S = T$, a morphism is a bijection of the set S to itself, so M_n is a representation of the symmetric group S_n.

Given an $\mathbf{FI}_d$-module M, define an A-module structure on $\bigoplus_{n\geq 0} M_n$ as follows. A basis element of $A_n = E^{\otimes n} \cong (\mathbf{k}^d)^{\otimes n}$ can be identified with a function $[n] \to [d]$, and we define multiplication by this basis element to be the action of the morphism (f, g) where $f(i) = i + n$ and g is the function $[n] \to [d]$. We omit the (routine) check that this is well-defined.

PROPOSITION 3.1. *The functor just defined gives an equivalence between the category of $\mathbf{FI}_d$-modules (over $\mathbf{k}$) and the category of A-modules.*

While the categorical model $\mathbf{FI}_d$ is much easier to work with, we have given the general definition of tca since it suggests a more general algebraic framework. In particular, the category $\mathbf{FI}$ gives an algebraic structure on the cohomology of configuration spaces.

Let X be a topological space. Given a finite set S, let $\mathrm{Conf}_S(X)$ be the set of injective functions $S \to X$, topologized as a subspace of the set of all functions $S \to X$ (a direct product of $|S|$ copies of X). Given an injective function $f\colon S \to T$, there is an induced map $\mathrm{Conf}_T(X) \to \mathrm{Conf}_S(X)$ given by composition

$(T \to X) \mapsto (S \xrightarrow{f} T \to X)$. By contravariance, for any $i \geq 0$, and coefficient ring $\mathbf{k}$, one also gets a $\mathbf{k}$-linear map

$$f_*\colon \mathrm{H}^i(\mathrm{Conf}_S(X); \mathbf{k}) \to \mathrm{H}^i(\mathrm{Conf}_T(X); \mathbf{k}).$$

An alternative way to phrase this is that each choice of $X, i, \mathbf{k}$ as above gives a functor $\mathbf{FI} \to \mathrm{Mod}_{\mathbf{k}}$ which sends S to $\mathrm{H}^i(\mathrm{Conf}_S(X); \mathbf{k})$ and $f\colon S \to T$ to f_*. In the next section, we will see how to exploit this algebraic structure.

REMARK 3.2. The category **FI** was introduced and studied in [**CEF**] over a field of characteristic 0, and also in a completely different language, in [**SS2**]. See those papers for more applications. □

REMARK 3.3. There is no particular reason to favor symmetric groups over other sequences of groups, such as general linear groups. In fact, "linear versions" of tca's are studied in [**PS**]. These are not spelled out as such, but rather the categorical models (categories **VI**, **VIC**, and variants) of the simplest cases are the objects of study. □

3.3. Secant varieties. The structure behind the example of secant varieties is more involved, so we define the relevant category in the special case where $B = \mathbf{k}[x_1, \ldots, x_r]$ is a polynomial ring. See [**Sa2**] for more details and motivation.

Define

$$(\mathbf{Z}^r_{\geq 0})_d = \{(x_1, \ldots, x_r) \in \mathbf{Z}^r_{\geq 0} \mid x_1 + \cdots + x_r = d\}.$$

We think of this as a basis for $\mathrm{Sym}^d\, \mathbf{k}^r$, and we define addition componentwise

$$+\colon (\mathbf{Z}^r_{\geq 0})_d \times (\mathbf{Z}^r_{\geq 0})_e \to (\mathbf{Z}^r_{\geq 0})_{d+e}.$$

DEFINITION 3.4. Define the **Veronese category** $\mathcal{V}_r$ as follows. The objects of $\mathcal{V}_r$ are pairs $(d, m) \in \mathbf{Z}^2_{\geq 0}$ and a morphism $\alpha\colon (d, m) \to (e, n)$ consists of the following data:

- An order-preserving injection $\alpha_1\colon [m] \to [n]$,
- A function $\alpha_2\colon [n] \setminus \alpha_1([m]) \to (\mathbf{Z}^r_{\geq 0})_e$,
- A function $\alpha_3\colon [m] \to (\mathbf{Z}^r_{\geq 0})_{e-d}$.

In particular, $\mathrm{Hom}_{\mathcal{V}_r}((d, m), (e, n)) = \varnothing$ if $d > e$. Given another morphism $\beta\colon (e, n) \to (f, p)$, the composition $\beta \circ \alpha = \gamma\colon (d, m) \to (f, p)$ is defined by

- $\gamma_1 = \beta_1 \circ \alpha_1$,
- $\gamma_2\colon [p] \setminus \gamma_1([m]) \to (\mathbf{Z}^r_{\geq 0})_f$ is defined by:
  - if $i \in [p] \setminus \beta_1([n])$, then $\gamma_2(i) = \beta_2(i)$, and
  - if $i \in \beta_1([n] \setminus \alpha_1([m]))$, then $\gamma_2(i) = \alpha_2(i') + \beta_3(i')$ where i' is the unique preimage of i under β_1.
- $\gamma_3\colon [m] \to (\mathbf{Z}^r_{\geq 0})_{f-d}$ is defined by $\gamma_3(i) = \alpha_3(i) + \beta_3(\alpha_1(i))$. □

Given $c \in (\mathbf{Z}^r_{\geq 0})_d$, let x^c be the corresponding monomial in $x_1, \ldots, x_r$. The basic example of a functor on $\mathcal{V}_r$ is defined by $(d, m) \mapsto (\mathrm{Sym}^d\, \mathbf{k}^r)^{\otimes m}$. To a morphism $\alpha\colon (d, m) \to (e, n)$, we define

$$(\mathrm{Sym}^d\, \mathbf{k}^r)^{\otimes m} \to (\mathrm{Sym}^e\, \mathbf{k}^r)^{\otimes n}$$
$$f_1 \otimes \cdots \otimes f_m \mapsto g_1 \otimes \cdots \otimes g_n$$

where $g_i = f_j x^{\alpha_2(i)}$ if $i = \alpha(j)$, and $g_i = x^{\alpha_2(i)}$ if $i \notin \alpha([m])$.

The keypoint is that this functor encodes certain operations between all of the (non-commutative analogue of the) coordinate rings of the various Veronese embeddings of projective space. One can also work with the actual coordinate rings by introducing a symmetrized version of the Veronese category, but we will avoid discussing this technical point. Fortunately, the ideals of secant varieties (of a fixed secant degree) form a subrepresentation, and the point is to show that this subrepresentation is finitely generated. This will imply the desired result: suppose the subrepresentation is generated in bidegrees $(d_1, m_1), \dots, (d_k, m_k)$. Then for any e, the set of elements in bidegrees $(e, m_1), \dots, (e, m_k)$ that come from the generators will be generators for the ideal of the secant variety in the eth Veronese embedding; in particular, $\max(m_i)$ is the desired bound in Theorem 2.4.

REMARK 3.5. Modules of the Veronese category do not model modules over a twisted commutative algebra, but there is a relation. The full subcategory on objects of the form (d, m) for d fixed and m varying is the category $\mathbf{FI}_N$ where $N = |(\mathbf{Z}^r_{\geq 0})_d|$. Hence, the Veronese category can be interpreted as a certain way of combining many of these categories together. □

4. Noetherian properties

We can formalize the situations above as follows. Given a category $\mathcal{C}$, a $\mathcal{C}$-module (over a ring $\mathbf{k}$) is a functor from $\mathcal{C}$ to the category of (left) $\mathbf{k}$-modules. In our cases, the isomorphism classes of objects of $\mathcal{C}$ are indexed by either $\mathbf{Z}_{\geq 0}$ or $\mathbf{Z}^2_{\geq 0}$ and so one gets a graded or bigraded sequence of $\mathbf{k}$-modules. In general, these sequences probably have no interesting properties (or at least no easily usable ones) at all. The key to getting interesting properties is to prove a finite generation statement.

We say that a $\mathcal{C}$-module M is **finitely generated** if there exist finitely many elements $m_1, \dots, m_r$ with $m_i \in M(x_i)$ for objects x_i such that each element of $M(y)$ for any object y can be written as a linear combination $\sum_{i=1}^r \alpha_i M(f)(m_i)$ for morphisms $f \colon x_i \to y$. Alternatively, the smallest subfunctor of M that contains the elements $m_1, \dots, m_r$ is M. The properties mentioned in §2.1 are, in fact, properties for general finitely generated $\mathbf{FI}$-modules.

THEOREM 4.1.

(1) *Let M be a finitely generated $\mathbf{FI}$-module over a field $\mathbf{k}$. The function*

$$n \mapsto \dim_{\mathbf{k}} M_n$$

is given by a polynomial function in n for $n \gg 0$.

(2) *If $\mathbf{k}$ is a field of characteristic 0, then $n \mapsto \dim_{\mathbf{k}} M_n^{S_n}$ is constant for $n \gg 0$.*

(3) *If $\mathbf{k}$ is a field of characteristic $p > 0$, for each i, the function $n \mapsto \dim_{\mathbf{k}} \mathrm{H}^i(S_n; M_n)$ is periodic with period a power of p.*

Property (1) immediately implies Theorem 2.1 once we know that the $\mathbf{FI}$-module $S \mapsto \mathrm{H}^i(\mathrm{Conf}_S(X); \mathbf{k})$ is finitely generated. Further given this, property (2) implies the first point of Theorem 2.2 since $\mathrm{H}^i(\mathrm{Conf}_n(X)/S_n; \mathbf{k}) = \mathrm{H}^i(\mathrm{Conf}_n(X); \mathbf{k})^{S_n}$. The second point does not quite follow from property (3), instead we refer the reader to [**Na**] for details.

So we see that being finitely generated is a desirable property that we want of the $\mathbf{FI}$-modules that we encounter in §2.1. One way to compute these $\mathbf{FI}$-modules is using a version of the Leray spectral sequence (see [**CEFN**, §4] for details). Under the assumptions made, the initial page of this spectral sequence consists of **FI**-modules which can be shown directly to be finitely generated. Hence, our desired **FI**-modules are subquotients of finitely generated ones. Finite generation is clearly inherited by quotient modules, but in complete generality, this need not be inherited by submodules.

We say that a module is **noetherian** if all of its submodules are finitely generated. The following result is exactly what is needed to finish the argument sketched above. It was first proven in the language of tca's over a field of characteristic 0 in [**Sn**], re-proven (for $d = 1$ over characteristic 0) in [**CEF**], proven in general (for $d = 1$) in [**CEFN**], and finally in general in [**SS3**].

THEOREM 4.2. *If* $\mathbf{k}$ *is a noetherian ring, then every finitely generated* $\mathbf{FI}_d$*-module over* $\mathbf{k}$ *is noetherian.*

Given a category $\mathcal{C}$, we will say that the category of $\mathcal{C}$-modules is **locally noetherian** (over $\mathbf{k}$) if every finitely generated $\mathcal{C}$-module over $\mathbf{k}$ is noetherian. Which $\mathcal{C}$ are locally noetherian for every noetherian coefficient ring? To finish the discussion from before, we mention the Veronese category.

THEOREM 4.3 (Sam). *The category of modules of Veronese category over a noetherian ring* $\mathbf{k}$ *is locally noetherian. The same is true for the symmetrized version if we restrict to noetherian rings* $\mathbf{k}$ *which contain the rational numbers.*

4.1. Gröbner methods. A general setup for proving local noetherianity is developed in [**SS3**] using ideas from Gröbner bases and many examples of interest in representation stability can be found there.

We start by recalling a combinatorial proof of the Hilbert basis theorem (i.e., $\mathbf{k}[x_1, \ldots, x_n]$ is a noetherian ring when $\mathbf{k}$ is a field) using Gröbner bases. Pick an admissible order on the monomials, i.e., a well-order compatible with multiplication. Using the order, we can define initial ideals, and reduce the study of the ascending chain condition to monomial ideals. Now, the set of monomial ideals is naturally in bijection with the set of ideals in the poset $\mathbf{N}^r$. Thus noetherianity of $\mathbf{k}[x_1, \ldots, x_n]$ follows from noetherianity of the poset $\mathbf{N}^r$ (Dickson's lemma), which can be rephrased as the following combinatorial exercise: given infinitely many vectors $v_1, v_2, \ldots$ in $\mathbf{N}^r$, there exists $i < j$ with $v_i \leq v_j$, where $\leq$ means coordinate-by-coordinate comparison.

The same idea can be made to work to prove noetherianity of finitely generated $\mathcal{C}$-modules for various choices of $\mathcal{C}$. To illustrate this, we will discuss the case of $\mathbf{FI}_d$-modules. There are several reductions that one can make.

- Define the principal projective $\mathbf{FI}_d$-module generated in degree n to be the functor P_n given by $P_n(S) = \mathbf{k}[\mathrm{Hom}_{\mathbf{FI}_d}([n], S)]$, where $\mathbf{k}[X]$ means the free $\mathbf{k}$-module on the set X. The action of morphisms is by post-composition. By Yoneda's lemma, a morphism $P_n \to M$ is the same as choosing an element in $M([n])$. Hence, a module M is finitely generated if and only if it is a quotient of $P_{n_1} \oplus \cdots \oplus P_{n_r}$ for some finite r.
- If M and N are noetherian, then the same is true for $M \oplus N$.
- If M is a quotient of N and N is noetherian, then the same is true for M.
- Putting it all together, to prove that all finitely generated modules are noetherian, it suffices to prove this for P_n for all $n \geq 0$.

The P_n are the analogues of free modules, though they behave differently for different n (in the case of graded rings, P_n would simply be a shift of P_0, which is why it is sufficient in that case to prove noetherianity of just P_0).

To deal with this problem, one would like to reduce to working with *monomial subrepresentations* of P_n, i.e., those generated by the basis vectors corresponding to morphisms $[n] \to S$. An issue is that there are too few of these for a reduction to be possible: the presence of automorphisms in $\mathbf{FI}_d$ causes many monomials to generate each other. We won't detail this issue any further, but rather discuss how to resolve it.

Define an auxiliary category $\mathbf{OI}_d$ (O for ordered) as the category of totally ordered sets S where morphisms are pairs $(f, g)\colon S \to T$ as in the definition of $\mathbf{FI}_d$, except now $f\colon S \to T$ is required to be order-preserving. There is a forgetful functor $\Phi\colon \mathbf{OI}_d \to \mathbf{FI}_d$ and we can pullback $\mathbf{FI}_d$-modules to $\mathbf{OI}_d$-modules along Φ. Let P'_n be the corresponding principal projective for $\mathbf{OI}_d$: $P'_n(S) = \mathbf{k}[\mathrm{Hom}_{\mathbf{OI}_d}([n], S)]$. A key point is that $\Phi^*(P_n) \cong (P'_n)^{\oplus n!}$, where in the decomposition, an $\mathbf{FI}_d$-morphism f belongs to the copy indexed by the unique permutation σ of $[n]$ needed to make $f \circ \sigma$ an order-preserving map. Hence it will actually suffice to show that P'_n is a noetherian $\mathbf{OI}_d$-module.

To do this, one *can* reduce to the case of monomial subrepresentations. For this aspect, see [**SS3**, §4]. To deal with monomial subrepresentations, we can finally reduce to the following combinatorial problem. Put a partial ordering on the set $X_n = \amalg_{m \ge n} \mathrm{Hom}_{\mathbf{OI}_d}([n], [m])$ by $\varphi \le \varphi'$ if there exists ψ such that $\psi \circ \varphi = \varphi'$. We want to show that X_n is **well-ordered**, which means that given an infinite sequence $\varphi_1, \varphi_2, \dots$, there exists $i < j$ such that $\varphi_i \le \varphi_j$.

We encode morphisms as words in the alphabet $\Sigma = \{*, 1, 2, \dots, d\}$ as follows. Given a morphism $\varphi = (f, g)\colon [n] \to [m]$, create a word $w(\varphi)$ of length m whose ith letter is a $*$ if i is in the image of f, and is $g(i)$ otherwise. Then φ can be recovered from $w(\varphi)$, and furthermore, $\varphi \le \varphi'$ if and only if $w(\varphi)$ is a subsequence of $w(\varphi')$.

Given any set X, let $X^\star$ be the set of finite words in X, with the partial ordering given by the subsequence relation. Higman proved that this is well-ordered whenever X is finite (this can be generalized substantially, but we won't do it here):

THEOREM 4.4 (Higman's lemma). *If X is finite, then $X^\star$ is well-ordered.*

By the definition, a subposet of a well-ordered poset is itself also well-ordered, so this solves our problem.

This suggests using a language-theoretic approach to understanding morphisms in $\mathcal{C}$, and this is explored in [**SS3**, §5]. We will restrict ourselves to discussing $\mathbf{OI}_1 = \mathbf{OI}$ in this setting. Morphisms are encoded as words in $\{*, 1\}$ with n instances of $*$. The set of words that come from a subrepresentation can be shown to be a regular language (and hence the generating function that records the length of words is a rational function), but even further, what we call an *ordered language*, which implies that the function counting the number of words of a given length is eventually a polynomial in the length.

5. What next?

The big open problem is whether or not finitely generated tca's are noetherian. Some cases of tca's generated in degree 2 (and their skew-commutative analogues) are proved to be noetherian in [**NSS1**] and [**NSS2**] over a field of characteristic 0 (some related results can be found in [**Eg**]). An application of these algebras can be found in [**MW**].

A categorical model of one case can be given as follows. Let $\mathcal{C}$ be the category whose objects are sets and morphisms $(f, \Pi): S \to T$ consist of an injection $f: S \to T$ and a decomposition Π of $T \setminus f(S)$ into 2-element subsets (so $|S| \equiv |T| \pmod 2$). Then $\mathcal{C}$-modules are the same as modules over the free tca generated by a trivial element (with respect to the S_2-action) of degree 2.

One reason this case is so interesting is that it resists attempts to be solved using Gröbner methods. If we follow the outline above, we first define an ordered version $\mathcal{C}'$ of the category $\mathcal{C}$ by requiring that the sets are totally ordered and that f is order-preserving. It turns out the module category for $\mathcal{C}'$ is not locally noetherian! However, pullbacks of $\mathcal{C}$-modules to $\mathcal{C}'$ have much more structure which possibly could be utilized. One point here is that almost all known cases of successful Gröbner methods applications boil down to Higman's lemma or a variant of it, which might be considered a "linear" phenomenon or a "1-dimensional" phenomenon. The category $\mathcal{C}$ is "non-linear" in that it deals with graphs rather than words, but it's not clear how to make this more precise.

For any positive integer d, modules over the free tca generated by a trivial element of degree d can be modeled as functors on the following category which generalizes $\mathcal{C}$ above: the objects are finite sets S, and a morphism $S \to T$ consists of an injection $f: S \to T$ together with a decomposition of $T \setminus f(S)$ as a disjoint union of d-element subsets (so $|S| \equiv |T| \pmod d$ if a morphism exists). Almost nothing is known about these module categories when $d \geq 3$. However, over a field of characteristic 0, this can be transformed into a problem about the $\mathbf{GL}_\infty(\mathbf{C})$ action on $\mathrm{Sym}(\mathrm{Sym}^3 \mathbf{C}^\infty)$, in which case partial progress has been made in [**DES**]. In particular, [**DES**] shows that $\mathrm{Sym}(\mathrm{Sym}^3 \mathbf{C}^\infty)$ is "topologically noetherian", i.e., that all $\mathbf{GL}_\infty(\mathbf{C})$-equivariant ideals are finitely generated up to radical. The topological noetherianity result was recently extended to $\mathrm{Sym}(F(\mathbf{C}^\infty))$ for all finite length polynomial functors F by Draisma [**Dr2**]. For some applications of topological noetherianity to uniform bound problems in commutative algebra, see [**ESS**].

References

[Ba] Jörgen Backelin, *On the rates of growth of the homologies of Veronese subrings*, Algebra, algebraic topology and their interactions (Stockholm, 1983), Lecture Notes in Math., vol. 1183, Springer, Berlin, 1986, pp. 79–100, DOI 10.1007/BFb0075451. MR846440

[Ch] Thomas Church, *Homological stability for configuration spaces of manifolds*, Invent. Math. **188** (2012), no. 2, 465–504, DOI 10.1007/s00222-011-0353-4. MR2909770

[CEF] Thomas Church, Jordan S. Ellenberg, and Benson Farb, *FI-modules and stability for representations of symmetric groups*, Duke Math. J. **164** (2015), no. 9, 1833–1910, DOI 10.1215/00127094-3120274. MR3357185

[CEFN] Thomas Church, Jordan S. Ellenberg, Benson Farb, and Rohit Nagpal, *FI-modules over Noetherian rings*, Geom. Topol. **18** (2014), no. 5, 2951–2984, DOI 10.2140/gt.2014.18.2951. MR3285226

[DES] Harm Derksen, Rob H. Eggermont, Andrew Snowden, Topological noetherianity for cubic polynomials, `arXiv:1701.01849v1`.

[Dr1] Jan Draisma, *Noetherianity up to symmetry*, Combinatorial algebraic geometry, Lecture Notes in Math., vol. 2108, Springer, Cham, 2014, pp. 33–61, DOI 10.1007/978-3-319-04870-3_2. MR3329086

[Dr2] Jan Draisma, Topological Noetherianity of polynomial functors, `arXiv:1705.01419v1`.

[DE] Jan Draisma, Rob H. Eggermont, Plücker varieties and higher secants of Sato's Grassmannian, *J. Reine Angew. Math.*, to appear, `arXiv:1402.1667v3`.

[DK] Jan Draisma and Jochen Kuttler, *Bounded-rank tensors are defined in bounded degree*, Duke Math. J. **163** (2014), no. 1, 35–63, DOI 10.1215/00127094-2405170. MR3161311

[Eg] Rob H. Eggermont, *Finiteness properties of congruence classes of infinite-by-infinite matrices*, Linear Algebra Appl. **484** (2015), 290–303, DOI 10.1016/j.laa.2015.06.035. MR3385063

[ERT] David Eisenbud, Alyson Reeves, and Burt Totaro, *Initial ideals, Veronese subrings, and rates of algebras*, Adv. Math. **109** (1994), no. 2, 168–187, DOI 10.1006/aima.1994.1085. MR1304751

[ESS] Daniel Erman, Steven V Sam, Andrew Snowden, Generalizations of Stillman's conjecture via twisted commutative algebra; in preparation.

[Fa] Benson Farb, Representation stability, *Proceedings of the 2014 International Congress of Mathematicians. Volume II*, 1173–1196, `arXiv:1404.4065v1`.

[MW] Jeremy Miller, Jennifer C. H. Wilson, Higher order representation stability and ordered configuration spaces of manifolds, `arXiv:1611.01920v1`.

[Na] Rohit Nagpal, FI-modules and the cohomology of modular S_n-representations, `arXiv:1505.04294v1`.

[NSS1] Rohit Nagpal, Steven V. Sam, and Andrew Snowden, *Noetherianity of some degree two twisted commutative algebras*, Selecta Math. (N.S.) **22** (2016), no. 2, 913–937, DOI 10.1007/s00029-015-0205-y. MR3477338

[NSS2] Rohit Nagpal, Steven V Sam, Andrew Snowden, Noetherianity of some degree two twisted skew-commutative algebras, `arXiv:1610.01078v1`.

[Pe] Dan Petersen, *A spectral sequence for stratified spaces and configuration spaces of points*, Geom. Topol. **21** (2017), no. 4, 2527–2555, DOI 10.2140/gt.2017.21.2527. MR3654116

[PS] Andrew Putman, Steven V Sam, Representation stability and finite linear groups, `arXiv:1408.3694v2`.

[Sa1] Steven V. Sam, *Ideals of bounded rank symmetric tensors are generated in bounded degree*, Invent. Math. **207** (2017), no. 1, 1–21, DOI 10.1007/s00222-016-0668-2. MR3592755

[Sa2] Steven V. Sam, *Syzygies of bounded rank symmetric tensors are generated in bounded degree*, Math. Ann. **368** (2017), no. 3-4, 1095–1108, DOI 10.1007/s00208-016-1509-8. MR3673648

[SS1] Steven V Sam, Andrew Snowden, Introduction to twisted commutative algebras, `arXiv:1209.5122`.

[SS2] Steven V. Sam and Andrew Snowden, *GL-equivariant modules over polynomial rings in infinitely many variables*, Trans. Amer. Math. Soc. **368** (2016), no. 2, 1097–1158, DOI 10.1090/tran/6355. MR3430359

[SS3] Steven V. Sam and Andrew Snowden, *Gröbner methods for representations of combinatorial categories*, J. Amer. Math. Soc. **30** (2017), no. 1, 159–203, DOI 10.1090/jams/859. MR3556290

[Sn] Andrew Snowden, *Syzygies of Segre embeddings and Δ-modules*, Duke Math. J. **162** (2013), no. 2, 225–277, DOI 10.1215/00127094-1962767. MR3018955

[To] Philip Tosteson, Lattice spectral sequences and cohomology of configuration spaces, `arXiv:1612.06034v1`.

Department of Mathematics, University of Wisconsin, Madison, Wisconsin
Email address: `svs@math.wisc.edu`
URL: `http://math.wisc.edu/~svs/`

Contemporary Mathematics
Volume **705**, 2018
http://dx.doi.org/10.1090/conm/705/14193

Thick tensor ideals of right bounded derived categories of commutative rings

Ryo Takahashi

ABSTRACT. Let R be a commutative noetherian ring. Denote by $\mathsf{D}^-(R)$ the derived category of cochain complexes X of finitely generated R-modules with $\mathsf{H}^i(X) = 0$ for $i \gg 0$. Then $\mathsf{D}^-(R)$ has the structure of a tensor triangulated category with tensor product $\otimes^{\mathbf{L}}_R$ and unit object R. In this article, we study thick tensor ideals of $\mathsf{D}^-(R)$, i.e., thick subcategories closed under the tensor action by each object in $\mathsf{D}^-(R)$, and investigate the Balmer spectrum $\mathsf{Spc}\,\mathsf{D}^-(R)$ of $\mathsf{D}^-(R)$, i.e., the set of prime thick tensor ideals of $\mathsf{D}^-(R)$. The results of the author given in this article are all taken from [*Thick tensor ideals of right bounded derived categories*, arXiv:1611.02826v1].

CONTENTS

Introduction

This article makes a report of the series of lectures given by the author at the Workshop and International Conference on Representations of Algebras (ICRA 2016), which was held at Syracuse University in August, 2016. Those lectures are based on joint work with Hiroki Matsui. The complete proofs of the results in this article that are due to Matsui and the author are all stated in [**10**], together with more detailed information and other related results.

Tensor triangular geometry is a theory established by Balmer [**2**] at the beginning of the current century. Let $\mathcal{T} = (\mathcal{T}, \otimes, \mathbb{1})$ be an (essentially small) *tensor triangulated category*, that is, a triangulated category $\mathcal{T}$ equipped with symmetric

---

2010 *Mathematics Subject Classification.* Primary 13D09; Secondary 18D10, 18E30, 19D23.
Key words and phrases. (thick tensor) ideal, Balmer spectrum, derived category, specialization-closed subset, support.
The author was partly supported by JSPS Grants-in-Aid for Scientific Research 16K05098.

tensor product $\otimes$ and unit object $\mathbb{1}$. A (thick tensor) *ideal* of $\mathcal{T}$ is by definition a thick subcategory of $\mathcal{T}$ which is closed under the action of $\mathcal{T}$ by $\otimes$. A proper ideal $\mathcal{P}$ of $\mathcal{T}$ is called *prime* if it satisfies:

$$X \otimes Y \in \mathcal{P} \implies X \in \mathcal{P} \text{ or } Y \in \mathcal{P}.$$

Prime ideals of tensor triangulated categories turn out to behave similarly to prime ideals of commutative rings; both share a lot of analogous properties. Among other things, the *Balmer spectrum* $\mathsf{Spc}\,\mathcal{T}$ of $\mathcal{T}$, which is by definition the set of prime ideals of $\mathcal{T}$, has the structure of a topological space, corresponding to the fact that the Zariski spectrum $\operatorname{Spec} R$ of a commutative ring R has a Zariski topology. Tensor triangular geometry studies Balmer spectra and develops commutative-algebraic and algebro-geometric observations. It is related to a lot of areas of mathematics, including commutative algebra, algebraic geometry, stable homotopy theory, modular representation theory, motivic theory, noncommutative topology and symplectic geometry. As Balmer [**4**] addressed an invited lecture at the International Congress of Mathematicians (ICM) in 2010, tensor triangular geometry has been attracting a great deal of attention.

Let R be a commutative noetherian ring. Let $\mathsf{D}^-(R)$ be the right bounded derived category of finitely generated R-modules. It is then a routine to verify that

$$(\mathsf{D}^-(R), \otimes_R^{\mathbf{L}}, R)$$

is a tensor triangulated category. The main topics of the series of lectures at ICRA 2016 by the author concern the structure of the ideals of $\mathsf{D}^-(R)$ and the structure of the Balmer spectrum $\mathsf{Spc}\,\mathsf{D}^-(R)$ of $\mathsf{D}^-(R)$.

The organization of this article is as follows. In Section 1, we make a review of some general properties of tensor triangulated categories and their Balmer spectra, which have been obtained by Balmer. From Section 2 on, we study the tensor triangulated category $\mathsf{D}^-(R)$ for a commutative noetherian ring R. In Section 2, we completely classify the compactly generated thick tensor ideals and the cocompactly ones of $\mathsf{D}^-(R)$. In this section, it turns out that a thick tensor ideal is compactly generated if and only if it is cocompactly generated, and when this is the case, we simply call it compact. In Section 3, we explore the topological structure of the Balmer spectrum of $\mathsf{D}^-(R)$, and make correspondences among certain classes of thick tensor ideals and subsets of the Zariski and Balmer spectra. We introduce the notion of tame thick tensor ideals, and compare those ideals with compact and radical ones. In Section 4, we investigate what is going on in the case where R is a discrete valuation ring. As we observe later, everything is clarified when R is artinian. Hence, the case of discrete valuation rings should be the first nontrivial case, but in fact, it turns out that even in this case the structure of $\mathsf{D}^-(R)$ is highly complicated.

1. Tensor triangulated categories and Balmer spectra

In this section, we introduce some of Balmer's works on general tensor triangulated categories. All the materials in this section are taken from [**2–4**]. First of all, we recall the definition of a tensor triangulated category.

DEFINITION 1.1. A *tensor triangulated category* $(\mathcal{T}, \otimes, \mathbb{1})$ is a triangulated category $\mathcal{T}$ equipped with symmetric tensor product $\otimes$ and unit object $\mathbb{1}$. To be more precise, $\mathcal{T}$ is both a triangulated category and a symmetric monoidal category such that the triangulated and symmetric monoidal structures are compatible.

Here are several examples of a tensor triangulated category. Note that all of them are essentially small.

EXAMPLE 1.2.

(1) Let X be a (quasi-compact and quasi-separated) scheme. Denote by $\mathsf{D}^{\mathsf{perf}}(X)$ the derived category of perfect complexes of $\mathcal{O}_X$-modules. Then $(\mathsf{D}^{\mathsf{perf}}(X), \otimes^{\mathbf{L}}_{\mathcal{O}_X}, \mathcal{O}_X)$ is a tensor triangulated category.
(2) Let R be a commutative ring. Denote by $\mathsf{K}^{\mathsf{b}}(\mathsf{proj}\, R)$ the homotopy category of bounded complexes of finitely generated projective R-modules. Then $(\mathsf{K}^{\mathsf{b}}(\mathsf{proj}\, R), \otimes_R, R)$ is a tensor triangulated category. This is nothing but the affine case of (1).
(3) Let k be a field of positive characteristic, and G a finite group (scheme over k). Denote by $\underline{\mathsf{mod}}\, kG$ the stable category of finitely generated kG-modules. Then $(\underline{\mathsf{mod}}\, kG, \otimes_k, k)$ is a tensor triangulated category.
(4) Let k, G be as in (3). Denote by $\mathsf{D}^{\mathsf{b}}(\mathsf{mod}\, kG)$ the derived category of bounded complexes of finitely generated kG-modules. Then $(\mathsf{D}^{\mathsf{b}}(\mathsf{mod}\, kG), \otimes_k, k)$ is a tensor triangulated category.
(5) Let R be a commutative noetherian ring. Denote by $\mathsf{D}^{-}(\mathsf{mod}\, R)$ the derived category of homologically right bounded complexes of finitely generated R-modules. Then $(\mathsf{D}^{-}(\mathsf{mod}\, R), \otimes^{\mathbf{L}}_R, R)$ is a tensor triangulated category. This tensor triangulated category plays a main role in this article.

Next, we give the definitions of a (thick tensor) ideal and a Balmer spectrum. We recall here that a *thick* subcategory of a triangulated category is by definition a nonempty full subcategory which is closed under direct summands, shifts and cones.

DEFINITION 1.3. Let $\mathcal{T}$ be an essentially small tensor triangulated category.

(1) A thick subcategory $\mathcal{I}$ of $\mathcal{T}$ is a *(tensor) ideal* if it satisfies the following implication.
$$a \in \mathcal{T},\, x \in \mathcal{I} \implies a \otimes x \in \mathcal{I}.$$
This is an analogue of an ideal of a commutative ring.
(2) An ideal $\mathcal{I}$ of $\mathcal{T}$ is *radical* if $\mathcal{I} = \sqrt{\mathcal{I}}$, where
$$\sqrt{\mathcal{I}} := \{a \in \mathcal{T} \mid \underbrace{a \otimes \cdots \otimes a}_{n} \in \mathcal{I} \text{ for some } n > 0\}$$
is the *radical* of $\mathcal{I}$. These are analogues of a radical ideal and the radical of an ideal of a commutative ring, respectively.
(3) A proper ideal $\mathcal{P}$ of $\mathcal{T}$ is *prime* if it satisfies the following implication.
$$x \otimes y \in \mathcal{P} \implies x \in \mathcal{P} \text{ or } y \in \mathcal{P}.$$
This is an analogue of a prime ideal of a commutative ring.
(4) The *Balmer spectrum* of $\mathcal{T}$ is defined by:
$$\mathsf{Spc}\, \mathcal{T} = \{\text{Prime ideals of } \mathcal{T}\}.$$
This corresponds to the Zariski spectrum $\operatorname{Spec} R$ of a commutative ring R.

(5) The *Balmer support* of an object x of $\mathcal{T}$ is defined by:

$$\mathtt{supp}(x) = \{\mathcal{P} \in \mathtt{Spc}\,\mathcal{T} \mid x \notin \mathcal{P}\}.$$

This corresponds to the subset $\mathrm{V}(f) = \{\mathfrak{p} \in \operatorname{Spec} R \mid f \in \mathfrak{p}\}$ of $\operatorname{Spec} R$ for an element f of R. Note that the containment is opposite.

(6) We put

$$\mathtt{U}(x) := \mathtt{supp}(x)^{\mathtt{C}} = \{\mathcal{P} \in \mathtt{Spc}\,\mathcal{T} \mid x \in \mathcal{P}\}.$$

This corresponds to the subset $\mathrm{D}(f) = \{\mathfrak{p} \in \operatorname{Spec} R \mid f \notin \mathfrak{p}\}$ of $\operatorname{Spec} R$.

Throughout the rest of this article, we assume that all tensor triangulated categories are essentially small, so that we can always define their Balmer spectra.

We make the definitions of a maximal ideal and a minimal prime of a tensor triangulated category.

DEFINITION 1.4. Let $\mathcal{T}$ be a tensor triangulated category.

(1) An ideal of $\mathcal{T}$ is said to be a *maximal ideal* of $\mathcal{T}$ if it is a proper ideal of $\mathcal{T}$ which is maximal with respect to the inclusion relation. We denote by $\max(\mathtt{Spc}\,\mathcal{T})$ the set of maximal ideals of $\mathcal{T}$.
(2) An ideal of $\mathcal{T}$ is said to be a *minimal prime* of $\mathcal{T}$ if it is a prime ideal of $\mathcal{T}$ which is minimal with respect to the inclusion relation. We denote by $\min(\mathtt{Spc}\,\mathcal{T})$ the set of minimal primes of $\mathcal{T}$.

Each Balmer spectrum has the structure of a topological space such that the Balmer supports are closed subsets. We state this here together with several fundamental properties which will often be used later.

PROPOSITION 1.5 ([**2**]). *Let $\mathcal{T}$ be a tensor triangulated category.*

(1) $\mathtt{Spc}\,\mathcal{T}$ *is a topological space with an open basis* $\{\mathtt{U}(x)\}_{x\in\mathcal{T}}$.
(2) *Every proper ideal of $\mathcal{T}$ is contained in a maximal ideal.*
(3) *Maximal ideals of $\mathcal{T}$ are prime.*
(4) *Every prime ideal of $\mathcal{T}$ contains a minimal prime.*
(5) *For each $\mathcal{P} \in \mathtt{Spc}\,\mathcal{T}$ the closure $\overline{\{\mathcal{P}\}}$ of $\{\mathcal{P}\}$ is irreducible, and described as follows.*

$$\overline{\{\mathcal{P}\}} = \{\mathcal{Q} \in \mathtt{Spc}\,\mathcal{T} \mid \mathcal{Q} \subseteq \mathcal{P}\}. \tag{1.5.1}$$

Conversely, any nonempty irreducible closed subset of $\mathtt{Spc}\,\mathcal{T}$ *has this form.*
(6) *The open subset $\mathtt{U}(x)$ of $\mathtt{Spc}\,\mathcal{T}$ is quasi-compact for each $x \in \mathcal{T}$. Conversely, any nonempty quasi-compact open subset of* $\mathtt{Spc}\,\mathcal{T}$ *has this form.*
(7) *For an ideal $\mathcal{I}$ of $\mathcal{T}$ one has*

$$\sqrt{\mathcal{I}} = \bigcap_{\mathcal{I}\subseteq\mathcal{P}\in\mathtt{Spc}\,\mathcal{T}} \mathcal{P}.$$

The equality (1.5.1) corresponds to the equality

$$\overline{\{\mathfrak{p}\}} = \{\mathfrak{q} \in \operatorname{Spec} R \mid \mathfrak{q} \supseteq \mathfrak{p}\}$$

of subsets of $\operatorname{Spec} R$ for a commutative ring R and a prime ideal $\mathfrak{p}$ of R. Again, the containment is opposite.

Thus, ideals of tensor triangulated categories have a lot of similar properties to ideals of commutative rings.

For a full subcategory $\mathcal{X}$ of $\mathcal{T}$ and a subset S of $\mathsf{Spc}\,\mathcal{T}$, set

$$\mathsf{supp}\,\mathcal{X} = \bigcup_{x\in\mathcal{X}} \mathsf{supp}(x),$$

$$\mathsf{supp}^{-1} S = \{x \in \mathcal{T} \mid \mathsf{supp}(x) \in S\}.$$

The following theorem is a celebrated result due to Balmer [**2**, Theorem 4.10].

THEOREM 1.6 (Balmer (2005)). *Let $\mathcal{T}$ be a tensor triangulated category. Then there is a one-to-one correspondence*

$$\{\textit{Radical ideals of } \mathcal{T}\} \underset{\mathsf{supp}^{-1}}{\overset{\mathsf{supp}}{\underset{1-1}{\rightleftarrows}}} \{\textit{Thomason subsets of } \mathsf{Spc}\,\mathcal{T}\}.$$

Here, a subset A of a topological space X is said to be *Thomason* if one can write

$$A = \bigcup_{i\in I} B_i$$

for some family $\{B_i\}_{i\in I}$ of subsets of X such that $B_i^{\complement} = X \setminus B_i$ is a quasi-compact open subset. A subset C of X is said to be *specialization-closed* if it satisfies the implication

$$x \in C \implies \overline{\{x\}} \subseteq C.$$

We notice that this condition is equivalent to saying that C is a (possibly infinite) union of closed subsets. Therefore, a Thomason subset is always specialization-closed. The name of a Thomason subset comes from the fact that for a quasi-compact quasi-separated scheme X, Thomason [**13**] gives a complete classification of the ideals of $\mathsf{D}^{\mathsf{perf}}(X)$ in terms of the Thomason subsets of the underlying topological space of X.

Theorem 1.7 says that for a given tensor triangulated category $\mathcal{T}$ the understanding of the structure of the Balmer spectrum of $\mathcal{T}$ provides a complete classification of the radical ideals of $\mathcal{T}$. Since each ideal of $\mathcal{T}$ is the kernel of some tensor triangulated functor from $\mathcal{T}$ and vice versa, classifying ideals of $\mathcal{T}$ leads us to the understanding of the structure of tensor triangulated functors from $\mathcal{T}$. In this sense, the above theorem is quite meaningful.

For each tensor triangulated category $\mathcal{T}$ one can define the structure sheaf $\mathcal{O}_{\mathcal{T}}$ on $\mathcal{T}$, and then the Balmer spectrum $\mathsf{Spc}\,\mathcal{T}$ has the structure of a locally ringed space [**4**, Constructions 24 and 29]. More precisely, for each quasi-compact open subset U of $\mathsf{Spc}\,\mathcal{T}$ we define

$$\mathcal{T}(U) := (\mathcal{T}/\,\mathsf{supp}^{-1}(U^{\complement}))^{\natural},$$

where $(-)^{\natural}$ stands for the idempotent completion. Then it holds that

$$\mathsf{Spc}\,\mathcal{T}(U) \cong U.$$

The assignment $U \mapsto \mathrm{End}_{\mathcal{T}(U)}(\mathbb{1})$ induces a presheaf of commutative rings, and we define the structure sheaf $\mathcal{O}_{\mathcal{T}}$ on $\mathcal{T}$ as its sheafification. Thus we obtain a locally ringed space

$$\mathrm{Spec}\,\mathcal{T} := (\mathsf{Spc}\,\mathcal{T}, \mathcal{O}_{\mathcal{T}}).$$

The following theorem due to Balmer is also well-known. We refer the reader to [**2**, Theorem 6.3] and [**4**, Theorem 57]; see also [**3**, Proposition 6.11].

THEOREM 1.7 (Balmer (2005, 2010)).

(1) *Let X be a quasi-compact quasi-separated scheme. Then there is an isomorphism*

$$\operatorname{Spec} \mathsf{D}^{\mathsf{perf}}(X) \cong X$$

of locally ringed spaces.

(2) *Let k be a field of positive characteristic, and G a finite group (scheme over k). Then there are isomorphisms*

$$\operatorname{Spec} \mathsf{D}^{\mathsf{b}}(\mathsf{mod}\, kG) \cong \operatorname{Spec}^{\mathrm{h}} \mathsf{H}^{\bullet}(G,k),$$
$$\operatorname{Spec}(\underline{\mathsf{mod}}\, kG) \cong \operatorname{Proj} \mathsf{H}^{\bullet}(G,k)$$

of locally ringed spaces.

Here $\mathsf{H}^{\bullet}(G,k)$ stands for the group cohomology ring. For a graded-commutative ring A, we denote by $\operatorname{Spec}^{\mathrm{h}} A$ the set of homogeneous prime ideals of A, that is, homogeneous ideals $\mathfrak{p}$ of A such that for two homogeneous elements $x, y \in A$ one has

$$xy \in \mathfrak{p} \implies x \in \mathfrak{p} \text{ or } y \in \mathfrak{p}.$$

For a commutative nonnegatively graded ring R we denote by $\operatorname{Proj} R$ the set of homogeneous prime ideals of R that do not contain $R_+ = \bigoplus_{i>0} R_i$. Note that $\operatorname{Proj} \mathsf{H}^{\bullet}(G,k)$ is nothing but the (projective) support variety $\mathcal{V}_G(k)$.

The isomorphism in Theorem 1.7(1) says that a scheme X is reconstructed from its derived category $\mathsf{D}^{\mathsf{perf}}(X)$; see also [**1**]. This is actually because of the tensor structure of $\mathsf{D}^{\mathsf{perf}}(X)$. Indeed, only from the triangulated structure of $\mathsf{D}^{\mathsf{perf}}(X)$ the original scheme X cannot be reconstructed, since there are a lot of derived equivalences of nonsingular algebraic varieties (e.g. the Fourier–Mukai transformation).

The second isomorphism in Theorem 1.7(2) is obtained by restricting the first one. Key roles in the proof of Theorem 1.7 are played by the classification theorems of ideals due to Hopkins [**9**], Neeman [**11**], Thomason [**13**], Benson–Carlson–Rickard [**5**] and Friedlander–Pevtsova [**8**]; see also the works of Benson–Iyengar–Krause [**6**] and Benson–Iyengar–Krause–Pevtsova [**7**]. The Balmer spectra are described for some other tensor triangulated categories by several authors; details can be found in [**4**].

Let $(\mathcal{T}, \otimes, \mathbb{1})$ be a tensor triangulated category. Balmer [**3**] constructs a continuous map

$$\rho_{\mathcal{T}}^{\bullet} : \operatorname{Spc} \mathcal{T} \to \operatorname{Spec}^{\mathrm{h}} \mathrm{R}_{\mathcal{T}}^{\bullet},$$

which is given by

$$\rho_{\mathcal{T}}^{\bullet}(\mathcal{P}) := (f \in \mathrm{R}_{\mathcal{T}}^{\bullet} \mid \operatorname{cone}(f) \notin \mathcal{P}).$$

Here,

$$\mathrm{R}_{\mathcal{T}}^{\bullet} = \operatorname{Hom}_{\mathcal{T}}(\mathbb{1}, \Sigma^{\bullet}\mathbb{1})$$

is a graded-commutative ring.

It is seen that for $\mathcal{T} = \mathsf{K}^{\mathsf{b}}(\mathsf{proj}\, R)$ with R being a commutative ring we have $\mathrm{R}_{\mathcal{T}}^{\bullet} = R$, and it is also observed that for $\mathcal{T} = \mathsf{D}^{\mathsf{b}}(\mathsf{mod}\, kG)$ with k being a field k of positive characteristic and G being a finite group (scheme over k) we have $\mathrm{R}_{\mathcal{T}}^{\bullet} = \mathsf{H}^{\bullet}(G,k)$. It is shown by Balmer [**3**, Propositions 8.1 and 8.5] that the isomorphism in Theorem 1.7(1) in the affine case, and the first isomorphism in Theorem 1.7(2) are given by the map $\rho_{\mathcal{T}}^{\bullet}$ given above. Thus the following conjecture has been presented by Balmer [**4**, Conjecture 72] in his invited lecture at the International Congress of Mathematicians (ICM), which was held in 2010 at Hyderabad.

CONJECTURE 1.8 (Balmer, ICM 2010). The map $\rho^{\bullet}_{\mathcal{T}}$ is (locally) injective if $\mathcal{T}$ is algebraic as a triangulated category.[1]

Let $f : X \to Y$ be a continuous map of topological spaces. We say that f is *locally injective* at a point $x \in X$ if there exists a neighborhood N of x such that the restriction $f|_N$ of f on N is injective. The map f is called *locally injective* if for all points $x \in X$ it is locally injective at x. Also, recall that a triangulated category is called *algebraic* if it is described as the stable category of a Frobenius exact category.

It is known that the conjecture does not hold for a non-algebraic triangulated category; indeed, if $\mathcal{T}$ is the Spanier–Whitehead stable homotopy category $\mathrm{SH}^{\mathrm{fin}}$ of finite pointed CW-complexes, then $\rho^{\bullet}_{\mathcal{T}}$ is not injective; see [**4**, Theorem 51]. On the other hand, as we have seen above, the conjecture does hold for $\mathsf{K}^{\mathsf{b}}(\mathsf{proj}\, R)$ and $\mathsf{D}^{\mathsf{b}}(\mathsf{mod}\, kG)$.

Now we introduce some notation, which will be used throughout the rest of this article.

NOTATION 1.9.

(1) Let R be a commutative noetherian ring.
(2) We denote by $\operatorname{Spec} R$ the *Zariski spectrum* of R, namely, the set of prime ideals of R equipped with the Zariski topology.
(3) For an ideal I of R we define $\mathrm{V}(I)$ the set of prime ideals of R containing I, and put $\mathrm{D}(I) = \mathrm{V}(I)^{\complement} = \operatorname{Spec} R \setminus \mathrm{V}(I)$.
(4) The set of maximal ideals (respectively, minimal primes) of R is denoted by $\operatorname{Max} R$ (respectively, $\operatorname{Min} R$).
(5) We denote by $\mathsf{mod}\, R$ the category of finitely generated R-modules, and by $\mathsf{proj}\, R$ the full subcategory of $\mathsf{mod}\, R$ consisting of finitely generated projective R-modules.
(6) We denote by $\mathsf{D}^*(R)$ the derived category $\mathsf{D}^*(\mathsf{mod}\, R)$ of the abelian category $\mathsf{mod}\, R$, and by $\mathsf{K}^*(R)$ the homotopy category $\mathsf{K}^*(\mathsf{proj}\, R)$ of the additive category $\mathsf{proj}\, R$, where $* \in \{-, \mathrm{b}\}$. There are obvious inclusions
$$\mathsf{K}^{\mathsf{b}}(R) \subseteq \mathsf{D}^{\mathsf{b}}(R) \subseteq \mathsf{D}^{-}(R).$$
Taking projective resolutions induces an equivalence
$$\mathsf{D}^{-}(R) \cong \mathsf{K}^{-}(R)$$
of tensor triangulated categories. We will often identify $\mathsf{D}^{-}(R)$ with $\mathsf{K}^{-}(R)$ via this equivalence.

From the next section on, we will investigate the structure of $\mathsf{D}^{-}(R)$ as a tensor triangulated category. We close this section by giving several comments about how hard it is.

DIFFICULITIES FOR $\mathsf{D}^{-}(R)$. The tensor triangulated category $\mathsf{D}^{-}(R)$ possesses a lot of defects on its structure, compared with the other well-established tensor triangulated categories:

(1) $\mathsf{D}^{-}(R)$ does not have arbitrary products or coproducts. (However, it does have some specific infinite coproducts, which will somehow play a crucial role in the proofs of our results.)

---

[1]Precisely speaking, Balmer conjectures the (local) injectivity of $\rho^{\bullet}_{\mathcal{T}}$ if $\mathcal{T}$ is "algebraic enough", one of the candidates for which is that $\mathcal{T}$ is algebraic.

(2) $\mathsf{D}^-(R)$ is not closed under duals. For example, in the case where R is an algebra over a field k, $\mathsf{D}^-(R)$ is not closed under k-duals.
(3) In particular, $\mathsf{D}^-(R)$ is never rigid. Recall that a triangulated category $\mathcal{T}$ is called *rigid* if there exist an exact functor $D : \mathcal{T}^{\mathrm{op}} \to \mathcal{T}$ and a functorial isomorphism

$$\mathrm{Hom}_{\mathcal{T}}(a \otimes b, c) \cong \mathrm{Hom}_{\mathcal{T}}(a, D(b) \otimes c)$$

for $a, b, c \in \mathcal{T}$. In fact, $\mathsf{D}^-(R)$ is even never closed as a symmetric monoidal category. There are a lot of results on rigid tensor triangulated categories, but we cannot use them for $\mathsf{D}^-(R)$.
(4) One has

$$\mathsf{thick}_{\mathsf{D}^-(R)}\, R \neq \mathsf{D}^-(R).$$

Indeed, the left hand side coincides with $\mathsf{K}^{\mathsf{b}}(R)$. There are several results on tensor triangulated categories $(\mathcal{T}, \otimes, \mathbb{1})$ satisfying $\mathsf{thick}_{\mathcal{T}}\, \mathbb{1} = \mathcal{T}$, but they are not available for $\mathsf{D}^-(R)$.

Thus, results in the literature are quite limited on tensor triangulated categories that can be applied to our tensor triangulated category $\mathsf{D}^-(R)$.

2. Compactly and cocompactly generated thick tensor ideals of $\mathsf{D}^-(R)$

In this section, we classify compactly or cocompactly generated ideals of the tensor triangulated category $\mathsf{D}^-(R)$. We begin with recalling the definitions of compact and cocompact objects.

DEFINITION 2.1. Let $\mathcal{T}$ be a triangulated category.

(1) An object $M \in \mathcal{T}$ is called *compact* (respectively, *cocompact*) if the natural morphism

$$\bigoplus_{\lambda\in\Lambda} \mathrm{Hom}_{\mathcal{T}}(M, N_\lambda) \to \mathrm{Hom}_{\mathcal{T}}(M, \bigoplus_{\lambda\in\Lambda} N_\lambda)$$

$$\left(\text{respectively, } \bigoplus_{\lambda\in\Lambda} \mathrm{Hom}_{\mathcal{T}}(N_\lambda, M) \to \mathrm{Hom}_{\mathcal{T}}(\prod_{\lambda\in\Lambda} N_\lambda, M)\right)$$

is an isomorphism for all families $\{N_\lambda\}_{\lambda\in\Lambda}$ of objects of $\mathcal{T}$ such that the coproduct $\bigoplus_{\lambda\in\Lambda} N_\lambda$ (respectively, the product $\prod_{\lambda\in\Lambda} N_\lambda$) exists in $\mathcal{T}$.
(2) We denote by $\mathcal{T}^{\mathsf{c}}$ (respectively, $\mathcal{T}^{\mathsf{cc}}$) the full subcategory of $\mathcal{T}$ consisting of compact (respectively, cocompact) objects of $\mathcal{T}$.
(3) An ideal of $\mathcal{T}$ is said to be *compactly generated* (respectively, *cocompactly generated*) if it is generated by some compact (respectively, cocompact) objects of $\mathcal{T}$ as an ideal.

The following equalities hold for compactly and cocompactly generated ideals of $\mathsf{D}^-(R)$.

FACT 2.2. There are equalities

$$\mathsf{D}^-(R)^{\mathsf{c}} = \mathsf{K}^{\mathsf{b}}(R),$$
$$\mathsf{D}^-(R)^{\mathsf{cc}} = \mathsf{D}^{\mathsf{b}}(R).$$

The second equality is due to Oppermann–Stovicek [**12**, Theorem 18]. The first equality in the above fact is well-known, and actually proved along the same lines as in the proof of the fact that the compact objects of the unbounded derived category of all R-modules are the perfect complexes over R. More precisely, take a complex $X \in \mathsf{K}^-(R)^c$ and denote by d^n the n-th differential map for each $n \in \mathbb{Z}$. Let C^n be the cokernel of the map d^{n-1}. There is a natural morphism

$$X \to \bigoplus_{n \in \mathbb{Z}} C^n[-n],$$

which factors through a finite direct sum by the compactness of X. Thus $X \to C^n[-n]$ is a zero morphism for all $n \ll 0$, and X is quasi-isomorphic to a truncation of X.

Next, let us recall the definition of the (usual) support of a chain complex. Note that this notion is different from that of a Balmer support introduced in the previous section.

DEFINITION 2.3.

(1) Let $X \in \mathsf{D}^-(R)$ be a complex. The *support* of X is defined to be the union of the supports (as R-modules) of homologies of X. One has equalities

$$\begin{aligned} \operatorname{supph} X &= \bigcup_{i \in \mathbb{Z}} \operatorname{supph} \mathsf{H}^i(X) \\ &= \{\mathfrak{p} \in \operatorname{Spec} R \mid X_{\mathfrak{p}} \neq 0\} \qquad (2.3.1) \\ &= \{\mathfrak{p} \in \operatorname{Spec} R \mid \kappa(\mathfrak{p}) \otimes^{\mathbf{L}}_R X \neq 0\} \end{aligned}$$

of subsets of $\operatorname{Spec} R$, where $\kappa(\mathfrak{p})$ stands for the residue field $R_{\mathfrak{p}}/\mathfrak{p}R_{\mathfrak{p}}$ of the local ring $R_{\mathfrak{p}}$.

(2) For a full subcategory $\mathcal{X}$ of $\mathsf{D}^-(R)$, set

$$\operatorname{supph} \mathcal{X} = \bigcup_{X \in \mathcal{X}} \operatorname{supph} X.$$

It is easy to see that the following hold.
- $\operatorname{supph} \mathcal{X}$ is a specialization-closed subset of $\operatorname{Spec} R$.
- There is an equality $\operatorname{supph} \mathcal{X} = \operatorname{supph}(\mathsf{thick}^{\otimes} \mathcal{X})$.

Here, $\mathsf{thick}^{\otimes} \mathcal{X}$ stands for the ideal generated by $\mathcal{X}$, that is, the smallest ideal of $\mathsf{D}^-(R)$ containing $\mathcal{X}$.

(3) For a subset S of $\operatorname{Spec} R$, set

$$\langle S \rangle = \mathsf{thick}^{\otimes}_{\mathsf{D}^-(R)} \{R/\mathfrak{p} \mid \mathfrak{p} \in S\}.$$

The second equality in (2.3.1) holds even for unbounded complexes of non-finitely generated R-modules, while the third equality only holds for complexes in $\mathsf{D}^-(R)$. To see this, let X be a complex in $\mathsf{D}^-(R)$ and $\mathfrak{p}$ a prime ideal of R. Then, by virtue of Nakayama's lemma, we have

$$\begin{aligned} \sup(\kappa(\mathfrak{p}) \otimes^{\mathbf{L}}_R X) &= \sup(\kappa(\mathfrak{p}) \otimes^{\mathbf{L}}_{R_{\mathfrak{p}}} X_{\mathfrak{p}}) \\ &= \sup \kappa(\mathfrak{p}) + \sup X_{\mathfrak{p}} \\ &= \sup X_{\mathfrak{p}}. \end{aligned}$$

This proves the third equality in (2.3.1). (For a complex X, the *supremum* $\sup X$ of X is by definition the supremum of integers i such that $\mathsf{H}^i(X) \neq 0$.)

The following theorem is the first main result of this article.

THEOREM 2.4 ([**10**, Theorem 2.12]). *There is a one-to-one correspondence*

$$\left\{\begin{matrix}\textit{Cocompactly generated}\\ \textit{ideals of } \mathsf{D}^-(R)\end{matrix}\right\} \underset{\langle\rangle}{\overset{\text{supph}}{\underset{\longleftarrow}{\overset{\longrightarrow}{\scriptstyle 1-1}}}} \left\{\begin{matrix}\textit{Specialization-closed}\\ \textit{subsets of } \operatorname{Spec} R\end{matrix}\right\}.$$

Thus the cocompactly generated ideals of $\mathsf{D}^-(R)$ are completely classified.

In fact, this one-to-one correspondence is not just a bijection of sets. For ideals $\mathcal{X}, \mathcal{Y}$ of $\mathsf{D}^-(R)$, define $\mathcal{X} \wedge \mathcal{Y}$ and $\mathcal{X} \vee \mathcal{Y}$ by:

$$\begin{cases}\mathcal{X} \wedge \mathcal{Y} = \mathsf{thick}^{\otimes}\{X \otimes_R^{\mathbf{L}} Y \mid X \in \mathcal{X},\ Y \in \mathcal{Y}\},\\ \mathcal{X} \vee \mathcal{Y} = \mathsf{thick}^{\otimes}(\mathcal{X} \cup \mathcal{Y}).\end{cases}$$

It is then seen that for specialization-closed subsets A, B of $\operatorname{Spec} R$ there are equalities

$$\begin{cases}\langle A\rangle \wedge \langle B\rangle = \langle A \cap B\rangle,\\ \langle A\rangle \vee \langle B\rangle = \langle A \cup B\rangle.\end{cases}$$

Using these equalities, one can show that the set of cocompactly generated ideals of $\mathsf{D}^-(R)$ forms a lattice with join $\vee$ and meet $\wedge$, and that the bijections in the theorem are lattice isomorphisms; see [**10**, Proposition 2.17].

On the other hand, using the above theorem, we observe that the assignments $\mathcal{X} \mapsto \mathcal{X} \cap \mathsf{K}^{\mathsf{b}}(R)$ and $\mathsf{thick}^{\otimes}\,\mathcal{Y} \leftarrow\!\shortmid\, \mathcal{Y}$ make a one-to-one correspondence

$$\{\text{Cocompactly generated ideals of } \mathsf{D}^-(R)\} \rightleftarrows \{\text{Thick subcategories of } \mathsf{K}^{\mathsf{b}}(R)\};$$

see [**10**, Corollary 2.14].

To prove the theorem, we need to extend the Hopkins–Neeman smash nilpotence theorem as follows; see [**10**, Theorem 2.7].

LEMMA 2.5 (Generalized smash nilpotence). *Let $f : X \to Y$ be a morphism in $\mathsf{K}^-(R)$ such that $Y \in \mathsf{K}^{\mathsf{b}}(R)$. If $f \otimes_R \kappa(\mathfrak{p}) = 0$ for all prime ideals $\mathfrak{p}$ of R, then $f^{\otimes t} = 0$ for some integer $t > 0$.*

We do not state the proof of this lemma. Instead, we give several comments on the proof.

REMARK 2.6.

(1) If we assume further that $X \in \mathsf{K}^{\mathsf{b}}(R)$, then the assertion of the lemma is nothing but the original smash nilpotence due to Hopkins [**9**, Theorem 10] and Neeman [**11**, Theorem 1.1]. In the proof of the original smash nilpotence, one can reduce to the case where $X = R$ by replacing the morphism $f : X \to Y$ with a morphism $f' : R \to \mathbf{R}\mathrm{Hom}_R(X, Y)$ via the isomorphim

$$\mathrm{Hom}_{\mathsf{K}^{\mathsf{b}}(R)}(R, \mathbf{R}\mathrm{Hom}_R(X, Y)) \cong \mathrm{Hom}_{\mathsf{K}^{\mathsf{b}}(R)}(X, Y).$$

Thanks to this reduction, one can identify the morphism $f \in \mathrm{Hom}_{\mathsf{K}^{\mathsf{b}}(R)}(R, Y)$ with the element $f(1) \in \mathsf{H}^0 Y$, which plays a key role in the proof of the original smash nilpotence.

(2) We show and use the following statements; see [**10**, Lemmas 2.5 and 2.6].
   (a) Let $\mathcal{T}$ be a tensor triangulated category. Let f, g be a morphism in $\mathcal{T}$, and let $\mathcal{X}, \mathcal{Y}$ be full subcategories of $\mathcal{T}$. If $f \otimes \mathcal{X} = 0$ and $g \otimes \mathcal{Y} = 0$, then $(f \otimes g) \otimes (\mathcal{X} * \mathcal{Y}) = 0$.

(b) Let $\boldsymbol{x} = x_1, \ldots, x_n$ be a sequence of elements of R. Let f be a morphism in $\mathsf{K}^{-}(R)$. If $f \otimes_R R/(\boldsymbol{x}) = 0$, then $f^{\otimes 2^n} \otimes_R \mathrm{K}(\boldsymbol{x}) = 0$.

Here, $\mathcal{X} * \mathcal{Y}$ stands for the full subcategory of $\mathsf{K}^{-}(R)$ consisting of objects E such that there exists an exact triangle

$$X \to E \to Y \rightsquigarrow$$

in $\mathcal{T}$ with $X \in \mathcal{X}$ and $Y \in \mathcal{Y}$, and $\mathrm{K}(\boldsymbol{x})$ stands for the Koszul complex of R with respect to $\boldsymbol{x}$. The statement (b) is deduced by using (a).

(3) We need the assumption that $Y \in \mathsf{K}^{\mathrm{b}}(R)$ to have the equality

$$\mathrm{ann}_{R_\mathfrak{p}}(f_\mathfrak{p}) = \mathrm{ann}_R(f)_\mathfrak{p}$$

for all prime ideals $\mathfrak{p}$ of R. Here, the *annihilator* of a morphism $f : X \to Y$ in $\mathsf{D}^{-}(R)$ is defined by

$$\mathrm{ann}_R(f) := \{a \in R \mid af = 0 \text{ in } \mathsf{D}^{-}(R)\},$$

which is nothing but the kernel of the morphism $R \to \mathrm{Hom}_{\mathsf{D}^{-}(R)}(X, Y)$ given by $a \mapsto af$.

By virtue of the generalized smash nilpotence, we can prove the following key proposition. For an object X of $\mathsf{D}^{-}(R)$ we define the *annihilator* $\mathrm{ann}\, X$ of X as the annihilator of the identity morphism of X.

PROPOSITION 2.7 ([**10**, Proposition 2.9]). *Let $X, Y \in \mathsf{D}^{-}(R)$ be complexes. Then the implication*

$$\mathrm{V}(\mathrm{ann}\, X) \subseteq \mathrm{supph}\, Y \implies X \in \mathsf{thick}^{\otimes} Y$$

holds true.

Again, we do not state the proof of this proposition, and instead, we give some comments on it.

REMARK 2.8.

(1) For every $X \in \mathsf{D}^{-}(R)$ one has

$$\mathrm{V}(\mathrm{ann}\, X) \supseteq \mathrm{supph}\, X.$$

The equality holds if $X \in \mathsf{D}^{\mathrm{b}}(R)$.

(2) The original statement that is due to Hopkins and Neeman and corresponds to the proposition asserts that for perfect complexes X, Y over R the implication

$$\mathrm{supph}\, X \subseteq \mathrm{supph}\, Y \implies X \in \mathsf{thick}^{\otimes} Y$$

holds true; see [**11**, Lemma 1.2].

(3) Proposition 2.7 does not hold if $\mathrm{V}(\mathrm{ann}\, X)$ is replaced with $\mathrm{supph}\, X$ or if $\mathrm{supph}\, Y$ is replaced with $\mathrm{V}(\mathrm{ann}\, Y)$; we will see this in Remarks 3.17 and 4.2.

(4) In the proof of the proposition, we first take a truncation $Y' \in \mathsf{K}^{\mathrm{b}}(R)$ of Y such that $\mathrm{V}(\mathrm{ann}\, X)$ is contained in $\mathrm{supph}\, Y'$. Then we consider the morphism $R \to \mathrm{Hom}_R(Y', Y)$ sending $1 \in R$ to the inclusion morphism $Y' \to Y$. The stream of the proof is similar to [**11**, Lemma 1.2], but we need to make various modifications.

(5) In the proposition, we can replace the object Y of $\mathsf{D}^-(R)$ with any full subcategory $\mathcal{Y}$ of $\mathsf{D}^-(R)$. Indeed, we find an object $Y \in \mathcal{Y}$ such that $\operatorname{supph} Y$ contains all the prime ideals (minimally) containing $\operatorname{ann} X$. Then $\mathrm{V}(\operatorname{ann} X)$ is contained in $\operatorname{supph} Y$, and we can reduce to the case where the subcategory $\mathcal{Y}$ consists only of Y.

As a corollary of Proposition 2.7 we have the following result. This result will be used in the proof of Theorem 2.4, and several other places.

COROLLARY 2.9 ([**10**, Corollary 2.11 and Proposition 4.11]).

(1) *Let X be a complex in $\mathsf{D}^-(R)$. Then it holds that*

$$\operatorname{supph} X = \operatorname{Spec} R \iff \mathsf{thick}^{\otimes} X = \mathsf{D}^-(R).$$

(2) *Let I be an ideal of R, and let $\mathcal{X}$ be an ideal of $\mathsf{D}^-(R)$. Take a system of generators $\boldsymbol{x} = x_1, \dots, x_n$ of I. Then it holds that*

$$\mathrm{V}(I) \subseteq \operatorname{supph} \mathcal{X} \iff R/I \in \mathcal{X} \iff \mathrm{K}(\boldsymbol{x}) \in \mathcal{X}.$$

PROOF. (1) The implication ($\Leftarrow$) follows from the equalities

$$\operatorname{supph} X = \operatorname{supph}(\mathsf{thick}^{\otimes} X) = \operatorname{supph} \mathsf{D}^-(R) = \operatorname{Spec} R.$$

As for the implication ($\Rightarrow$), for all objects $M \in \mathsf{D}^-(R)$ one has that $\mathrm{V}(\operatorname{ann} M)$ is contained in $\operatorname{supph} X$. Hence M belongs to $\mathsf{thick}^{\otimes} X$ by Proposition 2.7.

(2) We have

$$\operatorname{supph} R/I = \mathrm{V}(\operatorname{ann} R/I) = \mathrm{V}(I) = \mathrm{V}(\operatorname{ann} \mathrm{K}(\boldsymbol{x})) = \operatorname{supph} \mathrm{K}(\boldsymbol{x}).$$

Using Proposition 2.7 completes the proof of the assertion. □

Now we can obtain the proof of the main result of this section.

Proof of Theorem 2.4. Let $\mathcal{X}$ be a cocompactly generated ideal of $\mathsf{D}^-(R)$. Then one can write $\mathcal{X} = \mathsf{thick}^{\otimes} \mathcal{C}$ for some full subcategory $\mathcal{C}$ of $\mathsf{D}^{\mathsf{b}}(R)$. What we want to show is the equality $\mathcal{X} = \langle \operatorname{supph} \mathcal{X} \rangle$. As to the inclusion ($\supseteq$), Corollay 2.9(2) implies that $R/\mathfrak{p}$ is in $\mathcal{X}$ for all $\mathfrak{p} \in \operatorname{supph} \mathcal{X}$. As for the inclusion ($\subseteq$), it suffices to show that $\mathcal{C}$ is contained in $\langle \operatorname{supph} \mathcal{X} \rangle = \langle \operatorname{supph} \mathcal{C} \rangle$. Pick an object $M \in \mathcal{C}$. Then M is a bounded complex of finitely generated R-modules, whence it belongs to $\mathsf{thick}\{R/\mathfrak{p} \mid \mathfrak{p} \in \operatorname{supph} M\}$. Now we are done. □

As a corollary of Theorem 2.4 we have the following result.

COROLLARY 2.10 ([**10**, Corollary 2.15]). *The following are equivalent for an ideal $\mathcal{X}$ of $\mathsf{D}^-(R)$.*

(1) *$\mathcal{X}$ is compactly generated.*
(2) *$\mathcal{X}$ is cocompactly generated.*

When this is the case, we simply say that $\mathcal{X}$ is compact.

PROOF. Since $\mathsf{K}^{\mathsf{b}}(R)$ is contained in $\mathsf{D}^{\mathsf{b}}(R)$, compact generation implies cocompact generation. Therefore (1) implies (2). Let us show that (2) implies (1). Let W be a specialization-closed subset of $\operatorname{Spec} R$. Put

$$\mathcal{A} := \mathsf{thick}^{\otimes}\{R/\mathfrak{p} \mid \mathfrak{p} \in W\},$$
$$\mathcal{B} := \mathsf{thick}^{\otimes}\{\mathrm{K}(\boldsymbol{x}) \mid \boldsymbol{x}R \in W\}.$$

Then $\mathcal{A}$ is cocompactly generated, while $\mathcal{B}$ is compactly generated. We see that $\operatorname{supph} \mathcal{A} = \operatorname{supph} \mathcal{B} = W$. Using Theorem 2.4, we obtain $\mathcal{A} = \mathcal{B}$. □

As another corollary of Proposition 2.7, we get the following result.

COROLLARY 2.11 ([**10**, Corollary 2.19]). *If R is artinian, then all ideals of $\mathsf{D}^-(R)$ are compact. Therefore one has a one-to-one correspondence*

$$\{\textit{Ideals of } \mathsf{D}^-(R)\} \rightleftarrows \{\textit{Subsets of } \operatorname{Spec} R\}.$$

PROOF. The second assertion is deduced from the first one. To prove that all ideals $\mathcal{X}$ of $\mathsf{D}^-(R)$ are compact, it suffices to verify that the equality $\mathcal{X} = \langle \operatorname{supph} \mathcal{X} \rangle$ holds. The inclusion ($\supseteq$) follows from Corollary 2.9(2). To show the inclusion ($\subseteq$), we may assume that $\mathcal{X}$ consists of a single object X of $\mathsf{D}^-(R)$. Write the set of maximal ideals of R as follows.

$$\operatorname{Max} R = \{\underbrace{\mathfrak{m}_1, \dots, \mathfrak{m}_s}_{\in \operatorname{supph} X}, \underbrace{\mathfrak{m}_{s+1}, \dots, \mathfrak{m}_n}_{\notin \operatorname{supph} X}\}.$$

Letting t be the Loewy length of R, we have $\mathfrak{m}_1^t \cdots \mathfrak{m}_n^t = 0$. The Chinese remainder theorem then implies that there is an isomorphism

$$R \cong R/\mathfrak{m}_1^t \oplus \cdots \oplus R/\mathfrak{m}_n^t.$$

Taking the tensor product of X with this, we get an isomorphism

$$\begin{aligned} X \cong (X \otimes_R^{\mathbf{L}} R/\mathfrak{m}_1^t) &\oplus \cdots \oplus (X \otimes_R^{\mathbf{L}} R/\mathfrak{m}_s^t) \\ &\oplus (X \otimes_R^{\mathbf{L}} R/\mathfrak{m}_{s+1}^t) \oplus \cdots \oplus (X \otimes_R^{\mathbf{L}} R/\mathfrak{m}_n^t). \end{aligned}$$

We have the equality $\operatorname{supph}(X \otimes_R^{\mathbf{L}} R/\mathfrak{m}_i^t) = \operatorname{supph} X \cap \{\mathfrak{m}_i\}$ for each i, which shows that $X \otimes_R^{\mathbf{L}} R/\mathfrak{m}_i^t = 0$ for all $s+1 \le i \le n$. Hence X is isomorphic to the direct sum $(X \otimes_R^{\mathbf{L}} R/\mathfrak{m}_1^t) \oplus \cdots \oplus (X \otimes_R^{\mathbf{L}} R/\mathfrak{m}_s^t)$, and it is obtained that

$$\begin{aligned} X &\in \mathsf{thick}^{\otimes}\{R/\mathfrak{m}_1^t, \dots, R/\mathfrak{m}_s^t\} \\ &\subseteq \mathsf{thick}^{\otimes}\{R/\mathfrak{m}_1, \dots, R/\mathfrak{m}_s\} = \langle \operatorname{supph} X \rangle. \end{aligned}$$

Here, the inclusion relation follows from the fact that for each maximal ideal $\mathfrak{m}_i$ there exists an exact sequence

$$0 \to \mathfrak{m}_i^k/\mathfrak{m}_i^{k+1} \to R/\mathfrak{m}_i^{k+1} \to R/\mathfrak{m}_i^k \to 0$$

for all $k \ge 0$, and $\mathfrak{m}_i^k/\mathfrak{m}_i^{k+1}$ is a direct sum of copies of $R/\mathfrak{m}_i$. Now the proof of the corollary is completed. □

3. The Balmer spectrum of $\mathsf{D}^-(R)$ and classifications of thick tensor ideals

In this section, we consider the structure of the Balmer spectrum of $\mathsf{D}^-(R)$, and make correspondences among some classes of ideals of $\mathsf{D}^-(R)$ and subsets of $\operatorname{Spec} R$ and $\mathsf{Spc}\,\mathsf{D}^-(R)$. The section consists of three subsections.

3.1. The structure of $\mathsf{Spc}\,\mathsf{D}^-(R)$.

We investigate the structure of the Balmer spectrum of $\mathsf{D}^-(R)$ as a topological space, comparing it with the Zariski spectrum of R. We start by defining a tame ideal of $\mathsf{D}^-(R)$.

DEFINITION 3.1.

(1) For a subset S of $\operatorname{Spec} R$, we define the full subcategory $\operatorname{supph}^{-1} S$ of $\mathsf{D}^-(R)$ by

$$\operatorname{supph}^{-1} S = \{X \in \mathsf{D}^-(R) \mid \operatorname{supph} X \subseteq S\}.$$

One easily sees that $\operatorname{supph}^{-1} S$ is an ideal of $\mathsf{D}^-(R)$, and furthermore, the following equalities hold.
- $\operatorname{supph}^{-1} S = \operatorname{supph}^{-1} S_{\mathsf{spcl}}$.
- $\operatorname{supph}(\operatorname{supph}^{-1} S) = S_{\mathsf{spcl}}$.

Here, S_{spcl} stands for the largest specialization-closed subset of $\operatorname{Spec} R$ contained in S. (This is the spcl-*interior* of S in $\operatorname{Spec} R$ if we use the terminology in the next Subsection 3.2.)

(2) An ideal $\mathcal{X}$ of $\mathsf{D}^-(R)$ is called *tame* if $\mathcal{X} = \operatorname{supph}^{-1} S$ for some subset S of $\operatorname{Spec} R$. We set

$$^{\mathsf{t}}\mathsf{Spc}\,\mathsf{D}^-(R) = \{\text{tame prime ideals of } \mathsf{D}^-(R)\}.$$

One can construct the following correspondence between $\operatorname{Spec} R$ and $\mathsf{Spc}\,\mathsf{D}^-(R)$; see [**10**, Propositions 3.4 and 3.7].

PROPOSITION 3.2.

(1) *For* $\mathfrak{p} \in \operatorname{Spec} R$, *the full subcategory*

$$\mathcal{S}(\mathfrak{p}) := \{X \in \mathsf{D}^-(R) \mid X_{\mathfrak{p}} = 0\}$$

of $\mathsf{D}^-(R)$ *is a prime ideal of* $\mathsf{D}^-(R)$

(2) *For* $\mathcal{P} \in \mathsf{Spc}\,\mathsf{D}^-(R)$, *the set*

$$\{I \subseteq R \mid R/I \notin \mathcal{P}\}$$

of ideals of R *has a unique maximal element* $\mathfrak{s}(\mathcal{P})$ *with respect to the inclusion relation, which is a prime ideal of* R

Concerning the correspondence constructed in the above proposition, the following statements hold.

THEOREM 3.3 ([**10**, Theorems 3.9, 4.5, 4.7, 4.12 and 4.14]).

(1) *One has the order-reversing maps*

$$\mathcal{S} : \operatorname{Spec} R \rightleftarrows \mathsf{Spc}\,\mathsf{D}^-(R) : \mathfrak{s}$$

such that

$$\begin{cases} \mathfrak{s} \cdot \mathcal{S} = 1, \\ \mathcal{S} \cdot \mathfrak{s} = \operatorname{supph}^{-1} \operatorname{supph}. \end{cases}$$

In particular, the inequality

$$\dim(\mathsf{Spc}\,\mathsf{D}^-(R)) \geq \dim R$$

between the Krull dimensions holds.

(2) *The subset* ${}^{\mathsf{t}}\mathsf{Spc}\,\mathsf{D}^{-}(R)$ *of* $\mathsf{Spc}\,\mathsf{D}^{-}(R)$ *is dense. Moreover, there is a commutative diagram*

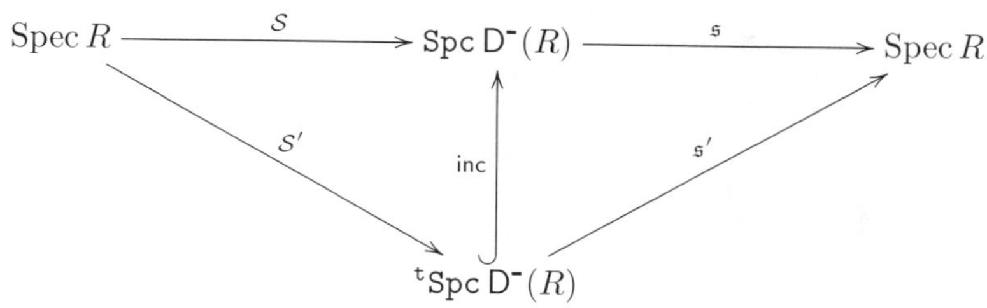

such that $\mathcal{S}'$ *is an open bijection,* $\mathfrak{s}'$ *is a continuous bijection and* $\mathfrak{s}$ *is a continuous map. In particular, the image of* $\mathcal{S}$ *coincides with* ${}^{\mathsf{t}}\mathsf{Spc}\,\mathsf{D}^{-}(R)$.

(3) *There is a commutative diagram*

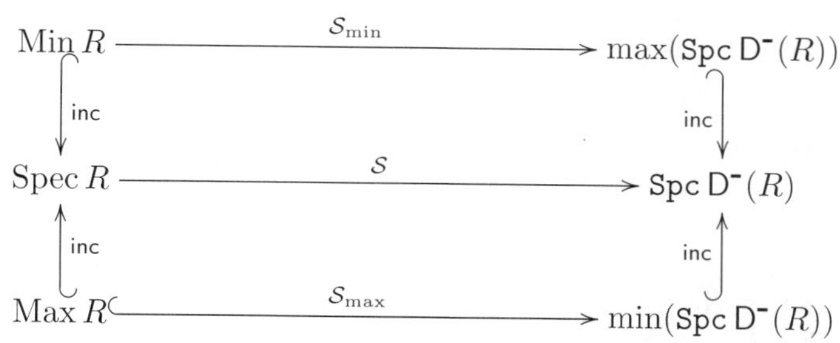

such that $\mathcal{S}_{\min}$ *is a homeomorphism, and the injective map* $\mathcal{S}_{\max}$ *is also a homeomorphism if* R *is a semilocal ring.*

(4) *The following are equivalent.*
  (a) *The map* $\mathcal{S}$ *is continuous.*
  (b) *The map* $\mathcal{S}'$ *is homeomorphic.*
  (c) *The map* $\mathfrak{s}'$ *is homeomorphic.*
  (d) *The set* $\operatorname{Spec} R$ *is finite.*

Here are several comments on this theorem.

REMARK 3.4.

(1) Recall that for a topological space X the *Krull dimension* $\dim X$ of X is by definition the supremum of the lengths of chains of nonempty irreducible closed subsets of X. For a tensor triangulated category $\mathcal{T}$ we have

$$\begin{aligned}\dim(\mathsf{Spc}\,\mathcal{T}) &= \sup\{n \geq 0 \mid \exists\,\text{chain } \overline{\{\mathcal{P}_0\}} \subsetneq \cdots \subsetneq \overline{\{\mathcal{P}_n\}} \text{ of subsets of } \mathsf{Spc}\,\mathcal{T}\}\\ &= \sup\{n \geq 0 \mid \exists\,\text{chain } \mathcal{P}_0 \subsetneq \cdots \subsetneq \mathcal{P}_n \text{ of points of } \mathsf{Spc}\,\mathcal{T}\}.\end{aligned}$$

(2) Note that $\operatorname{Max} R$, $\operatorname{Min} R$, $\max(\mathsf{Spc}\,\mathcal{T})$ and $\min(\mathsf{Spc}\,\mathcal{T})$ are all T_1-spaces, and that, in general, any finite subset of a T_1-space is closed. Thus, to show Theorem 3.3(3), it is enough to check that the top and bottom horizontal maps are bijective and injective, respectively (after we verify that they are induced).

(3) The following are equivalent ([**10**, Lemma 4.6]).
  - All specialization-closed subsets of $\operatorname{Spec} R$ are closed.
  - There are only finitely many specialization-closed subsets of $\operatorname{Spec} R$.
  - There are only finitely many closed subsets of $\operatorname{Spec} R$.
  - There are only finitely many prime ideals of R.

Using this equivalences, we can deduce Theorem 3.3(4).

(4) More precisely than Theorem 3.3(1), we actually have a commutative diagram

$$\begin{array}{ccccc} & & {}^{t}\mathsf{Spc}\,\mathsf{D}^-(R) & & \\ & \nearrow^{\mathcal{S}'} & \theta\Big\downarrow \mathsf{inc} & \searrow^{\mathfrak{s}'} & \\ \operatorname{Spec} R & \xrightarrow{\ \mathcal{S}\ } & \mathsf{Spc}\,\mathsf{D}^-(R) & \xrightarrow{\ \mathfrak{s}\ } & \operatorname{Spec} R \\ & \searrow_{\widetilde{\mathcal{S}}} & \pi\Big\downarrow \mathsf{can} & \nearrow_{\widetilde{\mathfrak{s}}} & \\ & & \mathsf{Spc}\,\mathsf{D}^-(R)/\operatorname{supph} & & \end{array}$$

such that $\mathfrak{s}\mathcal{S}$ is identity, $\mathcal{S}', \widetilde{\mathcal{S}}, \mathfrak{s}', \widetilde{\mathfrak{s}}, \pi\theta$ are bijections, $\mathcal{S}', \widetilde{\mathcal{S}}$ are open and closed, and $\mathfrak{s}, \mathfrak{s}', \widetilde{\mathfrak{s}}$ are continuous ([**10**, Theorem 4.5]). Here, $\mathsf{Spc}\,\mathsf{D}^-(R)/\operatorname{supph}$ stands for the quotient topological space by the equivalence relation induced by taking $\operatorname{supph}(-)$, and π the canonical surjection. (To be more precise, we define a relation $\sim$ in $\mathsf{Spc}\,\mathsf{D}^-(R)$ by

$$\mathcal{P} \sim \mathcal{Q} \quad \Longleftrightarrow \quad \operatorname{supph}\mathcal{P} = \operatorname{supph}\mathcal{Q}$$

for $\mathcal{P}, \mathcal{Q} \in \mathsf{Spc}\,\mathsf{D}^-(R)$. Then $\sim$ is an equivalence relation in $\mathsf{Spc}\,\mathsf{D}^-(R)$. We denote by $\mathsf{Spc}\,\mathsf{D}^-(R)/\sim$ the set of equivalence classes, and by $\pi : \mathsf{Spc}\,\mathsf{D}^-(R) \to \mathsf{Spc}\,\mathsf{D}^-(R)/\sim$ the map sending each $\mathcal{P} \in \mathsf{Spc}\,\mathsf{D}^-(R)$ to its equivalence class $[\mathcal{P}] \in \mathsf{Spc}\,\mathsf{D}^-(R)/\sim$. The set $\mathsf{Spc}\,\mathsf{D}^-(R)/\sim$ is a topological space, where a subset S of $\mathsf{Spc}\,\mathsf{D}^-(R)/\sim$ is open if and only if $\pi^{-1}(S)$ is an open subset of $\mathsf{Spc}\,\mathsf{D}^-(R)$.)

(5) More precisely than Theorem 3.3(4), the following assertion holds true ([**10**, Theorem 4.7]). Consider the three conditions
   (a) The map $\widetilde{\mathcal{S}}$ is a homeomorphism,
   (b) The map $\widetilde{\mathfrak{s}}$ is a homeomorphism,
   (c) The map $\pi\theta$ is a homeomorphism.

   Then (a) is equivalent to (b), and (b)$\wedge$(c) is equivalent to the four conditions in Theorem 3.3(4).

(6) We do not know whether $\mathcal{S}_{\max}$ is surjective even if R is not semilocal.

Suppose that R is artinian. Then R is semilocal, has only finitely many prime ideals and satisfies $\operatorname{Max} R = \operatorname{Spec} R = \operatorname{Min} R$. Hence, applying Theorem 3.3, we obtain the following corollary.

COROLLARY 3.5. *Let R be an artinian ring. Then the following statements hold true.*

(1) *The maps* $\mathcal{S} : \operatorname{Spec} R \rightleftarrows \mathsf{Spc}\,\mathsf{D}^-(R) : \mathfrak{s}$ *are mutually inverse homeomorphisms.*
(2) *One has* $\dim \mathsf{Spc}\,\mathsf{D}^-(R) = \dim R = 0 < \infty$.
(3) *All prime ideals of* $\mathsf{D}^-(R)$ *are tame.*

In fact, a more complete statement than the above corollary holds true; see Theorem 3.14.

Next, let us consider the locality of the Balmer spectrum of $\mathsf{D}^-(R)$ as a topological space.

DEFINITION 3.6.

(1) A topological space X is called *local* if for any open cover

$$X = \bigcup_{i \in I} U_i$$

of X there exists $t \in I$ such that $X = U_t$. In particular, a local topological space is quasi-compact.

(2) A tensor triangulated category $\mathcal{T}$ is said to be *local* if the topological space $\mathsf{Spc}\,\mathcal{T}$ is local.

Balmer [**3**, Proposition 4.2] obtains criteria for a given tensor triangulated category to be local.

PROPOSITION 3.7 (Balmer (2010)). *The following are equivalent for a tensor triangulated category $\mathcal{T}$.*

(1) *$\mathcal{T}$ is a local tensor triangulated category.*
(2) *$\mathcal{T}$ has a unique minimal prime.*
(3) *$\sqrt{\mathbf{0}}$ is a prime ideal of $\mathcal{T}$.*

If $\mathcal{T}$ is rigid, then (1)–(3) *are also equivalent to the following condition.*

(4) *$\mathbf{0}$ is a prime ideal of $\mathcal{T}$.*

Combining Theorem 3.3(3) and Proposition 3.7 yields that the tensor triangulated category $\mathsf{D}^-(R)$ is local if and only if the ring R is local. We should remark that $\mathsf{D}^-(R)$ is not rigid, as we stated in the previous section. Thus the last assertion of Proposition 3.7 is not available for $\mathcal{T} = \mathsf{D}^-(R)$, but we can directly show the corollary.

COROLLARY 3.8 ([**10**, Corollary 4.19]). *The following are equivalent.*

(1) *The tensor triangulated category $\mathsf{D}^-(R)$ is local.*
(2) *There exists a unique minimal prime of $\mathsf{D}^-(R)$.*
(3) *The radical $\sqrt{\mathbf{0}}$ of the zero ideal of $\mathsf{D}^-(R)$ is prime.*
(4) *The zero ideal $\mathbf{0}$ of $\mathsf{D}^-(R)$ is prime.*
(5) *The ring R is local.*

PROOF. (1) $\Leftrightarrow$ (2) $\Leftrightarrow$ (3): These equivalences follow from Proposition 3.7.

(2) $\Rightarrow$ (5): By Theorem 3.3(3) there is an inequality

$$\#\max(\operatorname{Spec} R) \leq \#\min(\mathsf{Spc}\,\mathsf{D}^-(R)) = 1,$$

which shows the implication.

(5) $\Rightarrow$ (4): Let $\mathfrak{m}$ be the unique maximal ideal of R. The zero ideal $\mathbf{0}$ of $\mathsf{D}^-(R)$ coincides with the prime ideal $\mathcal{S}(\mathfrak{m})$ of $\mathsf{D}^-(R)$.

(4) $\Rightarrow$ (3): It is easy to see that $\sqrt{\mathbf{0}} = \mathbf{0}$. □

As further observations on the topological structure of $\mathsf{Spc}\,\mathsf{D}^-(R)$, we can show that the topological space $\mathsf{Spc}\,\mathsf{D}^-(R)$ is irreducible if and only if so is $\operatorname{Spec} R$. Also, we can describe the Balmer spectrum of $\mathsf{D}^-(R)$ as follows.

$$\mathsf{Spc}\,\mathsf{D}^-(R) = \begin{cases} \mathtt{U}(R/\mathfrak{m}) \sqcup \{\mathbf{0}\} & \text{if } (R, \mathfrak{m}) \text{ is a local ring,} \\ \bigcup_{\mathfrak{m} \in \operatorname{Max} R} \mathtt{U}(R/\mathfrak{m}) & \text{if } R \text{ is a non-local ring.} \end{cases}$$

More strongly, for any two distinct maximal ideals $\mathfrak{m}, \mathfrak{n}$ of R the equality $\mathsf{D}^-(R) = \mathtt{U}(R/\mathfrak{m}) \cup \mathtt{U}(R/\mathfrak{n})$ holds. We refer the reader to [**10**, Corollary 4.13 and Proposition 4.20] for the details.

3.2. Classifications of ideals of $\mathsf{D}^-(R)$.
In this subsection, we consider making correspondences among compact, radical and tame ideals of $\mathsf{D}^-(R)$, and specialization-closed subsets of $\operatorname{Spec} R$, $\mathsf{Spc}\,\mathsf{D}^-(R)$ and ${}^{\mathsf{t}}\mathsf{Spc}\,\mathsf{D}^-(R)$. First of all, we explore the relationships among these three properties of ideals of $\mathsf{D}^-(R)$.

PROPOSITION 3.9 ([**10**, Lemma 5.7]). *Let $\mathcal{X}$ be an ideal of $\mathsf{D}^-(R)$.*

(1) *There are equalities*

$$\mathcal{X}_{\mathsf{cpt}} = \langle \operatorname{supph} \mathcal{X} \rangle, \quad \mathcal{X}^{\mathsf{rad}} = \sqrt{\mathcal{X}}, \quad \mathcal{X}^{\mathsf{tame}} = \operatorname{supph}^{-1} \operatorname{supph} \mathcal{X}$$

of ideals of $\mathsf{D}^-(R)$.

(2) *There are inclusions*

$$\mathcal{X}_{\mathsf{cpt}} \subseteq \mathcal{X} \subseteq \mathcal{X}^{\mathsf{rad}} \subseteq \mathcal{X}^{\mathsf{tame}}$$

of ideals of $\mathsf{D}^-(R)$, all of whose supports are equal. In particular, every tame ideal of $\mathsf{D}^-(R)$ is radical.

Here, $\mathcal{X}^{\mathbb{P}}$ (respectively, $\mathcal{X}_{\mathbb{P}}$) stands for the $\mathbb{P}$-*closure* (respectively, $\mathbb{P}$-*interior*) of $\mathcal{X}$, namely, the smallest (respectively, largest) $\mathbb{P}$-ideal containing (respectively, contained in) $\mathcal{X}$. Also, cpt and rad denote the compact and radical properties, respectively.

The assertion (1) of the above proposition is seen to hold just by checking the defintions. In relation to (2), the following statement holds: Let W be a specialization-closed subset of $\operatorname{Spec} R$. Then $\langle W \rangle$ (respectively, $\operatorname{supph}^{-1} W$) is the smallest (respectively, largest) ideal of $\mathsf{D}^-(R)$ whose support coincides with W; see [**10**, Theorem 6.6(2)].

To state the main result of this section, we introduce notation.

NOTATION 3.10. We use the following sets in the rest of this subsection.

$$\begin{aligned}
\mathbf{Rad} &= \{\text{Radical ideals of } \mathsf{D}^-(R)\},\\
\mathbf{Tame} &= \{\text{Tame ideals of } \mathsf{D}^-(R)\},\\
\mathbf{Cpt} &= \{\text{Compact ideals of } \mathsf{D}^-(R)\},\\
\mathbf{Spcl}(\mathrm{Spec}) &= \{\text{Specialization-closed subsets of } \operatorname{Spec} R\},\\
\mathbf{Spcl}({}^{\mathsf{t}}\mathsf{Spc}) &= \{\text{Specialization-closed subsets of } {}^{\mathsf{t}}\mathsf{Spc}\,\mathsf{D}^-(R)\},\\
\mathbf{Thom} &= \{\text{Thomason subsets of } \mathsf{Spc}\,\mathsf{D}^-(R)\}.
\end{aligned}$$

Proposition 3.9(2) implies the inclusion

$$\mathbf{Rad} \supseteq \mathbf{Tame}.$$

The main result of this section makes correspondences among the above six sets.

THEOREM 3.11. [**10**, Theorems 5.12 and 5.19] *One has the following diagram, which is naturally commutative. (More precisely, the diagram with sections and bijections and the diagram with retractions and bijections are commutative.)*

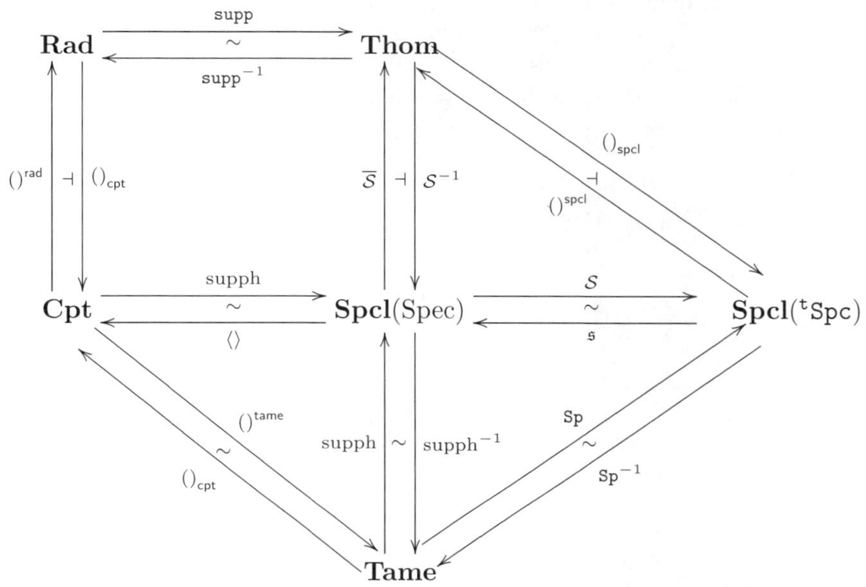

Here:

- $\begin{cases} f \sim g \iff gf = 1 \text{ and } fg = 1 \text{ (i.e. } (f,g) \text{ is a bijection pair)}, \\ f \dashv g \iff gf = 1 \text{ (i.e. } (f,g) \text{ is a section-retraction pair)}. \end{cases}$
- $\begin{cases} A_{\mathsf{spcl}} = \text{the } \mathsf{spcl}\text{-interior of } A \text{ in } {}^{\mathsf{t}}\mathsf{Spc}\,\mathsf{D}^{-}(R), \\ B^{\mathsf{spcl}} = \text{the } \mathsf{spcl}\text{-closure of } B \text{ in } \mathsf{Spc}\,\mathsf{D}^{-}(R). \end{cases}$
- $\begin{cases} \overline{\mathcal{S}}(W) = \bigcup_{\mathfrak{p}\in W} \overline{\{\mathcal{S}(\mathfrak{p})\}}, \\ \mathcal{S}^{-1}(A) = \{\mathfrak{p} \in \operatorname{Spec} R \mid \mathcal{S}(\mathfrak{p}) \in A\}, \\ \mathsf{Sp}(-) = \mathsf{supp}(-) \cap {}^{\mathsf{t}}\mathsf{Spc}\,\mathsf{D}^{-}(R), \\ \mathsf{Sp}^{-1}(B) = \{M \in \mathsf{D}^{-}(R) \mid \mathsf{Sp}\,M \subseteq B\}, \\ \mathcal{S}(W) = \{\mathcal{S}(\mathfrak{p}) \mid \mathfrak{p} \in W\}, \\ \mathfrak{s}(B) = \{\mathfrak{s}(\mathcal{P}) \mid \mathcal{P} \in B\}. \end{cases}$

Moreover, the following are equivalent.

(1) *The pair* $\mathcal{S} : \operatorname{Spec} R \rightleftarrows \mathsf{Spc}\,\mathsf{D}^{-}(R) : \mathfrak{s}$ *of maps is a one-to-one correspondence.*
(2) *The pair* $(()^{\mathsf{rad}}, ()_{\mathsf{cpt}})$ *of maps is a one-to-one correspondence.*
(3) *The pair* $(\overline{\mathcal{S}}, \mathcal{S}^{-1})$ *of maps is a one-to-one correspondence.*
(4) *The pair* $(()^{\mathsf{spcl}}, ()_{\mathsf{spcl}})$ *of maps is a one-to-one correspondence.*
(5) *The equality* **Rad** = **Tame** *holds.*

The spcl-*interior* A_{spcl} is the largest specialization-closed subset of ${}^{\mathsf{t}}\mathsf{Spc}\,\mathsf{D}^{-}(R)$ contained in A, while spcl-*closure* B^{spcl} is the smallest specialization-closed subset of $\mathsf{Spc}\,\mathsf{D}^{-}(R)$ containing B.

Here are some comments on the above theorem.

REMARK 3.12.

(a) The one-to-one correspondence **Rad** $\cong$ **Thom** in the diagram of Theorem 3.11 is nothing but Theorem 1.6 due to Balmer, while the one-to-one correspondence **Cpt** $\cong$ **Spcl**(Spec) is nothing but Theorem 2.4. Thus this diagram connects Theorems 1.6 and 2.4, and gives rise to several related correspondences.

(b) The proof of Theorem 3.11 proceeds step by step; for example, we show and use the equalities

$$\begin{cases} A_{\mathsf{spcl}} = A \cap {}^{\mathsf{t}}\mathsf{Spc}\,\mathsf{D}^{-}(R), \\ B^{\mathsf{spcl}} = \{\mathcal{P} \in \mathsf{Spc}\,\mathsf{D}^{-}(R) \mid \mathcal{P}^{\mathsf{tame}} \in B\} = \bigcup_{\mathcal{P} \in B^{\mathsf{spcl}}} \mathsf{supp}(R/\mathfrak{s}(\mathcal{P})). \end{cases}$$

(c) Theorem 3.11 yields a commutative diagram

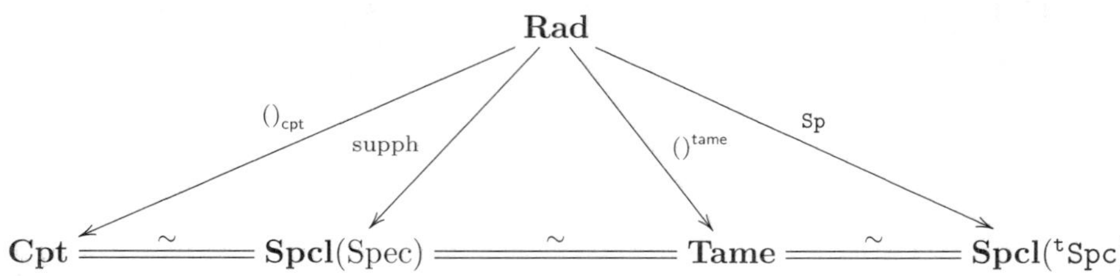

where the bottom bijections are the ones in the diagram of Theorem 3.11. Furthermore, the conditions (1)–(5) in Theorem 3.11 are also equivalent to the following three conditions.

(6) The map supph : **Rad** $\to$ **Spcl**(Spec) is a bijection.
(7) The map $()^{\mathsf{tame}}$: **Rad** $\to$ **Tame** is a bijection.
(8) The map Sp : **Rad** $\to$ **Spcl**($^{\mathsf{t}}\mathsf{Spc}$) is a bijection.

For the details, we refer the reader to [**10**, Corollary 5.20].

The corollary below is immediately obtained from the above theorem.

COROLLARY 3.13. *If every radical ideal of* $\mathsf{D}^{-}(R)$ *is compact, then* **Rad** $=$ **Tame**.

PROOF. For each radical ideal $\mathcal{X}$ of $\mathsf{D}^{-}(R)$ one has $\mathcal{X} = \mathcal{X}_{\mathsf{cpt}} = (\mathcal{X}_{\mathsf{cpt}})^{\mathsf{rad}}$. Hence

$$()^{\mathsf{rad}} : \mathbf{Cpt} \rightleftarrows \mathbf{Rad} : ()_{\mathsf{cpt}}$$

is a one-to-one correspondence. Theorem 3.11 implies **Rad** $=$ **Tame**. □

We are interested in what rings R are characterized by the eight conditions (1)–(8) appearing in Theorem 3.11 and Remark 3.12.

THEOREM 3.14 ([**10**, Theorem 6.5]). *The equivalent conditions* (1)–(8) *are also equivalent to the condition that*

(9) *the ring R is artinian.*

Furthermore, when this is the case, every ideal of $\mathsf{D}^{-}(R)$ *is compact, tame and radical.*

The most difficult part of the proof of this theorem is to show the necessity of the condition (9). Here, let us only check the last assertion of the theorem. Suppose that R is artinian, and pick any ideal $\mathcal{X}$ of $\mathsf{D}^{-}(R)$. Then Corollary 3.5(3) implies that $\mathcal{X}$ is compact, and that taking $\operatorname{supph}(-)$ makes an injective map. Hence the equality

$$\operatorname{supph}(\mathcal{X}) = \operatorname{supph}(\operatorname{supph}^{-1} \operatorname{supph} \mathcal{X})$$

implies that $\mathcal{X}$ coincides with $\operatorname{supph}^{-1} \operatorname{supph} \mathcal{X}$, which shows that $\mathcal{X}$ is tame. In general, a tame ideal of $\mathsf{D}^-(R)$ is radical, and hence $\mathcal{X}$ is radical.

Using Theorem 3.14 and Corollary 3.13, we immediately obtain the following.

COROLLARY 3.15. *Suppose that R is not artinian. Then there exists a non-compact radical ideal of* $\mathsf{D}^-(R)$.

In relation to this result, we consider the existence of compact prime ideals of $\mathsf{D}^-(R)$. If R is a local ring with maximal ideal $\mathfrak{m}$, then

$$\mathbf{Cpt} \cap \mathfrak{s}^{-1}(\mathfrak{m}) = \{\mathbf{0}\}.$$

In particular, there exists a compact prime ideal of $\mathsf{D}^-(R)$. On the other hand, if R is a non-local semi-local domain of positive dimension, then $\mathsf{D}^-(R)$ admits no compact prime ideal. We refer the reader to [**10**, Proposition 6.2]. If R is a discrete valuation ring, we can show

$$\mathbf{Cpt} \cap \operatorname{\mathsf{Spc}} \mathsf{D}^-(R) = \{\mathbf{0}, \operatorname{\mathsf{thick}}^{\otimes} k\},$$

where k is the residue field of R. This follows from Theorems 3.11 and 4.1. See also [**10**, Example 7.16].

3.3. On Balmer's conjecture for $\mathsf{D}^-(R)$.

From now on, we consider Balmer's conjecture stated in Section 1 for our tensor triangulated category $\mathsf{D}^-(R)$. First of all, we investigate the difference between radical and tame ideals of $\mathsf{D}^-(R)$. We have already learned that the following holds.

$$\mathcal{X}^{\mathsf{rad}} \subseteq \mathcal{X}^{\mathsf{tame}}.$$

The following theorem says that if $\mathcal{X}$ is compact, then the equality does not hold under mild assumptions.

THEOREM 3.16 ([**10**, Theorem 6.6]). *Let W be a nonempty proper specialization-closed subset of* $\operatorname{Spec} R$, *and put $\mathcal{X} = \langle W \rangle$. Assume that R is either a domain or a local ring. Then*

$$\mathcal{X}^{\mathsf{rad}} \subsetneq \mathcal{X}^{\mathsf{tame}}.$$

PROOF. Since W is nonempty, it contains a prime ideal P. Take a system of generators $\boldsymbol{x} = x_1, \dots, x_r$ of P. It is essential to think of the following complex.

$$C := \bigoplus_{i>0} \mathrm{K}(\boldsymbol{x}^i)[i].$$

Thanks to the shifts, this infinite direct sum exists in our tensor triangulated category $\mathsf{D}^-(R)$. Since $\operatorname{supph} C = \mathrm{V}(P)$ is contained in W, the complex C is in $\operatorname{supph}^{-1} W = \mathcal{X}^{\mathsf{tame}}$ by Proposition 3.9(1).

Suppose $\mathcal{X}^{\mathsf{rad}} = \mathcal{X}^{\mathsf{tame}}$. Then C belongs to $\mathcal{X}^{\mathsf{rad}} = \sqrt{\mathcal{X}}$. Hence there is an integer $n > 0$ such that

$$C' := \underbrace{C \otimes_R^{\mathbf{L}} \cdots \otimes_R^{\mathbf{L}} C}_{n}$$

belongs to $\mathcal{X}$. Note that C' contains

$$D := \bigoplus_{i>0} \mathrm{K}(\boldsymbol{x}^i)[ni]$$

as a direct summand. Therefore D is in $\mathcal{X} = \langle W \rangle = \mathsf{thick}^{\otimes}\{R/\mathfrak{p} \mid \mathfrak{p} \in W\}$, and we find a finite number of prime ideals $\mathfrak{p}_1, \dots, \mathfrak{p}_m$ in W such that

$$\operatorname{ann} D \supseteq (\operatorname{ann} R/\mathfrak{p}_1) \cdots (\operatorname{ann} R/\mathfrak{p}_m) = \mathfrak{p}_1 \cdots \mathfrak{p}_m.$$

Krull's intersection theorem implies

$$\operatorname{ann} D = \bigcap_{i>0} \boldsymbol{x}^i R = 0,$$

and we have $\mathfrak{p}_1 \cdots \mathfrak{p}_m = 0$. Thus for every prime ideal $\mathfrak{p}$ of R there exists an integer $1 \le t \le m$ such that $\mathfrak{p}$ contains $\mathfrak{p}_t$. Since W is specialization-closed and contains $\mathfrak{p}_t$, the prime ideal $\mathfrak{p}$ belongs to W. This shows that $W = \operatorname{Spec} R$, contrary to the assumption on W. □

Using the above proof, we have an observation related to Proposition 2.7.

REMARK 3.17. We use the same notation as in the proof of Theorem 3.16.

(1) It holds that $\operatorname{supph} C$ is contained in $\operatorname{supph} R/P$, but C does not belong to $\mathsf{thick}^{\otimes} R/P$. Indeed, we have $\operatorname{supph} C = \mathrm{V}(P) = \operatorname{supph} R/P$. Assume that C is in $\mathsf{thick}^{\otimes} R/P$. Then

$$0 = \operatorname{ann} C \supseteq (\operatorname{ann} R/P)^u = P^u$$

for some integer $u > 0$. Hence the equality $\operatorname{Spec} R = \mathrm{V}(P)$ holds, which is contained in W since W is specialization-closed. Therefore W coincides with $\operatorname{Spec} R$, which is a contradiction.

(2) It holds that $\mathrm{V}(\operatorname{ann} R)$ is contained in $\mathrm{V}(\operatorname{ann} C)$, but R does not belong to $\mathsf{thick}^{\otimes} C$. In fact, we have $\mathrm{V}(\operatorname{ann} R) = \mathrm{V}(0) = \mathrm{V}(\operatorname{ann} C)$. As $\operatorname{supph} C = \mathrm{V}(P)$ is a proper subset of $\operatorname{Spec} R$, it is observed from Corollary 2.9(1) that R is not in $\mathsf{thick}^{\otimes} C$.

Now, we consider Balmer's conjecture (Conjecture 1.8) for our tensor triangulated category $\mathsf{D}^-(R)$. First of all, let us check that the triangulated category $\mathsf{D}^-(R)$ is algebraic. The category $\mathsf{C}^-(R)$ of right bounded complexes of finitely generated R-modules is a Frobenius exact category with respect to the split short exact sequences of complexes in $\mathsf{C}^-(R)$, and $\mathsf{K}^-(R)$ is the stable category of $\mathsf{C}^-(R)$. Thus $\mathsf{K}^-(R)$ is an algebraic triangulated category.

Recall that Conjecture 1.8 concerns the continuous map

$$\rho^{\bullet}_{\mathsf{D}^-(R)} : \mathsf{Spc}\, \mathsf{D}^-(R) \to \operatorname{Spec}^{\mathrm{h}} \mathrm{R}^{\bullet}_{\mathsf{D}^-(R)}.$$

One can actually observe that

(a) $\mathrm{R}^{\bullet}_{\mathsf{D}^-(R)} = \mathrm{R}^{0}_{\mathsf{D}^-(R)} = R$,
(b) $\operatorname{Spec}^{\mathrm{h}} \mathrm{R}^{\bullet}_{\mathsf{D}^-(R)} = \operatorname{Spec} R$, and
(c) $\rho^{\bullet}_{\mathsf{D}^-(R)} = \mathfrak{s}$.

Thus, Conjecture 1.8 for $\mathsf{D}^-(R)$ just claims the local injectivity of the map $\mathfrak{s}$.

We can show that under quite mild assumptions the algebraic tensor triangulated category $\mathsf{D}^-(R)$ does not satisfy Balmer's conjecture.

COROLLARY 3.18 ([**10**, Corollary 6.10]). *Assume that* $\dim R > 0$, *and that* R *is either a domain or a local ring. Then* $\mathfrak{s}$ *is not locally injective. Hence, Balmer's Conjecture* 1.8 *does not hold true for* $\mathsf{D}^-(R)$.

PROOF. By assumption we find a nonunit $x \in R$ such that the principal ideal xR of R has positive height. We apply Theorem 3.16 to $\mathcal{X} = \langle \mathrm{V}(x) \rangle$ to get

$$\bigcap_{\mathcal{X} \subseteq \mathcal{P} \in \mathsf{Spc}\,\mathsf{D}^-(R)} \mathcal{P} = \mathcal{X}^{\mathsf{rad}} \subsetneq \mathcal{X}^{\mathsf{tame}} = \bigcap_{\mathcal{X} \subseteq \mathcal{P} \in {}^{\mathsf{t}}\mathsf{Spc}\,\mathsf{D}^-(R)} \mathcal{P}.$$

Hence we can choose a prime ideal $\mathcal{P}$ of $\mathsf{D}^-(R)$ such that $\mathcal{X} \subseteq \mathcal{P} \subsetneq \mathcal{P}^{\mathsf{tame}}$.

Assume that $\mathfrak{s}$ is locally injective at the point $\mathcal{P}$. Then there exists an object $M \in \mathsf{D}^-(R)$ with $\mathcal{P} \in \mathsf{U}(M)$ such that $\mathfrak{s}|_{\mathsf{U}(M)}$ is injective. Then $\mathsf{U}(M)$ contains two distinct points $\mathcal{P}$ and $\mathcal{P}^{\mathsf{tame}}$, which are sent by $\mathfrak{s}$ to the same point in $\operatorname{Spec} R$. This contradicts the injectivity of the map $\mathfrak{s}|_{\mathsf{U}(M)}$. □

4. The case of discrete valuation rings and applications

In this section, we mainly consider the case where R is a discrete valuation ring, and make an application for the higher-dimensional case.

For a local ring R with maximal ideal $\mathfrak{m}$, we denote by $\mathsf{D}^-_{\mathsf{fl}}(R)$ the full subcategory of $\mathsf{D}^-(R)$ consisting of complexes with finite length homologies. Note that this is nothing but $\operatorname{supph}^{-1}\{\mathfrak{m}\}$, and in particular, it is an ideal of $\mathsf{D}^-(R)$.

When R is a discrete valuation ring, we can construct certain concrete prime ideals of $\mathsf{D}^-(R)$ by making use of the growth of the Loewy lengths of homologies, and moreover we can obtain their generators.

THEOREM 4.1 ([**10**, Propositions 7.7, 7.17 and Theorems 7.11, 7.14]). *Let* R *be a discrete valuation ring with maximal ideal* xR. *For an integer* $n \geq 0$, *we denote by* $\mathcal{P}_n$ *the full subcategory of* $\mathsf{D}^-_{\mathsf{fl}}(R)$ *consisting of complexes* X *such that there exists an integer* $t \geq 0$ *with*

$$\ell\ell(\mathsf{H}^{-i}X) \leq ti^n$$

for all $i \gg 0$, *where* $\ell\ell(-)$ *stands for the Loewy length. Then the following assertions hold true.*

(1) *One has* $\mathcal{P}_n = \mathsf{thick}^{\otimes} G_n$, *where*

$$G_n = (\cdots \xrightarrow{0} R/x^{3^n}R \xrightarrow{0} R/x^{2^n}R \xrightarrow{0} R/x^{1^n}R \to 0).$$

In particular, $\mathcal{P}_n$ *is an ideal of* $\mathsf{D}^-(R)$.

(2) *The ideal* $\mathcal{P}_n$ *is prime and non-tame. It is non-compact if* $n \geq 1$.

(3) *There is a strictly ascending chain*

$$\mathcal{P}_0 \subsetneq \mathcal{P}_1 \subsetneq \mathcal{P}_2 \subsetneq \cdots$$

of prime ideals of $\mathsf{D}^-(R)$. *Therefore* $\mathsf{Spc}\,\mathsf{D}^-(R)$ *has infinite Krull dimension.*

Using this theorem, we have an observation related to Proposition 2.7, whose context is the same as Remark 3.17.

REMARK 4.2. We use the same notation as in Theorem 4.1. We denote by k the residue field of R, that is, $k = R/xR$.

(1) It holds that $\operatorname{supph} G_n$ is contained in $\operatorname{supph} k$, but G_n is not in $\mathsf{thick}^{\otimes} k$ for all $n \geq 1$. In fact, we have $\operatorname{supph} G_n = \mathrm{V}(x) = \operatorname{supph} k$. Assume that G_n belongs to $\mathsf{thick}^{\otimes} k$. Then $\mathcal{P}_n = \mathsf{thick}^{\otimes} G_n$ is contained in $\mathsf{thick}^{\otimes} k = \mathcal{P}_0$, which gives rise to a contradiction.
(2) It holds that $\mathrm{V}(\operatorname{ann} R)$ is contained in $\mathrm{V}(\operatorname{ann} G_n)$, but R does not belong to $\mathsf{thick}^{\otimes} G_n$. Indeed, we have $\mathrm{V}(\operatorname{ann} R) = \mathrm{V}(0) = \mathrm{V}(\operatorname{ann} G_n)$. As $\operatorname{supph} G_n = \mathrm{V}(x)$ is a proper subset of $\operatorname{Spec} R$, we see from Corollary 2.9(1) that R is not in $\mathsf{thick}^{\otimes} C$.

The proof of Theorem 4.1 is rather technical, which uses the following fact and lemma. See [**10**, Propositions 7.1, 7.2, Corollary 7.3 and Lemma 7.8].

FACT 4.3. Let R be a discrete valuation ring.

(1) For each complex $M \in \mathsf{D}^-(R)$ there is an isomorphism
$$M \cong \mathsf{H}(M) = \bigoplus_{i \in \mathbb{Z}} \mathsf{H}^i(M)[-i],$$
since the ring R is hereditary. In particular, we may assume that each object of $\mathsf{D}^-(R)$ is a complex with zero differentials.
(2) The nonisomorphic indecomposable R-modules of finite length are precisely given by the list
$$\{R/x^iR\}_{i>0},$$
which is a consequence of the structure theorem for finitely generated modules over a principal ideal domain.

LEMMA 4.4. *Let* $\mathcal{X} = \mathsf{thick}^{\otimes}(\cdots \xrightarrow{0} X_2 \xrightarrow{0} X_1 \xrightarrow{0} X_0 \to 0)$ *be an ideal of* $\mathsf{D}^-(R)$.

(1) *One has*
$$\mathcal{X} = \mathsf{thick}^{\otimes}(\cdots \xrightarrow{0} X_2 \oplus X_1 \oplus X_0 \xrightarrow{0} X_1 \oplus X_0 \xrightarrow{0} X_0 \to 0).$$
(2) *The ideal* $\mathcal{X}$ *contains the complex*
$$(\cdots \to 0 \to X_2^{\oplus a_2} \to 0 \to X_1^{\oplus a_1} \to 0 \to X_0^{\oplus a_0} \to 0)$$
for all integers $a_i \geq 0$.
(3) *The complex*
$$(\cdots \xrightarrow{0} X_3^{\oplus a_3} \xrightarrow{0} X_2^{\oplus a_2} \xrightarrow{0} X_1^{\oplus a_1} \xrightarrow{0} X_0^{\oplus a_0} \to 0)$$
is in $\mathsf{thick}^{\otimes}\{X_{\mathrm{even}}, X_{\mathrm{odd}}\}$, *where*
$$\begin{cases} X_{\mathrm{even}} = (\cdots \xrightarrow{0} X_6 \xrightarrow{0} X_4 \xrightarrow{0} X_2 \xrightarrow{0} X_0 \to 0), \\ X_{\mathrm{odd}} = (\cdots \xrightarrow{0} X_7 \xrightarrow{0} X_5 \xrightarrow{0} X_3 \xrightarrow{0} X_1 \to 0). \end{cases}$$

Here are some questions naturally arisen from the above lemma.

QUESTION. Let k be a field, and let $R = k[[x]]$ be a formal power series ring.

(1) Does the ideal $\mathsf{thick}^{\otimes}(\cdots \xrightarrow{0} X_2 \xrightarrow{0} X_1 \xrightarrow{0} X_0 \to 0)$ contain the complex $(\cdots \xrightarrow{0} X_2^{a_2} \xrightarrow{0} X_1^{a_1} \xrightarrow{0} X_0^{a_0} \to 0)$ for all integers $a_i \geq 0$?
(2) Does the ideal $\mathsf{thick}^{\otimes}(\cdots \to 0 \to X_1 \to 0 \to X_0 \to 0)$ contain the complex $(\cdots \xrightarrow{0} X_1 \xrightarrow{0} X_0 \to 0)$?

(3) Can we characterize the complexes C belonging to the ideal

$$\mathsf{thick}^{\otimes}(\cdots \xrightarrow{0} R/x^{2!} \xrightarrow{0} R/x^{1!} \xrightarrow{0} R/x^{0!} \to 0)$$

in terms of the Loewy lengths of homologies of C?

Recall that a triangulated functor $F : \mathcal{T} \to \mathcal{T}'$ of tensor triangulated categories is called *tensor triangulated* if $F(\mathbb{1}) = \mathbb{1}$ and $F(x \otimes y) = F(x) \otimes F(y)$ for all $x, y \in \mathcal{T}$. A tensor triangulated functor $F : \mathcal{T} \to \mathcal{T}'$ induces a continuous map

$$\mathsf{Spc}\, F : \mathsf{Spc}\, \mathcal{T}' \to \mathsf{Spc}\, \mathcal{T},$$

which is given by $\mathcal{P} \mapsto \mathcal{P} \cap \mathcal{T} := F^{-1}(\mathcal{P})$. This is injective if F is essentially surjective; see [**2**, Proposition 3.6 and Corollary 3.8].

Note that if $\mathsf{Spc}\, F$ is injective, then it sends a chain

$$\cdots \subsetneq \mathcal{P} \subsetneq \mathcal{Q} \subsetneq \cdots$$

of prime ideals of $\mathcal{T}'$ to a chain

$$\cdots \subsetneq \ \mathcal{P} \cap \mathcal{T} \ \subsetneq \ \mathcal{Q} \cap \mathcal{T} \ \subsetneq \cdots$$

of prime ideals of $\mathcal{T}$. Hence the inequality

$$\dim(\mathsf{Spc}\, \mathcal{T}') \leq \dim(\mathsf{Spc}\, \mathcal{T})$$

of Krull dimensions holds.

This observation and Theorem 4.1(3) yield the following result; see [**10**, Corollary 7.13].

COROLLARY 4.5. *Assume that $R_\mathfrak{p}$ is regular for some $\mathfrak{p}$ with positive height (e.g. R is normal). Then*

$$\dim(\mathsf{Spc}\, \mathsf{D}^-(R)) = \infty.$$

PROOF. We may assume that $\mathfrak{p}$ has height 1. Localization at $\mathfrak{p}$ induces an essentially surjective tensor triangulated functor

$$\mathsf{D}^-(R) \to \mathsf{D}^-(R_\mathfrak{p}).$$

The above argument implies that

$$\dim \mathsf{Spc}\, \mathsf{D}^-(R_\mathfrak{p}) \leq \dim \mathsf{Spc}\, \mathsf{D}^-(R).$$

Combining this with Theorem 4.1(3), we see that $\mathsf{Spc}\, \mathsf{D}^-(R)$ has infinite Krull dimension. □

Recall that a topological space is said to be *noetherian* if any descending chain of closed subsets stabilizes. Corollary 4.5 naturally makes us to have the following question.

QUESTION 4.6. Let R satisfy the same assumption as in Corollary 4.5. Is then the Balmer spectrum $\mathsf{Spc}\, \mathsf{D}^-(R)$ non-noetherian?

Acknowledgments

The author thanks the referee for giving him useful comments.

References

[1] Paul Balmer, *Presheaves of triangulated categories and reconstruction of schemes*, Math. Ann. **324** (2002), no. 3, 557–580, DOI 10.1007/s00208-002-0353-1. MR1938458

[2] Paul Balmer, *The spectrum of prime ideals in tensor triangulated categories*, J. Reine Angew. Math. **588** (2005), 149–168, DOI 10.1515/crll.2005.2005.588.149. MR2196732

[3] Paul Balmer, *Spectra, spectra, spectra—tensor triangular spectra versus Zariski spectra of endomorphism rings*, Algebr. Geom. Topol. **10** (2010), no. 3, 1521–1563, DOI 10.2140/agt.2010.10.1521. MR2661535

[4] Paul Balmer, *Tensor triangular geometry*, Proceedings of the International Congress of Mathematicians. Volume II, Hindustan Book Agency, New Delhi, 2010, pp. 85–112. MR2827786

[5] D. J. Benson, Jon F. Carlson, and Jeremy Rickard, *Thick subcategories of the stable module category*, Fund. Math. **153** (1997), no. 1, 59–80. MR1450996

[6] David J. Benson, Srikanth B. Iyengar, and Henning Krause, *Stratifying modular representations of finite groups*, Ann. of Math. (2) **174** (2011), no. 3, 1643–1684, DOI 10.4007/annals.2011.174.3.6. MR2846489

[7] D. Benson; S. B. Iyengar; H. Krause; J. Pevtsova, Stratification for module categories of finite group schemes, arXiv:1510.06773v2.

[8] Eric M. Friedlander and Julia Pevtsova, *Π-supports for modules for finite group schemes*, Duke Math. J. **139** (2007), no. 2, 317–368, DOI 10.1215/S0012-7094-07-13923-1. MR2352134

[9] Michael J. Hopkins, *Global methods in homotopy theory*, Homotopy theory (Durham, 1985), London Math. Soc. Lecture Note Ser., vol. 117, Cambridge Univ. Press, Cambridge, 1987, pp. 73–96. MR932260

[10] H. Matsui; R. Takahashi, Thick tensor ideals of right bounded derived categories, Preprint (2016), arXiv:1611.02826v1.

[11] Amnon Neeman, *The chromatic tower for $D(R)$*, Topology **31** (1992), no. 3, 519–532, DOI 10.1016/0040-9383(92)90047-L. With an appendix by Marcel Bökstedt. MR1174255

[12] Greg Stevenson, *Duality for bounded derived categories of complete intersections*, Bull. Lond. Math. Soc. **46** (2014), no. 2, 245–257, DOI 10.1112/blms/bdt089. MR3194744

[13] R. W. Thomason, *The classification of triangulated subcategories*, Compositio Math. **105** (1997), no. 1, 1–27, DOI 10.1023/A:1017932514274. MR1436741

Graduate School of Mathematics, Nagoya University, Furocho, Chikusaku, Nagoya, Aichi 464-8602, Japan

Email address: takahashi@math.nagoya-u.ac.jp

URL: http://www.math.nagoya-u.ac.jp/~takahashi/

Contemporary Mathematics
Volume **705**, 2018
http://dx.doi.org/10.1090/conm/705/14201

Stability, shards, and preprojective algebras

Hugh Thomas

ABSTRACT. In this note, I explain how stability, in the sense of King, provides a geometrical picture of (some of) the extension-closed abelian subcategories of a finite-dimensional algebra. I also describe Nathan Reading's shards of a hyperplane arrangement, and explain their relevance to understanding the stability picture for Dynkin-type preprojective algebras.

1. Semistable subcategories

Let A be a finite-dimensional algebra over a field k. We will work with the category of left A-modules. Suppose that A has n pairwise non-isomorphic simple modules $S_1, \ldots, S_n$.

The Grothendieck group of A can be defined as the free abelian group on a set of generators corresponding to the simple modules:

$$K_0(A) = \bigoplus_{i=1}^{n} \mathbb{Z}[S_i].$$

For any A-module M, there is a corresponding class in $K_0(A)$, which we denote $[M]$, and which is equal to $\sum_i c_i[S_i]$, where c_i is the number of times S_i appears in a composition series for M.

We will be interested in linear functionals on $K_0(A)$. For convenience in drawing pictures, we will extend scalars to consider real-valued functionals.

$$K_0^*(A)_\mathbb{R} = \mathrm{Hom}_\mathbb{Z}(K_0(A), \mathbb{R})$$

Let $\phi \in K_0^*(A)_\mathbb{R}$. An A-module M is called semistable with respect to ϕ if $\phi([M]) = 0$ and $\phi([N]) \leq 0$ for any submodule N of M. This definition is due to King [**Ki**], who showed that it is equivalent to a notion of semistability coming from geometric invariant theory (which we shall not need in this note). There is also another reformulation in terms of semi-invariants (see, for example, [**DW**]), but we shall not need to refer to this perspective either.

We will write $(A\text{-mod})^\phi$ for the full subcategory of A-modules semistable with respect to ϕ. It was shown by King (and it is an easy exercise) that $(A\text{-mod})^\phi$ is an extension-closed, exact abelian subcategory of A-mod.

For example, we could consider a path algebra of type A_2, with simples S_1 being projective and S_2 being injective, as shown in Figure 1. This is a picture

---

2010 *Mathematics Subject Classification.* 16G20, 05E10, 20F55, 52C35.

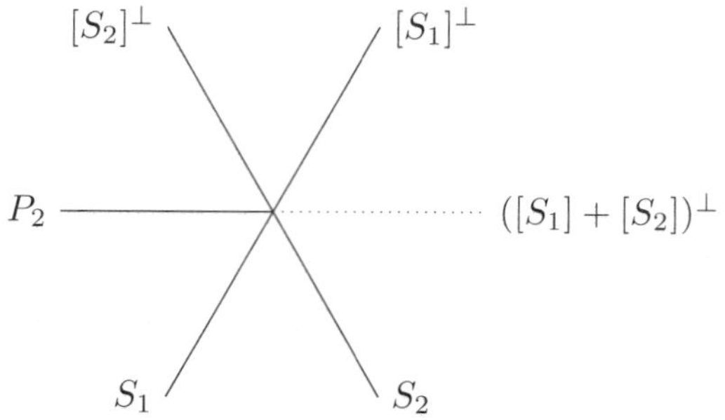

FIGURE 1. Regions of semistability for a path algebra of type A_2

of $K_0^*(A)_\mathbb{R} \cong \mathbb{R}^2$. Each point of the picture corresponds to a stability condition. There, $[S_1]^\perp$ designates the line consisting of elements of $K_0^*(A)_\mathbb{R}$ which vanish on $[S_1]$, and similarly for $[S_2]^\perp$ and $([S_1]+[S_2])^\perp$. We will consistently orient our stability diagrams so that the region where ϕ is positive on all the simples is at the bottom of the diagram. We have marked two lines and one half-line which are regions of semistability for the indecomposable modules of this algebra. S_1 is stable on the whole line $[S_1]^\perp$, and similarly for S_2. P_2, however, is only stable on half of the line $([S_1]+[S_2])^\perp$, the half that is drawn in solidly. On the other half, its submodule S_1 has $\phi([S_1]) > 0$, which causes it to become unstable. Note that the labels S_1, S_2, and P_2 in the diagram do not refer to specific points. Rather, they label their corresponding regions of stability (lines, or, in the case of P_2, a half-line). Generically, at a point not on any of the lines in the picture, the subcategory of semistable submodules is the zero category. Finally, at the origin, every module is semistable. Pictures like this, in the hereditary case, have been studied by [**IOTW**, **Ch**, **IT**, **IPT**].

Recall that an A-module is called a brick if its endomorphism ring is a division algebra. To understand the semistable subcategories of A-mod, it is sufficient to understand semistability of bricks, by the following lemma.

LEMMA 1. *An exact abelian extension-closed subcategory of A-mod is determined by the bricks it contains.*

PROOF. Let $\mathcal{C}$ be an abelian extension-closed subcategory of A-mod, and let $\mathcal{B}$ be the set of bricks it contains. The statement of the lemma follows from the fact which we shall establish that the objects of $\mathcal{C}$ are exactly those A-modules which admit a filtration by modules in $\mathcal{B}$.

Clearly, any module filtered by modules in $\mathcal{B}$ is contained in $\mathcal{C}$, because $\mathcal{C}$ is extension-closed. Conversely, let $X \in \mathcal{C}$. If X is a brick, it is in $\mathcal{B}$ and we are done. Otherwise, X admits a non-invertible endomorphism α. Now X is isomorphic to the extension of the image of α by its kernel, both of which have smaller total dimension than X, and both of which are in $\mathcal{C}$ because it is an exact abelian subcategory, so we are done by induction. □

Thus, if we want to understand the map from $K_0^*(A)_\mathbb{R}$ to semistable subcategories, it suffices to understand, for each brick of A-mod, the region of $K_0^*(A)_\mathbb{R}$ for which it is semistable. The category $(A\text{-mod})^\phi$ will consist of all modules filtered by the bricks that are semistable for ϕ.

Our goal in this paper is to describe this picture for finite-type preprojective algebras. This will require a detour into the theory of hyperplane arrangements.

2. Preprojective algebras

First, though, we introduce the finite-type preprojective algebras. Let Q be a simply-laced Dynkin quiver, with a set $Q_0 = \{1, \dots, n\}$ of vertices and a set Q_1 of arrows. Define $\overline{Q}$ to be the doubled quiver of Q, which is to say, for each arrow $a : i \to j$, we add an arrow $a^* : i \leftarrow j$. The preprojective algebra is then defined to be:

$$\Pi = k\overline{Q} / \sum_{a \in Q_1} (aa^* - a^*a).$$

This is a finite-dimensional self-injective algebra.

Preprojective algebras were originally introduced by Gelfand and Ponomarev [**GP**], and, in a version closer to the formulation which is now standard, by Dlab and Ringel [**DR**]. They arise naturally in geometric representation theoretic contexts, playing, for example, an essential role in Lusztig's definition of the semicanonical basis of the enveloping algebra of the positive part of a symmetric Kac-Moody Lie algebra [**Lu**]. For our purposes, we can just take them as an interesting class of algebras with a Dynkin classification; as we shall see, other elements of Dynkin diagram combinatorics will also turn out to be relevant to their analysis.

As a simple example, let us consider the preprojective algebra of type A_2. We have two vertices 1 and 2, an arrow a from 1 to 2, an arrow a^* from 2 to 1, and the relation $aa^* - a^*a$. Multiplying this relation on both sides by the idempotent at 1, and multiplying on both sides by the idempotent at 2, we deduce that the ideal generated by $aa^* - a^*a$ actually contains each of aa^* and a^*a, so in this case, we could have described the ideal of relations as being generated by aa^* and a^*a.

Either by noticing that this implies that the preprojective algebra of type A_2 happens to be a gentle algebra, or just by thinking about it, we determine that this algebra has four indecomposable modules: the simples at each vertex and the projectives at each vertex, which are of length two. All four of these modules are bricks. In keeping with the point of view developed in the previous section, we can ask ourselves where these bricks are semistable. The answer is given in Figure 2.

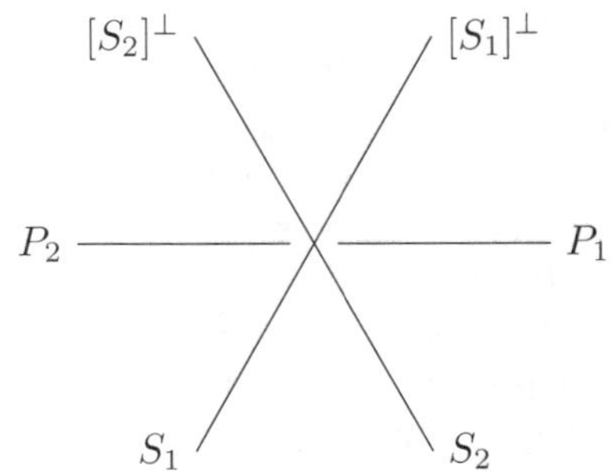

FIGURE 2. Regions of semistability for the preprojective algebra of type A_2

The verification that this picture is correct is essentially the same as for the hereditary example examined above. For clarity, the regions where P_1 and P_2 are semistable are drawn as if they don't quite touch the origin, but in fact they extend up to and include it.

We notice that if we consider the union of the lines and half-lines where at least one brick is semistable, this is a very symmetrical picture. As we shall see, this is no coincidence, but in order to make this notion precise, we shall have to introduce some further technology: specifically, we shall have to introduce the Weyl groups to provide the symmetries we want.

3. Weyl groups

Good general references for Weyl groups are [**Hu**, **BB**].

We want to define a bilinear form on $K_0(\Pi)$. This can be defined very explicitly by saying that $\langle [S_i], [S_i] \rangle = 2$, and for $j \neq i$, $\langle [S_i], [S_j] \rangle$ is minus the number of arrows between vertices i and j in Q.

A more conceptual definition is to consider the affine-type quiver $\hat{Q}$ which is obtained by adding a single vertex to Q. (There is a unique way to do this.) We can then define the corresponding preprojective algebra $\hat{\Pi}$, and the category of nilpotent $\hat{\Pi}$-modules, and consider its Grothendieck group and Euler form: for V and W nilpotent $\hat{\Pi}$-representations,

$$\begin{aligned} \langle [V], [W] \rangle &= \sum_{i=0}^{\infty} (-1)^i \dim \operatorname{Ext}^i(V, W) \\ &= \dim \operatorname{Hom}(V, W) - \dim \operatorname{Ext}^1(V, W) + \dim \operatorname{Ext}^2(V, W). \end{aligned}$$

Restricted to the subspace spanned by the classes of the simple modules of Π, we recover the form defined in the previous paragraph. Note that we cannot directly take the Euler form for Π because its global dimension is not finite, so $\sum_{i=0}^{\infty} (-1)^i \dim \operatorname{Ext}^i(V, W)$ is not well-defined.

The bilinear form on $K_0(\Pi)$ turns out to be positive definite. Since, in particular, it is non-degenerate, we can use it to identify $K_0(\Pi) \otimes \mathbb{R}$ with its dual, and thus define a positive definite symmetric bilinear form on $K_0^*(\Pi)_{\mathbb{R}}$.

We now want to define a group action on $V = K_0^*(\Pi)_{\mathbb{R}}$. Let $e_1, \ldots, e_n$ be the standard basis for V, with $e_i([S_j]) = \delta_{ij}$, where δ_{ij} is 1 if $i = j$ and 0 otherwise. Define a linear transformation s_i of V by

$$s_i(\phi) = \phi - \langle e_i, \phi \rangle e_i$$

Each s_i acts by reflecting in a hyperplane with respect to the bilinear form on V. We call the transformations s_i *simple reflections.*

The group generated by these n elements forms, by definition, the Weyl group associated to Q. We denote it by W, and we think of it as acting on V on the right.

For future use, for any $w \in W$, we define $\ell(w)$ to be the length of the shortest possible expression for w as a product of the simple reflections. The identity element is the unique element of length zero, and the simple reflections are exactly the elements of length one.

Define T to be the set of conjugates in W of $s_1, \ldots, s_n$. It is easy to see that all of these elements will also act by reflections. In fact, they are all the reflections in W.

By definition, the reflection arrangement $\mathcal{H}_\Pi$ associated to W consists of the collection of reflecting hyperplanes in V associated to the set T of reflections.

We can now state a rough version of our main result: the region in $K_0^*(\Pi)_\mathbb{R}$ where $(\Pi\text{-mod})^\phi \neq 0$ consists of exactly the union of the hyperplanes in $\mathcal{H}_\Pi$. This accounts for the regularity which we observed in the case of the A_2 preprojective algebra. In order to refine this result to get a picture like Figure 2, which reflects where each brick is semistable, we will need some way to divide the reflecting hyperplanes up into pieces. It turns out that a natural way to do this was developed, for superficially different purposes, by Reading [**Re1**], as we now explain.

4. Shards

We must now take a detour into the theory of hyperplane arrangements and, in particular, the poset structure on the poset of regions defined by a hyperplane arrangement. The key results we need are to be found in [**Re1**]. [**Re2**] is an exposition which provides further context.

Let $\mathcal{H}$ be a hyperplane arrangement in $\mathbb{R}^n$, by which we mean a collection of finitely many linear hyperplanes in $\mathbb{R}^n$.

$\mathcal{H}$ defines a set of *chambers*, which are the connected components of $\mathbb{R}^n \setminus \bigcup_{H\in\mathcal{H}} H$.

There is a natural graph structure associated to $\mathcal{H}$, which we denote $G(\mathcal{H})$. The vertices are the chambers, and two chambers are adjacent if their closures intersect in a codimension-one region in $\mathbb{R}^n$.

We shall define a poset structure on the set of chambers by specifying its *cover relations*, that is to say, the pairs E, F such that $E < F$ and there is no G with $E < G < F$. We write $E \lessdot F$ for a cover relation in a poset. The Hasse diagram of a poset is the directed graph on the elements of the poset, whose edges (E, F) are exactly the cover relations $E \lessdot F$ of the poset.

Choose a base chamber and call it C. Define the chamber poset $P(\mathcal{H}, C)$ on the set of chambers by imposing that $E \lessdot F$ if and only if E and F are adjacent and E lies on the same side as C of the hyperplane defined by the intersection of the closures of E and F.

The Hasse diagram of $P(\mathcal{H}, C)$ is then an orientation of the graph $G(\mathcal{H})$. The chamber C is the unique source, corresponding to the fact that it is the minimum element of the poset, and the chamber $-C$ is the maximum element of the poset. (For any chamber E, note that $-E$ also forms a chamber.) This poset was introduced by Edelman [**Ed**].

A chamber is called *simplicial* if it consists of positive linear combinations of n linearly independent vectors. $\mathcal{H}$ is called simplicial if all its chambers are simplicial. If $n \leq 2$, all hyperplane arrangements are simplicial, but this is not true for larger n.

An important source of simplicial arrangements are the *reflection arrangements*. The reflection arrangement associated to Π, which we have already introduced, is an example, but any finite Coxeter group yields a reflection arrangement in the same way. The reader who is unfamiliar with Coxeter groups may simply take our statements about reflection arrangements as applying to the reflection arrangements we have already introduced, with no loss.

Let $\mathcal{H}$ be a reflection arrangement. W acts on the set of chambers simply-transitively, so, after identifying the base chamber C with the identity element of

W, we can identify the chambers with the elements of W. The poset $P(\mathcal{H}, C)$ is then a well-known poset on W, known as *(right) weak order*, in which the cover relations are given by $v \lessdot vs_i$ if $\ell(vs_i) = \ell(v) + 1$.

A poset is called a *lattice* if any pair of elements E, F has a unique greatest lower bound, denoted $E \wedge F$ (the *meet* of E and F), and a unique least upper bound, denoted $E \vee F$ (the *join* of E and F).

THEOREM 1 ([**BEZ**, Theorem 3.4]). *If $\mathcal{H}$ is a simplicial arrangement, then the poset $P(\mathcal{H}, C)$ is a lattice.*

Simplicialness is not necessary for $P(\mathcal{H}, C)$ to be a lattice, see [**Re2**]. However, since our eventual application will be to reflection arrangements, we may as well not seek the greatest possible generality.

A lattice L is called semi-distributive if for E, F, G in L such that $E \vee F = E \vee G$ it follows that this element also equals $E \vee (F \wedge G)$, and dually if $E \wedge F = E \wedge G$, then this element also equals $E \wedge (F \vee G)$. We have the following result:

THEOREM 2 ([**Re2**, Corollary 9-3.9]). *If $\mathcal{H}$ is a simplicial arrangement, then the lattice $P(\mathcal{H}, C)$ is semidistributive.*

It is an immediate consequence of semidistributivity that if $G > E$ then there is a unique minimum element among all elements F such that $E \vee F = G$.

An element E of a lattice L is called *join-irreducible* if it is not the minimum element of the lattice, and it cannot be written as $E = F \vee G$ with $F, G < E$. The following lemma is an easy exercise.

LEMMA 2. *If L is a finite lattice, then E is join-irreducible in L if and only if E covers exactly one element.*

If E is join-irreducible, we write E_* for the unique element which it covers.

In a finite lattice, every element can be written as a join of join-irreducible elements, so they have an obviously important structural role. (This role is shared with the meet-irreducible elements, which are defined dually, and can be studied in the same way as we are doing for join-irreducible elements.)

There is a natural labelling of the edges of the Hasse diagram of $P(\mathcal{H}, C)$ by join-irreducible elements, as follows: define the join-irreducible label $j(E \gtrdot F)$ to be the minimum G such that $E \vee G = F$. By semidistributivity, this is well-defined, and it is clear that it must be join-irreducible.

Given the importance of join-irreducible elements of a lattice, it is natural to ask how to see the join-irreducible elements of $P(\mathcal{H}, C)$ in terms of the geometry of $\mathcal{H}$.

Each join-irreducible of $P(\mathcal{H}, C)$ is naturally associated to a particular hyperplane. Namely, if E is join-irreducible, then by Lemma 2, it covers a unique other chamber F, and by the definition of the cover relations in $P(\mathcal{H}, C)$, the span of intersection of the closures of E and F defines a hyperplane of $\mathcal{H}$.

This map from join-irreducible elements of $P(\mathcal{H}, C)$ to $\mathcal{H}$ is not a bijection, as demonstrated by Figure 3, which show the Hasse diagram of the poset of regions of a two-dimensional hyperplane arrangement superimposed over the hyperplane arrangement. We always draw the base chamber at the bottom. The join-irreducible elements are marked with black dots, and the arrows indicate the map from join-irreducibles to hyperplanes. We see that there are two join-irreducible elements which are associated to the hyperplanes H_2 and H_3.

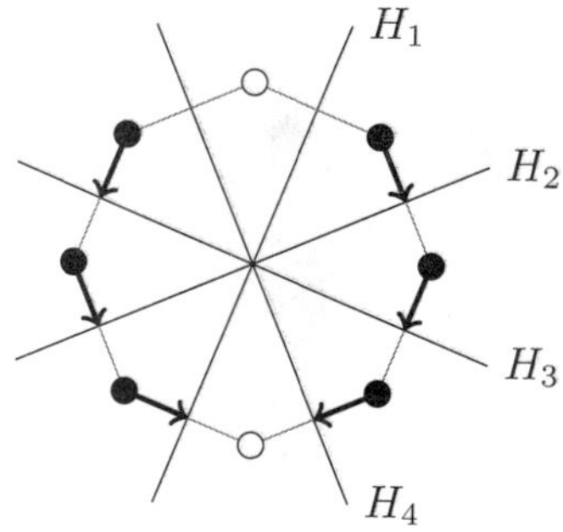

FIGURE 3. Poset of regions of a two-dimensional hyperplane arrangement

To define a bijection from join-irreducible elements to something geometric, Reading was impelled to split some of the hyperplanes in two, as in Figure 4. Now, each join-irreducible element (black dot) has a distinct hyperplane or half-hyperplane directly below it.

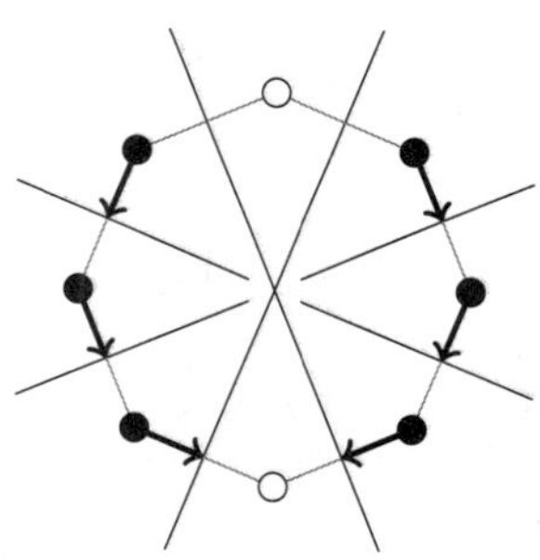

FIGURE 4. Splitting hyperplanes when $n = 2$

More formally, when $n = 2$, Reading splits in two the hyperplanes which are not adjacent to the base chamber, and calls this set of hyperplanes and half-hyperplanes the *shards* of $\mathcal{H}$. Now, if E is a join-irreducible element of $P(\mathcal{H}, e)$, we see that the facet of E corresponding to the unique cover $E > F$ lies in a well-defined shard of $\mathcal{H}$, and this gives us a bijection from join-irreducible elements to shards.

The fact that this works for $n = 2$ is a rather trivial observation. The surprising fact is that this simple strategy of splitting up hyperplanes is exactly what is needed in general.

To define the general strategy, we need to introduce some further notation. Let $\mathcal{H}^{(2)}$ be the set of codimension-two intersections of hyperplanes from $\mathcal{H}$, i.e.,

$$\mathcal{H}^{(2)} = \{H \cap K \mid H, K \in \mathcal{H}, H \neq K\}.$$

For each $X \in \mathcal{H}^{(2)}$, consider the hyperplanes in $\mathcal{H}$ containing X. Note that since C is a chamber of our original arrangement, and we are now considering the sub-arrangement of just those hyperplanes that contain X, the chamber C is located in a particular chamber of this sub-arrangement. Number the hyperplanes containing X cyclically as $H_1, H_2, \dots, H_r$ so that C is between H_1 and H_r, as in Figure 5.

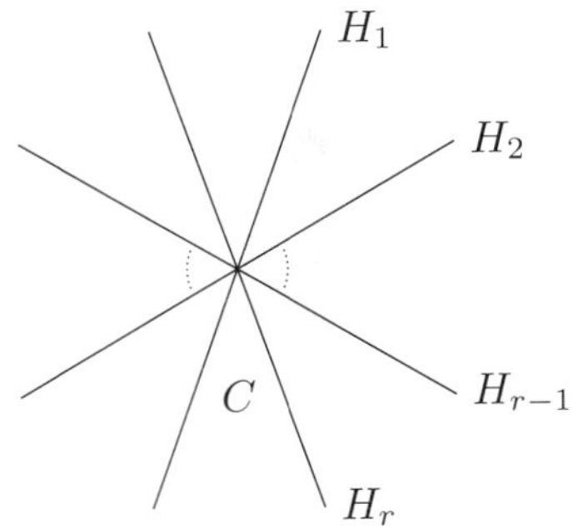

FIGURE 5. Numbering hyperplanes around a codimension 2 intersection

The idea is that, around X, we will split the hyperplanes as in the $n = 2$ situation previously discussed. We want X to split the hyperplanes which are not adjacent to C, so we define $\mathrm{Split}(X) = \{H_2, \dots, H_{r-1}\}$.

Now, a hyperplane $H \in \mathcal{H}$ is split into a set of shards, which we denote $Ш_H$, by defining

$$Ш_H = \text{ the components of } \left(H \setminus \bigcup_{X \mid H \in \mathrm{Split}(X)} X \right)$$

and

$$Ш(\mathcal{H}) = \bigcup_{H \in \mathcal{H}} Ш_H.$$

Given a cover relation $E \gtrdot F$ in $P(\mathcal{H}, C)$, the intersection of the two chambers E and F (which is a cone in the hyperplane separating them) lies entirely in one shard. We can therefore define $Ш(E \gtrdot F)$ to be this shard.

Reading proved:

THEOREM 3. [**Re1**, Proposition 3.3] *The map from join-irreducible elements of $P(\mathcal{H}, e)$ to shards, sending a join-irreducible G to $Ш(G \gtrdot G_*)$, is a bijection.*

Further, we have the following theorem (closely related to statements in [**Re1**, **Re2**], but expressed in a way that is convenient for us):

THEOREM 4. *The map sending G to $Ш(G \gtrdot G_*)$, sends the label $j(E \gtrdot F)$ to the label $Ш(E \gtrdot F)$ for any $E \gtrdot F$.*

PROOF. Let G be the join-irreducible corresponding to the shard separating E and F. Since E is above that shard, we have $G \leq E$ by [**Re1**, Lemma 3.5]. Thus $G \vee F = E$. Any element below G is below the hyperplane separating E and F. Thus G is a minimal element among those which join with F to give E. Since $P(\mathcal{H}, e)$ is semidistributive, G must be the minimum element, and thus $G = j(E \gtrdot F)$, and we have that the shard associated to $j(E \gtrdot F)$ is indeed $Ш(E \gtrdot F)$. □

5. Join-irreducibles of W and bricks of Π

In [**IRRT**], we constructed a bijection between join-irreducible elements of W and bricks of Π. The simplest way to state it is as follows. Let e_i be the idempotent of Π corresponding to the vertex i. Define the two-sided ideal $I_i = \Pi(1 - e_i)\Pi$.

Consider a word $\underline{w} = (i_1, \dots, i_r)$ with each $i_j \in \{1, \dots, n\}$. Define $I_{\underline{w}} = I_{i_1} \dots I_{i_r}$. We say that $(i_1, \dots, i_r)$ is a reduced word for $w \in W$ if $w = s_{i_1} \dots s_{i_r}$ and this is an expression for w of the minimum possible length.

PROPOSITION 1 ([**IR**]). *If $\underline{w}_1$ and $\underline{w}_2$ are reduced words for w, then $I_{\underline{w}_1} = I_{\underline{w}_2}$.*

We can therefore define I_w to be the ideal $I_{\underline{w}}$ where $\underline{w}$ is any reduced word for w.

Let $w \gtrdot u$ be a cover relation in weak order. Following [**IRRT**], we define the brick label for $B(w \gtrdot u)$ to be I_u/I_w. This module turns out to be, indeed, a brick.

In [**IRRT**], we also consider the join-irreducible labelling $j(w \gtrdot u)$. The definition used there is not the same as the one given here, but they are equivalent by [**IRRT**, Proposition 2.1].

One of the main results of [**IRRT**] can be stated as follows:

THEOREM 5 ([**IRRT**, Theorem 1.3]). *The map from join-irreducibles of W to bricks of Π sending w to $B(w \gtrdot w_*)$ is a bijection which transforms the join-irreducible labelling into the brick labelling.*

We can now state the main theorem of this note:

THEOREM 6. *For w a join-irreducible of W, the region where the brick $B(w \gtrdot w_*)$ is semistable is the closure of the shard $\text{Ш}(w \gtrdot w_*)$.*

6. Technical Lemmas

Before we begin the proof of the main theorem, we need a few technical lemmas.

LEMMA 3. *Let M be a Π-module such that $\operatorname{Hom}(S_i, M) = 0$. Then $[I_i \otimes M] = s_i([M])$.*

PROOF. If $\operatorname{Hom}(S_i, M) = 0$, then $I_i \otimes M$ is isomorphic to the result of applying a certain spherical twist functor to M, where M is thought of in the derived category of the corresponding affine-type preprojective algebra $\hat{\Pi}$. (This part of the conclusion of [**IRRT**, Proposition 3.2(b)] follows if we assume only that $\operatorname{Hom}(S_i, M) = 0$, although there, an additional homological assumption on M is made.) Spherical twists act like reflections on the level of the Grothendieck group. (See for example [**AIRT**, Lemma 2.6].) □

LEMMA 4. *Let $w \gtrdot u$ be a cover in weak order on W. Let i be such that $\ell(s_i w) > \ell(w)$. Then $\operatorname{Hom}(S_i, B(w \gtrdot u)) = 0$.*

PROOF. The Weyl group element u determines a torsion class $T_u = \operatorname{Fac} I_u$, and a corresponding torsion-free class F_u. Because $\ell(s_i u) > \ell(u)$, $I_{s_i u}$ is properly contained in I_u, and thus S_i is in the top of I_u, and in particular, $S_i \in T_u$. On the other hand, by [**IRRT**, Theorem 4.5], $B(w \gtrdot u) \in F_u$. Thus $\operatorname{Hom}(S_i, B(w \gtrdot u)) = 0$. □

We remark that under the hypotheses of Lemma 4, since $B(s_i w \gtrdot s_i u) \cong I_i \otimes B(w \gtrdot u)$, what Lemma 4 says is that Lemma 3 applies, so that $[B(s_i w \gtrdot s_i u)] = s_i([B(w \gtrdot u)])$. This is part of what is implied by Theorem 6; see also [**AIRT**, Theorem 2.7(1)].

LEMMA 5. *Let N be a submodule of M, and suppose that $\operatorname{Hom}(S_i, M) = 0$. Then the kernel of the induced map from $I_i \otimes N$ to $I_i \otimes M$ is a sum of some number (possibly zero) of copies of S_i.*

PROOF. From the short exact sequence

$$0 \to N \to M \to M/N \to 0$$

we obtain

$$\mathrm{Tor}_1(I_i, M/N) \to I_i \otimes N \to I_i \otimes M \to I_i \otimes M/N \to 0$$

To evaluate $\mathrm{Tor}_1(I_1, M/N)$, we can take

$$0 \to I_i \to \Pi \to S_i \to 0$$

and tensor by M/N, obtaining that $\mathrm{Tor}_1(I_i, M/N) \cong \mathrm{Tor}_2(S_i, M/N)$. As a Π-module, $\mathrm{Tor}_2(S_i, M/N)$ is congruent to a sum of some number of copies of S_i. □

7. Proof of Main Theorem

PROOF OF THEOREM 6. Let $w = us_i \gtrdot u$ be a cover in weak order on W. We will prove by reverse induction on the length of w that $B(w \gtrdot u)$ is semistable on the (closed) facet of the Coxeter fan corresponding to $w \gtrdot u$.

There is a unique element of W of maximal length, usually denoted w_0, and the chamber corresponding to it is $-C$. The hyperplanes that bound it are perpendicular to the simple roots, and the modules corresponding to the covers are the simple modules, each of which is semistable on its entire perpendicular hyperplane. This establishes the base case of the induction.

Now suppose that $w < w_0$. Let s_j be a simple reflection such that $\ell(s_j w) > \ell(w)$. Let $w' = s_j w$, $u' = s_j u$, $B' = B(w' \gtrdot u')$. B' is related to B by $B' = I_j \otimes B$. By Lemma 4 and Lemma 3, we have that $[B'] = s_i([B])$.

Suppose that B is not semistable for some ϕ in the facet corresonding to $w \gtrdot u$. This must be because of some subobject E of B such that $\phi([E]) > 0$. Define $\phi' = s_i(\phi)$. It falls on the facet $s_j w \gtrdot s_j u$. We want to conclude that there is a corresponding destabilizing subobject of B' for ϕ', which would contradict our induction hypothesis.

By Lemma 4, $\mathrm{Hom}(S_i, B) = 0$. It therefore follows that $\mathrm{Hom}(S_i, E) = 0$, so we can apply Lemma 3 to conclude that $[I_i \otimes E] = s_i([E])$. Therefore $\phi'([I_i \otimes E]) = \phi([E]) > 0$. Let E' be the image of $I_i \otimes E$ in B'. The kernel of the natural map from $I_i \otimes E$ to E' is a sum of copies of S_i by Lemma 5. Since $\ell(s_i u) > \ell(u)$, the chamber of u lies on the opposite side from C of the hyperplane perpendicular to $[S_i]$. Thus, $\phi'([S_i]) \leq 0$, so $\phi'([E']) \geq \phi'([I_i \otimes E]) > 0$. It follows that E' is destabilizing for B' with respect to ϕ', which is contrary to our induction hypothesis. Therefore $B(w \gtrdot u)$ is semistable with respect to weights on the facet corresponding to $w \gtrdot u$, as desired.

Now we prove the opposite direction, namely, that a brick must be unstable outside the closure of the corresponding shard. Let w be a join-irreducible of W. Let $Ш = Ш(w \gtrdot w_*)$, $B = B(w \gtrdot w_*)$. Consider a facet X of $Ш$. By the construction of shards, the span of X is a codimension two intersection in $\mathcal{H}^{(2)}$, and around X we have a picture with four shards, as shown in Figure 6. As always, the base chamber C is at the bottom.

By [**IRRT**, Proposition 4.3], if E and F are the bricks associated to the shards as in the picture, then there is a short exact sequence:

$$0 \to E \to B \to F \to 0.$$

For $\phi \in C$, we have that $\phi([S_i]) > 0$ for all i. Let $\theta \in Ш$. Since $Ш$ is on the opposite side of $[E]^{\perp}$ from E, $\theta([E]) \leq 0$. This is consistent with the fact which

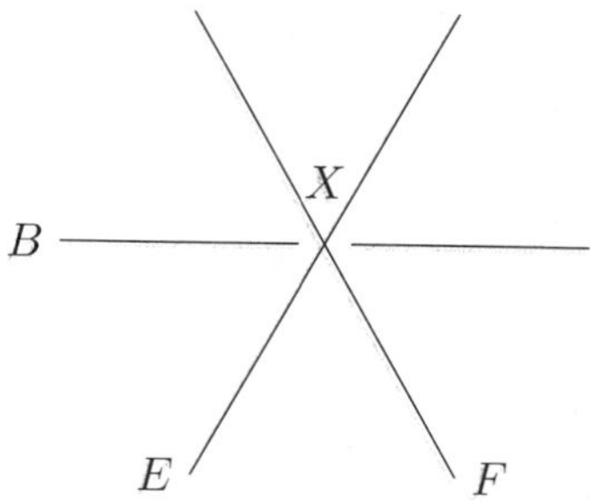

FIGURE 6. The local picture around X

we have already established that B is θ-semistable. However, if θ is strictly on the opposite side of the hyperplane in $[B]^{\perp}$ defined by the span of X, then $\theta([E]) > 0$, so B is not θ-semistable. This establishes the theorem. $\square$

8. Connection to other work

Baumann, Kamnitzer, and Tingley [**BKT**] study the representation theory of preprojective algebras of affine type. Many of the ideas from this note could also be extracted from their work, but they do not discuss shards, so the combinatorics we present here is less explicitly developed.

Crawley-Boevey establishes a result about the existence of representations of deformed preprojective algebras [**CB2**, Theorem 1.2] which implies that $(\Pi\text{-mod})^{\phi} \neq 0$ iff ϕ lies on a reflecting hyperplane by [**CB**, Lemma 3]. However, the argument to pass from the deformed preprojective algebra to semistable representations of the usual preprojective algebras depends on an assumption that the ground field is the complex numbers.

Acknowledgements

I would like to thank David Speyer, who gave me the suggestion that there should be a connection between shards and stability. Discussions with him were crucial to the development of my understanding of this topic. The work presented here is part of an ongoing larger joint project. I would also like to thank Osamu Iyama, Nathan Reading, and Idun Reiten; this paper also draws greatly on things I learned from our collaboration. I am grateful to the organizing committee of ICRA 2016 for having invited me to give the talk the details of which are fleshed out here, and to the referee for his or her comments. I would also like to record my gratitude to the Université Paris VII for the excellent working conditions in which I finished writing up this paper. I gratefully acknowledge financial support from NSERC and the Canada Research Chairs program.

References

[AIRT] Claire Amiot, Osamu Iyama, Idun Reiten, and Gordana Todorov, *Preprojective algebras and c-sortable words*, Proc. Lond. Math. Soc. (3) **104** (2012), no. 3, 513–539, DOI 10.1112/plms/pdr020. MR2900235

[BKT] Pierre Baumann, Joel Kamnitzer, and Peter Tingley, *Affine Mirković-Vilonen polytopes*, Publ. Math. Inst. Hautes Études Sci. **120** (2014), 113–205, DOI 10.1007/s10240-013-0057-y. MR3270589

[BB] Anders Björner and Francesco Brenti, *Combinatorics of Coxeter groups*, Graduate Texts in Mathematics, vol. 231, Springer, New York, 2005. MR2133266

[BEZ] Anders Björner, Paul H. Edelman, and Günter M. Ziegler, *Hyperplane arrangements with a lattice of regions*, Discrete Comput. Geom. **5** (1990), no. 3, 263–288, DOI 10.1007/BF02187790. MR1036875

[Ch] Calin Chindris, *Cluster fans, stability conditions, and domains of semi-invariants*, Trans. Amer. Math. Soc. **363** (2011), no. 4, 2171–2190, DOI 10.1090/S0002-9947-2010-05184-0. MR2746679

[CB] William Crawley-Boevey, *On the exceptional fibres of Kleinian singularities*, Amer. J. Math. **122** (2000), no. 5, 1027–1037. MR1781930

[CB2] William Crawley-Boevey, *Geometry of the moment map for representations of quivers*, Compositio Math. **126** (2001), no. 3, 257–293, DOI 10.1023/A:1017558904030. MR1834739

[DW] Harm Derksen and Jerzy Weyman, *The combinatorics of quiver representations* (English, with English and French summaries), Ann. Inst. Fourier (Grenoble) **61** (2011), no. 3, 1061–1131, DOI 10.5802/aif.2636. MR2918725

[DR] Vlastimil Dlab and Claus Michael Ringel, *The preprojective algebra of a modulated graph*, Representation theory, II (Proc. Second Internat. Conf., Carleton Univ., Ottawa, Ont., 1979), Lecture Notes in Math., vol. 832, Springer, Berlin-New York, 1980, pp. 216–231. MR607155

[Ed] Paul H. Edelman, *A partial order on the regions of* $\mathbf{R}^n$ *dissected by hyperplanes*, Trans. Amer. Math. Soc. **283** (1984), no. 2, 617–631, DOI 10.2307/1999150. MR737888

[GP] I. M. Gel′fand and V. A. Ponomarev, *Model algebras and representations of graphs* (Russian), Funktsional. Anal. i Prilozhen. **13** (1979), no. 3, 1–12. MR545362

[Hu] James E. Humphreys, *Reflection groups and Coxeter groups*, Cambridge Studies in Advanced Mathematics, vol. 29, Cambridge University Press, Cambridge, 1990. MR1066460

[IOTW] Kiyoshi Igusa, Kent Orr, Gordana Todorov, and Jerzy Weyman, *Cluster complexes via semi-invariants*, Compos. Math. **145** (2009), no. 4, 1001–1034, DOI 10.1112/S0010437X09004151. MR2521252

[IT] Colin Ingalls and Hugh Thomas, *Noncrossing partitions and representations of quivers*, Compos. Math. **145** (2009), no. 6, 1533–1562, DOI 10.1112/S0010437X09004023. MR2575093

[IPT] Colin Ingalls, Charles Paquette, and Hugh Thomas, *Semi-stable subcategories for Euclidean quivers*, Proc. Lond. Math. Soc. (3) **110** (2015), no. 4, 805–840, DOI 10.1112/plms/pdv002. MR3335288

[IRRT] O. Iyama, N. Reading, I. Reiten, and H. Thomas. Lattice structure of Weyl groups via representation theory of preprojective algebras. arXiv:1604.08401.

[IR] Osamu Iyama and Idun Reiten, *Fomin-Zelevinsky mutation and tilting modules over Calabi-Yau algebras*, Amer. J. Math. **130** (2008), no. 4, 1087–1149, DOI 10.1353/ajm.0.0011. MR2427009

[Lu] G. Lusztig, *Semicanonical bases arising from enveloping algebras*, Adv. Math. **151** (2000), no. 2, 129–139, DOI 10.1006/aima.1999.1873. MR1758244

[Ki] A. D. King, *Moduli of representations of finite-dimensional algebras*, Quart. J. Math. Oxford Ser. (2) **45** (1994), no. 180, 515–530, DOI 10.1093/qmath/45.4.515. MR1315461

[Re1] Nathan Reading, *Noncrossing partitions and the shard intersection order*, J. Algebraic Combin. **33** (2011), no. 4, 483–530, DOI 10.1007/s10801-010-0255-3. MR2781960

[Re2] N. Reading, *Lattice theory of the poset of regions*, Lattice theory: special topics and applications. Vol. 2, Birkhäuser/Springer, Cham, 2016, pp. 399–487. MR3645055

Université du Québec à Montréal, Montréal, Québec, Canada

Contemporary Mathematics
Volume **705**, 2018
http://dx.doi.org/10.1090/conm/705/14192

Computations and applications of some homological constants for polynomial representations of GL_n

Antoine Touzé

ABSTRACT. In this paper, we review the applications of the homological constants for polynomial representations of GL_n defined in [*Connectedness of cup products for polynomial representations of* GL_n, Annals of K-theory, to appear]. We also give new applications of these constants, in particular to the cohomology of classical groups. We make further progress on the problem of computing these constants for polynomial modules of interest.

1. Introduction

This article deals with polynomial representations of the general linear group over a field $\Bbbk$ of positive characteristic p, and with their restrictions to the classical matrix subgroups (i.e. symplectic and orthogonal groups). We let $\mathrm{Pol}_{n,d}$ be the category of homogeneous polynomial representations of degree d of GL_n, see Green's book [**8**] or [**7**, Section 3] for basic definitions. Such polynomial representations also belong to the realm of representations of finite dimensional $\Bbbk$-algebras. Indeed, one of the first basic results is the equivalence between $\mathrm{Pol}_{n,d}$ and the category of modules[1] over the Schur algebra $S(n,d)$.

Simple objects in $\mathrm{Pol}_{n,d}$ are indexed by partitions λ in at most n parts, of size $|\lambda| = \sum_{0\le i\le n} \lambda_i = d$. We fix a simple object L_λ for each partition λ and we let P_λ, resp. J_λ, be its projective cover, resp. its injective hull. Recall that a partition in at most n parts is called p^r-restricted if $\lambda_n < p^r$ and for all $i < n$, $\lambda_i - \lambda_{i+1} < p^r$. Such partitions appear naturally when studying the structure of polynomial modules, e.g. in the statement of the Steinberg tensor product theorem [**9**, II.3.17]. We introduced the following definition in [**20**].

DEFINITION. Given a polynomial representation M and a nonnegative integer r, we let $p(M,r) \in \mathbb{N} \cup \{+\infty\}$ be the maximal integer k such that M admits a projective resolution

$$\cdots \to P_i \to P_{i-1} \to \cdots \to P_0 \to M \to 0$$

2010 *Mathematics Subject Classification.* Primary 20G10, Secondary 18G15.
This work was supported in part by the Labex CEMPI (ANR-11–LABX–0007-01).

[1]Unlike [**8**] or [**7**], we do not assume here that our modules have finite dimension, but for our purposes it is a rather cosmetic change. Indeed, since the Schur algebra is finite dimensional, infinite dimensional modules have nice finiteness properties (they are direct limits of their finite dimensional submodules, they have finite socle length,...). Thus, most of our arguments working for finite dimensional modules equally work with infinite dimensional ones.

whose first k terms $P_0, \dots, P_{k-1}$ are direct sums of indecomposable projectives P_λ with p^r-restricted λ. Similarly, we let $i(M,r)$ be the maximal integer k such that M has an injective resolution whose first k terms $J^0, \dots, J^{k-1}$ are products of J_λ with p^r-restricted λ.

We are not aware that the homological constants $i(M,r)$ and $p(M,r)$ have been considered before. In [**20**], we found several applications of these cohomological constants. They are related to the behavior of cup products in the cohomology of GL_n, to generalizations and variants of Steinberg tensor product theorem, as well as to the homological behaviour of the Schur functor which compares the cohomology of GL_n with that of $\mathfrak{S}_n$. We also established basic computations for these cohomological constants. The purpose of this paper is to continue the work started in [**20**]. Namely, we give new applications of the constants $i(M,r)$ and $p(M,r)$, and we make further progress on the problem of computing these constants for concrete polynomial modules.

The paper is organized as follows. Section 2 concentrates on the applications of our cohomological constants. We review the applications already proved in [**20**] and we prove additional applications. The most noteworthy new result here is theorem 2.1, which shows that the surprising behavior of cup products proved in [**20**] is not specific to GL_n, but also holds for other classical matrix groups (othogonal or symplectic type). Section 3 concentrates on the problem of computing these constants for polynomial modules of interest. We prove a new characterization of $i(M,r)$ and $p(M,r)$ in proposition 3.4, related to the (derived) adjoint of the tensor product by $\Bbbk^{n\,(r)}$ (i.e. the r-th Frobenius twist of the defining representation of GL_n). With this new tool at our disposal, we undertake to study $i(M,r)$ and $p(M,r)$ for some new families of examples, including tilting modules and costantard modules associated to hooks or to some thin shaped partitions.

Thus, this article reviews some known results and establishes new ones. In order to make a clear distinction between what material is new and what is not, only new theorems and propositions are numbered.

Review of strict polynomial functors. Our applications of the homological constants $p(M,r)$ and $i(M,r)$ all assume that the polynomial modules in play are *stable* polynomial modules, i.e. they work for objects of $\mathrm{Pol}_{n,d}$ for $n \geq d$. The natural home for stating and proving theorems involving stable polynomial modules is the category of strict polynomial functors. Indeed, the categories $\mathrm{Pol}_{n,d}$ for $n \geq d$ are all equivalent to one another, and strict polynomial functors provide a model for these categories where the parasite integer n does not appear.

The remainder of the article is written in the language of strict polynomial functors, so we end this introduction by reviewing briefly some points of this theory. The reader may have full details and additional references by reading the seminal article of Friedlander and Suslin [**7**, Section 2], or [**12**] or [**20**, Section 2].

We denote by $\mathcal{P}_d$ the category of homogeneous strict polynomial functors of degree d over $\Bbbk$ (with values in arbitrary $\Bbbk$-vector spaces), and we let $\mathcal{P} = \bigoplus_{d\geq 0} \mathcal{P}_d$ be the category of strict polynomial functors of bounded degree. If $\Bbbk$ is an infinite field, a strict polynomial functor of degree d is simply a functor F from finite dimensional vector spaces to all vector spaces, such that the function $\mathrm{Hom}_\Bbbk(U,V) \to \mathrm{Hom}_\Bbbk(F(U),F(V))$ determined by F is a polynomial of degree d. In characteristic zero these functors appear in MacDonald's classical book [**13**], where they are called 'polynomial functors'. As already mentioned above, over any field $\Bbbk$, $\mathcal{P}_d$ is equivalent to the category $S(n,d)-\mathrm{Mod}$ of modules over the Schur algebra, or to the category $\mathrm{Pol}_{n,d}$, for any $n \geq d$. The equivalence is obtained by evaluating a functor F on $\Bbbk^n$ [**7**, Thm 3.2].

Many usual functors of $\Bbbk$-vector spaces are strict polynomial functors. To fix notations, we now provide a list of examples of strict polynomial functors which will be considered in this article.

- We let Γ^d be the d-th divided power functor, that is $\Gamma^d(V) = (V^{\otimes d})^{\mathfrak{S}_d}$ is the subspace of invariants of $V^{\otimes d}$ under the action of $\mathfrak{S}_d$ which permutes the factors of the tensor product (in particular Γ^0 is the constant functor with value $\Bbbk$). We let S^d be the d-th symmetric power functor, that is $S^d(V) = (V^{\otimes d})_{\mathfrak{S}_d}$ is the quotient space of coinvariants. More generally, if $\lambda = (\lambda_1, \dots, \lambda_n)$ is any n-tuple of nonnegative integers of weight $|\lambda| = \sum \lambda_i = d$, we let

$$\Gamma^\lambda = \Gamma^{\lambda_1} \otimes \cdots \otimes \Gamma^{\lambda_n}\ , \qquad \text{and} \qquad S^\lambda = S^{\lambda_1} \otimes \cdots \otimes S^{\lambda_n}\ .$$

  The functors Γ^λ, resp. S^λ, for all tuples λ of weight d, form a projective generator, resp. injective cogenerator, of $\mathcal{P}_d$. We also use a similar notation Λ^λ indexed by tuples λ for tensor products of exterior power functors. If $(1^{\{d\}}) = (1, \dots, 1)$ (d terms equal to one), we will denote by $\otimes^d$ the tensor product

$$\otimes^d := \Gamma^{(1^{\{d\}})} = \Lambda^{(1^{\{d\}})} = S^{(1^{\{d\}})}\ .$$

- $\mathcal{P}_d$ is a highest weight category [**4**] with simple objects indexed by the poset of partitions λ of weight (or size) $|\lambda| = d$, equipped with the dominance order $\trianglerighteq$. In particular we have the following functors attached to a partition λ. In the first four examples, λ is called the *highest weight* of the functor F_λ, because the evaluation $F_\lambda(\Bbbk^n)$ with $n \geq d$ is a polynomial module with highest weight λ.
  (1) We let L_λ be the simple functor indexed by λ.
  (2) We let S_λ be the costandard object indexed by λ. This functor is nothing but the Schur functor defined by Akin Buchsbaum and Weyman [**2**]. The notation used in [**2**] is different from ours, see [**17**, Section 6.1.1] for the conversion between notations. In particular $S_{(1^{\{d\}})} = \Lambda^d$ and $S_{(d)} = S^d$.
  (3) We let W_λ be the standard object indexed by λ. The functors W_λ are called Weyl functors and can be obtained from the Schur functors by duality: $W_\lambda = S_\lambda^\sharp$, where the dual $F^\sharp$ of a functor F satisfies $F^\sharp(V) = F(V^\vee)^\vee$ with '$\vee$' denoting the duality of $\Bbbk$ vector spaces. In particular $W_{(1^{\{d\}})} = \Lambda^d$ and $W_{(d)} = \Gamma^d$.
  (4) We let T_λ be the indecomposable tilting object indexed by λ. Each T_λ is characterized by indecomposability, self-duality, and by the existence of a filtration whose subquotients are Schur functors S_μ satisfying $\mu \trianglelefteq \lambda$, such that S_λ appears exactly once as a subquotient. The functor T_λ is a direct summand of $\Lambda^{\widetilde{\lambda}}$, where $\widetilde{\lambda}$ is the conjugate partition of λ.
  (5) We let P_λ, resp. J_λ, be the projective cover, resp. injective envelope, of L_λ. Then P_λ is a direct summand of Γ^λ and J_λ is a direct summand of S^λ.

- Finally, given a nonnegative integer r, we let $I^{(r)}$ be the r-th Frobenius twist. Thus $I^{(r)}(V)$ is the subspace of $S^{p^r}(V)$ generated by the elements of the form v^{p^r}, $v \in V$. The functor $I^{(0)} = S^1 = \Lambda^1 = \Gamma^1$ is often simply denoted by I (it is the identity functor of $\Bbbk$-vector spaces). For an arbitrary F, we denote by $F^{(r)}$ the composition $F \circ I^{(r)}$.

REMARK 1.1. In our notations, upper partitions or tuples mean tensor products, while lower partitions mean an index related to the highest weight category structure. For example if $\lambda = (3, 2, 2)$, then $S^\lambda = S^3 \otimes S^2 \otimes S^2$ is a tensor product of symmetric powers, while S_λ is the costandard (Schur) functor with highest weight $(3, 2, 2)$.

The homological constants $i(M, r)$ and $p(M, r)$ have the following alternative definition. We say that a tuple of nonnegative integers $\lambda = (\lambda_1, \dots, \lambda_n)$ is p^r-bounded if for all i, $\lambda_i < p^r$.

PROPOSITION ([**20**, Prop 4.1]). *Let $F \in \mathcal{P}_d$. Then $p(F, r)$ is the maximal (possibly infinite) integer k such that F admits a projective resolution*

$$\cdots \to P_i \to P_{i-1} \to \cdots \to P_0 \to F \to 0$$

in which the first k terms $P_0, \dots, P_{k-1}$ are direct sums of functors Γ^λ with p^r-bounded λ. Similarly, $i(F, r)$ is the maximal (possibly infinite) integer k such that F admits a injective resolution in which the first k terms $J^0, \dots, J^{k-1}$ are products of functors S^λ with p^r-bounded λ.

2. Applications of the constants $i(F, r)$ and $p(F, r)$

2.1. Stable cup products for GL_n and applications. Let us say that a quadruple of homogeneous functors (F, F', G, G') satisfies the Künneth condition if F has values in finite dimensional vector spaces and F' or G also has values in finite dimensional vector spaces. This technical condition ensures that the tensor product induces an isomorphism

$$\mathrm{Hom}_{\mathcal{P}}(F, G) \otimes \mathrm{Hom}_{\mathcal{P}}(F', G') \simeq \mathrm{Hom}_{\mathcal{P}(2)}(F \boxtimes F', G \boxtimes G')$$

where $\mathcal{P}(2)$ refers to the category of strict polynomial bifunctors. It is automatically satisfied if all the functors have values in finite dimensional vector spaces, as for the examples of strict polynomial functors given in the introduction. Our first use of the homological constants $i(F, r)$ and $p(F, r)$ is the following result on cup products.

THEOREM ([**20**, Thm 3.6]). *Let (F, G, F', G') be a quadruple of homogeneous strict polynomial functors satisfying the Künneth condition, and let $r \geq 0$. The cup product induces a graded injective map:*

$$\mathrm{Ext}^*_{\mathcal{P}_\Bbbk}(F, G) \otimes \mathrm{Ext}^*_{\mathcal{P}_\Bbbk}(F'^{(r)}, G'^{(r)}) \hookrightarrow \mathrm{Ext}^*_{\mathcal{P}_\Bbbk}(F \otimes F'^{(r)}, G \otimes G'^{(r)}) .$$

Moreover, this graded injective map is an isomorphism in degree k in the following situations.

(1) *When* $\deg F < \deg G$, *and* $k < i(G, r)$.
(2) *When* $\deg F > \deg G$, *and* $k < p(F, r)$.
(3) *When* $\deg F = \deg G$, *and* $k < p(F, r) + i(G, r)$.

This theorem may seem surprising. Indeed, it does not hold in the finite group (scheme) cohomology setting. For example, the cohomological support variety of a representation M of a finite group scheme G is the variety defined from the kernel of the cup product with $\mathrm{Id}_M \in \mathrm{Ext}^0_G(M,M)$:

$$-\cup \mathrm{Id}_M : \mathrm{Ext}^*_G(\Bbbk,\Bbbk) \to \mathrm{Ext}^*_G(M,M)\;.$$

What makes cohomological support non trivial is precisely the fact that cup products are not injective in this case.

Since precomposition by $I^{(r)}$ induces an isomorphism in Ext-degrees zero and one, we can even remove the Frobenius twist when working in these degrees. We then obtain the following corollary. This corollary allows to reduce some Ext^1 computations between simple functors to Ext^1-computations between simple functors with p-restricted highest weight. This should be compared to [**9**, II 10.16 and 10.17].

COROLLARY ([**20**, Cor 3.7]). *Let F and G be two homogeneous functors of the same degree with values in finite dimensional vector spaces. Assume that both the head of F and the socle of G are direct sums of functors L_λ with λ p^r-restricted. There are isomorphisms:*

$$\mathrm{Hom}_{\mathcal{P}}(F,G)\otimes \mathrm{Hom}_{\mathcal{P}}(F',G') \simeq \mathrm{Hom}_{\mathcal{P}}(F\otimes F'^{(r)}, G\otimes G'^{(r)})\;, \tag{2.1}$$

$$\begin{array}{c}\mathrm{Hom}_{\mathcal{P}_\Bbbk}(F,G)\otimes \mathrm{Ext}^1_{\mathcal{P}_\Bbbk}(F',G')\\ \oplus\, \mathrm{Ext}^1_{\mathcal{P}_\Bbbk}(F,G)\otimes \mathrm{Hom}_{\mathcal{P}_\Bbbk}(F',G')\end{array} \simeq \mathrm{Ext}^1_{\mathcal{P}_\Bbbk}(F\otimes F'^{(r)}, G\otimes G'^{(r)})\;. \tag{2.2}$$

In [**20**], we use this corollary to investigate the structure of functors or the form $F\otimes G^{(r)}$ where all the composition factors of F are p^r-restricted. For example the subfunctor lattice of such tensor products is essentially determined by the subfunctor lattice of F and the subfunctor lattice of G [**20**, Cor 5.12 and 5.13]. Another example is the next corollary, which can be thought of as a categorical version of the Steinberg tensor product theorem.

COROLLARY ([**20**, Cor 5.14]). *Let λ be a p^r restricted partition, and denote by $L_\lambda\otimes\mathcal{P}^{(r)}$ the full subcategory of $\mathcal{P}$ whose objects are tensor products of the form $L_\lambda\otimes F^{(r)}$ for any F. Then $L_\lambda\otimes\mathcal{P}^{(r)}$ is a localizing and colocalizing subcategory of $\mathcal{P}$. Moreover, the functor $\mathcal{P}\to L_\lambda\otimes\mathcal{P}^{(r)}$, $F\mapsto L_\lambda\otimes F^{(r)}$ is an equivalence of categories.*

2.2. Stable cup products for other classical types. In this section, we show that the property of cup products of polynomial representations described in section 2.1 has an analogue for the other classical matrix groups. To be more specific, if $G=Sp_{2n}\subset GL_{2n}$ or $G=SO_n\subset GL_n$, any polynomial representation of GL_n restricts to a representation of G. If M and N are polynomial representations, and $r\geq 0$, there is a cup product

$$H^*(G,M)\otimes H^*(G,N^{(r)})\to H^*(G,M\otimes N^{(r)})\;.$$

We will show that when the rank of G is big enough with respect to the degree of the polynomial representations in play, this cup product is injective and it is an isomorphism in low degrees. The result is stated in theorem 2.1.

As with cup products for GL_n, the natural home for stating and proving theorem 2.1 is the category of strict polynomial functors. So we first recall the connection between classical groups and strict polynomial functors proved in [**16**] (and improved in [**19**, section 7.4]). Let us fix a strict polynomial functor X. Then for all F, we let $H^*_X(F)$ be the extension groups:

$$H^i_X(F) := \bigoplus_{k\geq 0} \mathrm{Ext}^i_{\mathcal{P}}(\Gamma^k \circ X, F)\ .$$

For all k,ℓ, there is a natural inclusion $\Gamma^{k+\ell}(V) \to \Gamma^k(V)\otimes\Gamma^\ell(V)$. Replacing V by $X(-)$, we obtain a morphism of strict polynomial functors:

$$\Delta_{k,\ell} : \Gamma^{k+\ell}\circ X \to (\Gamma^k\circ X)\otimes(\Gamma^\ell\circ X)\ .$$

We can now define an associative cup product as the composite

$$H^i_X(F)\otimes H^j_X(G) \xrightarrow{\otimes} \bigoplus_{k,\ell\geq 0} \mathrm{Ext}^{i+j}_{\mathcal{P}}((\Gamma^k\circ X)\otimes(\Gamma^\ell\circ X), F\otimes G) \xrightarrow{\Delta^*_{k,\ell}} H^{i+j}_X(F\otimes G)\ .$$

The next theorem gives the link with the cohomology of symplectic groups.

THEOREM ([**16**, Thm 3.17], [**19**, Thm 7.24]). *There is a graded map, which is natural with respect to $F\in\mathcal{P}_{\Bbbk}$ and compatible with cup products:*

$$\phi_{F,2n} : H^*_{\Lambda^2}(F) \to H^*(Sp_{2n}, F(\Bbbk^{2n}))\ .$$

Moreover it is a graded isomorphism if $2n\geq \deg F$.

There are similar results for orthogonal groups and special orthogonal groups, with Λ^2 replaced by S^2. See [**16**, Thm 3.24] and [**19**, Thm 7.27 and Cor 7.31]. We can now describe the behavior of cup products for symplectic and orthogonal types.

THEOREM 2.1. *Let $X=\Lambda^2$ or $X=S^2$, let F and G be two homogeneous strict polynomial functors and let $r\geq 0$. The cup product*

$$\cup\ :\ H^*_X(F)\otimes H^*_X(G^{(r)}) \to H^*_X(F\otimes G^{(r)})$$

is injective. Moreover, it is an isomorphism in degree k if $k < i(F,r)$.

In the remainder of the section, we outline the proof of theorem 2.1. This proof is a modification of the proof for GL_n given in [**20**]. We will treat only the case when F and G both have even degrees. If one of the two functors has odd degree, then the proof is similar, and actually easier. For example, if F or G has odd degree, then the source of the cup product map in theorem 2.1 is zero for degree reasons, so injectivity is trivial.

So in the remainder of section 2.2 we assume that F has degree $2d$ and G has degree $2e$. The proof decomposes in several steps.

Step 1: reduction to a connectedness statement. Injectivity of the cup product was already proved in [**16**], and we recall the proof here. For all vector spaces V,W and all nonnegative integers n we have a canonical decomposition

$$\Gamma^n(X(V\oplus W)) \simeq \bigoplus_{k+\ell+m=n} \Gamma^k(X(V))\otimes\Gamma^\ell(V\otimes W)\otimes\Gamma^m(X(W))\ .$$

Thus $\Gamma^d(X(V))\otimes\Gamma^e(X(W))$ is a direct summand on $\Gamma^{d+e}(X(V\oplus W))$, and there is a commutative diagram, in which $\mathcal{P}(V,W)$ denotes the category of strict polynomial *bifunctors* with variables V and W (see e.g. [**20**, Section 2], [**16**, Section 2]), the top vertical arrow is induced by the canonical projection of $\Gamma^{d+ep^r}(X(V\oplus W))$

onto $\Gamma^d(X(V)) \otimes \Gamma^{ep^r}(X(W))$, and the bottom vertical arrow is induced by the sum diagonal adjunction map

$$\begin{array}{ccc}
\begin{array}{c}\mathrm{Ext}^*_{\mathcal{P}}(\Gamma^d \circ X, F) \\ \otimes \mathrm{Ext}^*_{\mathcal{P}}(\Gamma^{ep^r} \circ X, G^{(r)})\end{array} & \xrightarrow[\simeq]{\otimes} & \mathrm{Ext}^*_{\mathcal{P}(V,W)}(\Gamma^d(X(V)) \otimes \Gamma^{ep^r}(X(W)), F(V) \otimes G^{(r)}(W)) \\
& & \downarrow \\
& \searrow^{\cup} & \mathrm{Ext}^*_{\mathcal{P}(V,W)}(\Gamma^{d+ep^r}(X(V \oplus W)), F(V) \otimes G^{(r)}(W)) \\
& & \downarrow \simeq \\
& & \mathrm{Ext}^*_{\mathcal{P}}(\Gamma^{d+ep^r} \circ X, F \otimes G^{(r)})
\end{array}$$

Thus, the cup product is injective. Moreover, the cokernel of the cup product map can be identified from the diagram above. Namely, it is isomorphic to:

$$\mathrm{Ext}^*_{\mathcal{P}(V,W)}\left(\bigoplus \Gamma^k(X(V)) \otimes \Gamma^\ell(V \otimes W) \otimes \Gamma^m(X(W)), F(V) \otimes G^{(r)}(W)\right),$$

where the sum is taken over all triples (k, ℓ, m) satisfying $k + \ell + m = d + ep^r$ and $k > 0$, $m > 0$. Since there is no nontrivial extensions between homogeneous bifunctors of different bidegrees, it is also isomorphic to the direct sum of the graded vector spaces for $i > 0$:

$$\mathrm{Ext}^*_{\mathcal{P}(V,W)}\left(\Gamma^{d-i}(X(V)) \otimes \Gamma^{2i}(V \otimes W) \otimes \Gamma^{ep^r - i}(X(W)), F(V) \otimes G^{(r)}(W)\right).$$

We denote for short by $E^*(i, F, G^{(r)})$ these graded vector spaces. So to prove theorem 2.1, it remains to prove the following connectedness statement:

$$E^k(i, F, G^{(r)}) = 0 \text{ for } i > 0 \text{ and } k < i(F, r). \qquad (*)$$

Step 2: proof of assertion $(*)$ in a special case. We now assume that $F = S^\mu$ for a p^r-bounded tuple μ, and $G = S^\nu$ for an arbitrary tuple ν. We observe that $S^\mu(V \oplus V')$ can be written as a direct sum of terms of the form $S^{\mu^1}(V) \otimes S^{\mu^2}(V')$ where both μ^1 and μ^2 are p^r bounded. Thus, using sum-diagonal adjunction (see e.g. [**20**, Section 2] or [**16**, Section 5]), we obtain that $E^*(i, S^\mu, S^{\nu\,(r)})$ is isomorphic to a finite direct sum of terms of the form

$$\mathrm{Ext}^*_{\mathcal{P}(V,W,V',W')}\Big(\Gamma^{d-i}(X(V)) \otimes \Gamma^{2i}(V' \otimes W') \otimes \Gamma^{ep^r-i}(X(W')),$$
$$S^{\mu^1}(V) \otimes S^{\mu^2}(V') \otimes S^{\nu^1\,(r)}(W) \otimes S^{\nu^2\,(r)}(W')\Big),$$

where μ^1 and μ^2 are p^r-bounded, and $\mathcal{P}(V, W, V', W')$ is the category of strict polynomial quadrifunctors with variables V, W, V', W'. By the Künneth theorem, such an Ext is isomorphic to $E_1^* \otimes E_2^* \otimes E_3^*$ with

$$E_1^* = \mathrm{Ext}^*_{\mathcal{P}}(\Gamma^{d-i} \circ X, S^{\mu^1}),$$
$$E_2^* = \mathrm{Ext}^*_{\mathcal{P}(V',W')}(\Gamma^{2i}(V' \otimes W'), S^{\mu^2}(V') \otimes S^{\nu^1\,(r)}(W')),$$
$$E_3^* = \mathrm{Ext}^*_{\mathcal{P}}(\Gamma^{ep^r-i} \circ X, S^{\nu^2\,(r)}).$$

We claim that E_2^* is zero in all degrees. Indeed, replacing V' by $U^\vee$, where U is a finite dimensional vector space and '$\vee$' denote $\Bbbk$-linear duality, yields an equivalence of categories between $\mathcal{P}(V', W')$ and $\mathrm{Bif}(U, W')$ where the latter denotes the category of strict polynomial bifunctors, contravariant in the first variable U and covariant with the second variable W'. With this substitution, E_2^* is isomorphic to the graded vector space:

$$\mathrm{Ext}^*_{\mathrm{Bif}(U,W')}(\Gamma^{2i}(\mathrm{Hom}_{\Bbbk}(U,W')), \mathrm{Hom}_{\Bbbk}(\Gamma^{\mu^2}(U), S^{\nu^1\,(r)}(W')) \ .$$

By [**6**, Thm 1.5], the latter is isomorphic to $\mathrm{Ext}^*_{\mathcal{P}}(\Gamma^{\mu^2}, S^{\nu^1\,(r)})$, which is zero in positive degrees because Γ^{μ^2} is projective, and which is zero in degree zero by [**20**, Lm 3.10]. Thus $E_1^* \otimes E_2^* \otimes E_3^*$ is zero, which implies assertion $(*)$ in our special case.

Step 3: proof of assertion $(*)$ **in general.** We now return to the general case. The functor F has an injective coresolution J_F whose first $i(F,r)$ terms are products of functors S^μ for p^r-bounded tuples μ. Let J_G be an injective coresolution of G. Then $J_F(V) \otimes J_G^{(r)}(W)$ is a coresolution of $F(V) \otimes G(W)$, whose first $i(F,r)$ terms are finite direct sums of functors of the form $(\prod S^{\mu^i}(V)) \otimes (\prod S^{\nu^j\,(r)}(W))$ where the μ^i are p^r-bounded tuples of integers.

LEMMA 2.2. *Let μ^i be p^r-bounded tuples of integers and let ν^j be arbitrary tuples of integers. Then $(\prod S^{\mu^i}(V)) \otimes (\prod S^{\nu^j}(W))$ is a direct summand of a product of bifunctors of the form $S^\mu(V) \otimes S^\nu(W)$ for p^r-bounded tuples μ and arbitrary tuples ν.*

PROOF. First, by [**20**, Lm 3.3(iii)] $(\prod S^{\mu^i})$ is a direct summand of a direct sum $\bigoplus_k S^{\alpha^k}$ for p^r-bounded tuples α^k. Similarly, $\prod S^{\nu^j}$ is a direct summand of a direct sum of a direct sum $\bigoplus_\ell S^{\beta^\ell}$. So $(\prod S^{\mu^i}(V)) \otimes (\prod S^{\nu^j}(W))$ is a direct summand of $\bigoplus_{k,\ell} S^{\alpha^k}(V) \otimes S^{\beta^\ell}(W)$. The latter is an injective bifunctor, hence by injectivity it is a direct summand of $\prod_{k,\ell} S^{\alpha^k}(V) \otimes S^{\beta^\ell}(W)$. □

Using lemma 2.2 and applying the r-th Frobenius twist on the variable W, we obtain that in the complex $J_F(V) \otimes J_G^{(r)}(W)$, the terms $(J_F(V) \otimes J_G^{(r)}(W))^s$ with $0 \le s < i(F,r)$ are direct summands of products of bifunctors of the form $S^\mu(V) \otimes S^{\nu\,(r)}(W)$ for p^r-bounded tuples μ and arbitrary tuples ν. Thus using the special case of assertion $(*)$ proved in step 2, together with the spectral sequence

$$E_1^{s,t} = \mathrm{Ext}^t_{\mathcal{P}(V,W)}(B(V,W),(J_F(V) \otimes J_G^{(r)}(W))^s) \\ \Longrightarrow \mathrm{Ext}^{s+t}_{\mathcal{P}(V,W)}(B(V,W), F(V) \otimes G^{(r)}(W))$$

which we apply to the bifunctor

$$B(V,W) = \Gamma^{d-i}(X(V)) \otimes \Gamma^{2i}(V \otimes W) \otimes \Gamma^{ep^r-i}(X(W))$$

we obtain assertion $(*)$ in the general case. This finishes the proof of theorem 2.1.

2.3. On the closed monoidal structure of $\mathcal{P}_d$. As explained in [**12**], $\mathcal{P}_d$ has a closed symmetric monoidal structure. We denote by $\underline{\otimes}$ the internal tensor product, and by $\underline{\mathrm{Hom}}$ the internal tensor product. Since $\mathcal{P} = \bigoplus_{d \ge 0} \mathcal{P}_d$, we formally extend this symmetric monoidal structure to the whole category $\mathcal{P}$ by letting for all F, G with homogeneous components F^d, G^d of degree d :

$$\underline{\mathrm{Hom}}(F,G) = \bigoplus_{d \ge 0} \underline{\mathrm{Hom}}(F^d, G^d) \ , \qquad F \underline{\otimes} G = \bigoplus_{d \ge 0} F^d \underline{\otimes} G^d \ .$$

The properties of cup products described in section 2.1 have consequences for the symmetric monoidal structure of $\mathcal{P}_d$. This is what is behind the proof of [**20**, Prop 6.1], and we explain everything in more details here.

The internal Hom can be concretely described in terms of parametrizations. To be more specific, we denote by F^V and F_V the strict polynomial functors $U \mapsto F(\mathrm{Hom}_{\Bbbk}(V,U)$, $U \mapsto F(U \otimes V)$. Then one has isomorphisms of functors of the variable V, natural with respect to F, G:

$$\mathrm{Hom}_{\mathcal{P}}(F^V, G) \simeq \underline{\mathrm{Hom}}(F,G)(V) \simeq \mathrm{Hom}_{\mathcal{P}}(F, G_V)\;.$$

We let $\underline{\mathrm{Ext}}^*(F,G)$ be the value on G of the derived functor of $\underline{\mathrm{Hom}}(F,-) : \mathcal{P} \to \mathcal{P}$. Since parametrization is exact and preserves projectives and injectives, one has

$$\mathrm{Ext}^*_{\mathcal{P}}(F^V, G) \simeq \underline{\mathrm{Ext}}^*(F,G)(V) \simeq \mathrm{Ext}^*_{\mathcal{P}}(F, G_V).$$

Such parametrized extension groups appear naturally in computations of strict polynomial functors, see e.g. [**17**, **18**]. It is easy to see [**20**, Prop 7.3 (a)] that for all V one has $p(F,r) = p(F^V,r)$ and $i(G,r) = i(G_V,r)$. Thus we have a parametrized version of [**20**, Thm 3.6].

Proposition 2.3. *Let (F,G,F',G') be a quadruple of homogeneous strict polynomial functors satisfying the Künneth condition, and let $r \geq 0$. Cup product induces a graded injective map:*

$$\underline{\mathrm{Ext}}^*(F,G) \otimes \underline{\mathrm{Ext}}^*(F'^{(r)}, G'^{(r)}) \hookrightarrow \underline{\mathrm{Ext}}^*(F \otimes F'^{(r)}, G \otimes G'^{(r)})\;.$$

Moreover, this graded injective map is an isomorphism in degree k in the following situations.

1. *When* $\deg F < \deg G$, *and* $k < i(G,r)$.
2. *When* $\deg F > \deg G$, *and* $k < p(F,r)$.
3. *When* $\deg F = \deg G$, *and* $k < p(F,r) + i(G,r)$.

We let $\underline{\mathrm{Tor}}_*(F,G)$ be the value on G of the derived functors of $F \underline{\otimes} - : \mathcal{P} \to \mathcal{P}$. Internal Tor are related to internal Ext by isomorphisms of functors, natural with respect to F, G:

$$\underline{\mathrm{Tor}}_i(F,G)^\sharp \simeq \underline{\mathrm{Ext}}^i(F, G^\sharp)\;,$$

where $^\sharp$ refers to Kuhn (or contragedient) duality: $F^\sharp(V) = F(V^\vee)^\vee$ with $^\vee$ denoting the duality of $\Bbbk$-vector spaces. Internal Tor are not equipped with a cup product but rather with a coproduct. The construction of the coproduct in degree zero is explained in [**20**, Section 6.1], and after deriving, this coproduct induces a graded map:

$$\underline{\mathrm{Tor}}_*(F \otimes F', G \otimes G') \to \underline{\mathrm{Tor}}_*(F,G) \otimes \underline{\mathrm{Tor}}_*(F', G')\;.$$

One has the following connectedness property for internal Tor.

Proposition 2.4. *Let F, G, F', G' be homogeneous strict polynomial functors and let $r \geq 0$. The coproduct induces a graded surjective map:*

$$\underline{\mathrm{Tor}}_*(F \otimes F'^{(r)}, G \otimes G'^{(r)}) \twoheadrightarrow \underline{\mathrm{Tor}}_*(F,G) \otimes \underline{\mathrm{Tor}}_*(F'^{(r)}, G'^{(r)})\;.$$

Moreover, this graded surjective map is an isomorphism in degree k in the following situations.

1. *When* $\deg F < \deg G$, *G is finite and* $k < p(G,r)$.
2. *When* $\deg F > \deg G$, *F is finite and* $k < p(F,r)$.
3. *When* $\deg F = \deg G$, *F and G are finite and* $k < p(F,r) + p(G,r)$.

Proof. We first prove the statement when F, F', G, G' are finite (i.e their values are finite dimensional vector spaces). As noted in the proof of [**20**, Prop 6.1], in this case the coproduct is just the dual (for the duality $^\sharp$) of the cup product for internal Ext for the quadruple $(F, F'^{(r)}, G^\sharp, G'^{(r)\sharp})$. Since $G'^{(r)\sharp} \simeq (G'^\sharp)^{(r)}$, and $i(G^\sharp, r) = p(G, r)$, the statement follows from proposition 2.3. For the general case, we use that each functor is the filtered colimit of its finite subfunctors, and that tensor products and internal Tor commute with filtered colimits. □

Remark 2.5. The finiteness hypotheses in proposition 2.4 can actually be removed, but then our short proof using dualization does not work anymore. One rather needs to redo all the calculations of the proof of [**20**, Thm 3.6], replacing the Hom spaces by tensor products over the source category or by its parametrized variant $\underline{\otimes}$.

Proposition 2.4 is already interesting in the cases where one has isomorphism in degree zero. For example, the following application is analogous to the Steinberg tensor product theorem in that it allows to reduce the computation of the internal tensor product between simples to that of internal tensor products between simples with p-restricted highest weights.

Theorem ([**20**, Thm 6.2, Cor 6.5]). *Let $\lambda^0, \dots, \lambda^r$ and $\mu^0, \dots, \mu^s$ be p-restricted partitions, and let $\lambda = \sum p^i \lambda^i$ and $\mu = \sum p^i \mu^i$.*

(1) *If $r = s$ and μ^i and λ^i have the same weight for all i, then $L_\lambda \underline{\otimes} L_\mu$ is nonzero and there is an isomorphism:*

$$L_\lambda \underline{\otimes} L_\mu \simeq (L_{\lambda^0} \underline{\otimes} L_{\mu^0}) \otimes (L_{\lambda^1} \underline{\otimes} L_{\mu^1})^{(1)} \otimes \cdots \otimes (L_{\lambda^r} \underline{\otimes} L_{\mu^r})^{(r)} .$$

(2) *Otherwise, $L_\lambda \underline{\otimes} L_\mu$ is zero.*

2.4. Connectivity of the Schur functor. Given a degree d homogenous strict polynomial functor F, we let $f_d(F) = \mathrm{Hom}_{\mathcal{P}}(\otimes^d, F)$. The action of the symmetric group $\mathfrak{S}_d$ on $\otimes^d$ makes $f_d(F)$ into a $\Bbbk\mathfrak{S}_d$-module. The functor f_d is known as the Schur functor, and it is part of a recollement of abelian categories:

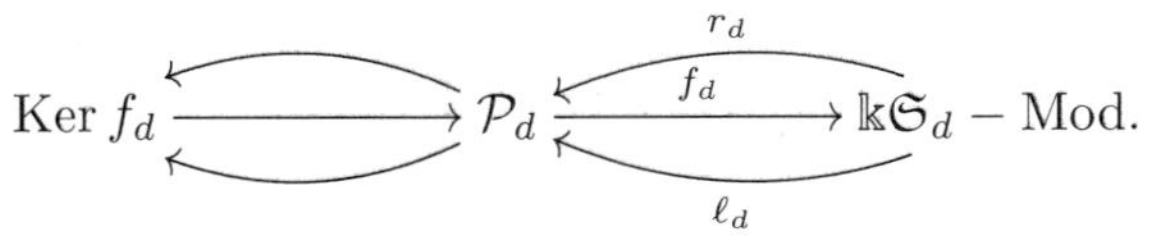

The interaction between cohomology of the symmetric group and cohomology of GL_n has been studied by the means of the Grothendieck spectral sequences associated to the composition of f_d and its adjoints [**5, 10**]. We observe that $i(F, 1)$ can be understood as the maximal integer k such that F has an injective resolution J such that the first k terms of the complex $f_d(J)$ are injective $\Bbbk\mathfrak{S}_d$-modules (and $p(F, r)$ has a similar description). So we deduce the following comparison result.

Theorem ([**20**, Thm 8.2]). *Let F and G be homogeneous functors of degree d. The map induced by the Schur functor:*

$$\mathrm{Ext}^k_{\mathcal{P}}(F, G) \to \mathrm{Ext}^k_{\Bbbk\mathfrak{S}_d}(f_d(F), f_d(G))$$

is an isomorphism in degrees $k < p(F, 1) + i(G, 1) - 1$, and it is injective in degree $k = p(F, 1) + i(G, 1) - 1$.

For some classes of functors, it is not to hard to compute lower bounds for $p(F,1)$ or $i(G,1)$, and we recover in this way various known results on the cohomology of symmetric groups.

- The right adjoint r_d of f_d is left exact and sends injective $\Bbbk\mathfrak{S}_d$-modules to p-bounded injectives, hence one has $i(r_d M,1) \geq 2$ for all $\Bbbk\mathfrak{S}_d$-modules M. This remark together with the previous theorem encompasses most of the Ext^1-computations and comparison results between representations of Schur algebras and symmetric groups obtained in [**5**].
- Using Ringel duality for strict polynomial functors we proved [**20**, Prop 7.6] that $i(W^d_\lambda, 1) \geq p-1+i(S_{\widetilde{\lambda}},1)$ where W_λ, resp. $S_{\widetilde{\lambda}}$ denotes the Weyl functor associated to λ, resp. the Schur functor associated to the conjugate partition $\widetilde{\lambda}$. This computation and the previous theorem encompass all the applications in [**10**, Section 6], except [**10**, Theorem 6.1].
- We compute in corollary 3.12 below that $i(S_\lambda, 1) \geq 2$ when λ is a partition with $\lambda_1 \leq p-2$. Thus for all p-restricted partitions μ such that $\mu \not\trianglerighteq \lambda$ one has (the first equality comes from highest weight category theory):

$$0 = \mathrm{Ext}^1_{\mathcal{P}}(L_\mu, S_\lambda) \simeq \mathrm{Ext}^1_{\Bbbk\mathfrak{S}_d}(f_d L_\mu, f_d S_\lambda)\ .$$

This result is similar (but not equivalent) to the main result of [**11**].

3. Computations of $i(F,r)$ and $p(F,r)$

The results of section 2 motivates us to investigate how to compute the homological constants $i(F,r)$ and $p(F,r)$. In this section, we give basic tools to compute $i(F,r)$ and $p(F,r)$, and partial results regarding the values of these constants for simples L_λ, Schur functors S_λ, Weyl functors W_λ and tilting functors T_λ.

3.1. General computation rules and reductions. We first recall some general facts established in [**20**, Section 7].

PROPOSITION. *Let F and G be two strict polynomial functors, let λ be a partition and $\widetilde{\lambda}$ its conjugate partition. The following holds.*

(3.1) $$p(F,r) = i(F^\sharp, r)\ ,$$

(3.2) $$i(F,r) = i(F^{(s)}, r+s)\ ,$$

(3.3) $$i(F\otimes G, r) = \min\{\, i(F,r)\, ,\, i(G,r)\,\}\ ,$$

(3.4) $$i(W_\lambda, r) \geq p^r - 1 + i(S_{\widetilde{\lambda}}, r)\ ,$$

(3.5) $$i(\Lambda^d, r) = p^r - 1 \text{ if } d \geq p^r,$$

(3.6) $$i(\Gamma^d, r) = 2p^r - 2 \text{ if } d \geq p^r,$$

(3.7) $$i(F,r) = +\infty \text{ if } \deg F < p^r.$$

In view of these results we will restrict ourselves to understand $i(F,r)$ for the following functors in the sequel.

- Functors $F = L_\lambda$ with p-restricted highest weight λ. Indeed, the value for arbitrary λ can then be obtain by equality (3.2) and (3.3) and the Steinberg tensor product theorem.
- Functors $F = S_\lambda$, with λ arbitrary. Indeed, we then obtain lower bounds for W_λ by inequality (3.4). Note that (3.5) and (3.6) show that inequality

(3.4) is an equality for $\lambda = (1, \dots, 1)$ and for $\lambda = (d)$. We don't know if equality holds in general.
- Functors $F = T_\lambda$ with λ arbitrary.

3.2. The integer $i(F, r)$ as a connectedness bound. We now recall a homological criterion to compute $i(F, r)$. Given a tuple of nonnegative integers $(d_0, \dots, d_k)$ we let

$$T^{(d_0,\dots,d_k)} := (I^{(0)})^{\otimes d_0} \otimes (I^{(1)})^{\otimes d_1} \otimes \cdots \otimes (I^{(k)})^{\otimes d_k} .$$

Thus $T^{(d_0,\dots,d_k)}$ is a homogeneous strict polynomial functor of degree $\sum_{i=0}^{k} p^i d_i$. We let $T(d, r)$ be the (finite) direct sum of all the $T^{(d_0,\dots,d_k)}$ of degree d which have at least one factor $I^{(s)}$ with $s \geq r$ (in particular $T(d, r) = 0$ as soon as $d < p^r$). The next proposition shows that we can interpret $i(F, r)$ as a connectedness bound. For brevity we will use the following notation.

Definition 3.1 (Notation). If E^* is a graded vector space, or a graded functor, we let

$$\operatorname{conn} E^* = \inf \left\{ k \in \mathbb{N} \mid E^k \neq 0 \right\} .$$

In particular $\operatorname{conn} E^* = +\infty$ if and only if $E^* = 0$.

Proposition ([**20**, Prop 7.1]). *For all F homogeneous of degree d:*

$$i(F, r) = \operatorname{conn} \operatorname{Ext}^*_{\mathcal{P}}(T(d, r), F) .$$

We will now prove another characterization of $i(F, r)$ which will be easier to check in practice. This characterization, given in proposition 3.4 below, uses the right adjoint of the functor $- \otimes I^{(r)} : \mathcal{P} \to \mathcal{P}$. This adjoint exists for formal reasons [**14**, Chap V], but we need an explicit formula. The reasonning works for the adjoint of $- \otimes F$ for any degree d homogeneous strict polynomial functor F, so we present the results in this generality. Given a strict polynomial functor G, we form the bifunctor $(V, W) \mapsto G(V \oplus W)$. As for every strict polynomial bifunctor, there is a decomposition

$$G(V \oplus W) = \bigoplus_{k \geq 0} G^{(k,*)}(V, W)$$

where each $G^{(k,*)}(V, W)$ is homogeneous of degree k with respect to the variable V. We let $[F : -] : \mathcal{P} \to \mathcal{P}$ be the functor defined by:

$$[F : G](W) = \operatorname{Hom}_{\mathcal{P}(V)} \left(F(V), G^{(d,*)}(V, W) \right) ,$$

where $\operatorname{Hom}_{\mathcal{P}(V)}(-, -)$ means that we are computing Hom between strict polynomial functors of the variable V.

In the decomposition of $G(V \oplus W)$, the direct summands $G^{(k,*)}(V, W)$ have degree less or equal to $\deg G - k$ as strict polynomial functors of the variable W and they are zero if $\deg G - k$ is negative. Thus $\deg[F : G] \leq \deg G - d$ and $[F : G]$ is zero if $\deg G - d$ is negative. Moreover $G^{(\deg G,*)}(V, W) = G(V)$, hence if $\deg G = d$ then $[F : G]$ is a constant functor with value $\operatorname{Hom}_{\mathcal{P}}(F, G)$.

Lemma 3.2. *The functor $[F : -] : \mathcal{P} \to \mathcal{P}$ is right adjoint to $- \otimes F : \mathcal{P} \to \mathcal{P}$.*

PROOF. We have a chain of natural isomorphisms:

$$\begin{aligned}\mathrm{Hom}_{\mathcal{P}}(F\bar{\otimes}H,G) &\simeq \mathrm{Hom}_{\mathcal{P}(V,W)}(F(V)\otimes H(W),G(V\oplus W))\\ &\simeq \mathrm{Hom}_{\mathcal{P}(V,W)}(F(V)\otimes H(W),G^{(d,*)}(V,W))\\ &\simeq \mathrm{Hom}_{\mathcal{P}(W)}(H(W)\,,\,\mathrm{Hom}_{\mathcal{P}(V)}(F(V),G^{(d,*)}(V,W))\,)\\ &= \mathrm{Hom}_{\mathcal{P}(W)}(H(W),[F:G](W))\,.\end{aligned}$$

The first isomorphism is by sum-diagonal adjunction, the second one follows from the fact that there is no Hom between homogeneous strict polynomial bifunctors of different bidegrees, the third one is [**14**, Prop 1 p. 37]. □

Next we consider the right derived functors $\mathbf{R}^i[F:-]:\mathcal{P}\to\mathcal{P}$. If $\deg G<d$ then $\mathbf{R}^*[F:G]$ is zero, and if $\deg G=d$ then $\mathbf{R}^*[F:G]$ is a constant functor with value $\mathrm{Ext}^*_{\mathcal{P}}(F,G)$. The next lemma gives basic rules to compute $\mathbf{R}^*[F:G]$ when the degree of G is greater than d.

LEMMA 3.3. *Assume that F is homogeneous of degree d with finite dimensional values. For all i, there is a functor $\Phi^i:\mathcal{P}^{\mathrm{op}}\times\mathcal{P}\times\mathcal{P}\to\mathcal{P}$ and a canonical decomposition, natural with respect to F,G,H*

$$\mathbf{R}^i[F:G\otimes H]\simeq G\otimes\mathbf{R}^i[F:H]\ \oplus\ \Phi^i(F,G,H)\ \oplus\ \mathbf{R}^i[F:G]\otimes H\,.$$

Moreover, for all $r\geq 0$, $\Phi^i(I^{(r)},G,H)=0$ for all G, H. Finally, let $E^*=S^*$, Λ^*, or any exponential functor with each E^k homogeneous of degree k. Let $\Delta_{k,\ell}:E^{k+\ell}\to E^k\otimes E^\ell$ and $m_{k,\ell}:E^k\otimes E^\ell\to E^{k+\ell}$ be the components of the multiplication and comultiplications of E^*. Then for all $k\geq 0$ the following compositions (with $E^j=0$ for $j<0$) are isomorphisms

$$E^{k-d}\otimes\mathbf{R}^i[F:E^d]\hookrightarrow\mathbf{R}^i[F:E^{k-d}\otimes E^d]\xrightarrow{\mathbf{R}^i[F:m_{k-d,\ell}]}\mathbf{R}^i[F:E^k]\,,$$

$$\mathbf{R}^i[F:E^k]\xrightarrow{\mathbf{R}^i[F:\Delta_{k-d,\ell}]}\mathbf{R}^i[F:E^{k-d}\otimes E^d]\twoheadrightarrow E^{k-d}\otimes\mathbf{R}^i[F:E^d]\,.$$

PROOF. We have a direct sum decomposition:

$$(G\otimes H)^{(d,*)}(V,W)=\bigoplus_{i=0}^{d}G^{(i,*)}(V,W)\otimes H^{(d-i,*)}(V,W)\,.\qquad(*)$$

The summands indexed by $i=0$ and $i=d$ are respectively equal to

$$G(W)\otimes H^{(d,*)}(V,W)\ \text{and}\ G^{(d,*)}(V,W)\otimes H(W)\,.$$

The direct sum decomposition is obtained by applying $\mathrm{Ext}^i_{\mathcal{P}(V)}(F(V),-)$ to this decomposition, and using the canonical isomorphisms (which hold since F has finite dimensional values hence it has a resolution by finitely generated projectives):

$$\begin{aligned}\mathrm{Ext}^i_{\mathcal{P}(V)}(F(V),G(W)\otimes H^{(d,*)}(V,W))&\simeq G(W)\otimes\mathrm{Ext}^i_{\mathcal{P}(V)}(F(V),H^{(d,*)}(V,W))\,,\\ \mathrm{Ext}^i_{\mathcal{P}(V)}(F(V),G^{(d,*)}(V,W)\otimes H(W))&\simeq\mathrm{Ext}^i_{\mathcal{P}(V)}(F(V),G^{(d,*)}(V,W))\otimes H(W)\,.\end{aligned}$$

If $F=I^{(r)}$, Pirashvili's vanishing lemma [**7**, Thm 2.13] shows that the summands of $(*)$ indexed by $i=0$ and $i=d$ are the only ones which may have a nonzero contribution after applying $\mathrm{Ext}^i_{\mathcal{P}(V)}(I^{(r)}(V),-)$. Finally, the last statement follows from the fact that there is an isomorphism:

$$(E^k)^{(d,*)}(V,W)\xrightarrow{\simeq}E^d(V)\otimes E^{k-d}(W)$$

induced by the comultiplication of E, with inverse induced by the multiplication of E. □

PROPOSITION 3.4. *For all F homogeneous of degree d one has*

$$i(F,r) = \min_{r\le s\le \log_p(d)} \operatorname{conn} \mathbf{R}^*[I^{(s)} : F]\ .$$

PROOF. For all G, there is a Grothendieck spectral sequence

$$E_2^{k,\ell} = \mathrm{Ext}^k_{\mathcal{P}}(G, \mathbf{R}^\ell[I^{(s)} : F]) \Longrightarrow \mathrm{Ext}^{k+\ell}_{\mathcal{P}}(G\otimes I^{(s)}, F)\ .$$

Let E_2^* denote the second page with total grading. Then

$$\operatorname{conn} \mathrm{Ext}^*_{\mathcal{P}}(G\otimes I^{(s)}, F) \ge \operatorname{conn} E_2^* \ge \operatorname{conn} \mathbf{R}^*[I^{(s)} : F]\ . \quad (*)$$

Moreover, if $c = \operatorname{conn} \mathbf{R}^*[I^{(s)} : F]$ and $\mathrm{Hom}_{\mathcal{P}}(G, \mathbf{R}^c[I^{(s)} : F]) \ne 0$, then $E_2^{0,c} = E_\infty^{0,c}$ for degree reasons, so the inequalities in $(*)$ are equalities.

This being said, let m denote the right hand side of the equality of proposition 3.4. Since any direct summand $T^{(d_0,\dots,d_k)}$ of $T(d,r)$ is of the form $G\otimes I^{(s)}$ for some $s\in[r,\log_p(d)]$, inequality $(*)$ implies that $i(F,r)\ge m$. Let t be such that $\operatorname{conn} \mathbf{R}^*[I^{(t)} : F] = m$. Since all simple functors are quotients of some $T^{(d_0,\dots,d_\ell)}$ (this is a consequence of the Steinberg tensor product theorem and Clausen and James' theorem, see e.g [**20**, Cor 4.3]), one may find a tuple $(d_0,\dots,d_\ell)$ such that $\mathrm{Hom}_{\mathcal{P}}(T^{(d_0,\dots,d_\ell)}, \mathbf{R}^m[I^{(t)} : F])$ is nonzero. Thus we have

$$i(F,r) \le \operatorname{conn} \mathrm{Ext}^*_{\mathcal{P}}(T^{(d_0,\dots,d_\ell)}\otimes I^{(t)}, F) = \operatorname{conn} \mathbf{R}^*[I^{(t)} : F] = m\ .$$

This finishes the proof of proposition 3.4. □

3.3. Computations for tilting functors.

LEMMA 3.5. *Let λ be a partition of weight d. The following assertions are equivalent.*

(i) *λ is p^r-restricted,*
(ii) *$\mathrm{Hom}_{\mathcal{P}}(T(d,r), J_\lambda)$ is zero,*
(iii) *$\mathrm{Hom}_{\mathcal{P}}(\otimes^{d-p^r}\otimes I^{(r)}, J_\lambda)$ is zero.*

PROOF. (i)⇒(ii) since no composition factor of $T(d,r)$ is p^r-restricted, and (ii)⇒(iii) since $\otimes^{d-p^r}\otimes I^{(r)}$ is a direct summand of $T(d,r)$. It remains to prove (iii)⇒(i). But (iii) implies that L_λ is not a composition factor of $\otimes^{d-p^r}\otimes I^{(r)}$. Thus, to prove (i), it suffices to prove that all L_μ with μ of weight d and not p^r-restricted appear as composition factors of $\otimes^{d-p^r}\otimes I^{(r)}$. If μ is not p^r-restricted then by Clausen and James' theorem and the Steinberg tensor product theorem L_μ is a quotient of a tensor product $T^{(d_0,\dots,d_k)}$ with $d_k>0$ for some $k\ge r$. Observe that for all s, $I^{(s+1)}$ is a subfunctor of $S^{p^{s+1}}$, hence a subquotient of $(I^{(s)})^{\otimes p}$. Using this observation repeatedly, we find that $T^{(d_0,\dots,d_k)}$ is a subquotient of $\otimes^{d-p^r}\otimes I^{(r)}$, hence that L_μ is a composition factor of $\otimes^{d-p^r}\otimes I^{(r)}$. □

PROPOSITION 3.6. *Let λ be an arbitrary partition. Then $i(T_\lambda, r)\in\{p^r-1,\infty\}$. Moreover $i(T_\lambda,r)=+\infty$ if and only if $\widetilde{\lambda}$ is p^r-restricted.*

PROOF. We use Ringel duality, in the same way as in the proof of [**20**, Lemma 7.5]. To be more specific, Ringel duality is an equivalence of categories $\Theta : D(\mathcal{P}) \to D(\mathcal{P})$, which sends $J_{\widetilde{\lambda}}$ to T_λ and $T^{(d_0,\dots,d_k)}$ to $T^{(d_0,\dots,d_k)}[-s]$, where $s = \sum_{i=0}^k d_i(p^i - 1)$. In particular

$$\mathrm{Ext}^*_{\mathcal{P}}(T^{(d_0,\dots,d_k)}, J_{\widetilde{\lambda}}) \simeq \mathrm{Ext}^{*+s}_{\mathcal{P}}(T^{(d_0,\dots,d_k)}, T_\lambda)\,. \qquad (*)$$

Since $J_{\widetilde{\lambda}}$ is injective, the left hand side of $(*)$ is zero in positive degrees. Now if $\widetilde{\lambda}$ is p^r-restricted, then $\mathrm{Hom}_{\mathcal{P}}(T(d,r), J_{\widetilde{\lambda}})$ is zero by lemma 3.5, hence $\mathrm{Ext}^*_{\mathcal{P}}(T(d,r), T_\lambda)$ is zero by $(*)$. If $\widetilde{\lambda}$ is not p^r-restricted, then $\mathrm{Hom}_{\mathcal{P}}(\otimes^{d-p^r} \otimes I^{(r)}, J_\lambda)$ is nonzero by lemma 3.5. Since the degree shift s is minimal for the summand $\otimes^{d-p^r} \otimes I^{(r)}$ of $T(d,r)$, we obtain

$$i(T_\lambda, r) = \operatorname{conn} \mathrm{Ext}^*_{\mathcal{P}}(\otimes^{d-p^r} \otimes I^{(r)}, T_\lambda) = p^r - 1\,.$$

□

3.4. Computations for Schur functors indexed by hooks. Schur functors indexed by hooks are often quite easy to deal with. Indeed they appear as the kernels of some Koszul complexes which have a very simple description. To be more specific, for $d \geq 1$ we let κ^d be the complex:

$$\Lambda^d \to \Lambda^{d-1} \otimes S^1 \to \cdots \to \Lambda^1 \otimes S^{d-1} \to S^d$$

where each $\Lambda^{d-i} \otimes S^i$ is placed in homological degree $d-i$, and the differential $d : \Lambda^{d-i} \otimes S^i \to \Lambda^{d-i-1} \otimes S^{i+1}$ is the composition (recall that $S^1 = \Lambda^1 = I$):

$$\Lambda^{d-i} \otimes S^i \xrightarrow{\mathrm{comult}\otimes S^i} \Lambda^{d-i-1} \otimes I \otimes S^i \xrightarrow{\Lambda^{d-i-1}\otimes \mathrm{mult}} \Lambda^{d-i-1} \otimes S^{i+1}\,.$$

Then κ^d is an exact complex, and for all i one has (see e.g. [**3**, p. 80]):

$$S_{(i,1^{\{d-i\}})} \;=\; \ker\left[\Lambda^{d-i} \otimes S^i \to \Lambda^{d-i-1} \otimes S^{i+1}\right]\,.$$

All of what is recalled above is valid for $d > 0$. If $d = 0$, we let κ^0 be the complex equal to $\Bbbk$ concentrated in degree zero.

For all nonnegative i, we may apply the functor $\mathbf{R}^i[I^{(r)} : -]$ termwise to the Koszul complex κ^d to obtain another complex of strict polynomial functors, which we denote by $\mathbf{R}^i[I^{(r)} : \kappa^d]$. In the next lemma we compute these complexes. As usual, the shift $C[s]$ of a complex C is a complex satisfying $C[s]_n = C_{s+n}$.

LEMMA 3.7. *Let $d \geq p^r$. We have*

$$\mathbf{R}^i[I^{(r)} : \kappa^d] \simeq \begin{cases} \kappa^{d-p^r} & \text{if } i = 0,\\ \kappa^{d-p^r}[p^r] & \text{if } i = p^r - 1,\\ 0 & \text{else.}\end{cases}$$

PROOF. As computed in [**7**, (4.5.1) p. 241], $\mathbf{R}^i[I^{(r)} : \Lambda^{p^r}] = \mathrm{Ext}^i_{\mathcal{P}}(I^{(r)}, \Lambda^{p^r})$ equals $\Bbbk$ if $i = p^r - 1$ and zero if $i \neq p^r - 1$. Also, $\mathbf{R}^i[I^{(r)} : S^{p^r}] = \mathrm{Ext}^i_{\mathcal{P}}(I^{(r)}, S^{p^r})$ is zero in positive degrees and equals $\Bbbk$ in degree zero. Thus, by lemma 3.3, $\mathbf{R}^i[I^{(r)} : \kappa^d]$ is zero for $i \notin \{0, p^r - 1\}$. Also, we have the following commutative diagram:

$$\begin{array}{ccccc} \mathbf{R}^0[I^{(r)}, \Lambda^{d-i} \otimes S^i] & \longrightarrow & \mathbf{R}^0[I^{(r)}, \Lambda^{d-i-1} \otimes I \otimes S^i] & \longrightarrow & \mathbf{R}^0[I^{(r)}, \Lambda^{d-i-1} \otimes S^{i+1}] \\ \simeq\uparrow & & \simeq\uparrow & & \simeq\uparrow \\ \Lambda^{d-i} \otimes S^{i-p^r} \otimes \Bbbk & \longrightarrow & \Lambda^{d-i-1} \otimes I \otimes S^{i-p^r} \otimes \Bbbk & \longrightarrow & \Lambda^{d-i-1} \otimes S^{i+1-p^r} \otimes \Bbbk \end{array}$$

where the horizontal maps are induced by the comultiplication of the exterior algebra and the multiplication of the symmetric algebra, and the vertical isomorphisms are compositions of the form (for suitable values of G, j) as in lemma 3.3:

$$G \otimes S^j \otimes \Bbbk = G \otimes S^j \otimes \mathbf{R}^0[I^{(r)}, S^{p^r}] \simeq G \otimes \mathbf{R}^0[I^{(r)}, S^{j+p^r}] \to \mathbf{R}^0[I^{(r)}, G \otimes S^{j+p^r}] \,.$$

Thus $\mathbf{R}^0[I^{(r)} : \kappa^d] \simeq \kappa^{d-p^r}$. One identifies $\mathbf{R}^{p^r-1}[I^{(r)} : \kappa^d]$ similarly. □

LEMMA 3.8. *Let $\lambda = (i, 1^{\{d-i\}})$ be a hook-shaped partition of weight d and let $1 \le r \le \log_p(d)$. If $d < p^r$ then $\mathbf{R}^*[I^{(r)}, S_\lambda]$ is zero. If $d = p^r$ then*

$$\mathbf{R}^k[I^{(r)}, S_\lambda] = \begin{cases} \Bbbk & \text{if } k = (d-i), \\ 0 & \text{else.} \end{cases}$$

If $d > p^r$ then

$$\mathbf{R}^k[I^{(r)}, S_\lambda] = \begin{cases} S_{(i-p^r, 1^{\{d-i\}})} & \text{if } k = 0 \text{ and } i > p^r, \\ S_{(i, 1^{\{d-i-p^r\}})} & \text{if } k = p^r - 1 \text{ and } d-i \ge p^r, \\ 0 & \text{else.} \end{cases}$$

PROOF. If $d < p^r$ vanishing occurs for degree reasons ($[I^{(r)} : F]$ is zero on functors of degree less than p^r). So we may assume that $d \ge p^r$. Let $\kappa^d_{\le d-i}$ be the stupid truncation of κ^d:

$$\kappa^d_{\le d-i} = \Lambda^{d-i} \otimes S^i \to \cdots \to \Lambda^1 \otimes S^{d-1} \to S^d \,.$$

The homology of $\kappa^d_{\le d-i}$ is $S_{(i,1^{\{d-i\}})}$ in degree $d-i$ and zero in other degrees. Thus, we have a hypercohomology spectral sequence:

$$E_1^{-s,t} = \mathbf{R}^t[I^{(r)} : \Lambda^s \otimes S^{d-s}] \Longrightarrow \mathbf{R}^{t-s+(d-i)}[I^{(r)} : S_{(i,1^{\{d-i\}})}] \,.$$

By lemma 3.7 there are only two nonzero rows in the first page, the row $t = 0$ and the row $t = p^r - 1$, and the complex $(E_1^{*,0}, d_1)$ is isomorphic to $\kappa^{d-p^r}_{\le d-i}$ while the complex (E_1^{*,p^r-1}, d_1) is isomorphic to $\kappa^{d-p^r}_{\le d-i-p^r}[p^r]$. Thus we can compute explicitly page E_2 of the spectral sequence. Assume first that $d = p^r$. Then $E_2^{0,0} = \Bbbk$ is the only nonzero term of the second page. Thus $E_2^{*,*} = E_\infty^{*,*}$ and we obtain $\mathbf{R}^*[I^{(r)} : S_\lambda]$ as described in the lemma. Assume now that $d > p^r$. Then the only possibly nonzero terms of page E_2 are $E_2^{-(d-i),0}$ and $E_2^{-(d-i),p^r-1}$. Thus

$$\mathbf{R}^*[I^{(r)} : S_{(i,1^{\{d-i\}})}] = E_\infty^{-(d-i),*} = E_2^{-(d-i),*},$$

and we obtain $\mathbf{R}^*[I^{(r)} : S_\lambda]$ as described in the lemma. □

We are now ready to prove the main result of this section.

PROPOSITION 3.9. *Let $\lambda = (i, 1^{\{d-i\}})$ be a hook-shaped partition, and let $r \ge 1$. The following holds.*

(i) *If $d < p^r$, then $i(S_\lambda, r) = +\infty$.*
(ii) *If $d = p^r$, then $i(S_\lambda, r) = (d-i)$.*
(iii) *If $p^r < d < 2p^r$ then*

$$i(S_\lambda, r) = \begin{cases} p^r - 1 & \text{if } i \le d - p^r, \\ +\infty & \text{if } d - p^r < i \le p^r, \\ 0 & \text{if } p^r < i. \end{cases}$$

(iv) *If* $2p^r \le d$, *then* $i(S_\lambda, r) = \begin{cases} p^r - 1 & \text{if } i \le p^r, \\ 0 & \text{if } i > p^r. \end{cases}$

PROOF. Statement (i) follows from general rules of computations recalled in section 3.1. For (ii), (iii) and (iv) we use that $i(S_\lambda, r)$ is the minimal value of the connectedness of $\mathbf{R}^*[I^{(s)}, S_\lambda]$ with $r \le s \le \log_p(d)$ by proposition 3.4. If $p^r \le d < 2p^r$ one only has to consider the connectedness of $\mathbf{R}^*[I^{(s)}, S_\lambda]$ for $s = r$, which is directly given by lemma 3.8, and one obtains (ii) and (iii). For (iv) there might be several integers s to consider. As $p^r \le d - p^r$ we have by lemma 3.8:

$$\operatorname{conn} \mathbf{R}^*[I^{(r)}, S_\lambda] = \begin{cases} p^r - 1 & \text{if } i \le p^r, \\ 0 & \text{if } i > p^r. \end{cases}$$

To prove (iv), it remains to prove that if $s > r$, then $\operatorname{conn} \mathbf{R}^*[I^{(s)}, S_\lambda]$ is greater or equal to $\operatorname{conn} \mathbf{R}^*[I^{(r)}, S_\lambda]$. It suffices to check it when $i \le p^r$. If $s > r$ and $d = p^s$ then $\operatorname{conn} \mathbf{R}^*[I^{(s)}, S_\lambda] = p^s - i$ by lemma 3.8. But $i \le p^r$ implies that $p^s - i \ge p^s - p^r > p^r - 1$ and we are done. If $s > r$ and $d \ne p^s$ then $i \le p^r$ implies that $i \le p^s$, hence by lemma 3.8 $\operatorname{conn} \mathbf{R}^*[I^{(r)}, S_\lambda] \in \{p^s - 1, +\infty\}$ is greater than $p^r - 1$ and we are done. □

3.5. Thin partitions. In this section we do not compute exact values for $i(S_\lambda, r)$ or $i(L_\lambda, r)$, but we rather provide lower bounds for these integers. The philosophy of the results obtained can be roughly explained as follows. First if $\lambda = (1^{\{d\}})$, then $S_\lambda = L_\lambda = \Lambda^d$ and $i(\Lambda^d, r)$ are high numbers ($p^r - 1$ or $+\infty$). Now if λ is a thin partition, that is if λ resembles $(1^{\{d\}})$ then S_λ and L_λ should resemble Λ^d, hence $i(S_\lambda, r)$ and $i(L_\lambda, r)$ should be high numbers, too.

LEMMA 3.10. *Let* $\lambda = (\lambda_1, \dots, \lambda_n)$ *be a partition of* p^r *in exactly* n *parts. Then*

$$i(S_\lambda, r) = \operatorname{conn} \operatorname{Ext}^*_{\mathcal{P}}(I^{(r)}, S_\lambda) \ge n - 1\,.$$

PROOF. The equality follows from proposition 3.4 and the fact that λ has weight p^r. By the Pieri rule [**1**, Thm (3)], one has a short exact sequence

$$0 \to S_\lambda \to S^{\lambda_1} \otimes S_{(\lambda_2, \dots, \lambda_n)} \to C_\lambda \to 0$$

and C_λ has a finite length filtration whose layers are Schur functors indexed by partitions μ of p^r in n or $n - 1$ parts, which all strictly dominate λ. Since $\operatorname{Ext}^*(I^{(r)}, S^{\lambda_1} \otimes S_{(\lambda_2, \dots, \lambda_n)})$ is zero by Pirashvili vanishing lemma [**7**, Thm 2.13], the associated long exact sequence of Ext-groups tells us that

$$i(S_\lambda, r) \ge \min_\mu \, i(S_\mu, r) + 1 \qquad (*)$$

where the minimum is taken over the layers S_μ of the Pieri filtration of $S^{\lambda_1} \otimes S_{(\lambda_2, \dots, \lambda_n)}$. Now we prove lemma 3.10 inductively. For $n = 1$ there is nothing to prove. Assume that lemma 3.10 holds for partitions of p^r in at most n parts. Then if λ is a partition of p^r in $(n+1)$ parts which is maximal among those partitions with respect to the dominance order, the filtration of C_λ has only layers of the form S_μ for partitions of μ in n parts. Thus, by using $(*)$ lemma 3.10 holds for S_λ. We then propagate the statement of lemma 3.10 by downwards induction on the finite poset (with respect to the dominance order) of partitions of p^r in $(n+1)$ parts. Thus we obtain that lemma 3.10 holds for all partitions of p^r into $(n+1)$ parts. □

PROPOSITION 3.11. *Let λ be a partition. Assume that λ contains no partition α of p^r in less or equal to n parts with trivial p-core. Then $i(S_\lambda, r) \geq n$.*

PROOF. By [**2**, Thm II.4.11], $S_\lambda(V \oplus W)$ has a filtration whose layers have the form $S_\alpha(V) \otimes S_{\lambda/\alpha}(W)$ for $\alpha \subset \lambda$. Let $s \geq r$. Restricting to the direct summand of $S_\lambda(V \oplus W)$ which has degree p^s with respect to V, one obtains

$$\operatorname{conn} \mathbf{R}^*[I^{(s)} : S_\lambda] \geq \min_\alpha \left\{ \operatorname{conn} \left(\operatorname{Ext}^*_{\mathcal{P}}(I^{(s)}, S_\alpha) \right) \right\} , \qquad (*)$$

where the minimum is taken over the partitions α of p^s which are contained in λ. If λ contains no partition α of p^r in less or equal to n parts with trivial p-core, then it contains no partition of α of p^s in less or equal to n parts with trivial p-core. Thus by lemma 3.10 the right hand side of $(*)$ is greater or equal to n, so that $\operatorname{conn} \mathbf{R}^*[I^{(s)} : S_\lambda] \geq n$. By proposition 3.4, this implies that $i(S_\lambda, r) \geq n$. □

The hypothesis in proposition 3.11 is not very simple. We give a simpler equivalent formulation for $r = 1$ in corollary 3.12, and a simpler but weaker condition for $r \geq 1$ in corollary 3.13.

COROLLARY 3.12. *If λ satisfies $\lambda_1 \leq p - n$, then $i(S_\lambda, 1) \geq n$.*

PROOF. The partitions of p with trivial p-core are the hooks $(p - i, 1^{\{i\}})$. □

COROLLARY 3.13. *If λ satisfies $\lambda_1 + \cdots + \lambda_n < p^r$, then $i(S_\lambda, r) \geq n$.*

PROOF. The condition implies that λ is too thin to contain any partition of p^r into less or equal to n parts. □

PROPOSITION 3.14. *Let λ be an arbitrary partition. The following holds.*

(1) *If λ satisfies $\lambda_1 \leq p - n$ then $i(L_\lambda, 1) \geq n$.*
(2) *If λ satisfies $\lambda_1 + \cdots + \lambda_n < p^r$, then $i(L_\lambda, r) \geq n$.*

PROOF. We will use the short exact sequences (coming from the highest weight structure of S_λ)

$$0 \to L_\lambda \to S_\lambda \to D_\lambda \to 0 , \qquad (*)$$

where $D_\lambda := S_\lambda / L_\lambda$ has a finite filtration whose layers are L_μ for μ dominated by λ. We prove both statements of proposition 3.14 simultaneously, by induction on the poset of partitions (with respect to the dominance order). First, if λ is minimal in this poset, then $L_\lambda = S_\lambda$ and the result is given by corollaries 3.12 and 3.13. Now let λ be a partition and assume that proposition 3.14 holds for all partitions dominated by λ. The key observation is that if μ is dominated by λ and λ satisfies one of the conditions of proposition 3.14, then μ also satisfies the same condition. In particular, we obtain that $i(D_\lambda, r) \geq n$ by the induction hypothesis. We also have $i(S_\lambda, r) \geq n$ by corollary 3.12 or 3.13. Thus the $\operatorname{Ext}^*_{\mathcal{P}}(T(d, r), -)$-long exact sequence associated to $(*)$ implies that $i(L_\lambda, r) \geq n$. □

References

[1] Kaan Akin and David A. Buchsbaum, *Characteristic-free representation theory of the general linear group*, Adv. in Math. **58** (1985), no. 2, 149–200, DOI 10.1016/0001-8708(85)90115-X. MR814749

[2] Kaan Akin, David A. Buchsbaum, and Jerzy Weyman, *Schur functors and Schur complexes*, Adv. in Math. **44** (1982), no. 3, 207–278, DOI 10.1016/0001-8708(82)90039-1. MR658729

[3] Giandomenico Boffi and David A. Buchsbaum, *Threading homology through algebra: selected patterns*, Oxford Mathematical Monographs, The Clarendon Press, Oxford University Press, Oxford, 2006. Oxford Science Publications. MR2247272

[4] E. Cline, B. Parshall, and L. Scott, *Finite-dimensional algebras and highest weight categories*, J. Reine Angew. Math. **391** (1988), 85–99. MR961165

[5] Stephen R. Doty, Karin Erdmann, and Daniel K. Nakano, *Extensions of modules over Schur algebras, symmetric groups and Hecke algebras*, Algebr. Represent. Theory **7** (2004), no. 1, 67–100, DOI 10.1023/B:ALGE.0000019454.27331.59. MR2046956

[6] Vincent Franjou and Eric M. Friedlander, *Cohomology of bifunctors*, Proc. Lond. Math. Soc. (3) **97** (2008), no. 2, 514–544, DOI 10.1112/plms/pdn005. MR2439671

[7] Eric M. Friedlander and Andrei Suslin, *Cohomology of finite group schemes over a field*, Invent. Math. **127** (1997), no. 2, 209–270, DOI 10.1007/s002220050119. MR1427618

[8] J. A. Green, *Polynomial representations of* GL_n, Second corrected and augmented edition, Lecture Notes in Mathematics, vol. 830, Springer, Berlin, 2007. With an appendix on Schensted correspondence and Littelmann paths by K. Erdmann, Green and M. Schocker. MR2349209

[9] Jens Carsten Jantzen, *Representations of algebraic groups*, 2nd ed., Mathematical Surveys and Monographs, vol. 107, American Mathematical Society, Providence, RI, 2003. MR2015057

[10] Alexander S. Kleshchev and Daniel K. Nakano, *On comparing the cohomology of general linear and symmetric groups*, Pacific J. Math. **201** (2001), no. 2, 339–355, DOI 10.2140/pjm.2001.201.339. MR1875898

[11] A. S. Kleshchev and J. Sheth, *On extensions of simple modules over symmetric and algebraic groups*, J. Algebra **221** (1999), no. 2, 705–722, DOI 10.1006/jabr.1998.8038. MR1728406

[12] Henning Krause, *Koszul, Ringel and Serre duality for strict polynomial functors*, Compos. Math. **149** (2013), no. 6, 996–1018, DOI 10.1112/S0010437X12000814. MR3077659

[13] I. G. Macdonald, *Symmetric functions and Hall polynomials*, 2nd ed., Oxford Mathematical Monographs, The Clarendon Press, Oxford University Press, New York, 1995. With contributions by A. Zelevinsky; Oxford Science Publications. MR1354144

[14] Saunders MacLane, *Categories for the working mathematician*, Springer-Verlag, New York-Berlin, 1971. Graduate Texts in Mathematics, Vol. 5. MR0354798

[15] Stuart Martin, *Schur algebras and representation theory*, Cambridge Tracts in Mathematics, vol. 112, Cambridge University Press, Cambridge, 1993. MR1268640

[16] Antoine Touzé, *Cohomology of classical algebraic groups from the functorial viewpoint*, Adv. Math. **225** (2010), no. 1, 33–68, DOI 10.1016/j.aim.2010.02.014. MR2669348

[17] Antoine Touzé, *Ringel duality and derivatives of non-additive functors*, J. Pure Appl. Algebra **217** (2013), no. 9, 1642–1673, DOI 10.1016/j.jpaa.2012.12.007. MR3042627

[18] Antoine Touzé, *Bar complexes and extensions of classical exponential functors* (English, with English and French summaries), Ann. Inst. Fourier (Grenoble) **64** (2014), no. 6, 2563–2637, DOI 10.5802/aif.2921. MR3331175

[19] A. Touzé, Foncteurs strictement polynomiaux et applications, Habilitation Thesis, 2014.

[20] A. Touzé, Connectedness of cup products for polynomial representations of GL_n and applications, to appear in Annals of K-theory.

UNIVERSITÉ LILLE 1 - SCIENCES ET TECHNOLOGIES, LABORATOIRE PAINLEVÉ, CITÉ SCIENTIFIQUE - BÂTIMENT M2, F-59655 VILLENEUVE D'ASCQ CEDEX, FRANCE

Email address: `antoine.touze@math.univ-lille1.fr`

Selected Published Titles in This Series

705 **Graham J. Leuschke, Frauke Bleher, Ralf Schiffler, and Dan Zacharia, Editors,** Representations of Algebras, 2018

701 **Joan-Carles Lario and V. Kumar Murty, Editors,** Number Theory Related to Modular Curves, 2018

700 **Alexandre Girouard, Dmitry Jakobson, Michael Levitin, Nilima Nigam, Iosif Polterovich, and Frédéric Rochon, Editors,** Geometric and Computational Spectral Theory, 2017

699 **Mark L. Agranovsky, Matania Ben-Artzi, Catherine Bénéteau, Lavi Karp, Dmitry Khavinson, Simeon Reich, David Shoikhet, Gilbert Weinstein, and Lawrence Zalcman, Editors,** Complex Analysis and Dynamical Systems VII, 2017

698 **Alexander M. Blokh, Leonid A. Bunimovich, Paul H. Jung, Lex G. Oversteegen, and Yakov G. Sinai, Editors,** Dynamical Systems, Ergodic Theory, and Probability: in Memory of Kolya Chernov, 2017

697 **Fabrizio Broglia, Françoise Delon, Max Dickmann, Danielle Gondard-Cozette, and Victoria Ann Powers, Editors,** Ordered Algebraic Structures and Related Topics, 2017

696 **Ara S. Basmajian, Yair N. Minsky, and Alan W. Reid, Editors,** In the Tradition of Ahlfors–Bers, VII, 2017

695 **Katrina Barron, Elizabeth Jurisich, Antun Milas, and Kailash Misra, Editors,** Lie Algebras, Vertex Operator Algebras, and Related Topics, 2017

694 **Manjul Bhargava, Robert Guralnick, Gerhard Hiss, Klaus Lux, and Pham Huu Tiep, Editors,** Finite Simple Groups: Thirty Years of the Atlas and Beyond, 2017

693 **Michael Cwikel and Mario Milman, Editors,** Functional Analysis, Harmonic Analysis, and Image Processing, 2017

692 **Anatole Katok, Yakov Pesin, and Federico Rodriguez Hertz, Editors,** Modern Theory of Dynamical Systems, 2017

691 **Farrell Brumley, Maria Paula Gómez Aparicio, and Alberto Mínguez, Editors,** Around Langlands Correspondences, 2017

690 **Andrés Eduardo Caicedo, James Cummings, Peter Koellner, and Paul B. Larson, Editors,** Foundations of Mathematics, 2017

689 **Erica Flapan, Allison Henrich, Aaron Kaestner, and Sam Nelson, Editors,** Knots, Links, Spatial Graphs, and Algebraic Invariants, 2017

688 **Jeffrey Bergen, Stefan Catoiu, and William Chin, Editors,** Groups, Rings, Group Rings, and Hopf Algebras, 2017

687 **Fernanda Botelho, Raena King, and T. S. S. R. K. Rao, Editors,** Problems and Recent Methods in Operator Theory, 2017

686 **Alp Bassa, Alain Couvreur, and David Kohel, Editors,** Arithmetic, Geometry, Cryptography and Coding Theory, 2017

685 **Heather A. Harrington, Mohamed Omar, and Matthew Wright, Editors,** Algebraic and Geometric Methods in Discrete Mathematics, 2017

684 **Anna Beliakova and Aaron D. Lauda, Editors,** Categorification in Geometry, Topology, and Physics, 2017

683 **Anna Beliakova and Aaron D. Lauda, Editors,** Categorification and Higher Representation Theory, 2017

682 **Gregory Arone, Brenda Johnson, Pascal Lambrechts, Brian A. Munson, and Ismar Volić, Editors,** Manifolds and K-Theory, 2017

681 **Shiferaw Berhanu, Nordine Mir, and Emil J. Straube, Editors,** Analysis and Geometry in Several Complex Variables, 2017

For a complete list of titles in this series, visit the
AMS Bookstore at **www.ams.org/bookstore/conmseries/**.